Phoebus J. Dhrymes

Introductory
Econometrics

Springer-Verlag

New York   Heidelberg   Berlin

Phoebus J. Dhrymes

Department of Economics
Columbia University
New York, New York 10027
USA

Library of Congress Cataloging in Publication Data

Dhrymes, Phoebus J., 1932–
  Introductory econometrics.

  Bibliography: p.
  Includes index.
  1. Econometrics. I. Title.
HB139.D483   330'.01'82   78-18907

© 1978 by Springer-Verlag New York Inc.

Printed in the United States of America.

9 8 7 6 5 4 3 2 1

ISBN 0-387-90317-8 Springer-Verlag New York
ISBN 3-540-90317-8 Springer-Verlag Berlin Heidelberg

To B & P & P

Preface

This book has taken form over several years as a result of a number of courses taught at the University of Pennsylvania and at Columbia University and a series of lectures I have given at the International Monetary Fund.

Indeed, I began writing down my notes systematically during the academic year 1972–1973 while at the University of California, Los Angeles. The diverse character of the audience, as well as my own conception of what an introductory and often terminal acquaintance with formal econometrics ought to encompass, have determined the style and content of this volume.

The selection of topics and the level of discourse give sufficient variety so that the book can serve as the basis for several types of courses. As an example, a relatively elementary one-semester course can be based on Chapters one through five, omitting the appendices to these chapters and a few sections in some of the chapters so indicated. This would acquaint the student with the basic theory of the general linear model, some of the problems often encountered in empirical research, and some proposed solutions. For such a course, I should also recommend a brief excursion into Chapter seven (logit and probit analysis) in view of the increasing availability of data sets for which this type of analysis is more suitable than that based on the general linear model.

A more ambitious two-semester course would attempt to cover the totality of the text with the exception of a few sections, which the instructor may find too idiosyncratic for his tastes. Some parts of the appendix to Chapter three easily fall in this category.

The book assumes that the student is familiar with the basic theory of inference, but a number of extremely useful results are given in Chapter eight. Of the material in Chapter eight, Sections one and two and a part of Section three deal with very basic topics such as the multivariate normal distribution, point estimation, and properties of estimators, and thus ought to be an integral part of the student's preparation. If students are lacking in this respect, then I have always found it a good procedure to review such material at the very beginning. Other more specialized results—also given in Chapter eight—may simply be referred to when the need for them arises.

Similarly, a knowledge of calculus, but more importantly of matrix algebra, is also assumed. However, nearly all the results from matrix algebra employed in the text may be found in the booklet *Mathematics for Econometrics*, appearing as an appendix at the end of this volume. In this sense the present volume is completely self contained. For students not overly well acquainted with matrix algebra I have found it useful to review the basic concepts in three or so lectures and let the student read the specialized results as the need for them arises.

There are a number of novel features for an introductory text. These include a discussion of logit models in some detail (Chapter seven); a discussion of power aspects of tests of hypotheses and consequently of noncentral chi square and F distributions; a discussion of the S-method for multiple comparison tests (both in the appendix to Chapter two); a discussion of the elementary aspects of Bayesian inference (Chapter eight and applications to the problem of multicollinearity and other aspects of the general linear model in Chapters two and four).

Finally, it is a pleasure to acknowledge my debt to the many individuals who have made this volume possible. Needless to say, my debt to the intellectual tradition in econometrics is immeasurable, and the individuals involved are simply too numerous to mention.

I also want to express my gratitude to successive generations of students who by their questions and puzzlement have made a substantial contribution to whatever clarity this volume might exhibit: to my present or former colleagues Michael McCarthy, Robert Shakotko, and John Taylor who read parts of the manuscript and offered suggestions and comments; to my student, Roman Frydman, who read this volume in its entirety in both manuscript and galley-proof form and thus prevented many misprints and incongruities. Last but not least, I express my gratitude to Columbia University for providing support for part of this undertaking and to Princeton University where I spent part of the calendar year 1977 completing this work.

New York, New York   Phoebus J. Dhrymes

May 1978

Contents

Chapter 8

The General Linear Model I 1

1 Introduction

1.1 Generalities

In many contexts of empirical research in economics we deal with situations in which one variable, say y, is determined by one or more other variables, say $\{x_i : i = 1, 2, \ldots, n\}$, without the former determining the latter. For example, suppose we are interested in the household demand for food. Typically, we assume that the household's operations in the market for food are too insignificant to affect the price in the market, so that our view of the household is that of an atomistic competitor. Thus, we take the price of food, as well as the other commodity prices, as given irrespective of the actions of the household. The household's income is, typically, determined independently of its food consumption activity—even if the household is engaged in agricultural pursuits. Here, then, we have a situation in which a variable of interest viz, the demand for food by a given household, is determined by its income and the prices of food and other commodities, while the latter group of variables is not influenced by the household's demand for food.[1]

[1] The perceptive reader would have, perhaps, noted some aspect of the fallacy of composition argument. If no household's consumption of food activity has *any* influence on the price of food, then by aggregation the entire collection of households operating in a given market has no effect on price. It is not clear, then, how price is determined in this market. Of course, the standard competitive model would have price determined by the interaction of (market) supply and demand. What is meant by the atomistic assumption is that an individual economic agent's activity has so infinitesimal an influence on price as to render it negligible for all practical purposes.

It is easy to give many other examples. Thus, the demand for medical services on the part of a household may also be, inter alia, a function of its demographic characteristics, its income, and the price of medical services. The latter, however, when the household possesses medical insurance, is a function of the deductible and coinsurance features of the (insurance) policy. Now, if we had a sample of n households that are asserted to be (parametrically) homogeneous, i.e., they differ in terms of their demographic characteristics and income but otherwise behave in similar fashion, then we can take this to be a sample of data generated from the hypothesized relationship. Consequently, the data may be utilized in order to make inferences about the parametric structure, characterizing the demand for food or medical services.

1.2 Models and Their Uses

The question may be asked why we should be interested in estimating the parameters of such hypothesized relationships. There are at least two responses to that. First, knowledge of the parametric structure may be important in its own right. Thus, a number of "theories" involving the firm in its pricing and production activities require the assumption of nonincreasing returns to scale, and involve the assertion of the existence of a "neoclassical" production function. It may, thus, be intrinsically interesting to determine whether firms in a given real context exhibit increasing returns to scale or not. Similarly, in the context of a system of demand equations describing the consumption behavior of a (group of) household(s) we may be interested in whether the income elasticity of the demand for food, say, is greater or less than one, i.e., whether food, however defined, is a luxury good or a necessity in the standard classification scheme. Frequently, the interest in a given parametric structure is a derived one, derived from a well-specified concern. For example, an oligopolistic firm may be interested, inter alia, in the price elasticity of demand for its output. While this may well be motivated simply by a desire to know, in fact it is more typically an interest that is derived from the following concern. Suppose the firm were to raise the price for its output somewhat; would this increase or reduce total revenue? The answer to that question depends crucially on the price elasticity of demand at current prices. Thus, knowledge of the parameters of the firm's demand function is extremely important in determining what course of action it ought to follow.

Second, and closely related to the first type of response, is the desire or requirement to "predict." Prediction means that, a relationship being known to exist between a given variable, called the *dependent* variable, and one or more other variables, called the *independent* or *explanatory* variables, it is desired, given knowledge of the independent variables at some future time, to say something about the behavior of the dependent variable. Prediction may be undertaken by someone who may not be closely concerned about the uses to which this information is to be put. Thus, e.g., there are, by now, a number of commercial firms that undertake to construct "models" of the United States economy and disseminate information regarding its course, in its various aspects, several quarters in advance. Those who generate these predictions are not closely concerned about the implications of such predictions. On the other hand, their clients are most interested, and for a variety of reasons. Large firms may be interested in the conditions they can expect to encounter with respect to the sectors that supply them with raw materials and/or the money market. Various agencies of government may be concerned about the course of the "price level," or of unemployment in various sectors, and so on.

Somewhat more formally we may state that the two main "reasons" one may be interested in the estimation of economic relationships are *structural identification, prediction, and control.* Accurate "knowledge" of the parametric structure is basic to both. Prediction alone, however, places relatively less stringent requirements on what we need to know about the phenomenon under consideration.

Since many economic phenomena have been modeled on the basis of the general linear model, we shall examine the estimation theory pertinent to the latter and its many variations.

2 Model Specification and Estimation

2.1 Notation, Basic Concepts, and Assumptions

The considerations of the preceding section lead, in the simplest possible case, to the formulation

$$y_t = \sum_{i=0}^{n} \beta_i x_{ti}.$$

In the above, y is the dependent variable of interest and the x_i, $i = 0, 1, 2, \ldots, n$, are the determining variables. In concrete terms, the reader may think of y_t as expenditures for medical services on the part of the tth household, and of the x_{ti} as pertaining to the household's demographic characteristics, its income, and various aspects of the household's insurance status. If the model were specified as above, and asserted to apply to a given set of households, the implication would be that given the characteristics of the household its expenditure on medical services is uniquely determined so that, for example, two households having roughly similar characteristics would have roughly similar expenditures for medical services. In the equation above, the β_i, $i = 0, 1, 2, \ldots, n$, are considered to be fixed, but unknown, constants and are said to be the *parameters* (or *parametric structure*) of the relation exhibited there. If the parametric structure is given and if two households' characteristics are identical, then any difference in the medical expenditures of the two households would immediately lead to the rejection of that relationship as a description of how households in a given set determine their expenditures for medical services. Typically, in econometrics we do not construct models that exhibit such stringent characteristics. While we might claim that, on the "average," households determine their expenditures for medical services in accordance with the relation above, we do not necessarily want to eliminate the possibility that basically similar households may, on occasion, differ appreciably in the size of their expenditures for medical services. This leads us to the distinction between *deterministic* and *stochastic* models. We have:

Definition 1. Consider an economic phenomenon of interest and denote by y the variable corresponding to this phenomenon. Let $\{x_i : i = 1, 2, \ldots, n\}$ be *observable* nonstochastic variables and suppose that for a real-valued function $f(\cdot)$, we specify

$$y = f(x), \qquad x = (x_1, x_2, \ldots, x_n)'$$

as a description of how x determines y. The relationship above is said to represent a *deterministic model*.

Remark 1. A deterministic model is one in which the interaction among the specified (and *observable*) variables of interest is uniquely determined, i.e., in the relation above the magnitude of y is uniquely determined through the function $f(\cdot)$ once the values of the elements of the vector x are specified. This is to be contrasted to a specification in which even if x and $f(\cdot)$ are completely specified it is not necessarily true that

$$y = f(x),$$

although it may be true that the relation above holds in some "average" sense.

Definition 2. Consider again the context of Definition 1 and suppose that

$$y = g(x, u),$$

where $g(\cdot, \cdot)$ is some real-valued function and u is an *unobservable* random variable. The model above is said to be a *stochastic model*. If $g(\cdot, \cdot)$ has the special form

$$g(x, u) = h(x) + s(u)$$

the model is said to be *additively stochastic*; if it has the form

$$g(x, u) = r(x)m(u)$$

the model is said to be *multiplicatively stochastic*. In the above $h(\cdot), s(\cdot), r(\cdot)$ and $m(\cdot)$ are real-valued functions.

Remark 2. Although we have specified in the definitions that x is a non-stochastic vector, this is not, strictly speaking, necessary. *The essential distinction is that x is observable while u is unobservable.*

EXAMPLE 1. Suppose we wish to specify the process by which a firm combines primary factors of production, say, capital K and labor L, in order to produce output Q. We may do this through the "production function" of economic theory. Thus

$$Q = F(K, L)$$

where $F(\cdot, \cdot)$ is a suitable function, as required by economic theory. The relation above would indicate that if capital and labor are specified then output would be uniquely determined. We may not, however, wish to operate in this stringent context—which at any rate is belied by observations on the triplet (Q, K, L). We may, instead, wish to replace the deterministic model above by the (multiplicatively) stochastic model.

$$Q = F(K, L)U$$

where U is a nonnegative random variable. The model above indicates that if (K, L) are specified, then output is not completely determined but is still subject to some variation depending on the random variable U. If $F(K, L)$ is the usual Cobb–Douglas function, the deterministic model is

$$Q = AK^{\alpha}L^{1-\alpha}, \qquad A > 0, 0 < \alpha < 1$$

and the (multiplicatively) stochastic model is

$$Q = AK^{\alpha}L^{1-\alpha}e^{u},$$

where we have written $U = e^{u}$. The additively stochastic model is

$$Q = AK^{\alpha}L^{1-\alpha} + U.$$

Here it is, perhaps, more convenient to operate with the multiplicatively stochastic model.

Similarly, if $F(\cdot, \cdot)$ is the constant elasticity of substitution production function, the variants noted above become

$$Q = A[a_1 K^{\alpha} + a_2 L^{\alpha}]^{1/\alpha}, \qquad A, a_1, a_2 > 0, \alpha \in (-\infty, 1)$$

or

$$Q = A[a_1 K^{\alpha} + a_2 L^{\alpha}]^{1/\alpha}e^{u}, \qquad Q = A[a_1 K^{\alpha} + a_2 L^{\alpha}]^{1/\alpha} + U.$$

Remark 3. What interpretation does one give to the random variables appearing in stochastic models? A rationalization often employed is the following. In addition to the determining variables (capital K and labor L in the example above) actually enumerated in a given model, there may well be (and typically are) a number of other, individually infinitesimal, factors affecting the phenomenon under consideration. These factors are too numerous and individually too insignificant to take into account explicitly. Collectively, however, they may exert a perceptible influence, which may be represented by the random variable of the model. Frequently, a loose form of a central limit theorem is invoked in order to justify the claim that the random variable in question is normally distributed. The reader should bear in mind that such considerations represent merely a rationalization for stochastic models. They are not to be considered an intrinsic justification of the procedure. Whether such models are useful descriptions of the aspect of the world economists wish to study would depend on whether or not such models (once their parameters have been estimated) can generate predictions of future behavior that are in accord with actual (future) observations.

Remark 4. Since we are discussing the taxonomy of models, perhaps a few words are in order regarding the philosophy of empirical work in economics and the nature and function of models in econometric work. The term "model" is, perhaps, best left without formal definition. A model usually refers to a set of conceptualizations employed by the economist to capture the essential features of the phenomenon under consideration. An econometric model, specifically, involves a mathematical formulation of the conceptual scheme referred to above; the mathematical form is rather specific and an attempt has been made to estimate (make inferences about) the model's parameters. If in Example 1 we claim that factor employment is determined by the usual (marginal) productivity conditions and that a set of observations are available on factor prices, output, capital, and labor, we can estimate the parameters of the Cobb–Douglas or (more generally) the *constant elasticity of substitution* (CES) production function. We would then have an econometric model of the production process of the firm that would exhibit the following characteristics: the firm is assumed to be perfectly competitive; the production function exhibits constant returns to scale; the firm's factor employment policies are always at equilibrium; the production function is of the CES variety.

In the context of the model and the estimates obtained (of the production model with a CES function), we can test the hypothesis that the elasticity of substitution is unity (Cobb–Douglas case) or is less (or greater) than unity. Having done this and having concluded that the CES parameter is, say, greater than unity, what can we claim for the model? Can we, for instance, make the claim that we know beyond question that the production process is as described in the model? Unfortunately, there is no basis for such a claim. We can only state that on the basis of the evidence (sample) at hand the assertion that the production process is as described above is not contradicted, i.e., the assertions are compatible with the evidence. As time passes more observations will become available. If, given the parametric structure of the model as estimated, we generate predictions and the model-based predictions are compatible with the observations then the degree of belief or confidence in the model is strengthened. If, on the other hand, as new data becomes available we observe incompatibilities between the model-based predictions and the observations then we conclude that the model is not a sufficiently accurate representation of the production process and we may wish to revise or altogether reject it. So, generally, econometric models help to describe concretely our conceptualizations about economic phenomena. No model can be asserted to represent absolute or ultimate "truth." The degree of cogency with which we can urge it as a faithful representation of that aspect of reality to which it is addressed would depend on the validity of predictions generated by the model. It will be the task of this and subsequent chapters to elaborate and make more precise the procedure by which inferences are made regarding a model's parameters or the validity of a model as a description of a phenomenon.

2.2 The General Linear Model Defined

The preceding discussion has set forth in general and somewhat vague terms the nature of econometric models and has alluded to procedures by which inferences are made regarding their parametric configuration and the degree of belief one may have in such models. Here we shall approach these topics more systematically, even though the context of the discussion will be some- what abstract. We have:

Definition 3. The stochastic model

$$y = \sum_{i=0}^{n} \beta_i x_i + u, \tag{1}$$

where $\{x_i : i = 0, 1, 2, \ldots, n\}$ is a set of nonstochastic observable variables, u is an unobservable random variable, and the $\{\beta_i : i = 0, 1, 2, \ldots, n\}$ are fixed but unknown constants, is said to be the *general linear model* (GLM).

Remark 5. The term *linear* refers to the manner in which the unknown parameters enter the model, not to the way in which the $\{x_i : i = 0, 1, 2, \ldots, n\}$ appear in the model. Thus, e.g., if $x_2 = x_1^2$, $x_3 = \ln x_4$, we are still dealing with a GLM since the parameters β_i enter linearly.

EXAMPLE 2. Let z be medical services expenditure by the ith household, let I be the household's income; and let c be the coinsurance feature of the house- hold's insurance policy. This means that if \$1 of medical expenses is incurred, the household pays the proportion c, while the household's insurance policy pays 1-c. Let n be the number of members of the household (perhaps standard- ized for age and state of health). A possible formulation of the "demand function" for medical services may be

$$\left(\frac{z}{n}\right) = \beta_0 + \beta_1\left(\frac{I}{n}\right) + \beta_2 \ln\left(\frac{I}{n}\right) + \beta_3 \frac{1}{c} + u. \tag{2}$$

Putting $y = (z/n)$, $x_0 = 1$, $x_1 = (I/n)$, $x_2 = \ln(I/n)$, $x_3 = (1/c)$, we can write the model above as

$$y = \beta_0 x_0 + \beta_1 x_1 + \beta_2 x_2 + \beta_3 x_3 + u.$$

This is still an instance of a GLM.

If, on the other hand, (with the same variables) the model were formulated as

$$y = \beta_0 x_0 + \beta_1 x_1 + \beta_2 x_2 + \frac{1}{\beta_3 - e^{-c}} + u$$

then it is no longer a general linear model, since β_3 enters the model non-linearly.

EXAMPLE 3. Frequently a model will become linear in the parameters upon a simple transformation. Thus, suppose we conducted an experiment in which we varied the quantity of capital and labor services available to a firm and observed the resultant output. Suppose, further, we hypothesized the production relation to be

$$Q = AK^\alpha L^\beta e^u. \tag{3}$$

Without loss of generality, we can take $A = 1$ (why?). The model in (3) is *not linear* in the parameters. On the other hand

$$\ln Q = \ln A + \alpha \ln K + \beta \ln L + u = \alpha \ln K + \beta \ln L + u.$$

Take $y = \ln Q$, $\ln K = x_0$, $\ln L = x_1$, and observe that we are dealing with the special case of Definition 3 in which $n = 1$.

2.3 Estimation of Parameters: Assumptions

Suppose we are dealing with the general linear model of the preceding section and we have T observations relating to it, the tth observation being

$$y_t = \sum_{i=0}^{n} \beta_i x_{ti} + u_t, \qquad t = 1, 2, \ldots, T. \tag{4}$$

A few remarks are now in order: x_{ti}, is the tth observation on the ith variable; the $\{x_{ti}: i = 1, 2, \ldots, n\}$ are said to be the *explanatory* or *regressor variables*—sometimes also called the *independent variables*; y_t is said to be the *dependent variable*, or *the regressand*, or *the variable to be explained*. In the context of this discussion we take the $\{x_{ti}: i = 1, 2, \ldots, n\}$ as nonstochastic for every t so that the dependent variable is stochastic by virtue of u_t and inherits all the stochastic properties of the latter, except possibly for its mean.

The usual set of assumptions under which estimation is carried out is as follows:

(A.1) the matrix $X = (x_{ti})$, $t = 1, 2, \ldots, T$, $i = 0, 1, 2, \ldots, n$, is of rank $n + 1$ (which implies $T \geq n + 1$);

(A.2) $\lim_{T \to \infty} (X'X/T)$ exists as a nonsingular matrix;

(A.3) $\{u_t : t = 1, 2, \ldots\}$ is a sequence of i.i.d. random variables with mean zero and variance σ^2.

The role played by these assumptions in determining the properties of the resulting estimators will become apparent as they are invoked in order to establish the various properties. At any rate the model may be written, in matrix form, as

$$y = X\beta + u,$$
$$y = (y_1, y_2, \ldots, y_T)', \qquad u = (u_1, u_2, \ldots, u_T)', \qquad \beta = (\beta_0, \beta_1, \ldots, \beta_n)'. \tag{5}$$

The problem, then, is to process the observable variables in (y, X) so as to make inferences regarding the unknown but constant parameter vector β, taking into account the properties of the unobservable random vector u. A number of methods exist for estimating the parameters of the GLM. Here we shall deal with the method of *least squares*, sometimes also called *ordinary least squares* (OLS). We shall not attempt to motivate this procedure. For the moment we shall let its justification be the properties of the resulting estimators.

We obtain estimators by the method of OLS through minimizing

$$S(b) = (y - Xb)'(y - Xb) \tag{6}$$

with respect to b. The minimizing value, say $\hat{\beta}$, serves as the estimator of β.

Remark 6. One may look on (6) as follows. If β is estimated by b and this is used in "predicting" y within the sample, the vector of errors committed is

$$y - Xb.$$

Sometimes this is also called the *vector of residuals* or *residual errors*. *The method of OLS then requires that the estimator of β be such that it minimizes the sum of the squares of the residual errors over the sample.* We have:

Proposition 1. *The* OLS *estimator of the parameter β of the model in (5) is the vector minimizing*

$$S(b) = (y - Xb)'(y - Xb)$$

with respect to b, and is given by

$$\hat{\beta} = (X'X)^{-1}X'y.$$

PROOF. We may rewrite the minimand as

$$S(b) = y'[I - X(X'X)^{-1}X']y$$
$$+ [b - (X'X)^{-1}X'y]'X'X[b - (X'X)^{-1}X'y], \tag{7}$$

which is to be minimized with respect to b. But the first term does not contain b; thus we can only operate on the second term. The latter, however, is a quadratic form in the positive definite matrix $X'X$. Thus, it is bounded below by zero and so the smallest value it can assume is zero. This occurs when we select, for b, the vector $(X'X)^{-1}X'y$.   q.e.d.

Remark 7. An alternative approach in determining the OLS estimator would have been to differentiate $S(b)$ with respect to b, set the derivative equal to zero, and solve the resulting equation. This approach was not followed for two reasons. First, it would have required some explanation of differentiation of a scalar function with respect to a vector (of arguments); second (and perhaps more importantly), *when solving from the first-order conditions we have no assurance that we are locating the global minimum.* From (7), however, it is obvious that, no matter what b is chosen,

$$S(b) \geq y'[I - X(X'X)^{-1}X']y.$$

Since

$$S(\hat{\beta}) = y'[I - X(X'X)^{-1}X']y$$

it is clear that $\hat{\beta}$ assigns to the minimand the smallest value it could possibly assume.

Remark 8. Notice that in order for the OLS estimator of β to be uniquely defined we must have that the inverse of $X'X$ exists. This is ensured by assumption (A.1) and we now see the role played by this assumption. The process of obtaining the estimator $\hat{\beta}$ is often referred to as *regressing y on x.*

2.4 Properties of the OLS Estimator of β

As we have noted earlier the basic random variable in the model is the error process $\{u_t: t = 1, 2, \ldots\}$; consequently, all stochastic properties of the variables of the model or of estimators must ultimately be derived from properties of the error process. Thus, in order to study the properties of the estimator of the preceding section we must first express it in terms of the error process. We have, substituting from (5),

$$\hat{\beta} = (X'X)^{-1}X'y = (X'X)^{-1}X'[X\beta + u] = \beta + (X'X)^{-1}X'u \qquad (8)$$

and we see that the estimator consists of the parameter we wish to estimate plus a linear function of the model's error process. This linear combination is often referred to as the *sampling variation.*

Remark 9. The representation in (8) makes clear why estimators, whatever their properties may be, can depart appreciably from the parameters they seek to estimate. For any given sample the estimate (i.e., the value of the estimator evaluated at the particular sample) will depart from the parameter it seeks to estimate, the extent of the departure depending *on the unobservable vector u and the values assumed by the observable explanatory variables* $\{x_{ti}: t = 1, 2, \ldots, T, i = 0, 1, 2, \ldots, n\}.$

EXAMPLE 4. Suppose we are given the sample $y = (2, 19, 10, 11, 9)'$, $x = (2, 8, 7, 3, 2)'$, and we are also given that it was generated by a process

$$y_t = \beta_0 + \beta_1 x_t + u_t, \qquad t = 1, 2, 3, 4, 5,$$

with $\beta_0 = 1$, $\beta_1 = 2$. To make this conformable with the practice employed earlier we define the fictitious variable $x_{t0} = 1$ for all t. The X matrix of the preceding discussion is, thus,

$$X = \begin{bmatrix} 1 & 2 \\ 1 & 8 \\ 1 & 7 \\ 1 & 3 \\ 1 & 2 \end{bmatrix}.$$

The matrix $X'X$ is

$$X'X = \begin{bmatrix} 5 & 22 \\ 22 & 130 \end{bmatrix}, \qquad |X'X| = 166,$$

and thus

$$(X'X)^{-1} = \frac{1}{166} \begin{bmatrix} 130 & -22 \\ -22 & 5 \end{bmatrix},$$

$$\hat{\beta} = \frac{1}{166} \begin{bmatrix} 130 & -22 \\ -22 & 5 \end{bmatrix} \begin{pmatrix} 51 \\ 277 \end{pmatrix} \approx \begin{pmatrix} 3.23 \\ 1.58 \end{pmatrix}.$$

As a matter of fact the values of the dependent variables were obtained by adding to $1 + 2x_t$, the x_t being, as defined at the beginning of the example, drawings from a population that is normal with mean zero and variance 4. The error values so generated were $u = (-3, 2, -5, 4, 4)$—rounding off to the nearest integer. It is clear that in this example the "sampling variation" is considerable, being 2.23 for β_0 and -0.42 for β_1.

Even though in the preceding example the estimates depart considerably from the underlying parameters, is there any sense in which the OLS estimator yields, on the average, correct inferences regarding the underlying parameters? This is, indeed, the case since we can show that it is an *unbiased estimator*. To be precise, let us prove the more general

Proposition 2. *Consider the GLM*

$$y_t = \sum_{i=0}^{n} \beta_i x_{ti} + u_t, \qquad t = 1, 2, \ldots, T,$$

such that:

(i) *the rank of the matrix* $X = (x_{ti})$, $t = 1, 2, \ldots, T$, $i = 0, 1, 2, \ldots, n$, *is* $(n + 1)$ *and* $T > n$, *the elements of* X *being nonstochastic*;
(ii) $\{u_t : t = 1, 2, \ldots\}$ *is a sequence of i.i.d. random variables with mean zero and variance* $\sigma^2 < \infty$;
(iii) $\lim_{T \to \infty} (X'X/T) = P$ *exists as a nonstochastic, nonsingular matrix (with finite elements)*.

Then, the OLS *estimator*

$$\hat{\beta} = (X'X)^{-1}X'y$$

has the following properties:

(a) *it is unbiased*;
(b) *it is consistent*;
(c) *within the class of linear unbiased estimators of* β *it is efficient (in the sense that if* $\tilde{\beta}$ *is any other linear unbiased estimator of* β, $\text{Cov}(\tilde{\beta}) - \text{Cov}(\hat{\beta})$ *is positive semidefinite)*.

PROOF. Substituting for y in the expression defining $\hat{\beta}$ we have, upon taking expectations,

$$E(\hat{\beta}) = \beta + E[(X'X)^{-1}X'u] = \beta + (X'X)^{-1}X'E(u) = \beta, \qquad (9)$$

which immediately shows unbiasedness.

To show consistency it is sufficient to show that the covariance of the estimator converges to the zero matrix as the sample size tends to infinity. This is a consequence of Proposition 19 of Chapter 7. Since

$$\hat{\beta} - \beta = (X'X)^{-1}X'u$$

we see that

$$\text{Cov}(\hat{\beta}) = E[(X'X)^{-1}X'uu'X(X'X)^{-1}] = \sigma^2(X'X)^{-1}.$$

The operations above use conditions (i) and (ii) of the proposition. We note that

$$\lim_{T \to \infty} (X'X)^{-1} = \lim_{T \to \infty} \frac{1}{T}\left(\frac{X'X}{T}\right)^{-1} = \lim_{T \to \infty} \frac{1}{T} P^{-1} = 0. \qquad (10)$$

Incidentally, in terms of Proposition 19 of Chapter 8 Equations (9) and (10) imply that $\hat{\beta}$ converges to β in quadratic mean and not only in probability.

To prove the last part of the proposition, let $\tilde{\beta}$ be any other linear unbiased estimator of β. Since it is linear (in the dependent variable) it has the representation

$$\tilde{\beta} = Hy \qquad (11)$$

where H does not depend on y.

But then we have, by unbiasedness,

$$E(\tilde{\beta}) = E[HX\beta + Hu] = HX\beta = \beta \qquad (12)$$

and the equation above is to be satisfied for all parameter vectors β. But this is easily seen to imply

$$HX = I. \qquad (13)$$

Now, define a matrix C by

$$H = (X'X)^{-1}X' + C \qquad (14)$$

and observe that (13) implies

$$CX = 0. \qquad (15)$$

Thus,

$$\tilde{\beta} = Hy = (X'X)^{-1}X'y + Cy = \hat{\beta} + Cy = \hat{\beta} + Cu \qquad (16)$$

and we see that

$$\text{Cov}(\tilde{\beta}) = \text{Cov}(\hat{\beta}) + \sigma^2 CC', \qquad (17)$$

which shows that if $\tilde{\beta}$ is *any* linear unbiased estimator of β then

$$\text{Cov}(\tilde{\beta}) - \text{Cov}(\hat{\beta}) = \sigma^2 CC',$$

the difference being positive semidefinite.   q.e.d.

Remark 10. The rank condition in (i) is used in uniquely determining the OLS estimator, for if the inverse $(X'X)^{-1}$ fails to exist the estimator is not uniquely defined. The zero mean assumption of (ii) and the nonstochastic character of the explanatory variables in (i) are used in (9) to obtain unbiasedness. The existence of the nonsingular limit in (iii) is used in (10) to establish the convergence, in quadratic mean and hence in probability, of $\hat{\beta}$ to β. The independent identically distributed aspect of (ii) is used in deriving (17).

Remark 11. The assumptions employed in Proposition 2 are actually overly strong, from the point of view of establishing the validity of its conclusions. In fact, conclusions (a) and (c) can be obtained on a weaker set of assumptions, the conclusions being, in such a case, conditional on the explanatory variables. It may be shown that even if the elements of the matrix $X = (x_{ti})$, $t = 1, 2, \ldots, T$, $i = 0, 1, 2, \ldots, n$, are *stochastic* it is sufficient to assert that the *conditional distribution* of the vector u given the elements of X has mean zero and covariance matrix $\sigma^2 I$ (which is considerably weaker than assumption (ii)) in order for conclusions (a) and (c) to hold. Typically, this restricted result is referred to as the *Gauss–Markov theorem* and characterizes the OLS estimator as BLUE (best linear unbiased estimator). The consistency conclusion of the proposition can be obtained from the assumption that

$$\plim_{T \to \infty} \frac{X'X}{T}$$

exists as a nonsingular nonstochastic matrix and that

$$\operatorname*{plim}_{T \to \infty} \frac{X'u}{T} = 0.$$

So the validity of the proposition's conclusions can be established on the weaker set of assumptions:

(a') the elements $\{x_{ti}: t = 1, 2, \ldots, T, i = 0, 1, 2, \ldots, n\}$ are stochastic such that $T > n$ and

$$\operatorname*{plim}_{T \to \infty} \frac{X'X}{T}$$

exists as a nonstochastic nonsingular matrix, where $X = (x_{ti})$;
(b') the vector $u = (u_1, u_2, \ldots, u_T)'$ has the properties

$$\mathrm{E}(u \,|\, X) = 0, \qquad \mathrm{Cov}(u \,|\, X) = \sigma^2 I, \qquad \operatorname*{plim}_{T \to \infty} \frac{X'u}{T} = 0.$$

We shall not, however, employ this weaker set of assumptions in the current discussion.

2.5 Estimation of σ^2

The remaining unknown parameter of the GLM may be estimated by

$$\tilde{\sigma}^2 = \frac{1}{T} \hat{u}'\hat{u}, \qquad \hat{u} = y - X\hat{\beta}. \tag{18}$$

This is an intuitively appealing estimator since we can think of $\hat{u}$ as a vector of "observations" on the error term of the model. The expression in (18) is, thus, akin to the "sample variance," which we would expect to yield a consistent estimator for the variance parameter σ^2. *But whatever the intuitive appeal of the expression in (18), what is important is the set of properties it may be shown to possess.* To investigate its properties we express $\hat{u}$, more explicitly, in terms of u. Thus, using (8) we have

$$\hat{u} = y - X\hat{\beta} = X\beta + u - X\beta - X(X'X)^{-1}X'u = [I - X(X'X)^{-1}X']u.$$

Since $I - X(X'X)^{-1}X'$ is an idempotent matrix we can rewrite (18) as

$$\tilde{\sigma}^2 = \frac{1}{T} u'[I - X(X'X)^{-1}X']u. \tag{19}$$

Using the results of Problem 6, at the end of this chapter, we immediately conclude that

$$\mathrm{E}(\tilde{\sigma}^2) = \frac{1}{T} \sigma^2 \operatorname{tr}[I - X(X'X)^{-1}X'] = \frac{T - n - 1}{T} \sigma^2,$$

which shows that if an *unbiased estimator* of σ^2 is desired we should operate
with

$$\hat{\sigma}^2 = \frac{1}{T - n - 1} \hat{u}'\hat{u} \tag{20}$$

which is, indeed, an unbiased estimator of σ^2.

Under the assumptions of Proposition 2 the estimators in (19) and (20) can
both be shown to be consistent. To see that, write

$$\tilde{\sigma}^2 = \frac{1}{T} u'u - \frac{1}{T} u'X \left(\frac{X'X}{T}\right)^{-1} \frac{X'u}{T}.$$

The first term consists of $(1/T) \sum_{t=1}^{T} u_t^2$. But $\{u_t^2 : t = 1, 2, \ldots\}$ is a sequence of
independent identically distributed random variables with mean σ^2. Thus by
Khinchine's theorem (Proposition 21 of Chapter 8)

$$\plim_{T \to \infty} \frac{1}{T} u'u = \sigma^2.$$

In addition,

$$\mathrm{E}\left(\frac{X'u}{T}\right) = 0 \quad \text{and} \quad \mathrm{Cov}\left(\frac{X'u}{T}\right) = \frac{\sigma^2}{T} \frac{X'X}{T},$$

which imply

$$\plim_{T \to \infty} \frac{u'X}{T} \left(\frac{X'X}{T}\right)^{-1} \frac{X'u}{T} = 0.$$

Thus, we conclude

$$\plim_{T \to \infty} \tilde{\sigma}^2 = \sigma^2.$$

Since

$$\hat{\sigma}^2 = \frac{T}{T - n - 1} \tilde{\sigma}^2$$

it is obvious that

$$\plim_{T \to \infty} \hat{\sigma}^2 = \plim_{T \to \infty} \tilde{\sigma}^2.$$

The preceding has established

Proposition 3. *Consider the GLM in the context of Proposition 2; then*

$$\tilde{\sigma}^2 = \frac{1}{T} \hat{u}'\hat{u}, \qquad \hat{\sigma}^2 = \frac{1}{T - n - 1} \hat{u}'\hat{u}$$

are both consistent estimators of σ^2, of which the first is biased (but asymptotically unbiased) and the second is unbiased.

Remark 12. The preceding shows that there may be more than one consistent estimator of the same parameter. Indeed, suppose that $\hat{\theta}_T$ is a consistent estimator of a parameter θ. Then

$$\hat{\phi}_T = \alpha_T \hat{\theta}_T$$

is also a consistent estimator of θ, provided

$$\operatorname*{plim}_{T \to \infty} \alpha_T = 1.$$

3 Goodness of Fit

3.1 Properties of the Vector of Residuals; the Coefficient of Determination of Multiple Regression

In the previous sections we operated in a fairly rigidly specified context. Thus, we assumed that the model that had generated the sample data was known, the only ambiguity being connected with the numerical magnitude of the elements of the vector β and σ^2. We also asserted that we knew that the determining variables on which we had observations $\{x_{ti} : t = 1, 2, \ldots, T, i = 0, 1, 2, \ldots, n\}$ are, indeed, the proper ones.

Quite frequently, however, what in the exposition above is a firmly established assertion is, in practice, only a conjecture. Thus, after the data have been processed we would like to know how well the "model fits the data." At an intuitive level this consideration may be formulated as follows: without benefit of explanatory variables we can "predict" the dependent variable by its mean. The variation of the observations (data) about the mean is given by the *sample variance*. Upon the formulation of the model we "predict" the dependent variable by a *linear combination* of the explanatory variables. The variation of the observations (on the dependent variable) about the *regression plane* (i.e., the linear combination of the explanatory variables) is what remains after the influence of the explanatory variables has been taken into account. It is only natural to look upon the difference between the variation about the mean and the variation about the regression plane as a measure of the "goodness of fit" of the model to the data. In a very basic sense, what empirical econometricians seek to do is to find "models" that "explain" certain economic phenomena. To do this they formulate certain hypotheses on the basis of which a model is constructed, data are gathered, and parameters are estimated. How they gather and how they process the data depends on what their underlying conceptual framework is, including the probability structure of the error term—when the model is the GLM. Even though in the process of constructing the model and processing the data one must be precise and rigid in one's specifications, when the process is ended one would always want to ask: after all this was the work fruitful? Does the model fit the data? How much of an explanation of the phenomenon do we obtain through our model? Could not the phenomenon be "explained" in simpler terms? We shall provide a means of answering these questions at an intuitive level now, postponing until the next chapter a formalization of these issues in the form of testing one or more hypotheses.

Before we tackle these problems let us elucidate some of the properties of the vector of residuals and the procedure by which we obtain the OLS estimator $\hat{\beta}$. We recall from the previous section that

$$\hat{u} = y - X\hat{\beta} = [I - X(X'X)^{-1}X']y$$

and consequently

$$X'\hat{u} = X'[I - X(X'X)^{-1}X']y = [X' - X']y = 0, \qquad (21)$$

so that *the vector of residuals is orthogonal to the matrix of explanatory variables.* Now, if the model contains *a constant term,* then one of the explanatory variables is the fictitious variable, say x_{t0}, all the observations on which are unity. Thus, the first column of X is the vector $e = (1, 1, \ldots, 1)'$, a T element vector all of whose elements are unity. But (21) then implies that

$$e'\hat{u} = \sum_{t=1}^{T} \hat{u}_t = 0,$$

so that the OLS residuals sum to zero over the sample.

When a constant term is included in the model, it is often convenient to measure all variables as deviations from their respective sample means. Thus, the dependent variable becomes

$$\left(I - \frac{ee'}{T}\right)y$$

and the matrix of observations on the explanatory variables

$$\left(I - \frac{ee'}{T}\right)X_1,$$

where X_1 is the matrix of observations on the bona fide explanatory variables excluding the fictitious variable unity, i.e., if X is the original data matrix then

$$X = (e, X_1).$$

To see exactly what is involved, consider

$$y_t = \sum_{i=0}^{n} \beta_i x_{ti} + u_t.$$

To obtain sample means simply sum and divide by T, bearing in mind that $x_{t0} = 1, t = 1, 2, \ldots, T$. We have

$$\bar{y} = \beta_0 + \sum_{i=1}^{n} \beta_i \bar{x}_i + \bar{u},$$

the overbars indicating sample means.

To center about sample means we consider

$$y_t - \bar{y} = \sum_{i=1}^{n} \beta_i (x_{ti} - \bar{x}_i) + u_t - \bar{u}$$

and we observe that *the centering operation has eliminated from consideration the constant term.* Thus, we need only deal with the bona fide variables, the fictitious variable x_{t0} having disappeared in the process.

Now, the OLS estimator of β obeys

$$X'X\hat{\beta} = X'y.$$

Separate $\hat{\beta}_0$ from the other coefficients by separating e from X_1. Thus

$$X'X = \begin{bmatrix} e'e & e'X_1 \\ X_1'e & X_1'X_1 \end{bmatrix}$$

and

$$X'X\hat{\beta} = \begin{bmatrix} e'e & e'X_1 \\ X_1'e & X_1'X_1 \end{bmatrix} \begin{pmatrix} \hat{\beta}_0 \\ \hat{\beta}_* \end{pmatrix} = \begin{pmatrix} e'y \\ X_1'y \end{pmatrix}$$

where

$$\hat{\beta}_* = (\hat{\beta}_1, \hat{\beta}_2, \dots, \hat{\beta}_n),$$

is the vector of estimated coefficients of the bona fide explanatory variables. The equations above may be written, more explicitly, as

$$e'e\hat{\beta}_0 + e'X_1\hat{\beta}_* = e'y,$$

$$X_1'e\hat{\beta}_0 + X_1'X_1\hat{\beta}_* = X_1'y.$$

Substituting from the first into the second set of equations, noting that

$$e'e = T, \qquad \frac{e'X_1}{T} = \bar{x}', \qquad \bar{x} = (\bar{x}_1, \bar{x}_2, \dots, \bar{x}_n)',$$

where the $\bar{x}_i$, $i = 1, 2, \dots, n$, are the sample means of the explanatory variables, we find

$$X_1'\left(I - \frac{ee'}{T}\right)X_1\hat{\beta}_* = X_1'\left(I - \frac{ee'}{T}\right)y, \qquad \hat{\beta}_0 = \bar{y} - \bar{x}'\hat{\beta}_*. \qquad (22)$$

Remark 13. Equation (22) suggests that the OLS estimator of β may be obtained in two steps. First, center all data about respective sample means; regress the centered dependent variable on the centered (bona fide) explanatory variables, obtaining

$$\hat{\beta}_* = \left[X_1'\left(I - \frac{ee'}{T}\right)X_1\right]^{-1} X_1'\left(I - \frac{ee'}{T}\right)y.$$

Second, estimate the "constant" term, by the operation

$$\hat{\beta}_0 = \bar{y} - \bar{x}'\hat{\beta}_*,$$

where $\bar{x}$ is the vector of sample means of the explanatory variables. The preceding discussion establishes that *whether we do or do not center the data the resulting coefficient estimates are numerically equivalent, provided the model contains a "constant" term. If the model does not contain a constant term then the statement above is false,* as will become apparent in subsequent discussion.

Let us now express the residual vector in a form suitable for answering the questions raised earlier regarding the "goodness of fit" criterion. In view of (22) we have

$$\hat{u} = y - X\hat{\beta} = y - e\hat{\beta}_0 - X_1\hat{\beta}_* = (y - e\bar{y}) - (X_1 - e\bar{x}')\hat{\beta}_*. \quad (23)$$

We also note that

$$y - e\bar{y} = X\hat{\beta} + \hat{u} - e\bar{y} = e\hat{\beta}_0 + X_1\hat{\beta}_* + \hat{u} - e\bar{y} = (X_1 - e\bar{x}')\hat{\beta}_* + \hat{u}. \quad (24)$$

But from (21) we note that

$$\hat{u}'(X_1 - e\bar{x}') = 0.$$

Consequently

$$(y - e\bar{y})'(X_1 - e\bar{x}')\hat{\beta}_* = \hat{\beta}'_*(X_1 - e\bar{x}')'(X_1 - e\bar{x}')\hat{\beta}_*, \quad (25)$$

and thus we can write

$$(y - e\bar{y})'(y - e\bar{y}) = \hat{\beta}'_*(X_1 - e\bar{x}')'(X_1 - e\bar{x}')\hat{\beta}_* + \hat{u}'\hat{u}. \quad (26)$$

The first term in the right side of (26) reflects the influence of the bona fide explanatory variables in determining the variation of the dependent variable (about its sample mean), while the second term gives the residual variation, i.e., the variation of the dependent variable about the regression plane or the part of the dependent variable's variation that is not accounted for by the (bona fide) explanatory variables.

Following the comments made at the beginning of this section we would want to define a measure of goodness of fit based on the magnitude of the variation (of the dependent variable) accounted for by the explanatory variables. This leads us to consider

$$(y - e\bar{y})'(y - e\bar{y}) - \hat{u}'\hat{u}.$$

However, the quantity above ranges, in principle, from zero to infinity. As such it does not easily lend itself to simple interpretation. In addition, its magnitude depends on the units of measurement of the dependent and explanatory variables. To avoid these problems we use a relative, instead of an absolute, measure of the reduction in variability through the use of the (bona fide) explanatory variables. In particular, we have:

Definition 4. Consider the standard GLM as set forth, say, in Proposition 2. A measure of goodness of fit is given by the *coefficient of determination of multiple regression* (unadjusted for degrees of freedom) R^2, which is defined by

$$R^2 = 1 - \frac{\hat{u}'\hat{u}}{(y - e\bar{y})'(y - e\bar{y})}.$$

The coefficient of determination of multiple regression (adjusted for degrees of freedom) $\bar{R}^2$ is given by

$$\bar{R}^2 = 1 - \frac{\hat{u}'\hat{u}/T - n - 1}{(y - e\bar{y})'(y - e\bar{y})/T - 1}.$$

Remark 14. The terminology "unadjusted" or "adjusted" for degrees of freedom will become more meaningful in the next chapter when the distributional aspects of the GLM will be taken up. The formal use of these quantities will become apparent at that stage as well. For the moment we can only look upon R^2 and $\bar{R}^2$ as intuitively plausible measures of the extent to which a given model "fits the data." (See also Problem 11 at the end of this chapter).

Another interpretation of R^2 is that it represents the square of the simple correlation between the actual sample observations on the dependent variable and the predictions, $\hat{y} = X\hat{\beta}$, generated by the OLS estimator of β, within the sample. To see this first note (see also Problem 12), that the (sample) mean of the "predictions" is the same as the sample mean of the actual observations. Thus, the correlation coefficient between the predicted and actual dependent variable within the sample is

$$r = \frac{(y - e\bar{y})'(\hat{y} - e\bar{y})}{\sqrt{(y - e\bar{y})'(y - e\bar{y})(\hat{y} - e\bar{y})'(\hat{y} - e\bar{y})}}.$$

We note

$$(y - e\bar{y})'(\hat{y} - e\bar{y}) = (\hat{y} + \hat{u} - e\bar{y})'(\hat{y} - e\bar{y}) = (\hat{y} - e\bar{y})'(\hat{y} - e\bar{y}),$$

so that

$$r^2 = \frac{(\hat{y} - e\bar{y})'(\hat{y} - e\bar{y})}{(y - e\bar{y})'(y - e\bar{y})}.$$

But

$$\hat{y} - e\bar{y} = X\hat{\beta} - e\bar{y} = e\hat{\beta}_0 + \sum_{i=1}^{n} \hat{\beta}_i x_{.i} - e\bar{y}$$

$$= \sum_{i=1}^{n} \hat{\beta}_i (x_{.i} - e\bar{x}_i) = \left(I - \frac{ee'}{T}\right) X_1 \hat{\beta}_*,$$

where $x_{.i}$ is the ith column of X. Consequently, using (26), we conclude that

$$r^2 = \frac{\hat{\beta}'_* X'_1 \left(I - \frac{ee'}{T}\right) X_1 \hat{\beta}_*}{(y - e\bar{y})'(y - e\bar{y})} = R^2. \tag{27}$$

We have therefore proved

Proposition 4. *Consider the GLM of Proposition 2 and the OLS estimator of its parameter vector β. Then the coefficient of correlation of multiple regression (unadjusted for degrees of freedom) R^2, as in Definition 4, has an interpretation as the (square of the) simple correlation coefficient between the actual and predicted values of the dependent variable over the sample.*

3.2 The GLM without a Constant Term

Frequently, models are specified that do not contain a constant term. When this is the case it is not appropriate to define R^2 in the same fashion as when the model does contain a constant term. This is but one aspect of the variation in results when the constant term is suppressed from the specification.

Thus, let us now ask: what modifications in the results, discussed in this chapter, are necessary if the GLM does *not* contain a constant term? The essential modifications are thus:

(i) the elements of the residual vector do not (necessarily) sum to zero;
(ii) the numerical values of the estimator of the coefficient vector, i.e., the estimates of β for any given sample, will vary according to whether we operate with data that are centered about sample means or not;
(iii) the coefficient of determination of multiple regression (unadjusted) should not be computed as in Definition 4, for if it is so computed it may well be negative!

To induce comparability with earlier discussion, write the (T) observations on the GLM as

$$y = X_1 \beta_* + u. \tag{28}$$

We remind the reader that the version considered earlier was

$$y = e\beta_0 + X_1 \beta_* + u.$$

The essential difference between the two is that in (28) we assert as true that $\beta_0 = 0$.

The OLS estimator of β_* in (28) is

$$\hat{\beta}_* = (X'_1 X_1)^{-1} X'_1 y. \tag{29}$$

If we *center the data* and *then* obtain the estimator of β_* we find

$$\tilde{\beta}_* = \left[X_1' \left(I - \frac{ee'}{T} \right) X_1 \right]^{-1} X_1' \left(I - \frac{ee'}{T} \right) y \qquad (30)$$

and it is clear by inspection that it is not necessarily true that, numerically, $\hat{\beta}_* = \tilde{\beta}_*$. This equality will hold, for example, when $e'X_1 = 0$; but otherwise it need not be true.

What are the differences between the estimators in (29) and (30)? We first observe that upon substitution for y we obtain

$$\tilde{\beta}_* = \beta_* + \left[X_1' \left(I - \frac{ee'}{T} \right) X_1 \right]^{-1} X_1' \left(I - \frac{ee'}{T} \right) u, \qquad (31)$$

whence unbiasedness follows quite easily. It is, moreover, easily established that

$$\mathrm{Cov}(\tilde{\beta}_*) = \sigma^2 \left[X_1' \left(I - \frac{ee'}{T} \right) X_1 \right]^{-1}. \qquad (32)$$

Recalling that $\hat{\beta}_*$ is also a linear unbiased estimator of β_* with covariance matrix $\sigma^2 (X_1' X_1)^{-1}$, we see that since

$$X_1' X_1 - X_1' \left(I - \frac{ee'}{T} \right) X_1 = \frac{1}{T} X_1' ee' X_1$$

is a positive semidefinite matrix so is

$$\mathrm{Cov}(\tilde{\beta}_*) - \mathrm{Cov}(\hat{\beta}_*).$$

Thus the estimator in (29) is efficient relative to that in (30) when, in fact, the GLM does not contain a constant term. *Thus, operating with data centered about their respective sample means entails loss of efficiency when the GLM does not contain a constant term.*

Remark 15. Although the preceding result is derived in connection with the constant term in a GLM, it is actually a particular instance of a more general proposition regarding the improvement of properties of estimators when more "information" is used in obtaining them. Thus in (29) the estimator takes into account the "information" that $\beta_0 = 0$, while in (30) this "information" is *not* utilized. Such problems will be taken up somewhat more fully in the next chapter.

In the preceding section we showed that the elements of the residual vector summed to zero. This, unfortunately, is not so here. Thus, *using* (29), we can write the vector of residuals

$$\hat{u} = y - X_1 \hat{\beta}_* = X_1 \beta_* + u - X_1 \beta_* - X_1 (X_1' X_1)^{-1} X_1' u$$
$$= [I - X_1 (X_1' X_1)^{-1} X_1'] u.$$

But it is obvious that while

$$X'_1 \hat{u} = 0,$$

in general

$$e' \hat{u} \neq 0.$$

The reader should also observe that *if* β_* is estimated according to (30), the residual vector

$$\tilde{u} = y - X_1 \tilde{\beta}_*$$

will no longer obey $X'_1 \tilde{u} = 0$; it will however, obey

$$X'_1 \left(I - \frac{ee'}{T} \right) \tilde{u} = 0.$$

Finally, if we compute R^2 in the usual way, i.e., as

$$\frac{(y - e\bar{y})'(y - e\bar{y}) - \hat{u}'\hat{u}}{(y - e\bar{y})'(y - e\bar{y})},$$

we cannot be certain that this will yield a number in $[0, 1]$. Thus, note that

$$y = X_1 \hat{\beta}_* + \hat{u}$$

and

$$y'y = \hat{\beta}'_* X'_1 X_1 \hat{\beta}_* + \hat{u}'\hat{u}.$$

Consequently,

$$(y - e\bar{y})'(y - e\bar{y}) - \hat{u}'\hat{u} = y'y - T\bar{y}^2 - (y'y - \hat{\beta}'_* X'_1 X_1 \hat{\beta}_*)$$
$$= \hat{\beta}'_* X'_1 X_1 \hat{\beta}_* - T\bar{y}^2$$

and we would have no assurance that in any given sample the last member of the relation above will be nonnegative. Moreover, since

$$\frac{(y - e\bar{y})'(y - e\bar{y}) - \hat{u}'\hat{u}}{(y - e\bar{y})'(y - e\bar{y})} = \frac{\hat{\beta}'_* X'_1 X_1 \hat{\beta}_* - T\bar{y}^2}{(y - e\bar{y})'(y - e\bar{y})}$$

the "coefficient of determination" thus computed could be either negative or positive but not greater than unity.

In such a context (i.e., when the GLM is specified as not containing a constant term) the coefficient of determination should be computed as

$$R^2 = 1 - \frac{\hat{u}'\hat{u}}{y'y}.$$

This quantity will have the usual properties of lying in $[0, 1]$. The adjusted coefficient would then be computed as

$$\bar{R}^2 = 1 - \frac{\hat{u}'\hat{u}/(T - n)}{y'y/T}.$$

Let us now summarize the development in this section.

(a) If we are dealing with the GLM for which the constant term is known to be zero, the efficient estimator of the coefficient vector is given by (29).

(b) If the coefficient vector is estimated according to (29) the elements of the residual vector do *not sum to zero*.

(c) If $\hat{u}$ is the residual vector as in (b) above the coefficient of determination should be computed as

$$R^2 = 1 - \frac{\hat{u}'\hat{u}}{y'y},$$

where $y = (y_1, y_2, \ldots, y_T)'$ is the vector of observations on the dependent variable. If it is computed in the standard way, i.e., as

$$R^2 = 1 - \frac{\hat{u}'\hat{u}}{(y - e\bar{y})'(y - e\bar{y})}$$

then we have no assurance that this quantity lies in $[0, 1]$; instead, it may be negative.

(d) If the estimator is obtained as in (30), then it is still unbiased and consistent. It is, however, less efficient than that in (29) and, of course, it will not generally coincide numerically for any given sample with the estimate obtained from (29).

(e) The residual vector, $\tilde{u}$, obtained from the estimator in (30) i.e.,

$$\tilde{u} = y - X_1\tilde{\beta}_*,$$

no longer has the property that

$$X_1'\tilde{u} = 0.$$

Nor, of course, is it true that

$$e'\tilde{u} = 0.$$

It does, however, obey

$$X_1'\left(I - \frac{ee'}{T}\right)\tilde{u} = 0.$$

The import of the preceding discussion is quite simple. *When the specification of the GLM states that $\beta_0 = 0$, do not center data in obtaining the OLS estimators of parameters. In computing the coefficient of determination use the variation of the dependent variable about zero and not about its sample mean, i.e., use $y'y$ and not $(y - e\bar{y})'(y - e\bar{y})$.*

QUESTIONS AND PROBLEMS

1. Prove that, in Equation (12), $HX\beta = \beta$ for all β implies $HX = I$. [*Hint*: take $\beta = (1, 0, 0, \ldots, 0)$, then $\beta = (0, 1, 0, \ldots, 0)$, and so on].

2. In Equation (16) show that $Cy = Cu$; also show that $\text{Cov}(\hat{\beta}, Cu) = 0$ and that CC' is positive semidefinite.

3. Show that for the validity of Equation (9) we need only $E(u|X) = 0$. (The last expression is read, "the conditional expectation of u given X is zero.")

4. Show that for the validity of the derivation of $Cov(\hat{\beta})$, in Proposition 2 we only need $Cov(u|X) = \sigma^2 I$. (The last expression is read, "the covariance matrix of the vector u given X is $\sigma^2 I$.")

5. Show that $E(\hat{\beta}) = \beta$ does not depend on the joint density of the elements of X. On the other hand, establish that $Cov(\hat{\beta})$ as derived in Proposition 2 is valid only conditionally on X if we operate with assumptions (a′) and (b′) of Remark 11.

6. Let x be an n-element vector of zero mean random variables such that $Cov(x) = \Sigma$. If A is any $n \times n$ nonstochastic matrix, show that $E[x'Ax] = \operatorname{tr} A\Sigma$.

7. If $(y_1, y_2, \ldots, y_T)$ are T observations on some random variable show that $\bar{y} = (1/T)e'y$ is a representation of the sample mean $(\bar{y})$, where $e = (1, 1, \ldots, 1)'$ and $Ts^2 = y'(I - ee'/T)y$, where s^2 is the sample variance. [*Hint:* observe that $I - (ee'/T)$ is an *idempotent* matrix.]

8. Let $X = (x_{ti})\, t = 1, 2, \ldots, T,\ i = 1, 2, \ldots, n$. Show that if we wish to express variables as deviations from sample means, i.e., if instead of x_{ti} we use $x_{ti} - \bar{x}_i$, then the matrix representation is $[I - (ee'/T)]X$. Show also that

$$\left(I - \frac{ee'}{T}\right)X = (X - e\bar{x}'), \qquad \bar{x}' = \frac{1}{T}e'X.$$

9. Show that the quantity

$$(y - e\bar{y})'(y - e\bar{y}) - \hat{u}'\hat{u}$$

derived from Equation (26) is nonnegative. Under what conditions, if ever, will it be zero? Show also that its magnitude will depend on the units in which the variables are measured.

10. Show that R^2 lies in the interval $[0, 1]$.

11. Let $X_k = (x_{.0}, x_{.1}, \ldots, x_{.k})$ and suppose we regress y on X_k, obtaining the coefficient of determination of multiple regression (unadjusted) R_k^2. Let $X_{k+1} = (x_{.0}, x_{.1}, \ldots, x_{.k}, x_{.k+1})$ and regress y on X_{k+1}, obtaining R_{k+1}^2. Show that it is *always true* that $R_{k+1}^2 \geq R_k^2$. Show also that it is not necessarily true that $\bar{R}_{k+1}^2 \geq \bar{R}_k^2$. What conclusions do you deduce from this regarding the *intuitive* usefulness of R^2 as a measure for judging whether one model (e.g., the one containing $k + 2$ variables above) is superior relative to another (e.g., the one containing $k + 1$ variables)?

12. In the context of the GLM show that if $\hat{y} = X\hat{\beta}$ then $e'\hat{y} = e'y$, where

$$y = (y_1, y_2, \ldots, y_T)', \qquad \hat{\beta} = (X'X)^{-1}X'y, \qquad e = (1, 1, \ldots, 1)', \qquad X = (e, X_1).$$

13. Under the conditions of Proposition 2—adapted to the model in (28)—show that the estimator in Equation (30) is consistent for β_*.

14. Show that the estimator $\tilde{\beta}_*$ of Equation (30) is a consistent estimator of β_*, even on the assumption that $\beta_0 = 0$. [*Hint:* on the assumption that Equation (28) is true and, otherwise, the conditions of Proposition 2 hold, show that

$$\lim_{T \to \infty} Cov(\tilde{\beta}_*) = 0.]$$

Appendix: A Geometric Interpretation of the GLM: the Multiple Correlation Coefficient

A.1 The Geometry of the GLM

For simplicity, consider the bivariate GLM

$$y_t = \beta_0 + \beta_1 x_t + u_t$$

and note that the conditional expectation of y_t given x_t is

$$E(y_t|x_t) = \beta_0 + \beta_1 x_t,$$

The model may be plotted in the y,x plane as density functions about the mean.

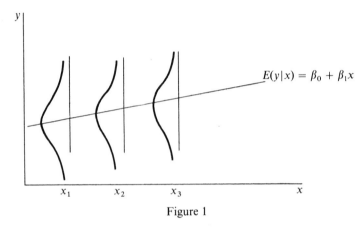

Figure 1

Specifically, in Figure 1 we have plotted the conditional mean of y given x as the straight line. Given the abscissa x, however, the dependent variable is not constrained to lie on the line. Instead, it is thought to be a random variable defined over the vertical line rising over the abscissa. Thus, for given x we can, in principle, observe a y that can range anywhere over the vertical axis. This being the conceptual framework, we would therefore not be surprised if in plotting a given sample in y,x space we obtain the disposition of Figure 2. In particular, even if the pairs $\{(y_t, x_t): t = 1, 2, \ldots, T\}$ have been generated by the process pictured in Figure 1 there is no reason why plotting the sample will not give rise to the configuration of Figure 2. A plot of the sample is frequently referred to as a *scatter diagram*. The least squares procedure is simply a method for determining a line through the scatter diagram such that for given abscissa (x) the square of the y distance between the corresponding point and the line is minimized.

In Figure 2 the sloping line is the hypothetical estimate induced by OLS. As such it represents an estimate of the unknown parameters in the conditional

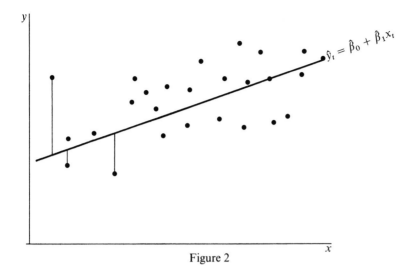

Figure 2

mean function. The vertical lines are the vertical distances (y distances) between the following two points: first, given an x that lies in the sample set of observations there corresponds a y that lies in the sample set of observations; second, given this same x there corresponds a y that lies on the sloping line. It is the sum of the squares of the distances between all points (such that the x component lies in the set of sample observations) that the OLS procedure seeks to minimize. In terms of the general results in the preceding discussion this is accomplished by taking

$$\hat{\beta}_0 = \bar{y} - \hat{\beta}_1\bar{x}, \qquad \hat{\beta}_1 = \frac{S_{yx}}{S_{xx}},$$

where

$$S_{yx} = \frac{1}{T}\sum_{t=1}^{T}(y_t - \bar{y})(x_t - \bar{x}), \qquad S_{xx} = \frac{1}{T}\sum_{t=1}^{T}(x_t - \bar{x})^2,$$

$$\bar{y} = \frac{1}{T}\sum y_t, \qquad \bar{x} = \frac{1}{T}\sum x_t.$$

The y distance referred to above is

$$y_t - \hat{\beta}_0 - \hat{\beta}_1 x_t = (y_t - \bar{y}) - \hat{\beta}_1(x_t - \bar{x}),$$

the square of which is

$$(y_t - \bar{y})^2 - 2\hat{\beta}_1(y_t - \bar{y})(x_t - \bar{x}) + \hat{\beta}_1^2(x_t - \bar{x})^2.$$

Notice, incidentally, *that to carry out an OLS estimation scheme we need only the sums and cross products of the observations.* Notice also that the variance of the slope coefficient is

$$\text{Var}(\hat{\beta}_1) = \frac{\sigma^2}{\sum_{t=1}^{T}(x_t - \bar{x})^2}.$$

Consequently, if we could design the sample by choosing the x coordinate we could further minimize the variance of the resulting estimator by choosing the x's so as to make $\sum_{t=1}^{T}(x_t - \bar{x})^2$ *as large as possible.* In fact, it can be shown that if the phenomenon under study is such that the x's are constrained to lie in the interval $[a, b]$ and we can choose the design of the sample, we should choose half the x's at a and half the x's at b. In this fashion we minimize the variance of $\hat{\beta}_1$. Intuitively, and in terms of Figures 1 and 2, the interpretation of this result is quite clear. By concentrating on two widely separated points in the x space we induce maximal discrimination between a straight line and a more complicated curve. If we focus on two x points that are very adjacent our power to discriminate is very limited, since over a sufficiently small interval all curves "look like straight lines." By taking half of the observations at one end point and half at the other, we maximize the "precision" with which we fix these two ordinates of the conditional mean function and thus fix the slope coefficient by the operation

$$\hat{\beta}_1 = \frac{\bar{y}^{(2)} - \bar{y}^{(1)}}{b - a}.$$

Above, $\bar{y}^{(2)}$ is the mean of the y observations corresponding to x's chosen at b and $\bar{y}^{(1)}$ is the mean of the y observations corresponding to the x's chosen at a.

In the multivariate context a pictorial representation is difficult; nonetheless a geometric interpretation in terms of vector spaces is easily obtained. The columns of the matrix of explanatory variables, X, are by assumption linearly independent. Let us initially agree that we deal with observations that are centered about respective sample means. Since we have, by construction, n such vectors, they span an n-dimensional subspace of the T-dimensional Euclidean space $\mathbb{R}_T$. We observe that $X(X'X)^{-1}X'$ is the matrix representation of a projection of $\mathbb{R}_T$ into itself. We recall that a projection is a linear idempotent transformation of a space into itself, i.e., if P represents a projection operator and $y_1, y_2 \in \mathbb{R}_T$, c being a real constant, then

$$P(cy_1 + y_2) = cP(y_1) + P(y_2), \qquad P[P(y_1)] = P(y_1),$$

where $P(y)$ is the image of $y \in \mathbb{R}_T$, under P.

We also recall that a projection divides the space $\mathbb{R}_T$ into two subspaces, say S_1 and S_2, where S_1 is the *range of the projection,* i.e.,

$$S_1 = \{z : z = P(y), y \in \mathbb{R}_T\},$$

while S_2 is the *null space of the projection,* i.e.,

$$S_2 = \{y : P(y) = 0, y \in \mathbb{R}_T\}.$$

We also recall that *any element of* $\mathbb{R}_T$ *can be written uniquely as the sum of two components, one from* S_1 *and one from* S_2.

The subspace S_2 is also referred to as the *orthogonal complement* of S_1, i.e., if $y_1 \in S_1$ and $y_2 \in S_2$ their inner product vanishes. Thus, $y_1'y_2 = 0$.

The application of these concepts to the regression problem makes the mechanics of estimation quite straightforward. What we do is to project the vector of observations y on the subspace of $\mathbb{R}_T$ spanned by the (linearly independent) columns of the matrix of observations on the explanatory variables X. The matrix of the projection is

$$X(X'X)^{-1}X',$$

which is an idempotent matrix of rank n. The orthogonal complement of the range of this projection is another projection, the matrix of which is

$$I - X(X'X)^{-1}X'.$$

It then follows immediately that we can write

$$y = \hat{y} + \hat{u},$$

where

$$\hat{y} = X(X'X)^{-1}X'y$$

is an element of the range of the projection defined by the matrix $X(X'X)^{-1}X'$, while

$$\hat{u} = [I - X(X'X)^{-1}X']y$$

and is an element of its orthogonal complement. Thus, mechanically, *we have decomposed y into $\hat{y}$, which lies in the space spanned by the columns of X, and $\hat{u}$, which lies in a subspace which is orthogonal to it.*

While the mechanics of regression become clearer in the vector space context above, it must be remarked that the context in which we studied the general linear model is by far the richer one in interpretation and implications.

A.2 A Measure of Correlation between a Scalar and a Vector

In the discussion to follow we shall draw an interesting analogy between the GLM and certain aspects of multivariate, and more particularly, multivariate *normal* distributions. To fix notation, let

$$x \sim N(\mu, \Sigma)$$

and partition

$$x = \begin{pmatrix} x^1 \\ x^2 \end{pmatrix}, \qquad \mu = \begin{pmatrix} \mu^1 \\ \mu^2 \end{pmatrix}, \qquad \Sigma = \begin{bmatrix} \Sigma_{11} & \Sigma_{12} \\ \Sigma_{21} & \Sigma_{22} \end{bmatrix}$$

such that x^1 has k elements, x^2 has $n - k$, μ has been partitioned conformably with x, Σ_{11} is $k \times k$, Σ_{22} is $(n - k) \times (n - k)$, Σ_{12} is $k \times (n - k)$, etc.

We recall that the conditional mean of x^1 given x^2 is simply

$$E(x^1 | x^2) = \mu^1 + \Sigma_{12}\Sigma_{22}^{-1}(x^2 - \mu^2).$$

If $k = 1$ then $x^1 = x_1$ and

$$E(x_1 | x^2) = \mu_1 + \sigma_1 . \Sigma_{22}^{-1}(x^2 - \mu^2) = \mu_1 - \sigma_1 . \Sigma_{22}^{-1}\mu^2 + \sigma_1 . \Sigma_{22}^{-1}x^2.$$

But, in the GLM we also have that

$$E(y | x) = \beta_0 + \sum_{i=1}^{n} \beta_i x_i$$

so that if we look upon $(y, x_1, x_2, \ldots, x_n)'$ as having a jointly normal distribution we can think of the "systematic part" of the GLM above as the conditional mean (function) of y given the x_i, $i = 1, 2, \ldots, n$.

In this context, *we might wish to define what is to be meant* by the correlation coefficient between a scalar and a vector. We have

Definition. Let x be an n-element random vector having (for simplicity) mean zero and covariance matrix Σ. Partition

$$x = \begin{pmatrix} x^1 \\ x^2 \end{pmatrix}$$

so that x^1 has k elements and x^2 has $n - k$ elements. Let $x_i \in x^1$. The correlation coefficient between x_i and x^2 is defined by

$$\max_{\alpha} \text{Corr}(x_i, \alpha'x^2) = \max_{\alpha} \frac{\text{Cov}(x_i, \alpha'x^2)}{[\sigma_{ii}\alpha' \, \text{Cov}(x^2)\alpha]^{1/2}}.$$

This is termed the *multiple correlation coefficient* and it is denoted by

$$R_{i \cdot k+1, k+2, \ldots, n}.$$

We now proceed to derive an expression for the multiple correlation coefficient in terms of the elements of Σ. To do so we require two auxiliary results. Partition

$$\Sigma = \begin{bmatrix} \Sigma_{11} & \Sigma_{12} \\ \Sigma_{21} & \Sigma_{22} \end{bmatrix}$$

conformably with x and *let σ_i. be the ith row of Σ_{12}.* We have:

Assertion A.1. *For $\gamma' = \sigma_i . \Sigma_{22}^{-1}$, $x_i - \gamma'x^2$ is uncorrelated with x^2.*

PROOF. We can write

$$\begin{pmatrix} x_i - \gamma'x^2 \\ x^2 \end{pmatrix} = \begin{bmatrix} e'_{\cdot i} & -\gamma' \\ 0 & I \end{bmatrix} \begin{pmatrix} x^1 \\ x^2 \end{pmatrix}$$

where $e_{\cdot i}$ is a k-element (column) vector all of whose elements are zero except the ith, which is unity. The covariance matrix of the left member above is

$$\begin{bmatrix} e'_{.i} & -\gamma' \\ 0 & I \end{bmatrix} \begin{bmatrix} \Sigma_{11} & \Sigma_{12} \\ \Sigma_{21} & \Sigma_{22} \end{bmatrix} \begin{bmatrix} e_{.i} & 0 \\ -\gamma & I \end{bmatrix}$$

$$= \begin{bmatrix} e'_{.i}\Sigma_{11}e_{.i} - 2e'_{.i}\Sigma_{12}\gamma + \gamma'\Sigma_{22}\gamma & e'_{.i}\Sigma_{12} - \gamma'\Sigma_{22} \\ \Sigma_{21}e_{.i} - \Sigma_{22}\gamma & \Sigma_{22} \end{bmatrix}$$

But

$$e'_{.i}\Sigma_{12} = \sigma_{i.}, \quad \gamma'\Sigma_{22} = \sigma_{i.}\Sigma_{22}^{-1}\Sigma_{22} = \sigma_{i.},$$

and the conclusion follows immediately.   q.e.d.

Assertion A.2. *The quantity*

$$\mathrm{Var}(x_i - \alpha'x^2)$$

is minimized for the choice $\alpha = \gamma$, γ *being as in Assertion* A.1.

PROOF. We may write, for any $(n - k)$-element vector α,

$$\mathrm{Var}(x_i - \alpha'x^2) = \mathrm{Var}[(x_i - \gamma'x^2) + (\gamma - \alpha)'x^2]$$
$$= \mathrm{Var}(x_i - \gamma'x^2) + \mathrm{Var}[(\gamma - \alpha)'x^2].$$

The last equality follows since the covariance between $x_i - \gamma'x^2$ and x^2 vanishes, by Assertion A.1.
Thus,

$$\mathrm{Var}(x_i - \alpha'x^2) = \mathrm{Var}(x_i - \alpha'x^2) + (\gamma - \alpha)'\Sigma_{22}(\gamma - \alpha),$$

which is (globally) minimized by the choice $\gamma = \alpha$ (why?).   q.e.d.

It is now simple to prove:

Proposition A.1. *Let* x *be as in Assertion* A.1, *and let* $x_i \in x^1$. *Then the (square of the) multiple correlation coefficient betweeen* x_i *and* x^2 *is given by*

$$R^2_{i\cdot k+1, k+2, \ldots, n} = \frac{\sigma_{i.}\Sigma_{22}^{-1}\sigma'_{i.}}{\sigma_{ii}}.$$

PROOF. For any $(n - k)$-element vector α and scalar c, we have by Assertion A.2

$$\mathrm{Var}(x_i - c\alpha'x^2) \geq \mathrm{Var}(x_i - \gamma'x^2).$$

Developing both sides we have

$$\sigma_{ii} - 2c\sigma_{i.}\alpha + c^2\alpha'\Sigma_{22}\alpha \geq \sigma_{ii} - 2\sigma_{i.}\gamma + \gamma'\Sigma_{22}\gamma.$$

This inequality holds, in particular, for

$$c^2 = \frac{\gamma'\Sigma_{22}\gamma}{\alpha'\Sigma_{22}\alpha}.$$

Substituting, we have

$$\sigma_{ii} - 2\left(\frac{\gamma'\Sigma_{22}\gamma}{\alpha'\Sigma_{22}\alpha}\right)^{1/2}\sigma_{i\cdot}\alpha + \gamma'\Sigma_{22}\gamma \geq \sigma_{ii} - 2\sigma_{i\cdot}\gamma + \gamma'\Sigma_{22}\gamma.$$

Cancelling σ_{ii} and $\gamma'\Sigma_{22}\gamma$, rearranging and multiplying both sides by $(\sigma_{ii}\gamma'\Sigma_{22}\gamma)^{-1/2}$, we find

$$\frac{\sigma_{i\cdot}\alpha}{(\sigma_{ii}\alpha'\Sigma_{22}\alpha)^{1/2}} \leq \frac{\sigma_{i\cdot}\gamma}{(\sigma_{ii}\gamma'\Sigma_{22}\gamma)^{1/2}}.$$

But

$$\frac{\sigma_{i\cdot}\alpha}{(\sigma_{ii}\alpha'\Sigma_{22}\alpha)^{1/2}} = \mathrm{Corr}(x_i, \alpha'x^2).$$

Consequently, we have shown that for every α

$$\mathrm{Corr}(x_i, \alpha'x^2) \leq \mathrm{Corr}(x_i, \gamma'x^2)$$

for $\gamma' = \sigma_{i\cdot}\Sigma_{22}$. Thus

$$R_{i\cdot k+1,k+2,\ldots,n} = \frac{\sigma_{i\cdot}\Sigma_{22}^{-1}\sigma_{i\cdot}'}{(\sigma_{ii}\sigma_{i\cdot}\Sigma_{22}^{-1}\sigma_{i\cdot}')^{1/2}} = \left(\frac{\sigma_{i\cdot}\Sigma_{22}^{-1}\sigma_{i\cdot}'}{\sigma_{ii}}\right)^{1/2} \qquad \text{q.e.d.}$$

Remark A.1. If, in addition, we assume that the elements of x are *jointly normal*, then the conditional distribution of x_i given x^2 is

$$N(\mu_i + \sigma_{i\cdot}\Sigma_{22}^{-1}(x^2 - \mu^2), \qquad \sigma_{ii} - \sigma_{i\cdot}\Sigma_{22}^{-1}\sigma_{i\cdot}').$$

The ratio of the conditional to the unconditional variance of x_i (given x^2) is given by

$$\frac{\sigma_{ii} - \sigma_{i\cdot}\Sigma_{22}^{-1}\sigma_i'}{\sigma_{ii}} = 1 - R_{i\cdot k+1,k+2,\ldots,n}^2.$$

Thus, $R_{i\cdot k+1,k+2,\ldots,n}^2$, measures the relative reduction in the variance of x_i between its marginal and conditional distributions (given $x_{k+1}, x_{k+2}, \ldots, x_n$).

The analogy between these results and those encountered in the chapter is now quite obvious. In that context, the role of x_i is played by the dependent variable, while the role of x^2 is played by the bona fide explanatory variables. If the data matrix is

$$X = (e, X_1),$$

where X_1 is the matrix of observations on the bona fide explanatory variables, then

$$\frac{1}{T}X_1'\left(I - \frac{ee'}{T}\right)y$$

plays the role of σ_i.. In the above, y is the vector of observations on the dependent variable and, thus, the quantity above is the *vector of sample covariances* between the explanatory and dependent variables. Similarly,

$$\frac{1}{T} X_1' \left(I - \frac{ee'}{T} \right) X_1$$

is the *sample covariance matrix* of the explanatory variables. The vector of residuals is analogous to the quantity $x_i - \gamma' x^2$, and Assertion A.1 corresponds to the statement that the vector of residuals in the regression of y on X is orthogonal to X, a result given in Equation (21). Assertion A.2 is analogous to the result in Proposition 1. Finally, the (square of the) multiple correlation coefficient is analogous to the (unadjusted) coefficient of determination of multiple regression. Thus, recall from Equation (26) that

$$R^2 = 1 - \frac{\hat{u}'\hat{u}}{y' \left(I - \dfrac{ee'}{T} \right) y}$$

$$= \frac{y' \left(I - \dfrac{ee'}{T} \right) X_1 \left[X_1' \left(I - \dfrac{ee'}{T} \right) X_1 \right]^{-1} X_1' \left(I - \dfrac{ee'}{T} \right) y}{y' \left(I - \dfrac{ee'}{T} \right) y},$$

which is the sample analog of the (square of the) multiple correlation coefficient between y and $x_1, x_2, \ldots, x_n$,

$$R_{y \cdot x_1, x_2, \ldots, x_n}^2 = \frac{\sigma_{y \cdot} \Sigma_{xx}^{-1} \sigma_{y \cdot}'}{\sigma_{yy}},$$

where

$$\Sigma = \text{Cov}(z) = \begin{bmatrix} \sigma_{yy} & \sigma_{y \cdot} \\ \sigma_{y \cdot}' & \Sigma_{xx} \end{bmatrix}, \qquad z = \begin{pmatrix} y \\ x \end{pmatrix}, \qquad x = (x_1, x_2, \ldots, x_n)',$$

i.e., it is the "covariance matrix" of the "joint distribution" of the dependent and bona fide explanatory variables.

2 The General Linear Model II

1 Generalities

In the preceding chapter we derived the OLS estimator of the (coefficient) parameters of the GLM and proved that a number of properties can be ascribed to it. In so doing, we have not assumed any specific form for the distribution of the error process. It was, generally, more than sufficient in that context to assert that the error process was one of i.i.d. random variables with zero mean and finite variance. However, even though unbiasedness, consistency, and efficiency could be proved, the distributional properties of such estimators could not be established. Consequently, tests of significance could not be formulated. In subsequent discussion we shall introduce an explicit assumption regarding the distribution of the error process, and determine what additional implications this might entail for the OLS estimators. In particular, recall that the assumptions under which estimation was carried out were:

(A.1) the explanatory variables are nonstochastic and linearly independent, i.e., if $X = (x_{ti})$, $t = 1, 2, \ldots, T$, $i = 0, 1, 2, \ldots, n$, is the matrix of observations on the explanatory variables then X is nonstochastic and rank $(X) = n + 1$;

(A.2) the limit

$$\lim_{T \to \infty} \frac{X'X}{T} = P$$

is well defined, i.e., the elements of P are nonstochastic finite quantities and P is *nonsingular*, i.e., the explanatory variables are *asymptotically linearly independent*.

34

(A.3) the error process $\{u_t : t = 1, 2, \ldots\}$ is one of i.i.d. random variables with mean zero and (finite) variance σ^2.

In the following we shall consider, in addition to the above,

(A.4) $u_t \sim N(0, \sigma^2)$ for all t.

Remark 1. Readers may think that (A.1) implies (A.2), and indeed such may have occurred to them in the discussion of the preceding chapter. Unfortunately, this is not the case, as an example will suffice to show. Take $x_{t0} = 1$ for all t, and $x_{t1} = \lambda^t$ where $|\lambda| < 1$. One sees that

$$X'X = \begin{bmatrix} T & \dfrac{\lambda - \lambda^{T+1}}{1 - \lambda} \\ \dfrac{\lambda - \lambda^{T+1}}{1 - \lambda} & \dfrac{\lambda^2 - \lambda^{2(T+1)}}{1 - \lambda^2} \end{bmatrix}.$$

Clearly, for every finite T, this is a *nonsingular matrix.* On the other hand,

$$\lim_{T \to \infty} \frac{X'X}{T} = \begin{bmatrix} 1 & 0 \\ 0 & 0 \end{bmatrix},$$

which is a singular matrix. Thus (A.1) does not imply (A.2).

The introduction of (A.4) opens the possibility of using *maximum likelihood* (ML) procedures in estimating the parameters of the GLM. It should be stressed that what allows us to do that *is not the normality aspect of* (A.4), but the fact that a specific distribution is postulated. *This allows us to write the likelihood function of a sample and, thus, to obtain* ML *estimators for the unknown parameters of the model.*

2 Distribution of the Estimator of β

2.1 Equivalence of OLS and ML Procedures

Consider again the standard GLM subject to the assumptions (A.1) *through* (A.4) of the preceding section. The sample may be written in matrix form as

$$y = X\beta + u, \tag{1}$$

and in view of (A.4) we have

$$u \sim N(0, \sigma^2 I).$$

The likelihood function, in terms of u, is nothing but the joint density of its elements, i.e.,

$$(2\pi)^{-T/2}(\sigma^2)^{-T/2} \exp\left\{ -\frac{1}{2\sigma^2} u'u \right\}.$$

Unfortunately, however, it is not written in terms of observable quantities and, thus, cannot possibly furnish us with any information regarding the unknown parameter vector β—*for that matter the function does not even contain β!* What can be done? Well, we can operate in terms of the observables, by viewing (1) as a transformation from u to y. To that effect we recall Proposition 4 of Chapter 8, which deals with the distribution of transformed variables. Let us state what is done more explicitly. Given the assumption (A.4) we can deduce the distribution of the dependent variable using Proposition 4 (or Proposition 14) of Chapter 8, and may then write the likelihood function of the sample in terms of the dependent (and independent) variables. This will enable us to use the observations (y, X) to make inferences about β, σ^2. Now, the likelihood function of the observations is simply their joint density. But this can be deduced from the joint density of the elements of the error vector, treating (1) as a transformation from u to y. The Jacobian matrix of the transformation is

$$\frac{\partial u}{\partial y} = I$$

and, thus, the *Jacobian is unity.*

Consequently, by Proposition 4 of Chapter 8 the likelihood function of the observations is

$$(2\pi)^{-T/2}(\sigma^2)^{-T/2} \exp\left\{ -\frac{1}{2\sigma^2}(y - X\beta)'(y - X\beta) \right\}.$$

The logarithm of this function is

$$L(\beta, \sigma^2; y, X) = -\frac{T}{2}\ln(2\pi) - \frac{T}{2}\ln\sigma^2 - \frac{1}{2\sigma^2}S(\beta) \qquad (2)$$

where

$$S(\beta) = (y - X\beta)'(y - X\beta).$$

The (log) likelihood function in (2) depends on σ^2 *and* β. Maximizing, we have the first-order conditions

$$\frac{\partial L}{\partial \sigma^2} = -\frac{T}{2}\frac{1}{\sigma^2} + \frac{1}{2\sigma^4}S(\beta) = 0,$$

$$\frac{\partial L}{\partial \beta} = -\frac{1}{2\sigma^2}\frac{\partial}{\partial \beta}S(\beta) = 0. \qquad (3)$$

It is easy to solve the first equation to obtain

$$\hat{\sigma}^2 = \frac{S(\beta)}{T}. \qquad (4)$$

It is also clear from the second equation of (3) that if a vector $\hat{\beta}$ can be found such that

$$\frac{\partial}{\partial \beta} S(\beta)|_{\beta = \hat{\beta}} = 0$$

then this vector satisfies the first equation of (3) as well. Thus, the problem is decomposable: first find a vector $\hat{\beta}$ satisfying the equation above; then evaluate $S(\hat{\beta})$ and estimate σ^2 from (4).

We observe that *this is equivalent to the following stepwise procedure.* First maximize (2) partially, with respect to σ^2. This yields the first equation of (3), which is solved by (4) for any admissible β. Insert this in (2) to obtain the concentrated likelihood function

$$L^*(\beta; y, X) = -\frac{T}{2}[\ln(2\pi) + 1] - \frac{T}{2}\ln\left[\frac{S(\beta)}{T}\right]$$

$$= -\frac{T}{2}[\ln(2\pi) + 1 - \ln T] - \frac{T}{2}\ln S(\beta). \tag{5}$$

The first term in the last member of (5) is a constant not depending on β. Consequently, *maximizing (5) with respect to β is equivalent to minimizing $S(\beta)$ with respect to β.* But the latter is, of course, the procedure by which we obtain the OLS estimator of β, and thus the ML and OLS estimators of β are, in this context, identical. We have therefore proved:

Proposition 1. *Consider the GLM subject to the assumptions (A.1) through (A.4) above. Then the OLS and ML estimators of β are identical. The maximum likelihood procedure suggests as the estimator of σ^2*

$$\hat{\sigma}^2 = \frac{1}{T}(y - X\hat{\beta})'(y - X\hat{\beta}),$$

*where $\hat{\beta}$ is the OLS **and** ML estimator of β.*

2.2 Distribution of the ML Estimator of β

The preceding section has established that the ML (and OLS) estimator of β, in the face of normality for the error process, is given by

$$\hat{\beta} = (X'X)^{-1}X'y = \beta + (X'X)^{-1}X'u. \tag{6}$$

It then follows immediately from Proposition 6 of Chapter 8 that

$$\hat{\beta} \sim N[\beta, \sigma^2(X'X)^{-1}]. \tag{7}$$

Tests of hypotheses regarding elements of β can, thus, be based on the distribution in (7). Unfortunately, however, unless σ^2 is known such tests cannot be carried out. If σ^2 were known, we observe that for testing the hypothesis

$$H_0: \quad \beta_i = \beta_i^0,$$

as against

$$H_1: \quad \beta_i \neq \beta_i^0,$$

where β_i^0 is a specified number, we can proceed as follows. Let q_{ii} be the ith diagonal element in $(X'X)^{-1}$. Then

$$\frac{\hat{\beta}_i - \beta_i^0}{\sqrt{\sigma^2 q_{ii}}} \sim N(0, 1).$$

The cumulative unit normal distribution, however, is tabulated, and thus a test of the hypothesis above can easily be carried out—in exactly the same way as we carry out a test on a (univariate) normal variable's mean *with known variance.*

However, typically, σ^2 is not known and thus the normal test given above is not widely applicable in empirical investigations. What we need is the test appropriate to the case where σ^2 is not known, but is estimated from the data.

2.3 Distribution of Quadratic Forms in Normal Variables[1]

In order to formulate tests in the case of unknown σ^2 and establish their properties, it is necessary, at least, to establish the distribution of

$$\hat{u}'\hat{u} = (y - X\hat{\beta})'(y - X\hat{\beta}).$$

The quantity above is basic to the various estimators of σ^2 we considered earlier. But it can be shown that the sum of the squared residuals is a quadratic form in the error process of the GLM. This provides the motivation for considering the distribution of quadratic forms in normal variables.

From previous work the reader no doubt recalls

Proposition 2. *Let* $x \sim N(0, I)$, x *being* $n \times 1$. *Then*

$$x'x \sim \chi_n^2$$

(which is read, "$x'x$ is chi square with n degrees of freedom").

PROOF. No formal proof will be given since it is assumed that the reader is basically familiar with this result. We point out that by definition a chi-square distribution with one degree of freedom is the distribution of the square of an

---

[1] The discussion in this section may be bypassed without loss of continuity. It represents a digression in several aspects of the distribution of quadratic forms. The reader need only know the conclusions of the various propositions. The proofs and ancillary discussion are not essential to the understanding of subsequent sections.

$N(0, 1)$ variable. The sum of n (independent) such variables has the chi-square distribution with n degrees of freedom. Now since the x_i, $i = 1, 2, \ldots, n$, are mutually independent and $x_i \sim N(0, 1)$ for all i, the conclusion of the proposition follows quite easily.   q.e.d.

A slight extension of the result above is given by

Proposition 3. *Let* $x \sim N(\mu, \Sigma)$, *where* x *is* $n \times 1$. *Then*

$$(x - \mu)'\Sigma^{-1}(x - \mu) \sim \chi_n^2.$$

PROOF. Since Σ (and thus Σ^{-1}) is positive definite, there exists a nonsingular matrix P such that

$$P'P = \Sigma^{-1}.$$

Consider, then,

$$y = P(x - \mu).$$

Using Proposition 6 of Chapter 8 we see that

$$y \sim N(0, I).$$

Using Proposition 2 of this chapter we conclude that

$$y'y \sim \chi_n^2$$

But

$$y'y = (x - \mu)'\Sigma^{-1}(x - \mu) \sim \chi_n^2. \quad \text{q.e.d.}$$

While the propositions above give the canonical form of the chi-square distribution, it would still be a very useful result to have a criterion for determining whether a given quadratic form is chi-square distributed or not. To partially answer this question we have

Proposition 4. *Let* $x \sim N(0, I)$, *where* x *is* $n \times 1$, *and let* A *be a symmetric matrix of rank* r. *Then*

$$x'Ax \sim \chi_r^2$$

if and only if A *is* **idempotent**.

PROOF. Suppose A is idempotent. Then, its roots are either zero or one. Hence the (diagonal) matrix of its characteristic vectors is

$$\begin{bmatrix} I_r & 0 \\ 0 & 0 \end{bmatrix}$$

where I_r is an identity matrix of order r.

Let Q be the (orthogonal) matrix of characteristic vectors. Then we can write

$$A = Q \begin{bmatrix} I_r & 0 \\ 0 & 0 \end{bmatrix} Q'.$$

Defining

$$y = Q'x$$

we note that

$$y \sim N(0, I)$$

Thus

$$x'Ax = x'Q \begin{bmatrix} I_r & 0 \\ 0 & 0 \end{bmatrix} Q'x = y' \begin{bmatrix} I_r & 0 \\ 0 & 0 \end{bmatrix} y \sim \chi_r^2.$$

The conclusion follows from Proposition 2.

Conversely, suppose

$$\phi = x'Ax \sim \chi_r^2. \tag{8}$$

Then its moment generating function is

$$M_\phi(t) = (1 - 2t)^{-r/2}. \tag{9}$$

On the other hand, since A is a symmetric matrix, it has real characteristic roots and its matrix of characteristic vectors can be chosen to be orthogonal. Thus, we can write

$$A = Q \Lambda Q'$$

where

$$\Lambda = \text{diag}(\lambda_1, \lambda_2, \ldots, \lambda_n),$$

the λ_i being the characteristic roots of A and Q being the matrix of the associated characteristic vectors.

Thus, an alternative representation of ϕ can be given, i.e., one that only utilizes the fact that A is symmetric. To be precise,

$$\phi = x'Ax = x'Q\Lambda Q'x = \sum_{i=1}^{n} \lambda_i y_i^2 \tag{10}$$

where, as before, $y = Q'x$ and thus the y_i^2 are, independently, chi square. We can compute the moment generating function of ϕ using the representation in (10) to obtain

$$M_\phi(t) = E[\exp(t \sum \lambda_i y_i^2)] = \prod_{i=1}^{n} E[\exp(t\lambda_i y_i^2)] = \prod_{i=1}^{n}(1 - 2t\lambda_i)^{-1/2}. \tag{11}$$

Comparing (9) to (11) we conclude that

$$(1 - 2t)^{-r/2} = \prod_{i=1}^{n} (1 - 2\lambda_i t)^{-1/2}. \tag{12}$$

For (12) to be valid, it can be shown that r of the λ_i must be unity and the remaining $n - r$ must be zero. But this shows that

$$A = Q \begin{bmatrix} I_r & 0 \\ 0 & 0 \end{bmatrix} Q'$$

and hence A is *idempotent*. q.e.d.

We next want to establish the conditions under which two quadratic forms in normal variables are mutually independent. This is accomplished by the following:

Proposition 5. *Let* $x \sim N(0, I)$ *and let* A, B *be two* $n \times n$ *symmetric matrices. If* $AB = 0$ *then* $x'Ax$ *and* $x'Bx$ *are mutually independent.*

PROOF. We observe that

$$(AB)' = B'A' = BA.$$

Since

$$AB = 0$$

we conclude

$$AB = BA.$$

Thus, by Proposition 53 of *Mathematics for Econometrics* there exists an orthogonal matrix P such that

$$P'AP = D_1, \qquad P'BP = D_2,$$

where D_1 and D_2 are diagonal matrices. We also observe that

$$AB = 0$$

implies

$$D_1 D_2 = 0 \tag{13}$$

so that, by rearrangement, if necessary, we can write

$$D_1 = \begin{bmatrix} D_{11} & 0 & 0 \\ 0 & 0 & 0 \\ 0 & 0 & 0 \end{bmatrix}, \qquad D_2 = \begin{bmatrix} 0 & 0 & 0 \\ 0 & D_{22} & 0 \\ 0 & 0 & 0 \end{bmatrix}, \tag{14}$$

where D_{11} and D_{22} are diagonal matrices containing, respectively, the nonnull elements of D_1 and D_2. The partition is dictated by (13), which states that if, in diagonal position i, D_1 has a nonnull element then the corresponding element in D_2 is null, and conversely. Without loss of relevance let D_{11} be $n_1 \times n_1$ and D_{22} be $n_2 \times n_2$, $n_1 + n_2 \leq n$, and define

$$y = Px = \begin{bmatrix} y^1 \\ y^2 \\ y^3 \end{bmatrix} \tag{15}$$

where y^1 is $n_1 \times 1$, y^2 is $n_2 \times 1$, and y^3 is $(n - n_1 - n_2) \times 1$. We see that

$$x'Ax = y^{1'}D_{11}y^1, \qquad x'Bx = y^{2'}D_{22}y^2. \tag{16}$$

We also note that

$$y \sim N(0, I) \tag{17}$$

so that y^1 is independent of y^2. We thus conclude that

$$x'Ax \quad \text{and} \quad x'Bx$$

are mutually independent.   q.e.d.

2.4 Tests of Significance in the GLM with Normal Errors

The preceding section has provided us with the essential results needed to complete the development of tests regarding the parameters of the GLM alluded to above. We recall that when the error process is assumed to be normal, the OLS and ML estimators of the coefficient vector β of the GLM coincide, and that, moreover, the estimator obeys

$$\hat{\beta} \sim N[\beta, \sigma^2(X'X)^{-1}].$$

The parameter σ^2 is typically unknown and its estimator is proportional to

$$\hat{u}'\hat{u} = (y - X\hat{\beta})'(y - X\hat{\beta}).$$

We also recall that

$$\hat{\beta} - \beta = (X'X)^{-1}X'u$$
$$\hat{u} = [I - X(X'X)^{-1}X']u. \tag{18}$$

We may now prove

Proposition 6. *Let A be a nonstochastic $s \times (n + 1)$ matrix $(s \leq n + 1)$ of rank s, and $\hat{\beta}$ the ML (or OLS) estimator of the parameter vector β in the context of the GLM subject to assumptions (A.1) through (A.4). Let $\hat{u}$ be the vector of residuals as exhibited in (18). Then, the following statements are true:*

$$\phi_1 = \frac{1}{\sigma^2}(\hat{\beta} - \beta)'A'[A(X'X)^{-1}A']^{-1}A(\hat{\beta} - \beta) \sim \chi_s^2;$$

$$\tag{19}$$

$$\phi_2 = \frac{1}{\sigma^2}\hat{u}'\hat{u} \sim \chi_{T-n-1}^2.$$

PROOF. By construction

$$A(\hat{\beta} - \beta) \sim N(0, \sigma^2 A(X'X)^{-1}A').$$

The truth of the first statement follows immediately from Proposition 3. For the second statement we note that

$$\frac{\hat{u}'\hat{u}}{\sigma^2} = \frac{u}{\sigma}[I - X(X'X)^{-1}X']\frac{u}{\sigma},$$

$$\frac{u}{\sigma} \sim N(0, I).$$

Since

$$\text{rank}[I - X(X'X)^{-1}X'] = T - n - 1$$

the truth of the second statement follows immediately from Proposition 4.

                                     q.e.d.

Proposition 7. *The two quadratic forms of Proposition 6 are mutually independent and*

$$\frac{\phi_1}{\phi_2}\frac{T - n - 1}{s} \sim F_{s, T-n-1}.$$

PROOF. Using the relations in (18) we have

$$\phi_1 = \frac{u'}{\sigma}[X(X'X)^{-1}A'[A(X'X)^{-1}A']^{-1}A(X'X)^{-1}X']\frac{u}{\sigma},$$

$$\phi_2 = \frac{u}{\sigma}[I - X(X'X)^{-1}X']\frac{u}{\sigma}.$$

The first part of the statement follows immediately by Proposition 5 upon noting that the matrices of the two quadratic forms are mutually orthogonal. The second part is obvious from the definition of a central F-variable with s and $T - n - 1$ degrees of freedom.   q.e.d.

Proposition 8. *Consider the GLM under the conditions set forth in Section 1. Let β_i be the ith element of the coefficient vector β and let $\hat{\beta}_i$ be its ML (or OLS) estimator. Let*

$$\hat{\sigma}^2 = \frac{1}{T - n - 1}\hat{u}'\hat{u}$$

be the unbiased estimator of σ^2. Then, a test of the null hypothesis

$$H_0: \quad \beta_i = \beta_i^0,$$

as against the alternative

$$H_1: \quad \beta_i \neq \beta_i^0,$$

can be carried out in terms of the test statistic

$$\frac{\hat{\beta}_i - \beta_i^0}{\sqrt{\hat{\sigma}^2 q_{ii}}},$$

which is t-distributed with $T - n - 1$ degrees of freedom, where q_{ii} is the ith diagonal element of $(X'X)^{-1}$.

PROOF. Obvious from Proposition 7 and the definition of a t-distributed random variable.

Remark 2. The preceding discussion suggests that the unbiased estimator of σ^2 is relatively more useful than the ML estimator—which is given by $\hat{u}'\hat{u}/T$. It also suggests that if one wants to test that a given coefficient is or is not equal to zero, one only has *to consider the ratio of the estimate of the coefficient divided by its standard error.* If the null hypothesis is true then such a ratio would have the t_{T-n-1}-distribution. Treating the particular statistic (i.e., the number resulting from this operation) as an observation from a population characterized by the t_{T-n-1}-distribution we ask: what is the probability that this observation has been generated by a process characterized by this density? If the probability is sufficiently low then we reject the null hypothesis, H_0, thus accepting the alternative, H_1. Pictorially, suppose Figure 1 represents a t_{T-n-1}-density where T and n have been specified. The

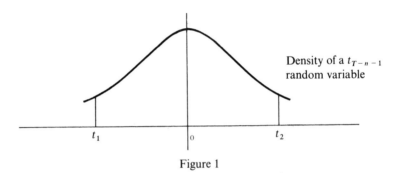

Density of a t_{T-n-1} random variable

t_1 0 t_2

Figure 1

integral (of this density) over (t_1, t_2) gives the probability that a t_{T-n-1}-variable assumes a value in the interval (t_1, t_2). Suppose the end points, t_1, t_2 are chosen from the appropriate t-distribution tables, so that this probability is, say, .95. Now, if the test statistic (observation) lies outside this interval we conclude that if the null hypothesis is correct the probability of obtaining the results obtained is $1 - .95 = .05$. This may lead us to reject the null hypothesis, thus accepting the alternative.

The preceding has shown how to carry out tests of significance on a single parameter. It would be natural now to extend our discussion to the case of tests of groups of parameters. Without loss of generality suppose we are interested in testing a hypothesis on the last k elements of the vector β. Partition

$$\beta = \begin{pmatrix} \beta^{(1)} \\ \beta^{(2)} \end{pmatrix} \tag{20}$$

so that $\beta^{(2)} = (\beta_{n-k+1}, \beta_{n-k+2}, \ldots, \beta_n)'$. Partition the matrix X conformably, i.e.,

$$X = (X^*, X_2), \tag{21}$$

where X_2 contains the last k columns of X, i.e., the variables corresponding to the coefficients in $\beta^{(2)}$. If $\hat{\beta}$ is the OLS (or ML) estimator of β, then in the context of this discussion we know that

$$\hat{\beta}^{(2)} \sim N(\beta^{(2)}, \sigma^2 R_2),$$

where

$$R_2 = [X_2'(I - M^*)X_2]^{-1}, \qquad M^* = X^*(X^{*\prime}X^*)^{-1}X^{*\prime}. \tag{22}$$

In order to carry out the desired test we must find some test statistic that contains all the elements of $\beta^{(2)}$. But from Proposition 3 we see immediately that

$$[\hat{\beta}^{(2)} - \beta^{(2)}]'\frac{R_2^{-1}}{\sigma^2}[\hat{\beta}^{(2)} - \beta^{(2)}] \sim \chi_k^2.$$

The problem is that even though the null hypothesis will specify, typically, the vector $\beta^{(2)}$, the quantity above still involves the unknown parameter σ^2. Thus, we must operate with an estimate of σ^2, and this inevitably will give rise to questions regarding the independence of the estimator of σ^2 and the quadratic form above.

But from the equations defining the OLS (or ML) estimator of β, after partitioning as in (20) and (21), we find

$$
\begin{aligned}
\hat{\beta}^{(2)} &= R_2 X_2'(I - M^*)y \\
&= R_2 X_2'(I - M^*)[X^*\beta^{(1)} + X_2\beta^{(2)} + u] \\
&= \beta^{(2)} + R_2 X_2'(I - M^*)u.
\end{aligned}
$$

Consequently, the quadratic form of interest is given by

$$v_1 = [\hat{\beta}^{(2)} - \beta^{(2)}]'\frac{R_2^{-1}}{\sigma^2}[\hat{\beta}^{(2)} - \beta^{(2)}] = \frac{u'}{\sigma}(I - M^*)X_2 R_2 X_2'(I - M^*)\frac{u}{\sigma}. \tag{23}$$

But it is easily verified that the matrix in the last member of (23) is symmetric idempotent. Moreover, consider again the quadratic form of Proposition 6, i.e.,

$$v_2 = \frac{\hat{u}'\hat{u}}{\sigma^2} = \frac{u'}{\sigma}(I - M)\frac{u}{\sigma}. \tag{24}$$

From Proposition 4 we know that these two quadratic forms (in Equations (23) and (24)) are chi-square distributed. From Proposition 5 we can conclude that they are also mutually independent *if we can show that the product of the two matrices is zero*. But, we note (see also Problem 3)

$$MM^* = M^*. \tag{25}$$

Consequently

$$(I - M)(I - M^*) = I - M^* - M + M^* = I - M$$

and (see also Problem 4)

$$(I - M)(I - M^*)X_2 = (I - M)X_2 = 0. \tag{26}$$

But this shows

$$(I - M)(I - M^*)X_2 R_2 X_2'(I - M^*) = 0,$$

which establishes the mutual independence of the two quadratic forms in (23) and (24).

We thus conclude that v_1 and v_2 (of (23) and (24) respectively) are mutually independent chi-square variables, the first with k degrees of freedom and the second with $T - n - 1$ degrees of freedom. Consequently,

$$\frac{v_1}{v_2} \frac{T - n - 1}{k}$$

has the $F_{k, T-n-1}$-distribution. But this means

$$\frac{1}{k} [\hat{\beta}^{(2)} - \beta^{(2)}]' R_2^{-1} [\hat{\beta}^{(2)} - \beta^{(2)}] \frac{1}{\hat{\sigma}^2} \sim F_{k, T-n-1}. \tag{27}$$

The above is, of course, a test statistic in that it does not contain unknown parameters, since H_0 will, typically, specify a particular value for $\beta^{(2)}$. The discussion above may be summarized in

Proposition 9. *Consider the GLM under the conditions set forth in Section 1. Partition the coefficient vector*

$$\beta = \begin{pmatrix} \beta^{(1)} \\ \beta^{(2)} \end{pmatrix}$$

so that $\beta^{(2)}$ contains k elements. A test of the null hypothesis

$$H_0: \quad \beta^{(2)} = \bar{\beta}^{(2)},$$

as against the alternative

$$H_1: \quad \beta^{(2)} \neq \bar{\beta}^{(2)},$$

can be carried out through the quantity

$$\frac{1}{k} [\hat{\beta}^{(2)} - \bar{\beta}^{(2)}]' \frac{R_2^{-1}}{\hat{\sigma}^2} [\hat{\beta}^{(2)} - \bar{\beta}^{(2)}],$$

whose distribution is $F_{k, T-n-1}$, i.e., it is F-distributed with k and $T - n - 1$ degrees of freedom.

Remark 3. It may be asked: why do we use the complicated procedure of Proposition 9 instead of using the procedure of Proposition 8 and applying it *seriatim* to each of the k elements of $\beta^{(2)}$? The answer, of course, is quite clear. Operating *seriatim* with each of the elements of $\beta^{(2)}$ means that we utilize the *marginal* distribution of the elements of $\hat{\beta}^{(2)}$. Operating in the context of Proposition 9 means that we utilize the *joint density* of the elements of $\hat{\beta}^{(2)}$. Thus, the latter procedure utilizes a more informative base on which to make inferences.

The test above has been derived from the joint distribution of the elements of $\hat{\beta}$, and follows exactly the same motivation as the test on a single coefficient. The test, however, appears unduly cumbersome since it would require the computation of R_2 as well as the quadratic form involving $\hat{\beta}^{(2)} - \bar{\beta}^{(2)}$. On the other hand, the ready availability of computer regression programs suggests an alternative approach. Thus, without loss of generality, take $\bar{\beta}^{(2)} = 0$ and consider the following motivation. In order to test whether $\beta^{(2)} = 0$ we first regress y on X^* and then y on X. If we compare the sum of the squared residuals in the two regressions we should be able to determine whether or not the hypothesis is to be accepted. We shall formulate this test precisely and show that it is equivalent, in fact identical, to the test outlined in Proposition 9.

The first regression yields

$$\tilde{\beta}^{(1)} = (X^{*\prime}X^*)^{-1}X^{*\prime}y$$

and the sum of the squared residuals under

$$H_0: \quad \beta^{(2)} = 0$$

is

$$Q_0 = (y - X^*\tilde{\beta}^{(1)})'(y - X^*\tilde{\beta}^{(1)}) = u'(I - M^*)u,$$
$$M^* = X^*(X^{*\prime}X^*)^{-1}X^{*\prime}$$

The sum of the squared residuals under the alternative

$$H_1: \quad \beta^{(2)} \neq 0$$

is simply

$$Q_1 = u'(I - M)u, \qquad M = X(X'X)^{-1}X'.$$

A test may be based on the difference $Q_0 - Q_1$ relative to Q_1. It would seem intuitively plausible that as the last k variables are introduced in the regression the explanatory power of the model, measured (inversely) by the sum of the squared residuals, must increase. The question is whether the increase is large enough to justify the implications of H_1.

Let us now consider the nature of the proposed test. We observe that

$$Q_0 - Q_1 = u'(M - M^*)u, \qquad Q_1 = u'(I - M)u.$$

Since both matrices are *idempotent* and since

$$(I - M)(M - M^*) = 0$$

we conclude that

$$\frac{(Q_0 - Q_1)/k}{Q_1/T - n - 1} \sim F_{k, T-n-1}. \tag{28}$$

The test statistic in (28) has *exactly the same distribution as the test statistic of Proposition* 9. Since

$$\frac{Q_1}{T - n - 1} = \hat{\sigma}^2,$$

in order to show that the two statistics are *identical* we need only show that under H_0, i.e., when $\bar{\beta}^{(2)} = 0$,

$$\begin{aligned} Q_0 - Q_1 &= u'(M - M^*)u \\ &= [\hat{\beta}^{(2)} - \bar{\beta}^{(2)}]' R_2^{-1} [\hat{\beta}^{(2)} - \bar{\beta}^{(2)}] \\ &= u'(I - M^*)X_2 R_2 X_2'(I - M^*)u. \end{aligned}$$

But this will be assured if we show that

$$X_2 R_2 X_2' - M^* X_2 R_2 X_2' - X_2 R_2 X_2' M^* + M^* X_2 R_2 X_2' M^* = M - M^*.$$

Utilizing Problems 6 and 7 at the end of this chapter, this reduces to showing that

$$M^* X_2 R_2 X_2' M^* = X^* R_1 X^{*\prime} - M^*,$$

whose validity follows from Problem 8.
    We have therefore established

Proposition 10. *The test described in Proposition* 10 *for testing*

$$H_0: \quad \beta^{(2)} = 0,$$

as against the alternative

$$H_1: \quad \beta^{(2)} \neq 0,$$

in the context of the GLM, where

$$X = (X^*, X_2), \qquad \beta = \begin{pmatrix} \beta^{(1)} \\ \beta^{(2)} \end{pmatrix}$$

(X, β being partitioned conformably), is equivalent (in fact identical) to the following procedure. Let Q_0 be the sum of the squared residuals in the regression of y on X^* and Q_1 the sum of the squared residuals in the regression of y on X. Define

$$\frac{(Q_0 - Q_1)/k}{Q_1/T - n - 1} \sim F_{k, T-n-1}$$

and use this statistic to test the hypothesis above. In particular, it is true that

$$\frac{Q_0 - Q_1}{Q_1} = \frac{v_1}{v_2}$$

where v_1 and v_2 are as defined in (23) and (24) respectively.

Remark 4. The virtue of the procedure in Proposition 10 is that it makes maximal use of commonly existing software to carry out such tests. Typically, the estimated covariance matrix of the coefficient estimators is not printed out in computer programs. Even if it were, the procedure of Proposition 9 would require the user to invert a submatrix thereof and compute the value of a certain quadratic form. The procedure of Proposition 10, on the other hand, requires the use—twice—of a regression program. It is, thus, a more capital intensive but also a more convenient procedure to employ.

2.5 Formal Tests of Goodness of Fit

In the preceding chapter we defined the (unadjusted) coefficient of determination of multiple regression R^2. It was advanced at the time as an intuitively appealing measure of the extent to which the model fits the data. It was, also, variously interpreted as a (relative) measure of the reduction in the variability of the dependent variable, when account is taken of the explanatory variables, or as the square of the (sample) correlation coefficient between actual and predicted dependent variables (within the sample). In this section we shall formalize the use of R^2 in goodness of fit considerations by showing that a simple transformation of R^2 will yield a test statistic for testing the hypothesis that the coefficients of all the bona fide variables are (simultaneously) equal to zero. If the hypothesis is rejected then we conclude that at least some of the explanatory variables exert some influence on the dependent variable of interest.

In the context of the GLM as developed in this chapter let it be desired to test the hypothesis

$$H_0: \quad \beta_* = 0,$$

as against the alternative

$$H_1: \quad \beta_* \neq 0,$$

where

$$\beta = (\beta_0, \beta_*')',$$

β_0 being the "constant term" of the equation.

Using the apparatus of Proposition 10 we carry out two regressions; one of y on $e = (1, 1, 1, \ldots, 1)'$, i.e., the fictitious explanatory variable corresponding to the constant term, and another of y on $X = (e, X_1)$, where X_1 is the matrix of observations on the explanatory variables corresponding to the coefficients in β_*. The sum of squares of the residuals from the regression carried out on the assumption that H_0 is true is given by

$$Q_0 = y'\left(I - \frac{ee'}{T}\right)y = u'\left(I - \frac{ee'}{T}\right)u.$$

The sum of the squared residuals obtained from the regression carried out under H_1 is

$$Q_1 = \hat{u}'\hat{u} = y'(I - M)y = u'(I - M)u.$$

The test statistic is then proportional to

$$\frac{Q_0 - Q_1}{Q_1} = \frac{u'\left(M - \dfrac{ee'}{T}\right)u}{u'(I - M)u}.$$

Since we are dealing with a special case of the discussion leading to Proposition 10 we already know that the numerator and denominator are independently chi-square distributed with

$$\mathrm{tr}\left(M - \frac{ee'}{T}\right) = n \quad \text{and} \quad \mathrm{tr}(I - M) = T - n - 1$$

degrees of freedom respectively. It remains now to connect the statistic above with the coefficient of determination R^2. But we note

$$R^2 = 1 - \frac{\hat{u}'\hat{u}}{y'\left(I - \dfrac{ee'}{T}\right)y} = \frac{y'\left(I - \dfrac{ee'}{T}\right)y - \hat{u}'\hat{u}}{y'\left(I - \dfrac{ee'}{T}\right)y} = \frac{Q_0 - Q_1}{Q_0}$$

and

$$1 - R^2 = \frac{Q_1}{Q_0}.$$

Consequently

$$\frac{R^2}{1 - R^2} = \frac{Q_0 - Q_1}{Q_1}$$

and we see immediately that

$$\frac{R^2}{1 - R^2} \frac{T - n - 1}{n} \sim F_{n, T-n-1}.$$

The preceding discussion may be summarized in

Proposition 11. *Consider the GLM*

$$y = e\beta_0 + X_1\beta_* + u$$

where $u \sim N(0, \sigma^2 I)$ *and let*

$$R^2 = 1 - \frac{\hat{u}'\hat{u}}{y'\left(I - \dfrac{ee'}{T}\right)y}$$

be the (unadjusted) coefficient of determination of multiple regression. A test of the "significance of R^2" or more formally a test of the null hypothesis

$$H_0: \quad \beta_* = 0,$$

as against the alternative

$$H_1: \quad \beta_* \neq 0,$$

may be based on the test statistic

$$\frac{R^2}{1 - R^2} \frac{T - n - 1}{n} \sim F_{n, T-n-1},$$

where T is the number of observations and n is the number of the bona fide explanatory variables, i.e., the number of the columns of X_1.

Remark 5. The null hypothesis at the level of significance α is accepted when the test statistic does not exceed a number F_α such that

$$\Pr\{F_{n, T-n-1} \leq F_\alpha\} = 1 - \alpha.$$

If the test statistic, based on a given sample, exceeds F_α then the alternative is accepted, i.e., we conclude that $\beta_* \neq 0$.

How do we interpret this conclusion? Does this mean that the evidence at hand supports the assertion that every element of β_* is different from zero? This is decidedly not the case. *Such a finding supports only the assertion that at least one element of β_* cannot be said to be zero. The finding as such does not even indicate which element it is that cannot be said to be zero.*

3 General Linear Restriction: Estimation and Tests

The logical structure of the discussion in the preceding two sections was roughly as follows. Beginning with the GLM

$$y = X\beta + u,$$

which incorporates the broadest statement we would wish to make about a given economic phenomenon, we place some restrictions on β. We then ask whether, on the basis of the sample evidence, we ought to accept the validity of these restrictions or not. The nature of the restrictions was of the general form

$$A\beta = 0$$

Thus, e.g., in Proposition 8 on the assumption that $\beta_i^0 = 0$ the restriction was

$$e_i'\beta = 0,$$

where e_i is an $(n + 1)$-element (column) vector all of whose elements are zero except the ith, which is unity. For the more general case, we would have

$$e_i'\beta = \beta_i^0.$$

Similarly, in Proposition 9 the nature of the restriction is

$$(0, I)\begin{pmatrix} \beta^{(1)} \\ \beta^{(2)} \end{pmatrix} = \bar{\beta}^{(2)}.$$

The essential feature of the restrictions above is that the matrix of the restrictions A has rows whose elements are zero except for one which is unity. Whether the restrictions are homogeneous or not is inconsequential. Since β_i^0 and $\bar{\beta}^{(2)}$ above are numerically specified, *we can always redefine the parameter vector so that the restrictions are of the homogeneous form*

$$A\beta = 0.$$

Given the statement of the restrictions as above, what we have done is to obtain the ML (or OLS) estimator of β, say $\hat{\beta}$, form the quantity

$$A\hat{\beta}$$

and determine its distribution. Based on the latter we have *tested the hypothesis*

$$A\beta = 0.$$

The question now arises: supposing the hypothesis is accepted or, better still, supposing it is given as a fact that such restrictions hold with respect to the parameter vector β, how should we estimate parameters in the face of such restrictions? In this context, let us take a somewhat broader point of view and consider (nonhomogeneous) restrictions of the form

$$A\beta = a \tag{29}$$

where A and a are some *known* matrix and vector respectively, not necessarily of the simple form employed in Propositions 8 and 9. In (29) we impose the condition

(A.5) A is an $r \times (n + 1)$ matrix of rank r.

Remark 6. Each row of A represents a restriction on the elements of β. Assuming that A is of rank r involves no loss of generality whatever—in this linear restriction context. It simply means that the r restrictions are *linearly independent*. If, for example, A were a $k \times (n + 1)$ matrix of rank $r < k$, *and if the k restrictions were not incompatible*, then by a series of elementary row operations we could reduce (29) to the equivalent system

$$\begin{pmatrix} A^* \\ 0 \end{pmatrix} \beta = \begin{pmatrix} a^* \\ 0 \end{pmatrix},$$

where A^* is $r \times (n + 1)$ of rank r. It is clear than the last $k - r$ equations of the system above imply no restrictions on the vector of coefficients β.

The estimation problem when (A.5) is added to the assumptions (A.1) through (A.4) depends on how we regard (A.5). If it is regarded as a *maintained hypothesis*[2] then we should estimate the parameters of the model subject to (29). On the other hand, if it is regarded as a *hypothesis to be tested*, then in estimation we ignore the restriction.

If (29) is taken as a maintained hypothesis, then (A.1) through (A.5) imply that we ought to maximize the (log) likelihood function subject to (29). Thus, we form the Lagrangian expression

$$F = -\frac{T}{2}\ln(2\pi) - \frac{T}{2}\ln\sigma^2 - \frac{1}{2\sigma^2}S(\beta) + \lambda'(a - A\beta) \qquad (30)$$

where λ is the vector of Lagrangian multipliers and

$$S(\beta) = (y - X\beta)'(y - X\beta).$$

As in Section 2.1, we may maximize[3] Equation (30) partially with respect to σ^2, obtaining for any admissible β

$$\hat{\sigma}^2 = \frac{1}{T}S(\beta).$$

Inserting into (30) we have the concentrated likelihood function

$$F^*(\beta, \lambda; y, X) = -\frac{T}{2}[\ln(2\pi) + 1 - \ln T] - \frac{T}{2}\ln S(\beta) + \lambda'(a - A\beta). \quad (31)$$

The first term in the right side of (31) is a constant depending neither on β nor λ. Thus, the first-order conditions for F^* as in (31) are

$$\frac{\partial F^*}{\partial \beta} = T\frac{X'(y - X\beta)}{S(\beta)} - A'\lambda = 0,$$

$$\frac{\partial F^*}{\partial \lambda} = a - A\beta = 0.$$

Rearranging the first set of equations above and multiplying through by

$$A(X'X)^{-1}$$

yields

$$A(X'X)^{-1}X'y - A\beta = \hat{\sigma}^2 A(X'X)^{-1}A'\lambda.$$

Using the second set of equations in the first-order conditions and eliminating the vector of Lagrangian multipliers λ (see also Problem 10) we find

$$\hat{\beta} = \tilde{\beta} + (X'X)^{-1}A'[A(X'X)^{-1}A']^{-1}(a - A\tilde{\beta}), \qquad (32)$$

---

[2] We remind the reader that a maintained hypothesis is one about whose validity we are certain or, at any rate, one whose validity we do not question—whether this is due to certainty or convenience is another matter.

[3] Strictly speaking we are seeking a saddle point of the Lagrangian, but commonly one speaks of "maximizing."

where

$$\tilde{\beta} = (X'X)^{-1}X'y \tag{33}$$

i.e., it is the estimator of β, which does not take into account the restrictions in (29).

Finally, the ML estimator of the variance parameter σ^2 is obviously given by

$$\hat{\sigma}^2 = \frac{1}{T} S(\hat{\beta}),$$

where $\hat{\beta}$ is as defined in (32).[4]

Remark 7. From (32) we see that the estimator of β obeying the restrictions in (29) may be expressed as the sum of two components; one is the estimator of β, ignoring the restrictions; and the other is a correction factor that is a linear transformation of the deviation of the unrestricted estimator from the restrictions imposed by (29).

Let us now establish the properties of the estimator in (32) when (A.5) is treated as a maintained hypothesis. To this effect it is notationally useful to define

$$C = (X'X)^{-1}A' \tag{34}$$

and, thus, to write (32) as

$$\hat{\beta} = \tilde{\beta} + C(C'X'XC)^{-1}(a - A\tilde{\beta}). \tag{35}$$

Substituting for y in (33) we have

$$(\hat{\beta} - \beta) = [(X'X)^{-1} - C(C'X'XC)^{-1}C']X'u.$$

It follows immediately that

$$E(\hat{\beta}) = \beta,$$

thus showing unbiasedness. Moreover, we see that

$$\text{Cov}(\hat{\beta}) = \sigma^2[(X'X)^{-1} - C(C'X'XC)^{-1}C'] = \sigma^2\Phi.$$

By (A.4) we thus conclude

$$\hat{\beta} \sim N(\beta, \sigma^2\Phi).$$

Since

$$(X'X)^{-1} - \Phi = C(C'X'XC)^{-1}C'$$

---

[4] Often the nature of the restrictions imposed by (29) will be sufficiently simple so that the estimator in (32) can be arrived at by first substituting from (29) in the model and then carrying out an ordinary (unrestricted) regression procedure.

is a positive semidefinite matrix we also conclude that the restricted estimator in (32) is efficient relative to the unrestricted estimator in (33). Noting that

$$C(C'X'XC)^{-1}C' = \frac{1}{T}\left[\left(\frac{X'X}{T}\right)^{-1}A'\left[A\left(\frac{X'X}{T}\right)^{-1}A'\right]^{-1}A\left(\frac{X'X}{T}\right)^{-1}\right]$$

(36)

we conclude (in view of (A.2)) that

$$\lim_{T\to\infty} C(C'X'XC)^{-1}C' = \lim_{T\to\infty}\frac{1}{T}[P^{-1}A'(AP^{-1}A')^{-1}AP^{-1}] = 0,$$

which shows $\hat{\beta}$ to converge to β in quadratic mean and hence in probability. The preceding discussion may be summarized in

Proposition 12. *Consider the GLM under assumptions (A.1) through (A.5) (of this chapter). Then the following statements are true:*

(i) *the ML (or OLS) estimator of β obeying the restrictions in (A.5) is given by $\tilde{\beta} = \hat{\beta} + C[C'X'XC]^{-1}(a - A\hat{\beta})$, where $\hat{\beta} = (X'X)^{-1}X'y$ is the estimator of β not necessarily obeying the restrictions in (A.5),*
(ii) *the distribution of $\tilde{\beta}$ is given by $\tilde{\beta} \sim N(\beta, \sigma^2\Phi)$, where $\Phi = (X'X)^{-1} - C[C'X'XC]^{-1}C'$;*
(iii) *the estimator $\tilde{\beta}$ is efficient relative to $\hat{\beta}$, i.e., $\mathrm{Cov}(\hat{\beta}) - \mathrm{Cov}(\tilde{\beta}) = \sigma^2 C[C'X'XC]^{-1}C'$ is positive semidefinite;*
(iv) *Φ is a singular matrix (see Problem 11).*

Remark 8. The result stated under item (iii) above is a generalization of a phenomenon noted at the end of the preceding chapter in connection with the constant term of the GLM. There we saw that if the model is specified *not* to contain a constant term and we obtain estimators of the unknown coefficients from centered data, such estimators are inefficient relative to estimators obtained from (raw) uncentered data. Proposition 12 generalizes this result by stating (in (iii)) that if restrictions on the vector β *are known to hold, estimators of β utilizing such restrictions are efficient relative to estimators that do not utilize them.* This may be paraphrased, somewhat loosely, as follows: other things being equal, the more information we utilize the more efficient the resulting estimators.

Remark 9. The singularity of Φ is an entirely expected result; since β has been estimated subject to a number of linear restrictions it is not surprising to find that the elements of the estimator $\tilde{\beta}$ are linearly dependent. In fact, it would have been very surprising if this linear dependency did not hold. Indeed, for the case where the constant term is known to be zero, the reader ought to verify that the procedure of Proposition 12 gives exactly the same results as we had obtained at the end of the preceding chapter. Problem 12 at the end of this chapter is instructive in this connection.

Let us now examine the problems posed when the relation

$$A\beta = a$$

is treated as a *hypothesis to be tested*. As before, A is an $r \times (n + 1)$ known matrix, and a is an $r \times 1$ vector of known constants (otherwise arbitrary). The problem may be posed as one involving an estimator for $A\beta$ and a test of the hypothesis that $A\beta = a$. Before we proceed to the solution of this problem we need the following useful proposition:

Proposition 13. *Let*

$$y = X\beta + u$$

be the standard GLM *subject to assumptions* (A.1) *through* (A.3) *of this chapter. Let A be an $r \times (n + 1)$ matrix of fixed constants of rank $r \leq n + 1$. Then the* BLUE *of $A\beta$ is $A\tilde{\beta}$ where $\tilde{\beta}$ is the* BLUE *of β.*

PROOF. The BLUE of β is the OLS estimator

$$\tilde{\beta} = (X'X)^{-1}X'y.$$

Let Hy be any other (i.e., other than $A\tilde{\beta}$) linear unbiased estimator of $A\beta$. By unbiasedness we have

$$E(Hy) = HX\beta = A\beta.$$

If we write

$$H = A(X'X)^{-1}X' + D$$

we note, by the unbiasedness condition, that $DX = 0$. Thus, we can write

$$Hy - A\beta = A(X'X)^{-1}X'u + Du$$

and

$$\text{Cov}(Hy) = \sigma^2 A(X'X)^{-1}A' + \sigma^2 DD' = \text{Cov}(A\tilde{\beta}) + \sigma^2 DD'.$$

Thus, *for any linear unbiased estimator, say Hy,*

$$\text{Cov}(Hy) - \text{Cov}(A\tilde{\beta}) = \sigma^2 DD',$$

which is positive semidefinite.   q.e.d.

In order to carry out a test we need the distribution of the statistics involved. Thus, suppose that, in addition to (A.1) through (A.3), (A.4) holds as well. In this context we conclude that

$$A\tilde{\beta} \sim N[A\beta, \sigma^2 A(X'X)^{-1}A'].$$

The application of Proposition 13 is now straightforward. Thus, take

$$H_0: \quad A\beta = a,$$

as against the alternative

$$H_1: \quad A\beta \neq a.$$

Under the null hypothesis

$$(A\tilde{\beta} - a) \sim N[0, \sigma^2 A(X'X)^{-1}A'],$$

and thus

$$\frac{[A\tilde{\beta} - a]'[A(X'X)^{-1}A']^{-1}[A\tilde{\beta} - a]}{\sigma^2} \sim \chi_r^2.$$

If we could show that

$$\frac{\tilde{u}'\tilde{u}}{\sigma^2} = \frac{1}{\sigma^2}(y - X\tilde{\beta})'(y - X\tilde{\beta})$$

is distributed independently of the preceding quadratic form we would complete the solution.

But we note that when H_0 is true

$$(A\tilde{\beta} - a)'(A(X'X)^{-1}A')^{-1}(A\tilde{\beta} - a)$$
$$= u'X(X'X)^{-1}A'[A(X'X)^{-1}A']^{-1}A(X'X)^{-1}X'u \qquad (37)$$
$$\tilde{u}'\tilde{u} = u'(I - X(X'X)^{-1}X')u$$

It is easily verified that the matrices of the two quadratic forms in (37) are *idempotent* and that *their product vanishes*. We, thus, conclude

$$\frac{(A\tilde{\beta} - a)'[A(X'X)^{-1}A']^{-1}(A\tilde{\beta} - a)}{\tilde{u}'\tilde{u}} \frac{T - n - 1}{r} \sim F_{r, T-n-1}. \qquad (38)$$

In the discussion culminating in Proposition 12 we have shown how to estimate the parameters of a GLM subject to the restriction

$$A\beta = a.$$

In the discussion just completed we have shown how, after estimating the parameters of a GLM without regard to any restrictions, we can test whether restrictions on the parameter vector, in the form given above, are supported by the sample evidence. Can any connection be made between the two procedures? This is, indeed, the case. In point of fact, *we shall show that the test statistic in (38) can be obtained from the sum of squares of the residuals from two regressions, one in which the restriction above is imposed and another in which the restriction is not imposed.*

To see this, consider again the estimator $\tilde{\beta}$ of item (i) of Proposition 12. On the assumption that the restriction is valid, the vector of residuals can be expressed as

$$\hat{u} = y - X\tilde{\beta} = (y - X\tilde{\beta}) - X(X'X)^{-1}A'[A(X'X)^{-1}A']^{-1}A(X'X)^{-1}X'u.$$

In view of the fact that

$$X'(y - X\tilde{\beta}) = 0$$

the squared residuals can be written as

$$\hat{u}'\hat{u} = \tilde{u}'\tilde{u} + u'X(X'X)^{-1}A'[A(X'X)^{-1}A']^{-1}A(X'X)^{-1}X'u$$

where, of course,

$$\tilde{u} = y - X\tilde{\beta}.$$

But $\hat{u}'\hat{u}$ is the sum of squares of the residuals from the regression in which the restrictions have been imposed, while $\tilde{u}'\tilde{u}$ is the sum of squares of the residuals from the regression in which the restrictions have not been imposed. Thus,

$$\frac{\hat{u}'\hat{u} - \tilde{u}'\tilde{u}}{\tilde{u}'\tilde{u}} \frac{T - n - 1}{r} \sim F_{r, T-n-1}. \tag{39}$$

Making use of the relation in (37) *we conclude that the two test statistics in* (38) *and* (39) *are identical.*

To recapitulate: in order to carry out a test of the hypothesis

$$H_0: \quad A\beta = a$$

it is sufficient to carry out two regressions. In one, we estimate β subject to the restriction imposed by H_0; let the sum of the squared residuals from this regression be $\hat{u}'\hat{u}$. In the other regression we estimate β without imposing H_0; let the sum of the squared residuals from this regression be $\tilde{u}'\tilde{u}$. The test statistic for H_0 is then given by (39).

The preceding discussion has established

Proposition 14. *Consider the* GLM

$$y = X\beta + u$$

under assumptions (A.1) *through* (A.4). *Suppose we consider, in addition*

$$A\beta = a, \tag{40}$$

where A is an $r \times (n + 1)$ *matrix of rank r. A test of the validity of* (40)—*treated as a testable hypothesis—can be carried out in the following two ways:*

(i) *Obtain the* BLUE *of* β, *say* $\tilde{\beta}$, *not subject to the restriction in* (40) *and consider the statistic*

$$\frac{(A\tilde{\beta} - a)'[A(X'X)^{-1}A']^{-1}(A\tilde{\beta} - a)}{\tilde{u}'\tilde{u}} \frac{T - n - 1}{r} \sim F_{r, T-n-1},$$

   where

$$\tilde{u} = y - X\tilde{\beta}.$$

(ii) *Obtain the estimator of β, say $\hat{\beta}$, subject to (40) and the estimator of β, say $\tilde{\beta}$, not subject to (40), and consider*

$$\hat{u} = y - X\hat{\beta}, \qquad \tilde{u} = y - X\tilde{\beta}.$$

Then

$$\frac{\hat{u}'\hat{u} - \tilde{u}'\tilde{u}}{\tilde{u}'\tilde{u}} \frac{T - n - 1}{r} \sim F_{r, T-n-1}, \qquad r = \text{rank}(A)$$

is a statistic for testing the statement in (40) treated as a testable hypothesis.

(iii) *The procedures in (i) and (ii) are equivalent not only in the sense of having the same distribution, but also in the sense that, apart from roundoff errors, they yield numerically the same statistic as well.*

Propositions 12 through 14 have wider implications than may be apparent from a casual reading. In part to stress and explore this aspect we give a number of examples illustrating some of their implications.

EXAMPLE 1. Consider the GLM under conditions (A.1) through (A.4). What this means is that the limit of what we may be prepared to state as true about an economic phenomenon, say y, is that it depends at most on $x_0, x_1, x_2, \ldots, x_n$ subject to the conditions (A.1) through (A.4). In the context of this information set we estimate β by

$$\tilde{\beta} = (X'X)^{-1}X'y.$$

Suppose we wish to test the hypothesis

$$H_0: \quad \beta_n = \beta_{n-1} = 0,$$

as against the alternative

$$H_1: \quad \beta_n \neq 0 \quad \text{or} \quad \beta_{n-1} \neq 0 \quad \text{or both.}$$

The preceding discussion states that the (generalized likelihood ratio) test statistic may be found by considering the regression of y on $x_0, x_1, x_2, \ldots, x_{n-2}$ and obtaining the sum of squared residuals, say Q_0. From the regression of y on $x_0, x_1, x_2, \ldots, x_n$ we have the sum of squared residuals

$$Q_1 = \tilde{u}'\tilde{u} = (y - X\tilde{\beta})'(y - X\tilde{\beta}).$$

The test statistic is then

$$\frac{Q_0 - Q_1}{Q_1} \frac{T - n - 1}{2} \sim F_{2, T-n-1}.$$

In addition, if it is *known* that

$$\beta_n = \beta_{n-1} = 0$$

then the estimator of $\beta_0, \beta_1, \ldots, \beta_{n-2}$ that utilizes the condition above is efficient when compared to the estimator of the same parameters not making use of this information. Of course, the distribution of the latter is obtained as the marginal distribution of the first $n - 1$ elements of $\hat{\beta}$ as given above.

EXAMPLE 2. Consider the production function

$$Q = \prod_{i=1}^{n} X_i^{\alpha_i} e^u.$$

Upon taking logarithms we have

$$\ln Q = y = \sum_{i=1}^{n} \alpha_i \ln X_i + u.$$

If the $\ln X_i$ can be taken to be fixed, independently of u, then we may wish to estimate the parameters α_i subject to the constraint

$$\sum_{i=1}^{n} \alpha_i = 1.$$

This is an instance of an application of Proposition 12.

EXAMPLE 3. Suppose in the GLM, as in Example 1, we wish to test the hypothesis

$$\beta_1 = \beta_3, \qquad \beta_2 = \beta_4.$$

This is accomplished by defining the matrix

$$A = \begin{bmatrix} 0 & 1 & 0 & -1 & 0 & \cdots & 0 \\ 0 & 0 & 1 & 0 & -1 & \cdots & 0 \end{bmatrix}.$$

If the model is estimated subject to

$$A\beta = 0$$

and not subject to this restriction, the relative difference in the sum of squared residuals (times an appropriate constant) gives the desired test statistic.

EXAMPLE 4. (Test of structure homogeneity). Consider a number, s, of GLM obeying (A.1) through (A.4), i.e., consider

$$y_{\cdot i} = X_i \beta_{\cdot i} + u_{\cdot i}, \qquad i = 1, 2, \ldots, s.$$

Here we assume, in addition, that the $u_{\cdot i}$ are mutually independent for $i = 1, 2, \ldots, s$ and, moreover, that

$$u_{\cdot i} \sim N(0, \sigma^2 I),$$

i.e., they have the same distribution.

We are only in doubt as to whether the s models are mean homogeneous, i.e., we wish to test

$$H_0: \quad \beta_{\cdot i} = \beta_*, \qquad i = 1, 2, \ldots, s,$$

as against the alternative

$$H_1: \quad \beta_{\cdot i} \neq \beta_* \quad \text{for at least one } i.$$

We observe that if we have T_i observations for the ith GLM, then each X_i is $T_i \times (n + 1)$, each $y_{\cdot i}$ is $T_i \times 1$, etc. Define

$$y = (y'_{\cdot 1}, y'_{\cdot 2}, \ldots, y'_{\cdot s})', \qquad X = \text{diag}(X_1, X_2, \ldots, X_s)$$
$$\beta = (\beta'_{\cdot 1}, \beta'_{\cdot 2}, \ldots, \beta'_{\cdot s}), \qquad u = (u'_{\cdot 1}, u'_{\cdot 2}, \ldots, u'_{\cdot s});$$

and write this compactly as

$$y = X\beta + u$$

where y is $T \times 1$, $T = \sum_{i=1}^{s} T_i$, X is $T \times s(n + 1)$, etc. In this context define the matrix

$$A = \begin{bmatrix} I & 0 & 0 & \cdots & & -I \\ 0 & I & & \cdots & & -I \\ \vdots & & & & & \vdots \\ 0 & & & \cdots & I & -I \end{bmatrix},$$

which is $(s - 1)(n + 1) \times s(n + 1)$, of rank $(s - 1)(n + 1)$. The null hypothesis can now be stated as

$$A\beta = 0.$$

Proposition 14, then, states that in order to carry out the test of this hypothesis it would be sufficient to carry out the following two regressions. First, obtain the estimator of β subject to the restrictions above, say $\hat{\beta}$, and compute the sum of squared residuals

$$\hat{u}'\hat{u} = (y - X\hat{\beta})'(y - X\hat{\beta}).$$

Second, obtain the estimator of β not subject to this restriction, say $\tilde{\beta}$; observe that in this case the sum of squared residuals may be written as

$$\tilde{u}'\tilde{u} = (y - X\tilde{\beta})'(y - X\tilde{\beta}) = \sum_{i=1}^{s} \tilde{u}'_{\cdot i} \tilde{u}_{\cdot i} = \sum_{i=1}^{s} (y_{\cdot i} - X_i \tilde{\beta}_{\cdot i})'(y_{\cdot i} - X_i \tilde{\beta}_{\cdot i}).$$

The desired test is then

$$\frac{\hat{u}'\hat{u} - \tilde{u}'\tilde{u}}{\tilde{u}'\tilde{u}} \frac{T - s(n + 1)}{(s - 1)(n + 1)} \sim F_{r, T - s(n+1)}, \qquad r = (s - 1)(n + 1).$$

Notice that, computationally, carrying out the test is not particularly difficult. We need to carry out the s separate regressions, obtaining in the process the $\tilde{\beta}_{\cdot i}$, $i = 1, 2, \ldots, s$, and hence $\tilde{\beta}$ and $\tilde{u}'\tilde{u}$. In this example the only additional calculation involved is that for obtaining $\hat{\beta}$; however, the latter can be obtained as

$$\hat{\beta} = (\hat{\beta}'_*, \hat{\beta}'_*, \ldots, \hat{\beta}'_*)',$$

i.e., as the vector $\hat{\beta}_*$ repeated s times. Thus, we need only compute $\hat{\beta}_*$ which in this case is given by

$$\hat{\beta}_* = \left[\sum_{i=1}^{s} X_i' X_i \right]^{-1} \left[\sum_{i=1}^{s} X_i' y_{\cdot i} \right],$$

and we see that at no time need we invert a matrix of higher dimension than $(n + 1)$.

Remark 10. When $s = 2$ and the two GLM are two subsamples, say for example pre- and post-World War II observations, *the procedure given in Example 4 is referred to as a Chow test.* As we pointed out earlier (and particularly in Problem 9) this is a special case of a generalized likelihood ratio test.

EXAMPLE 5. Consider again the s GLM of Example 4 and suppose the test of interest relates only to the coefficients of the bona fide variables, and not to the constant terms of the individual models. We can employ the procedure above as follows. Let I be an $n \times n$ identity matrix, and put

$$I^* = (0, I),$$

where 0 is an $n \times 1$ (column) vector.

Define

$$A = \begin{bmatrix} I^* & 0 & 0 & \cdots & & -I^* \\ 0 & I^* & 0 & \cdots & & -I^* \\ \vdots & & & & & \vdots \\ 0 & & & \cdots & I^* & -I^* \end{bmatrix}.$$

The desired test can be carried out by first estimating the unrestricted model, thus obtaining the sum of squared residuals $\tilde{u}'\tilde{u}$, and then obtaining the estimator of the vector β subject to $A\beta = 0$, thus obtaining the sum of the squared residuals $\hat{u}'\hat{u}$. Since here $\text{rank}(A) = (s - 1)n$, we conclude

$$\frac{\hat{u}'\hat{u} - \tilde{u}'\tilde{u}}{\tilde{u}'\tilde{u}} \frac{T - n - 1}{r} \sim F_{r, T-n-1}, \qquad r = \text{rank}(A) = n(s - 1).$$

4 Mixed Estimators and the Bayesian Approach

It may occur in the context of a GLM that some information is available on the parameters, but not in exact form. A way in which this might arise is through estimates obtained in previous studies. However, this facet is best handled in the context of Bayesian analysis, and this aspect will also be pursued.

We may express this inexact prior knowledge through

$$r = R\beta + v,$$

where r and R and a known vector and matrix, respectively, and v is a random variable such that

$$E(v) = 0, \qquad \text{Cov}(v) = \Omega.$$

We assume that R is $s \times (n + 1)$ of rank $s \leq n + 1$. Thus we have two sources from which to make inferences regarding β—the GLM represented in

$$y = X\beta + u$$

under the standard assumptions (A.1) through (A.3), and the inexact prior information above.

The situation is somewhat analogous to that explored in Problem 13, where we had two GLM and were interested in testing a hypothesis on mean homogeneity. Here, of course, it is given that the parameters characterizing the conditional mean of y and r are identical, and hence a question of testing does not arise. We may write the two sources of information on β as

$$w = Z\beta + u^* \tag{41}$$

where

$$w = \begin{pmatrix} y \\ r \end{pmatrix}, \qquad Z = \begin{pmatrix} X \\ R \end{pmatrix}, \qquad u^* = \begin{pmatrix} u \\ v \end{pmatrix}.$$

The OLS estimator of β in the context of (41) is given by

$$\hat{\beta} = (X'X + R'R)^{-1}[X'y + R'r]. \tag{42}$$

Substituting from (41) we see

$$\hat{\beta} = \beta + (X'X + R'R)^{-1}[X'u + R'v].$$

If it is additionally asserted that R is nonstochastic and that u and v are mutually independent, it immediately follows that

$$E(\hat{\beta}) = \beta,$$

$$\text{Cov}(\hat{\beta}) = (X'X + R'R)^{-1}(\sigma^2 X'X + R'\Omega R)(X'X + R'R)^{-1}.$$

It may also be verified that if we ignored the inexact prior information and based our inference regarding β only on the sample (y, X) the resulting estimator would be less efficient. By its very nature prior information is limited. It does not grow with sample size; hence, as the latter increases its informational content relative to prior information increases until, in the limit, the latter becomes insignificant. We should also point out that the covariance matrix of u^* is *not* of the form encountered when we examined the structure of the basic general linear model. Thus, strictly speaking, we do not know whether $\hat{\beta}$ in (42) is the most efficient estimator we can produce for β. We shall take up such aspects in later discussion.

A more satisfactory approach to the problem of prior information is through Bayesian analysis. In this context we would argue as follows. The GLM is

$$y = X\beta + u$$

with

$$u \sim N(0, \sigma^2 I).$$

Given the parameters, σ^2, β, the joint distribution of the observations is

$$p(y; \beta, \sigma^2, X) = (2\pi)^{-T/2} |\sigma^2 I|^{-1/2} \exp\left\{ -\frac{1}{2\sigma^2} (y - X\beta)'(y - X\beta) \right\}. \quad (43)$$

Now suppose, for simplicity, that σ^2 is *known* and that the prior density of β is given by

$$p(\beta) \sim N(b_0, h^{-2}I).$$

Then the joint density of the observations *and* the unknown parameters is

$$p(\beta, y; X) = (2\pi)^{-(T+n+1)/2} |\sigma^2 I|^{-1/2} |h^2 I|^{1/2}$$

$$\times \exp\left\{ -\frac{1}{2\sigma^2} (y - X\beta)'(y - X\beta) \right\} \exp\{ -\tfrac{1}{2} h^2 (\beta - b_0)'(\beta - b_0) \}.$$

$$(44)$$

We recall that the process of inference in Bayesian analysis consists of making the transition from the prior to the posterior distribution of the unknown parameters; the latter is nothing more than the conditional distribution of the parameters, given the observations. In order to facilitate this process we simplify the exponentials in (44). To this effect we recall from Equation (7) of Chapter 1 that

$$(y - X\beta)'(y - X\beta) = (\beta - b)'X'X(\beta - b) + y'(I - M)y$$

where

$$b = (X'X)^{-1}X'y, \qquad M = X(X'X)^{-1}X'. \quad (45)$$

Collecting terms in β and completing the quadratic form we can rewrite the exponentials as

$$\exp(-\tfrac{1}{2}Q)\exp\{ -\tfrac{1}{2}(\beta - b_1)'S_1(\beta - b_1) \},$$

where

$$Q = \frac{1}{\sigma^2} y'(I - M)y + \frac{b'X'Xb}{\sigma^2} + b_0'(h^2 I)b_0 - b_1'S_1 b_1,$$

$$(46)$$

$$S_1 = \frac{X'X}{\sigma^2} + h^2 I, \qquad b_1 = S_1^{-1}\left(\frac{X'X}{\sigma^2} b + (h^2 I)b_0 \right).$$

We note that Q *does not contain* β. Thus the joint density in (44) can be written in the more suggestive form

$$p(\beta, y; X) = (2\pi)^{-(n+1)/2}|S_1|^{1/2}\exp\{-\tfrac{1}{2}(\beta - b_1)'S_1(\beta - b_1)\} \cdot K, \quad (47)$$

where

$$K = (2\pi)^{-T/2}|\sigma^2 I|^{-1/2}|h^2 I|^{1/2}|S_1|^{-1/2}\exp(-\tfrac{1}{2}Q). \quad (48)$$

But the first component of the right-hand side of (47) (i.e., all terms exclusive of K) is recognized as a multivariate normal with mean b_1 and covariance matrix S_1^{-1}. Hence, the marginal density of y, which is obtained by integrating β out of (47), is

$$p(y; X) = (2\pi)^{-T/2}|\sigma^2 I|^{-1/2}|h^2 I|^{1/2}|S_1|^{1/2}\exp(-\tfrac{1}{2}Q) = K. \quad (49)$$

Thus, the posterior distribution of the unknown parameter—which is the conditional density of β given the data—is simply

$$p(\beta|y, X) = \frac{p(\beta, y; X)}{p(y; X)} = (2\pi)^{-(n+1)/2}|S_1|^{1/2}\exp\{-\tfrac{1}{2}(\beta - b_1)'S_1(\beta - b_1)\}. \quad (50)$$

Remark 11. Occasionally, *the density in (49), i.e., the marginal distribution of the observations once the unknown parameters (given an appropriate distribution) have been integrated out, is termed the predictive density.*

Remark 12. The density in (50) is the posterior density of the unknown parameters given the data. Notice *that the inverse of its covariance matrix, S_1, is the sum of the inverse of the covariance matrix of the prior distribution and that of the OLS estimator of β. Notice, further, that its mean b_1 is a weighted sum of the OLS estimator of β and the mean of the prior distribution.*

Remark 13. It is important, here, to point out the similarities and differences in the two approaches. In the case of mixed estimators we seek to estimate *a fixed but unknown parameter.* The estimate we obtain is a weighted sum of the estimate resulting when we use *only* the sample observations, and that resulting when we use *only* the "prior" information. The same may be said of the Bayesian procedure, provided by "estimate" we understand, in this case, the mean of the posterior distribution. The difference is that in the Bayesian case we are not constrained to choose any particular estimate. What we have is a density function—modified by the information contained in the sample—of the unknown parameter β. Moreover, in choosing a particular estimate we do so by minimizing an appropriate loss function. In the standard case the common practice is to obtain estimates by using least squares or maximum likelihood methods—although, of course, we are free to use other suitable methods if we desire.

The covariance matrix of the posterior distribution is given in (46) but this is not the appropriate matrix to compare with the covariance matrix of the mixed estimator. *The covariance matrix in* (46) *does not refer to the procedure by which an estimate is obtained by Bayesian methods.* It is important to recognize this since it has been the subject of considerable confusion. If it is desired that a comparison be carried out between the moments of the two procedures we *must select some estimator in the Bayesian context. We recall that an estimator is a function of the sample that assumes a specific numerical value when the sample observations are given. If the Bayesian procedure is terminated in* (50) *what we have then is the posterior density for the unknown parameter β, not an estimate on the basis of which we can act if the occasion requires.* As discussed in Chapter 8 the numerical estimate of a parameter is selected, in a Bayesian context, by minimizing an appropriate loss function. If the loss function is, say,

$$C(\beta, \gamma) = c - (\beta - \gamma)'\Phi(\beta - \gamma), \tag{51}$$

where Φ is a positive definite matrix, c a scalar constant, and γ the estimate to be chosen, then minimizing the expected value of (51)—using the posterior density of β—with respect to γ yields

$$\gamma = b_1 = S_1^{-1}\left[\frac{X'X}{\sigma^2}b + (h^2I)b_0\right]. \tag{52}$$

This would be a Bayesian "estimator" of β that is comparable to the one in (42). It is, indeed, an estimator in that it is a function of the sample such that when the observations are given (and in this case the prior information as well) it assumes a specific numerical value. A Bayesian practitioner, however, will not now proceed to derive its distribution, compute its moments, or test any hypotheses. The matter will be settled as follows. Given the prior and sample information and given the cost of choosing γ when the parameter is, in fact, β the choice in (52) is the one that minimizes expected cost. All other considerations are irrelevant. Nonetheless, if for comparability with the estimator in (42) we wish to compute the moments of (52) we can do so quite easily. Thus, substituting for b and y in (52) we find

$$b_1 = S_1^{-1}\left[\frac{X'X\beta}{\sigma^2} + (h^2I)b_0 + \frac{X'u}{\sigma^2}\right],$$

which yields, using the prior distribution[5] of β,

$$E(b_1) = b_0. \tag{53}$$

The results in (53) make it abundantly clear that the moments of "estimates" are rather irrelevant in the Bayesian context, since the mean of the posterior distribution is an "unbiased" estimator of the mean of the priori distribution, which is known!

---

[5] This amounts to considering what Raifa and Schleifer [27] have called the *preposterior distribution.*

Remark 14. In the preceding we have assumed that σ^2 is known. Of course, it is rare that this is so. For unknown σ^2 we have to deal with a joint prior on β *and* σ^2. This makes the manipulations a great deal more complicated but does not add to our understanding of what the Bayesian procedure is all about. For this reason we shall not deal with such a case.

QUESTIONS AND PROBLEMS

1. Verify that (4) locates the *global* maximum of (2), for any prespecified admissible vector β. [*Hint*: what is $\partial^2 L/\partial v^2$, where $v = \sigma^2$?]

2. Provide the details of the proof of Proposition 6 by showing that $x'Ax$ and $x'Bx$ contain distinct subsets of the vector y, the elements of which are mutually independent.

2a. Use Proposition 7 to derive the conclusion of Proposition 8. [*Hint*: take A to be a column vector all of whose elements are zero save the ith, which is unity. Also note that the square root of an $F_{1,k}$-variable is distributed as a t_k-variable.]

3. Provide the details for establishing the relation in (25). [*Hint*: notice that $X^* = X(\begin{smallmatrix}I\\0\end{smallmatrix})$, where I is an identity matrix of dimension equal to the number of columns in X^*.]

4. Provide the details in establishing the relation in Equation (26). [*Hint*: notice that $X_2 = X(\begin{smallmatrix}0\\I\end{smallmatrix})$ where I is of dimension equal to the number of columns in X_2.]

5. Show that the test given in Proposition 8 is a special case of that formulated in Proposition 9. [*Hint*: what is the square of the test statistic in Proposition 8? How does it compare with the test statistic in Proposition 9 for the special case $k = 1$?]

6. If $X = (X^*, X_2)$ show that

$$(X'X)^{-1} = \begin{bmatrix} R_1 & -(X^{*\prime}X^*)^{-1}X^{*\prime}X_2R_2 \\ -(X_2'X_2)^{-1}X_2'X^*R_1 & R_2 \end{bmatrix}$$

where $R_1 = [X^{*\prime}(I - M_2)X^*]^{-1}, R_2 = [X_2'(I - M^*)X_2]^{-1}, M_2 = X_2(X_2'X_2)^{-1}X_2'$, $M^* = X^*(X^{*\prime}X^*)^{-1}X^{*\prime}$.

7. Show that

$$M = X(X'X)^{-1}X' = X^*R_1X^{*\prime} - M_2X^*R_1X^{*\prime} - M^*X_2R_2X_2' + X_2R_2X_2'.$$

8. Show that $M^*X_2R_2X_2'M^* = X^*R_1X^{*\prime} - M^*$. [*Hint*: $X^*R_1X^{*\prime}M^* = X^*R_1X^{*\prime}$, $(X^{*\prime}X^*)^{-1}X^{*\prime}X_2R_2 = R_1X^{*\prime}X_2(X_2'X_2)^{-1}$.]

9. If $y = X\beta + u, u \sim N(0, \sigma^2 I)$, show that the procedure outlined in Proposition 10 for testing the hypotheses contained therein is a generalized likelihood ratio procedure, i.e., it is derived from considering

$$\max_{\beta\mid H_0} L(\beta, \sigma^2; y, X)/\max_{\beta\mid H_1} L(\beta, \sigma^2; y, X),$$

where L is the likelihood function of the observations and $\max_{\beta\mid H_0}$ means, maximize with respect to β given that H_0 is true."

10. In the context of the maximum problem in Equation (31) show

    (i) $A(X'X)^{-1}A'$ is an $r \times r$ matrix of rank r,
    (ii) $\hat{\sigma}^2\lambda = [A(X'X)^{-1}A']^{-1}[A\hat{\beta} - a]$, thus justifying Equation (32).

11. Show that the matrix Φ defined in item (ii) of Proposition 12 is singular. [*Hint*: show there exists a nonsingular matrix M such that

$$X'X = MM', \qquad A'[A(X'X)^{-1}A']^{-1}A = M\Lambda M',$$

    Λ being the diagonal matrix of the solutions to

$$|\lambda X'X - A'[A(X'X)^{-1}A']^{-1}A| = 0].$$

12. In item (iii) of Proposition 12, suppose $A = e_1 = (1, 0, 0, \ldots, 0)$, $a = 0$ (i.e., the restriction is that the constant term is zero). Show that the marginal distribution of the remaining elements of $\hat{\beta}$, i.e., of $\hat{\beta}_1, \hat{\beta}_2, \ldots, \hat{\beta}_n$, is $N[\beta_*, \sigma^2(X_1'X_1)^{-1}]$, where $\beta_* = (\beta_1, \beta_2, \ldots, \beta_n)'$ and X is partitioned as: $X = (e, X_1)$, where X_1 is the matrix of observations on the bona fide explanatory variable $e = (1, 1, 1, \ldots, 1)'$. [*Hint*: if P is a positive definite matrix, γ a conformable (column) vector, and α a scalar such that $1 + \alpha\gamma'P^{-1}\gamma \neq 0$, then

$$(P + \alpha\gamma\gamma')^{-1} = P^{-1} - \frac{\alpha}{1 + \alpha\gamma'P^{-1}\gamma}P^{-1}\gamma\gamma'P^{-1}.$$

    Also, for $X = (e, X_1)$, use the partitioned inverse form of $(X'X)^{-1}$ in Problem 6.]

13. In Example 4 consider the case $s = 2$. Verify that

$$\hat{\beta} = \begin{pmatrix} \hat{\beta}_* \\ \hat{\beta}_* \end{pmatrix}$$

    with $\hat{\beta}_*$ as defined in the last equation of that Example. [*Hint*: $[(X_1'X_1)^{-1} + (X_2'X_2)^{-1}]^{-1} = X_2'X_2(X_1'X_1 + X_2'X_2)^{-1}X_1'X_1.$]

14. Recall that the (adjusted) coefficient of determination of multiple regression is defined by

$$\bar{R}_n^2 = 1 - \frac{\hat{u}'\hat{u}/T - n - 1}{y'\left(1 - \dfrac{ee'}{T}\right)y/T - 1}$$

    where the dependent variable y has been regressed on $n + 1$ explanatory variables (including the fictitious variable corresponding to the constant term.) Occasionally, it is the practice of empirical researchers to introduce additional variables into the GLM so long as $\bar{R}^2$ keeps increasing. While this is admittedly a practice of doubtful validity, prove the following. Given that the dependent variable y has been regressed on n variables $(x_0, x_1, \ldots, x_{n-2})$, the introduction of x_n in the regression will increase $\bar{R}^2$ if the t-ratio of the coefficient of x_n exceeds unity (in absolute value). [*Hint*: use Proposition 10 with $k = 1$ and note that

$$\bar{R}_n^2 - \bar{R}_{n-1}^2 = c(T - n)[(Q_1/Q_0) - 1],$$

    where: R_n^2 is the adjusted coefficient in the regression of y on $x_0, x_1, x_2, \ldots, x_n$; Q_0 is the sum of squared residuals in this regression; Q_1 is the sum of squared residuals in the regression of y on $x_0, x_1, \ldots, x_{n-1}$; and c is a constant of proportionality not depending on the explanatory variables.]

15. The gamma function is defined by

$$\Gamma(\alpha) = \int_0^\infty e^{-u} u^{\alpha-1} \, du, \qquad \alpha > 0.$$

(a) Show that if α is an integer, then $\Gamma(\alpha) = (\alpha - 1)(\alpha - 2) \cdots 1 = (\alpha - 1)!$
(b) Show that for any $\alpha > 0$, $\Gamma(\alpha + 1) = \alpha \Gamma(\alpha)$.
(c) Show that $\Gamma(\frac{1}{2}) = \sqrt{\pi}$.

[*Hint*: for (a) and (b) use integration by parts; for (c) recall that an $N(0, 1)$-variable has density $(1/\sqrt{2\pi})\exp(-\frac{1}{2}x^2)$.]

16. If $a_1.$ is any nonnull (n-element) row vector, show that there exists an orthogonal matrix A with $a_1.$ as its first row. [*Hint*: choose vectors $a_{i.}^*, i = 2, 3, \ldots, n$, such that the set $\{a_1., a_{i.}^*: i = 2, 3, \ldots, n\}$ is linearly independent and use Gram–Schmidt orthogonalization.]

17. To evaluate

$$2 \int_0^1 t^{2r}(1 - t^2)^{(m/2)-1} \, dt$$

put $t^2 = s$

to obtain

$$\int_0^1 s^{r+(1/2)-1}(1 - s)^{(m/2)-1} \, ds.$$

This is the Beta integral and is denoted by $B(r + \frac{1}{2}, m/2)$. Show that this can be evaluated as

$$B\left(r + \frac{1}{2}, \frac{m}{2}\right) = \frac{\Gamma\left(r + \frac{1}{2}\right)\Gamma\left(\frac{m}{2}\right)}{\Gamma\left(\frac{m}{2} + r + \frac{1}{2}\right)}.$$

18. Suppose

$$X = (x._1, X_1), \qquad y = X\beta + u.$$

Show that the coefficient of $x._1$, say $\hat{\beta}_1$, in the regression of y on X can be expressed as $\hat{\beta}_1 = (s._1' s._1)^{-1} s._1' y$, where $s._1 = x._1 - X_1(X_1'X_1)^{-1}X_1'x._1$, and that its variance is $\sigma^2(s._1' s._1)^{-1}$. [*Hint*: use the results of Problem 6.]

Appendix

The discussion in this appendix has two objectives:

(i) to examine the power aspects of tests of significance, i.e., the probability of rejecting the null hypothesis when the alternative is true;

(ii) to present the multiple comparison test as a complement to the usual F-test for testing, simultaneously, a number of hypotheses.

We remind the reader that if we wish to test the hypothesis

$$H_0: \quad \beta = 0,$$

as against the alternative

$$H_1: \quad \beta \neq 0,$$

where β is, say, an $(n + 1)$-element vector, then:

If the hypothesis is accepted we "know" that, within the level of significance specified, all elements of β are zero;

If the hypothesis is rejected, however, then all we "know" is that at least one element of β is nonzero. It would be desirable to go beyond this and determine which of the elements are and which are not, zero.

It is this function that is performed by the multiple comparison test.

We recall that designing a test for a hypothesis is equivalent to specifying a critical region. The latter is a subset of the sample space such that, if the (sample) observations fall therein we reject the hypothesis and if they do not accept it. The probability assigned to the critical region (as a function of the unknown parameters of the underlying distribution) is called the *power function* associated with the test (or the critical region). The reader may have observed that when we discussed various tests in the preceding chapter we had always derived the distribution of the test statistic on the assumption that the null hypothesis is correct. The usefulness of tests in discriminating between true and false hypotheses is, in part, judged by the level of significance—which is the value assumed by the power function on the assumption that the null hypothesis represents the truth. This gives us some indication of the frequency with which such procedures will reject true null hypotheses. Another criterion, however, is the power of the test—which is the value assumed by the power function on the assumption that the alternative hypothesis represents the truth. In order to evaluate the latter, however, we need the distribution of the test statistic when the alternative represents the truth. The perceptive reader should have been impressed by the fact that (and possibly wondered why) in our discussions we had always derived the distribution of the various test statistics on the assumption that the null hypothesis is true. In part, this is consistent with the nature and inherent logic of the procedure as well as the presumption that an investigator will resort to a statistical test when his substantive knowledge of a phenomenon has progressed to the stage where, having mastered nearly all its ramifications, he has an inkling that his entire conception of the phenomenon is consistent with a specific set of parametric values. This he formulates as the null hypothesis; and, indeed he would be very surprised if the test were to lead to its rejection.[6] In part, however, the practice is also due to the fact

---

[6] At least this is the motivation inherited from the physical sciences.

that the distributions under the alternative hypothesis are appreciably more difficult to deal with.

Insofar as we have dealt earlier with the chi-square, t-, and F-distributions, we shall now deal with the *noncentral* chi-square, t-, and F-distributions. The latter are the distributions of the (appropriate) test statistics, dealt with previously, when the alternative hypothesis is true.

A.1 Noncentral Chi Square

We recall that the chi-square variable with m degrees of freedom (typically denoted by χ_m^2 and more appropriately called *central chi square*) has the density function

$$h_m(z) = \frac{e^{-(1/2)z}\left(\dfrac{z}{2}\right)^{(m/2)-1}}{2\Gamma\left(\dfrac{m}{2}\right)},$$

where $\Gamma(\cdot)$ is the gamma function (See Problem 15). We also recall that if $y \sim N(0, I)$, then

$$y'y = \sum_{i=1}^{m} y_i^2 \sim \chi_m^2.$$

The noncentral chi-square distribution arises when, in the relation above, the means of (some of) the basic normal variables are *nonzero*. Thus, the problem to be examined is as follows: if x is an n-element (column) random vector such that

$$x \sim N(\theta, I),$$

what can we say about the distribution of

$$z = x'x?$$

To handle this problem we employ a transformation that reduces the situation as closely as possible to the standard chi-square distribution — also called the *central chi-square* distribution.

We note that there exists *an orthogonal matrix A* such that its first row is (θ'/δ) where

$$\delta = (\theta'\theta)^{1/2}.$$

(In this connection see also Problem 16.) Now put

$$y = Ax$$

and observe that

$$y \sim N(\mu, I), \qquad \mu = (\delta, 0, 0, \ldots, 0)'.$$

Since

$$x'x = x'A'Ax = y'y = y_1^2 + u, \qquad u = \sum_{i=2}^{n} y_i^2,$$

we see that the desired random variable $(x'x)$ has been expressed as the sum of the square of an $N(\delta, 1)$ random variable (y_1^2) and a χ_m^2 $(m = n - 1)$ random variable (u). Moreover we know that y_1^2 and u are mutually independent.

To find the distribution of $x'x$ we can begin with the joint density of (y_1, u), which is

$$\frac{e^{-(1/2)(y_1 - \delta)^2}}{\sqrt{2\pi}} \cdot \frac{\left(\dfrac{u}{2}\right)^{(m/2) - 1} e^{-(1/2)u}}{2\Gamma\left(\dfrac{m}{2}\right)}, \qquad m = n - 1.$$

From this, applying the results of Proposition 4 of Chapter 8, we can hope to find the density of $y_1^2 + u$. Thus, consider the transformation

$$z = y_1^2 + u, \qquad z \in (0, \infty),$$

$$t = \frac{y_1}{z^{1/2}}, \qquad t \in (-1, 1).$$

The Jacobian of this transformation is $z^{1/2}$. Hence, the joint density of (z, t) is

$$\frac{e^{-(1/2)\delta^2} e^{\delta t z^{1/2}} e^{-(1/2)z} \left(\dfrac{z}{2}\right)^{(n/2) - 1} (1 - t^2)^{(m/2) - 1}}{\sqrt{\pi} \, 2\Gamma\left(\dfrac{m}{2}\right)}$$

$$= \frac{e^{-(1/2)\delta^2}}{2\sqrt{\pi}\,\Gamma\left(\dfrac{m}{2}\right)} \sum_{r=0}^{\infty} \frac{\delta^r}{r!} 2^{r/2} \left(\dfrac{z}{2}\right)^{[(n+r)/2] - 1} e^{-(1/2)z} t^r (1 - t^2)^{(m/2) - 1}.$$

The right member above is obtained by expanding

$$e^{\delta t z^{1/2}} = \sum_{r=0}^{\infty} \frac{\delta^r t^r z^{r/2}}{r!}.$$

To obtain the density of z we integrate out the irrelevant variable t, i.e., we obtain from the *joint density* of (z, t) the *marginal density* of z. The integration involved is

$$\int_{-1}^{1} t^r (1 - t^2)^{(m/2) - 1} \, dt.$$

and we observe that the integral vanishes for *odd r*. Hence in the series above we should replace r by $2r$, the range of r still being $\{0, 1, \ldots\}$. Now, for even powers we deal with

$$\int_{-1}^{1} t^{2r}(1 - t^2)^{(m/2)-1} \, dt = 2 \int_{0}^{1} t^{2r}(1 - t^2)^{(m/2)-1} \, dt$$

and the integral may be shown to be (see Problem 17)

$$\frac{\Gamma\!\left(r + \frac{1}{2}\right)\Gamma\!\left(\frac{m}{2}\right)}{\Gamma\!\left(\frac{n + 2r}{2}\right)}, \qquad n = m + 1.$$

Hence, remembering that terms containing *odd r vanish* and substituting the results above in the infinite sum, we find that the density function of z is given by

$$h(z) = \sum_{r=0}^{\infty} e^{-(1/2)\delta^2} \frac{\delta^{2r}}{(2r)!} \left[\frac{e^{-(1/2)z}\left(\dfrac{z}{2}\right)^{[(n+2r)/2]-1}}{2\Gamma\!\left(\dfrac{n+2r}{2}\right)} \right] \frac{\Gamma(r + \frac{1}{2})2^r}{\Gamma(\frac{1}{2})}$$

$$= \sum_{r=0}^{\infty} C_r h_{n+2r}(z),$$

where: $h_{n+2r}(\cdot)$ is the density function of a *central* chi-square variable with $n + 2r$ degrees of freedom;

$$C_r = e^{-(1/2)\delta^2} \frac{\delta^{2r}}{(2r)!} \frac{\Gamma(r + \frac{1}{2})2^r}{\Gamma(\frac{1}{2})} = e^{-(1/2)\delta^2}\left(\frac{\delta^2}{2}\right)^r \frac{1}{r!} \frac{2^r r!}{(2r)!} \frac{\Gamma(r + \frac{1}{2})2^r}{\Gamma(\frac{1}{2})};$$

and (see Problem 15)

$$\Gamma(\tfrac{1}{2}) = \sqrt{\pi}.$$

We shall now show that

$$\frac{2^{2r} r! \Gamma(r + \frac{1}{2})}{(2r)! \Gamma(\frac{1}{2})} = 1,$$

thus expressing the noncentral chi-square density as a weighted average of central chi-square densities (with degrees of freedom $n + 2r$). It is a *weighted average* since the weights $e^{-\lambda}(\lambda^r/r!)$ sum to unity, where $\lambda = \frac{1}{2}\delta^2$. In fact the weights are *simply the ordinates of the mass function of a Poisson distributed (discrete) random variable.*

Now, observe that

$$(2r)! = (2r)(2r - 1)2(r - 1)(2r - 3)2(r - 2)\cdots 1$$
$$= 2^r r!(2r - 1)(2r - 3)(2r - 5)\cdots 1.$$

Also

$$\Gamma(r + \tfrac{1}{2}) = (r - 1 + \tfrac{1}{2})(r - 2 + \tfrac{1}{2}) \cdots (r - r + \tfrac{1}{2})\Gamma(\tfrac{1}{2})$$

$$= \frac{1}{2^r}\,[(2r - 1)(2r - 3) \cdots 1]\Gamma(\tfrac{1}{2}).$$

Thus

$$\frac{2^{2r}r!\,\Gamma(r + \tfrac{1}{2})}{(2r)!\,\Gamma(\tfrac{1}{2})} = \frac{2^{2r}r!(1/2^r)[(2r - 1)(2r - 3) \cdots 1]\Gamma(\tfrac{1}{2})}{2^r r![(2r - 1)(2r - 3) \cdots 1]\Gamma(\tfrac{1}{2})} = 1$$

as claimed, and we can now write

$$h(z) = \sum_{r=0}^{\infty} e^{-\lambda}\frac{\lambda^r}{r!}\,h_{n+2r}(z), \qquad \lambda = \tfrac{1}{2}(\theta'\theta).$$

We, therefore, have the following:

Proposition A.1. *Let* $x \sim N(\theta, I)$, x *being* $n \times 1$; *the density function of*

$$z = x'x$$

is given by

$$h(z) = \sum_{r=0}^{\infty} e^{-\lambda}\frac{\lambda^r}{r!}\,h_{n+2r}(z),$$

where

$$\lambda = \tfrac{1}{2}\theta'\theta$$

and

$$h_{n+2r}(z) = \frac{e^{-(1/2)z}\left(\dfrac{z}{2}\right)^{[(n+2r)/2]-1}}{2\Gamma\left(\dfrac{n+2r}{2}\right)},$$

i.e., the density of z is a convex combination, with Poisson weights of central chi-square distributions with parameters $n + 2r$, $r = 0, 1, 2, \ldots$.

Remark A.1. The variable z above is said to have the *noncentral chi-square distribution* with parameters n and $\lambda = \tfrac{1}{2}\theta'\theta$. The latter is said to be the *noncentrality* parameter, while the former is called the *degrees of freedom* parameter. Such a variable is denoted by $\chi_n^2(\lambda)$. Note that $\chi_n^2(0)$ is the usual central chi-square variable.

A.2 Noncentral F-Distributions

We remind the reader that the (central) F-distribution with n_1, n_2 degrees of freedom is the distribution of the ratio

$$\frac{w_1}{w_2}\frac{n_2}{n_1}$$

where $w_i \sim \chi^2_{n_i}$, $i = 1, 2$, and the two random variables are mutually independent. We also recall that the (central) t-distribution is the distribution of the ratio

$$\frac{u}{\sqrt{\dfrac{w_2}{n_2}}}$$

where $u \sim N(0, 1)$, w_2 is $\chi^2_{n_2}$, and the two variables are mutually independent. Hence the square of a t-distributed variable with n_2 degrees of freedom is distributed according to the F-distribution with 1 and n_2 degrees of freedom. For symmetric t-tests, which are almost universal in applied econometric work, we may most conveniently employ the central F-distribution. Consequently, in the discussion to follow we shall only deal with noncentral F-distributions.

Thus, let w_i, $i = 1, 2$, be two mutually independent noncentral chi-square variables

$$w_i \sim \chi^2_{n_i}(\lambda_i), \qquad i = 1, 2.$$

We seek the distribution of

$$F = \frac{w_1}{w_2}\frac{n_2}{n_1}.$$

Instead of the distribution of F, however, we shall first find the distribution of

$$u = \frac{w_1}{w_1 + w_2}.$$

The reason why u is considered first instead of F is that there exist tabulations for the distribution of u (for selected parameter values) while no such tabulations exist for the distribution of F. Noting that

$$\Pr\{F \leq F_\alpha\} = \Pr\left\{u \leq \frac{n_1 F_\alpha}{n_2 + n_1 F_\alpha}\right\}$$

we see that operating with u is perfectly equivalent to operating with F.

To find the distribution of u we employ the same procedure as employed in the previous section. The joint density of w_1, w_2 is

$$e^{-(\lambda_1 + \lambda_2)} \sum_{r_1 = 0}^{\infty} \sum_{r_2 = 0}^{\infty} r_2 r_1 \frac{1}{4\Gamma\left(\dfrac{n_1 + 2r_1}{2}\right)\Gamma\left(\dfrac{n_2 + 2r_2}{2}\right)}$$

$$\times \left(\frac{w_1}{2}\right)^{[(n_1 + 2r_1)/2] - 1} \left(\frac{w_2}{2}\right)^{[(n_2 + 2r_2)/2] - 1} e^{-(1/2)(w_1 + w_2)}$$

where

$$r_2 = \frac{\lambda_1^{r_1}}{r_1!}, \qquad r_1 = \frac{\lambda_2^{r_2}}{r_2!}.$$

Use the transformation

$$u = \frac{w_1}{w_1 + w_2}, \qquad w = w_1, \qquad u \in (0, 1), \; w \in (0, \infty).$$

The Jacobian of the transformation is

$$\frac{w}{u^2}.$$

Upon substitution, the typical term of the infinite series above—apart from r_2, r_1—becomes

$$\frac{1}{C(r_1, r_2)} \left(\frac{w}{2}\right)^{[(n_1 + n_2 + 2r_1 + 2r_2)/2] - 1} e^{-(1/2)(w/u)} (1 - u)^{[(n_2 + 2r_2)/2] - 1} u^{-[(n_2 + 2r_2)/2] - 1}$$

where

$$C(r_1, r_2) = 2B(s_1, s_2)\Gamma\left(\frac{n_1 + n_2 + 2r_1 + 2r_2}{2}\right), \qquad s_1 = \frac{n_1 + 2r_1}{2},$$

$$s_2 = \frac{n_2 + 2r_2}{2},$$

and $B(s_1, s_2)$ is the beta function with parameters s_1 and s_2. To find the density of u we integrate out w; to this effect make the change in variable

$$w^* = \frac{w}{u}, \qquad w^* \in (0, \infty),$$

and observe that, apart from a factor of proportionality, we have to integrate with respect to w^*

$$\left[\frac{1}{2\Gamma\left(\dfrac{n_1 + n_2 + 2r_1 + 2r_2}{2}\right)} \left(\frac{w^*}{2}\right)^{[(n_1 + n_2 + 2r_1 + 2r_2)/2] - 1} e^{-(1/2)w^*} \right]$$

$$\times u^{s_1 - 1}(1 - u)^{s_2 - 1}.$$

The integral of the bracketed expression is unity, since it is recognized as the integral of the (central) chi-square density with $2(s_1 + s_2)$ degrees of freedom. Consequently, the density of u is given by

$$g(u; n_1, n_2, \lambda_1, \lambda_2) = e^{-(\lambda_1 + \lambda_2)} \sum_{r_1=0}^{\infty} \sum_{r_2=0}^{\infty} \frac{\lambda_1^{r_1} \lambda_2^{r_2}}{r_1! \, r_2!} \frac{1}{B(s_1, s_2)} u^{s_1 - 1} (1 - u)^{s_2 - 1},$$

which is recognized as a convex combination of beta distributions with parameters s_i, $i = 1, 2$, $s_i = (n_i + 2r_i)/2$. If we wish to obtain the density function of

$$F = \frac{w_1}{w_2} \frac{n_2}{n_1}$$

we need only observe that

$$u = \frac{n_1 F}{n_2 + n_1 F}.$$

The Jacobian of this transformation is

$$\left(\frac{n_1}{n_2}\right) \left(1 + \frac{n_1}{n_2} F\right)^{-2},$$

while

$$1 - u = \left(1 + \frac{n_1}{n_2} F\right)^{-1}.$$

Thus, substituting above, we have the density function of F,

$$h(F; n_1, n_2, \lambda_1, \lambda_2) = e^{-(\lambda_1 + \lambda_2)} \sum_{r_1=0}^{\infty} \sum_{r_2=0}^{\infty} \frac{\lambda_1^{r_1} \lambda_2^{r_2}}{r_1! \, r_2!} \frac{\left(\frac{n_1}{n_2}\right)}{B(s_1, s_2)} \frac{\left(\frac{n_1}{n_2} F\right)^{s_1 - 1}}{\left(1 + \frac{n_1}{n_2} F\right)^{s_2 + s_1}}.$$

We therefore have

Proposition A2. *Let*

$$w_i \sim \chi^2_{n_i}(\lambda_i), \qquad i = 1, 2,$$

be mutually independent. Then the density function of

$$F = \frac{w_1}{w_2} \frac{n_2}{n_1}$$

is given by

$$h(F; n_1, n_2, \lambda_1, \lambda_2), \qquad F \in (0, \infty).$$

The density function of

$$u = \frac{w_1}{w_1 + w_2}$$

is given by

$$g(u; n_1, n_2, \lambda_1, \lambda_2), \qquad u \in (0, 1).$$

In either case the density is uniquely determined by four parameters: n_1, n_2, *which are the degrees of freedom parameters, and* λ_1, λ_2, *which are the noncentrality parameters (of* w_1 *and* w_2 *respectively).*

The results of Proposition A.2 may be specialized to the various tests considered in the preceding chapter.

EXAMPLE A.1. Consider again the test given in Proposition 8. Suppose the null hypothesis is in fact false and

$$\beta_i = \bar{\beta}_i.$$

Define

$$\lambda = \bar{\beta}_i - \beta_i^0.$$

The square of the statistic developed there is

$$\frac{(\hat{\beta}_i - \beta_i^0)^2}{\hat{\sigma}^2 q_{ii}}$$

and is, thus, distributed as a noncentral F-distribution with parameters

$$n_1 = 1, \qquad \lambda_1 = \frac{\lambda^2}{2\sigma^2 q_{ii}}, \qquad n_2 = T - n - 1, \qquad \lambda_2 = 0.$$

Hence, given the critical region the power of the test may be evaluated on the basis of the noncentral F-distribution with parameters as above. Unfortunately, power calculations are not performed very frequently in applied econometric work, although the necessary tables do exist. (See e.g., [5].)

Remark A.2. The preceding example affords us an opportunity to assess, somewhat heuristically, the impact of multicollinearity on hypotheses testing. We shall examine collinearity in a subsequent chapter. For the moment it is sufficient to say that it is relevant to the extent that one or more of the explanatory variables in a GLM can be "explained" by the others.

In the context of this example note that if by $\hat{x}_{\cdot i}$ we denote that part of the variable $x_{\cdot i}$ that can be explained by (its regression on) the remaining explanatory variables and by $s_{\cdot i}$, the vector of residuals, we can write

$$x_{\cdot i} = \hat{x}_{\cdot i} + s_{\cdot i}.$$

Note that

$$s'_{.i}\hat{x}_{.i} = 0, \qquad s'_{.i}x_{.j} = 0, \qquad j \ne i,$$

the $x_{.j}$ being the observations on the remaining variables of the GLM.
    One can then show that

$$q_{ii} = (s'_{.i}s_{.i})^{-1}, \qquad \hat{\beta}_i = (s'_{.i}s_{.i})^{-1}s'_{.i}y,$$

where, of course,

$$s_{.i} = x_{.i} - X_1(X'_1X_1)^{-1}X'_1x_{.i}$$

and X_1 is the submatrix of X obtained when from the latter we suppress $x_{.i}$.
    If the hypothesis

$$\beta_i = \beta_i^0$$

is false then the quantity

$$\frac{\hat{\beta}_i - \beta_i^0}{\sqrt{\sigma^2 q_{ii}}}$$

has mean

$$\frac{\lambda}{\sqrt{\sigma^2 q_{ii}}}, \qquad \lambda = \beta_i - \beta_i^0.$$

Hence, the noncentrality parameter will be

$$\lambda_1 = \frac{1}{2}\frac{\lambda^2 s'_{.i}s_{.i}}{\sigma^2}.$$

If the sample is relatively collinear with respect to $x_{.i}$ then, even though λ^2 may be large, the noncentrality parameter λ_1 would tend to be "smaller" owing to the fact the sum of squared residuals, $s'_{.i}s_{.i}$, would tend to be (appreciably) smaller relative to the situation that would prevail if the sample were *not* collinear. Since, generally, we would expect the power of the test to increase with the noncentrality parameter it follows that collinearity would exert an undesirable effect in this context.

EXAMPLE A.2. A common situation is that of testing the returns to scale parameter in production function studies. If, for example, we deal with the Cobb–Douglas production function

$$AK^\alpha L^\beta U$$

(U representing the error process) we may estimate $\hat{\alpha} + \hat{\beta}$. If the variance of the sum is of the form $\hat{\sigma}^2 r$ then r will be known and $\hat{\sigma}^2$ would be proportional to a χ^2_{T-3}-variable. Thus, on the null hypothesis

$$H_0: \quad \alpha + \beta = 1,$$

we would have

$$\frac{(\hat{\alpha} + \hat{\beta} - 1)^2}{\hat{\sigma}^2 r} \sim F_{1, T-3}.$$

Suppose, however,

$$\alpha + \beta - 1 = \lambda \neq 0.$$

Then the statistic above is noncentral F with parameters

$$n_1 = 1, \qquad \lambda_1 = \frac{\lambda^2}{2\sigma^2 r}, \qquad n_2 = T - 3, \qquad \lambda_2 = 0.$$

Suppose that

$$\lambda_1 = .5, \qquad n_2 = 20.$$

From the relevant tables in [5] we see that the power of such a test (with level of significance .05) is approximately .3. But this means that the returns to scale parameter may be quite high and, depending on r, the probability of rejecting the null hypothesis will still only be .3. We will have such a situation if $\lambda = .7$ and $\sigma^2 r = .5$. Thus, such procedures lead to the acceptance of false null hypotheses alarmingly frequently. Of course, if λ_1 were even closer to zero the power would have been even lower. This is to warn the reader that *adjacent* hypotheses cannot be distinguished with great confidence when the sample on which the inference is based is not very large. Thus, in such procedures as that given in this example, if a given moderate-sized sample (say 30 or 40) leads us to accept the constant returns to scale hypothesis, the reader ought to bear in mind that in all likelihood it would also lead us to accept the null hypothesis

$$H_0: \quad \alpha + \beta = 1.1,$$

as against the alternative

$$H_1: \quad \alpha + \beta \neq 1.1.$$

The economic implications of constant, as against increasing, returns to scale, however, are very different indeed!

A.3 Multiple Comparison Tests

Consider the test of the hypothesis

$$H_0: \quad \beta_* = 0,$$

as against

$$H_1: \quad \beta_* = 0,$$

in the context of the GLM

$$y = X\beta + u,$$

where

$$\beta = (\beta_0, \beta_*')', \qquad X = (e, X_1),$$

in the usual notation of the chapter.

To carry out the test we proceed as follows. Letting $\sigma^2 Q_*$ be the covariance matrix of the OLS estimator of β_* we form the quantity

$$F = \frac{(\hat{\beta}_* - \beta_*)' Q_*^{-1} (\hat{\beta}_* - \beta_*)}{\hat{u}'\hat{u}} \cdot \frac{T - n - 1}{n}, \tag{A.1}$$

it being understood that there are T observations, that β_* contains n parameters, and that $\hat{u}$ is the vector of OLS residuals. Under the null hypothesis (A.1) becomes

$$F = \frac{\hat{\beta}_*' Q_*^{-1} \hat{\beta}_*}{\hat{u}'\hat{u}} \cdot \frac{T - n - 1}{n}. \tag{A.2}$$

The quantity in (A.2) is now completely determined, given the data, i.e., it is a statistic. The distribution of (A.2) under H_0 is central F with n and $T - n - 1$ degrees of freedom.

The mechanics of the test are these. From the tables of the $F_{n, T-n-1}$-distribution we determine a number, say F_α, such that

$$\Pr\{F_{n, T-n-1} \leq F_\alpha\} = 1 - \alpha,$$

where α is the level of significance, say $\alpha = .05$ or $\alpha = .025$ or $\alpha = .01$ or whatever.

The acceptance region is

$$F \leq F_\alpha,$$

while the rejection region is

$$F > F_\alpha.$$

The geometric aspects of the test are, perhaps, most clearly brought out if we express the acceptance region somewhat differently. Thus, we may write

$$(\hat{\beta}_* - \bar{\beta}_*)' Q_*^{-1} (\hat{\beta}_* - \bar{\beta}_*) \leq n\hat{\sigma}^2 F_\alpha, \tag{A.3}$$

where, evidently,

$$\hat{\sigma}^2 = \frac{\hat{u}'\hat{u}}{T - n - 1},$$

and where for the sake of generality we have written the hypothesis as

$$H_0: \quad \beta_* = \bar{\beta}_*,$$

as against

$$H_1: \quad \beta_* \neq \bar{\beta}_*.$$

In the customary formulation of such problems one takes

$$\bar{\beta}_* = 0.$$

In the preceding, of course, the elements of $\bar{\beta}_*$ are numerically specified. For greater clarity we may illustrate the considerations entering the test procedure, in the two-dimensional case, as in Figure A.1.

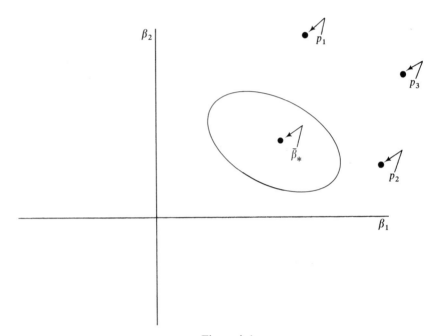

Figure A.1

The relation in (A.3) represents an ellipsoid with center at $\bar{\beta}_*$. If the statistic

$$(\hat{\beta}_* - \bar{\beta}_*)' Q_*^{-1} (\hat{\beta}_* - \bar{\beta}_*)$$

falls within the ellipsoid we accept H_0, while if it falls outside we accept H_1. If the statistic is represented by p_1 we will reject H_0, but it may well be that what is responsible for this is the fact that we "ought" to accept

$$\beta_1 = \bar{\beta}_{*1}, \qquad \beta_2 \neq \bar{\beta}_{*2}.$$

Similarly, if the statistic is represented by p_2 we will reject H_0 but it may well be that this is so because we "ought" to accept

$$\beta_1 \neq \bar{\beta}_{*1}, \qquad \beta_2 = \bar{\beta}_{*2}.$$

If the statistic is represented by p_3 then, perhaps, we "ought" to accept

$$\beta_1 \neq \bar{\beta}_{*1}, \qquad \beta_2 \neq \bar{\beta}_{*2}.$$

The F-test, however, does not give any indication as to which of these alternatives may be appropriate.

The configurations above belong to the class of functions of the form

$$h'\beta_* = 0. \tag{A.4}$$

It would be desirable to find a way in which tests for the hypotheses in (A.4) may be linked to the acceptance ellipsoid of the F-test. If this is accomplished, then perhaps upon rejection of a hypothesis by the F-test we may be able to find the "parameters responsible" for the rejection. Needless to say this will shed more light on the empirical implications of the test results.

The connection alluded to above is provided by the relation of the *planes of support* of an ellipsoid to the "supported" ellipsoid. Some preliminary geometric concepts are necessary before we turn to the issues at hand. In part because of this, the discussion will be somewhat more formal than in the earlier part of this appendix.

A.4 Geometric Preliminaries

Definition A.1. Let $x \in \mathbb{E}_n$, where $\mathbb{E}_n$ is the n-dimensional Euclidean space. The set of points

$$E = \{x : (x - a)'M(x - a) \le c\}$$

where c is a positive constant and M a positive definite matrix is said to be an ellipsoid with center at a.

Remark A.3. It entails no loss of generality to take $c = 1$. Thus, in the definition above, dividing through by c we have

$$E = \left\{ x : (x - a)'\left(\frac{M}{c}\right)(x - a) \le 1 \right\}.$$

If M is positive definite and $c > 0$ then, clearly,

$$\left(\frac{M}{c}\right)$$

is also a positive definite matrix. In the definitions to follow we will always take $c = 1$.

Remark A.4. The special case

$$M = \text{diag}(m_1, m_2, \ldots, m_n), \qquad a = 0,$$

is referred to as an ellipsoid in *canonical or standard form*.

Definition A.2. Let $x \in \mathbb{E}_n$. The set of points

$$S = \{x : (x - a)'(x - a) \le c\}$$

where $c > 0$ is said to be an n-dimensional sphere, with center at a. The special case

$$c = 1, \qquad a = 0,$$

is *referred to as a sphere in canonical or standard form, or simply a unit sphere.*

Remark A.5. Notice that an ellipsoid in canonical form is simply a unit sphere whose coordinates, x_i, $i = 1, 2, \ldots, n$, have been stretched or contracted, respectively, by the factors m_i, $i = 1, 2, \ldots, n$.

Lemma A.1. *Every ellipsoid*

(i) *can be put in canonical form, and*
(ii) *can be transformed into the unit sphere.*

PROOF. Let

$$E = \{x : (x - a)'M(x - a) \le 1\} \tag{A.5}$$

be an ellipsoid. Let R^*, Λ be, respectively, the matrices of characteristic vectors and roots of M. Put

$$y = R^{*\prime}(x - a)$$

and rewrite (A.5) as

$$E = \{y : y'\Lambda y \le 1\}. \tag{A.6}$$

This proves the first part of the lemma. For the second part, we put

$$z = \Lambda^{1/2}y, \qquad \Lambda^{1/2} = \mathrm{diag}(\lambda_1^{1/2}, \lambda_2^{1/2}, \ldots, \lambda_n^{1/2}),$$

and note that E is transformed to the unit sphere

$$S = \{z : z'z \le 1\}. \quad \text{q.e.d.}$$

Remark A.6. The relationship of the coordinates of the sphere (to which the ellipsoid is transformed) to the coordinates of the original ellipsoid is

$$z = \Lambda^{1/2}R^{*\prime}(x - a) = R'(x - a), \qquad R = R^*\Lambda^{1/2}$$

or

$$x = R'^{-1}z + a.$$

Definition A.3. Let $x \in \mathbb{E}_n$. The set of points

$$P = \{x : h'(x - a) = 0\}$$

is said to be a *plane through a, orthogonal to the vector h.*

Remark A.7. The plane above can be thought of as the set of vectors, measured from a, that are orthogonal to the vector h. It is obvious that $a \in P$.

Definition A.4. Let

$$P = \{x : h'(x - a) = 0\}$$

be the plane through a orthogonal to h, and put

$$h'a = c_0.$$

The planes described by

$$P_i = \{x : h'x = c_i\}$$

such that

$$c_i \neq c_0, \qquad c_i \neq c_j,$$

are said to be *parallel to P and to each other.*

Remark A.8. Parallel planes do not have points in common. Thus if P_1, P_2 are two parallel planes, let x_0 be a point on both of them. Since $x_0 \in P_1$ we have

$$h'x_0 = c_1$$

Since $x_0 \in P_2$ we have

$$h'x_0 = c_2.$$

But this implies $c_1 = c_2$, which is a contradiction.

Remark A.9. The planes through a and $-a, |a| \neq 0$, orthogonal to a vector h are parallel. Thus, the first plane is described by the equation

$$h'(x - a) = 0$$

while the second is described by

$$h'(x + a) = 0.$$

Rewriting slightly we have

$$h'x = h'a = c,$$
$$h'x = -h'a = -c.$$

Provided $c \neq 0$, it is evident that the two planes are parallel.

Definition A.5. Let $x \in \mathbb{E}_n$ and let E be the ellipsoid

$$E = \{x : (x - a)'M(x - a) \leq 1\}.$$

The plane (through the point x_0 and orthogonal to the vector h)

$$P = \{x : h'(x - x_0) = 0\}$$

is said to be a *plane of support of the ellipsoid* if

(a) E and P have one point in common, and
(b) E lies entirely on one side of P.

Remark A.10. To fully understand what "lies on one side of P" means, consider the special case of a line. Thus, if $n = 2$ and, for simplicity, $x_0 = 0$, we have

$$h_1 x_1 + h_2 x_2 = 0.$$

Notice that the equation above divides E_2 into three sets of points:

$$P = \{x : h'x = 0\}; \tag{A.7}$$

$$P_- = \{x : h'x < 0\}; \tag{A.8}$$

$$P_+ = \{x : h'x > 0\}. \tag{A.9}$$

The equation for the set given by (A.7) is

$$x_2 = -\frac{h_1}{h_2} x_1. \tag{A.10}$$

That for P_- is

$$x_2 = -\frac{c_1}{h_2} - \frac{h_1}{h_2} x_1, \tag{A.11}$$

and that for P_+ is

$$x_2 = \frac{c_1}{h_2} - \frac{h_1}{h_2} x_1. \tag{A.12}$$

Suppose, for definiteness in our discussion,

$$h_2 > 0.$$

The set P lies on the line in (A.10); the set P_- consists of all lines as in (A.11) with $c_1 > 0$; P_+ consists of all lines as in (A.12) with $c_1 > 0$. For any $c_1 > 0$ it is clear that the lines described in (A.11) lie below the line describing P, and the lines described by (A.12) lie above the line describing P. In this sense, P_- lies on one side of P (below) while P_+ lies on the other side (above). The directions "above" and "below" will of course be reversed for

$$h_2 < 0.$$

Now suppose we solve the problem of finding for a given ellipsoid the two parallel planes of support that are orthogonal to some vector h. By varying h it should be possible to describe the ellipsoid by its planes of support. This, then, will do exactly what we had asked at the beginning of this discussion, viz., produce a connection between the acceptance region of an F-test (an ellipsoid) and a number of linear hypotheses of the form $h'\beta_* = c$ (its planes of support). We have

Lemma A.2. *Let $x \in \mathbb{E}_n$, and let E be the ellipsoid*

$$E = \{x : (x - a)'M(x - a) \le 1\}$$

and h a given vector. Let x_0 be a point on the boundary of E, i.e., x_0 obeys

$$(x_0 - a)'M(x_0 - a) = 1.$$

Then the planes

$$P_1 = \{x: h'(x - a) = (h'M^{-1}h)^{1/2}\},$$
$$P_2 = \{x: h'(x - a) = -(h'M^{-1}h)^{1/2}\}$$

are parallel planes of support for E (at the points x_0 and $-x_0 + 2a$, respectively,) orthogonal to the vector h.

PROOF. Let E and h be given. By Lemma A.1, E can be transformed to the unit sphere

$$S = \{z: z'z \le 1\},$$

where

$$z = R'(x - a) \tag{A.13}$$

and R is such that

$$M = RR'. \tag{A.14}$$

The plane

$$P = \{z: z_0'(z - z_0) = 0\}$$

is a plane of support for S, where z_0 lies on the boundary of S and thus

$$z_0'z_0 = 1.$$

To see this, define

$$P_- = \{z: z_0'z - 1 < 0\}.$$

Clearly P and S have z_0 in common. Let $z \in S$. Then z obeys

$$z'z \le 1, \qquad |z| \le 1.$$

Since

$$|z_0'z| \le |z_0||z| \le |z| \le 1$$

we have

$$z_0'z - 1 \le 0,$$

so that if $z \in S$ then either $z \in P$ or $z \in P_-$; hence P is indeed a plane of support. Similarly, the plane

$$P^* = \{z: z_0'(z + z_0) = 0\}$$

is parallel to P and is a plane of support for S. First, $-z_0$ lies on the boundary of S since

$$(-z_0')(-z_0) = z_0'z_0 = 1;$$

moreover, $-z_0 \in P^*$ since

$$z_0'(-z_0 + z_0) = 0.$$

Second, define

$$P_+^* = \{z: z_0'z + 1 > 0\}$$

and note that if $z \in S$ then

$$|z| \le 1, \qquad |z_0'z| \le |z_0||z| \le 1.$$

Consequently,

$$z_0'z + 1 \ge 0,$$

which shows that S lies on one side of P_+^*; hence, the latter is a plane of support. The equation for P is, explicitly,

$$z_0'z = 1,$$

and that for P^* is

$$z_0'z = -1,$$

so that, indeed, they are parallel.

Let us now refer the discussion back to the original coordinates of the ellipsoid. From (A.13) we have that the equations for P and P^* are, respectively,

$$z_0'R'(x - a) = 1, \qquad z_0'R'(x - a) = -1. \tag{A.15}$$

Since we are *seeking planes that are orthogonal to a given vector h we must have*

$$rh = Rz_0,$$

where r is a constant. Alternatively,

$$z_0 = rR^{-1}h. \tag{A.16}$$

But since

$$1 = z_0'z_0 = r^2h'R'^{-1}R^{-1}h = r^2(h'M^{-1}h),$$

we conclude that

$$r = \pm(h'M^{-1}h)^{-1/2}$$

and the desired planes are

$$h'(x - a) = \pm(h'M^{-1}h)^{1/2} \tag{A.17}$$

We now show explicitly that

$$h'(x - a) = (h'M^{-1}h)^{1/2} \tag{A.18}$$

is a plane of support of E through the point

$$x_0 = a + R'^{-1}z_0 \tag{A.19}$$

orthogonal to the vector h.

Noting Equation (A.16) and related manipulations we can write (A.18) more usefully as

$$h'(x - a - R'^{-1}z_0) = 0 \qquad\qquad\qquad\text{(A.20)}$$

since

$$(h'M^{-1}h)^{1/2} = r(h'M^{-1}h) = h'R'^{-1}z_0. \qquad\qquad\text{(A.21)}$$

Now substituting from (A.19) in (A.20) we verify that x_0 lies on the plane described by (A.18). But x_0 lies also on E since

$$(x_0 - a)'M(x_0 - a) = z_0'z_0 = 1.$$

Moreover, if $x \in E$ then

$$|R'(x - a)| \leq 1$$

so that

$$|h'(x - a)| = \left| \frac{1}{r} z_0' R'(x - a) \right| \leq \frac{1}{r} = (h'M^{-1}h)^{1/2}.$$

Consequently,

$$h'(x - a) - (h'M^{-1}h)^{1/2} \leq 0,$$

which shows that (A.18) represents a plane of support for E. The plane parallel to that in (A.18) can be written, using the notation in (A.20), as

$$h'(x - a + R'^{-1}z_0) = 0. \qquad\qquad\qquad\text{(A.22)}$$

With x_0 as in (A.19) it is evident that

$$-x_0 + 2a = a - R'^{-1}z_0$$

also lies on the plane. Moreover,

$$(-x_0 + 2a - a)'M(-x_0 + 2a - a) = (x_0 - a)'M(x_0 - a) = 1,$$

so that it lies on E as well. It remains only to show that E lies on one side of the plane in (A.22). To see this let x be any point in E. As before, we have

$$|h'(x - a)| \leq (h'M^{-1}h)^{1/2}.$$

Consequently,

$$h'(x - a) + (h'M^{-1}h)^{1/2} \geq 0,$$

which shows that E lies on one side of the plane in (A.22).   q.e.d.

Corollary A.1. *The equation of the strip between the two parallel planes of support for E, say at x_0 and $-x_0 + 2a$, that are orthogonal to a vector h is given by*

$$-(h'M^{-1}h)^{1/2} \leq h'(x - a) \leq (h'M^{-1}h)^{1/2}.$$

PROOF. Obvious from the lemma.

Remark A.11. It is clear that the ellipsoid is contained in the strip above. Indeed, if we determine all strips between parallel planes of support, the ellipsoid E can be represented as the intersection of all such strips. Hence, a point x belongs to the ellipsoid E if and only if it is contained in the intersection.

Remark A.12. If the ellipsoid is centered at zero to begin with then

$$a = 0$$

and the results above are somewhat simplified. For such a case the strip is given by

$$-(h'M^{-1}h)^{1/2} \le h'x \le (h'M^{-1}h)^{1/2}$$

and the parallel planes of support have the points x_0 and $-x_0$, respectively, in common with the ellipsoid

$$x'Mx \le 1.$$

Multiple Comparison Tests—The S-Method

In this section we develop the S-method, first suggested by Scheffè (see [29] and [30]) and thus named after him. The method offers a solution to the following problem: upon rejection of a hypothesis on a set of parameters to find the parameter(s) responsible for the rejection.

Consider the GLM under the standard assumptions and further assume normality for the errors. Thus the model is

$$y = X\beta + u, \qquad u \sim N(0, \sigma^2 I),$$

and we have the OLS estimator

$$\hat{\beta} = \beta + (X'X)^{-1}X'u, \qquad \hat{\beta} \sim N(\beta, \sigma^2(X'X)^{-1}).$$

Let β_* be a subvector of β containing $k \le n + 1$ elements. Thus, in the obvious notation,

$$\hat{\beta}_* \sim N(\beta_*, \sigma^2 Q_*),$$

where Q_* is the submatrix of $(X'X)^{-1}$ corresponding to the elements of β_*. We are interested in testing, say,

$$H_0: \quad \beta_* = \bar{\beta}_*,$$

as against the alternative

$$H_1: \quad \beta_* \ne \bar{\beta}_*.$$

First, we recall that for the true parameter vector, say β_*^0,

$$\frac{1}{k}(\hat{\beta}_* - \beta_*^0)' \frac{Q_*^{-1}}{\hat{\sigma}^2} (\hat{\beta}_* - \beta_*^0) \sim F_{k, T-n-1},$$

where

$$\hat{\sigma}^2 = \frac{\hat{u}'\hat{u}}{T - n - 1}, \qquad \hat{u} = y - X\hat{\beta},$$

and $F_{k, T-n-1}$ is a central F-distributed variable with k and $T - n - 1$ degrees of freedom.

The mechanics of the test are as follows. Given the level of significance, say α, we find a number, say F_α, such that

$$\Pr\{F_{k, T-n-1} \leq F_\alpha\} = 1 - \alpha.$$

In the terminology of this appendix we consider the ellipsoid E with center $\hat{\beta}_*$;

$$E = \left\{ \beta_* : \frac{1}{k} (\beta_* - \hat{\beta}_*)' \frac{Q_*^{-1}}{\hat{\sigma}^2} (\beta_* - \hat{\beta}_*) \leq F_\alpha \right\}.$$

If the point specified by the null hypothesis lies in E, i.e., if

$$\bar{\beta}_* \in E$$

we accept H_0, while if

$$\bar{\beta}_* \notin E$$

we accept H_1.

Let us rewrite the ellipsoid slightly to conform with the conventions of this appendix. Thus

$$E = \{ \beta_* : (\beta_* - \hat{\beta}_*)'M(\beta_* - \hat{\beta}_*) \leq 1 \},$$

where

$$M = \frac{Q_*^{-1}}{k\hat{\sigma}^2 F_\alpha}.$$

The test then is as follows:

$$\text{accept } H_0 \quad \text{if } \bar{\beta}_* \in E;$$
$$\text{accept } H_1 \quad \text{if } \bar{\beta}_* \notin E.$$

In the previous discussion, however, we have established that a point belongs to an ellipsoid E if (and only if) it is contained in the intersection of the strips between all parallel planes of support. The strip between two parallel planes of support to E orthogonal to a vector h is described by

$$h'(\beta_* - \hat{\beta}_*) = \pm(h'M^{-1}h)^{1/2}.$$

Hence a point, say $\bar{\beta}_*$, obeys

$$\bar{\beta}_* \in E$$

if any only if for any vector $h \in \mathbb{E}_k$ it obeys

$$-(h'M^{-1}h)^{1/2} < h'(\bar{\beta}_* - \hat{\beta}^*) < (h'M^{-1}h)^{1/2}.$$

We are now in a position to prove

Theorem A.1. *Consider the GLM*

$$y = X\beta + u$$

under the standard assumptions, and suppose further that

$$u \sim N(0, \sigma^2 I).$$

Let

$$\hat{\beta} = \beta + (X'X)^{-1}X'u, \qquad \hat{\sigma}^2 = \frac{\hat{u}'\hat{u}}{T - n - 1}, \qquad \hat{u} = y - X\hat{\beta},$$

where β has $n + 1$ elements. Let β_* be a subvector of β containing $k \leq n + 1$ elements and $\hat{\beta}_*$ its OLS estimator, so that

$$\hat{\beta}_* \sim N(\beta_*, \sigma^2 Q_*),$$

where Q_* is the submatrix of $(X'X)^{-1}$ corresponding to the elements of β_*. Further, let there be a test of the hypothesis

$$H_0: \quad \beta_* = \bar{\beta}_*,$$

as against the alternative

$$H_1: \quad \beta_* \neq \bar{\beta}_*.$$

Then the probability is $1 - \alpha$ that simultaneously, for all vectors $h \in \mathbb{E}_k$, the intervals

$$(h'\hat{\beta}_* - S\hat{\sigma}_{\hat{\phi}}, \quad h'\hat{\beta}_* + S\hat{\sigma}_{\hat{\phi}})$$

will contain the true parameter point, where

$$S = (kF_\alpha)^{1/2}, \qquad \sigma_\phi^2 = \sigma^2 h' Q_* h = \text{Var}(h'\hat{\beta}_*), \qquad \hat{\sigma}_\phi^2 = \hat{\sigma}^2(h' Q_* h),$$

$$\phi = h'\beta_*,$$

and F_α is a number such that

$$\Pr\{F_{k, T-n-1} \leq F_\alpha\} = 1 - \alpha,$$

$F_{k, T-n-1}$ being a central F-distributed variable with k and $T - n - 1$ degrees of freedom.

PROOF. From the preceeding discussion we have determined that the mechanics of carrying out the F-test on the hypothesis above involves the construction of the ellipsoid E with center $\hat{\beta}_*$ obeying

$$E = \{\beta_* : (\beta_* - \hat{\beta}_*)' M(\beta_* - \hat{\beta}_*) \leq 1\},$$

where

$$M = \frac{Q_*^{-1}}{k\hat{\sigma}^2 F_\alpha}$$

and α is the specified level of significance. We accept

$$H_0: \quad \beta_* = \bar{\beta}_* \quad \text{if } \bar{\beta}_* \in E$$

and we accept

$$H_1: \quad \beta_* \neq \bar{\beta}_* \quad \text{if } \bar{\beta}_* \notin E.$$

Another implication of the construction above is that the ellipsoid E will contain the true parameter point with probability $1 - \alpha$. But a point lies in the ellipsoid above if it lies in the intersection of the strips

$$|h'(\beta_* - \hat{\beta}_*)| < (h'M^{-1}h)^{1/2} = S\hat{\sigma}_{\hat{\phi}}$$

for all $h \in \mathbb{E}_k$, where

$$\phi = h'\beta_*, \qquad \sigma_{\hat{\phi}}^2 = \text{Var}(h'\hat{\beta}_*).$$

Since the probability is $1 - \alpha$ that the ellipsoid contains the true parameter point, it follows that the probability is $1 - \alpha$ that the intersection of all strips

$$-S\hat{\sigma}_{\hat{\phi}} < h'(\beta_* - \hat{\beta}_*) < S\hat{\sigma}_{\hat{\phi}}$$

for $h \in \mathbb{E}_k$ will contain the true parameter point. Alternatively, we may say that the probability is $1 - \alpha$ that simultaneously, for all vectors $h \in \mathbb{E}_k$, the intervals

$$(h'\hat{\beta}_* - S\hat{\sigma}_{\hat{\phi}}, h'\hat{\beta}_* + S\hat{\sigma}_{\hat{\phi}})$$

will contain the true parameter point.   q.e.d.

Remark A.13. The result above is quite substantially more powerful than the usual F-test. If it is desired to test the hypothesis stated in the theorem we proceed to check whether the point $\bar{\beta}_*$ lies in the ellipsoid, i.e., whether

$$(\bar{\beta}_* - \hat{\beta}_*)'M(\bar{\beta}_* - \hat{\beta}_*) \leq 1.$$

If so we accept H_0; if not we accept H_1. In the latter case, however, we can only conclude that at least one element of β_* is different from the corresponding element in $\bar{\beta}_*$. But which, we cannot tell. Nor can we tell whether more than one such element differs. This aspect is perhaps best illustrated by a simple example. Consider the case where

$$\beta_* = (\beta_1, \beta_2)'.$$

If the F-test rejects H_0 we may still wish to ascertain whether we should accept:

$$\beta_1 = 0, \quad \beta_2 \neq 0;$$

or

$$\beta_2 = 0, \quad \beta_1 \neq 0;$$

or

$$\beta_1 \neq 0, \quad \beta_2 \neq 0, \quad \beta_1 \neq \beta_2.$$

The standard practice is to use the t-test on each of the relevant parameters. But proceeding in this sequential fashion means that the nominal levels of significance we claim for these tests are not correct. In particular, we shall proceed to test, e.g.,

$$\beta_1 = 0, \qquad \beta_2 \neq 0$$

only if the initial F-test leads us to accept

$$\beta_* \neq 0.$$

Then, the t-test above is a conditional test and the level of significance could not be what we would ordinarily claim. The theorem above ensures that we can carry out these tests simultaneously, at the α level of significance. Thus, consider the vectors

$$h_{.1} = (1, 0)', \qquad h_{.2} = (0, 1)', \qquad h_{.3} = (1, -1)'$$

and define

$$\phi_1 = \beta_1, \qquad \phi_2 = \beta_2, \qquad \phi_3 = \beta_1 - \beta_2,$$
$$Q_* = (q_{ij}), \qquad i, j = 1, 2.$$

If $\hat{\sigma}^2$ is the OLS induced estimate of σ^2 we have

$$\hat{\sigma}^2_{\hat{\phi}_1} = \hat{\sigma}^2 q_{11}, \qquad \hat{\sigma}^2_{\hat{\phi}_2} = \hat{\sigma}^2 q_{22}, \qquad \hat{\sigma}^2_{\hat{\phi}_3} = \hat{\sigma}^2 (q_{11} - 2q_{12} + q_{22}).$$

The intervals induced by the S-method are

$$(\hat{\beta}_1 - (2F_\alpha)^{1/2}\hat{\sigma}_{\hat{\phi}_1}, \hat{\beta}_1 + (2F_\alpha)^{1/2}\hat{\sigma}_{\hat{\phi}_1}), \quad (\hat{\beta}_2 - (2F_\alpha)^{1/2}\hat{\sigma}_{\hat{\phi}_2}, \hat{\beta}_2 + (2F_\alpha)^{1/2}\hat{\sigma}_{\hat{\phi}_2})$$
$$(\hat{\beta}_1 - \hat{\beta}_2 - (2F_\alpha)^{1/2}\hat{\sigma}_{\hat{\phi}_3}, \hat{\beta}_1 - \hat{\beta}_2 + (2F_\alpha)^{1/2}\hat{\sigma}_{\hat{\phi}_3}),$$

where F_α is a number such that

$$\Pr\{F_{2, T-n-1} \leq F_\alpha\} = 1 - \alpha.$$

Remark A.14. The common practice in testing the hypotheses above is to apply the t-test *seriatim*. Let t_α be a number such that

$$\Pr\{|t_{T-n-1}| \leq t_\alpha\} = 1 - \alpha,$$

where t_{T-n-1} is a central t-variable with $T - n - 1$ degrees of freedom. The intervals based on the t-statistic are

$$(\hat{\beta}_1 - t_\alpha\hat{\sigma}_{\hat{\phi}_1}, \hat{\beta}_1 + t_\alpha\hat{\sigma}_{\hat{\phi}_1}), \qquad (\hat{\beta}_2 - t_\alpha\hat{\sigma}_{\hat{\phi}_2}, \hat{\beta}_2 + t_\alpha\hat{\sigma}_{\hat{\phi}_2})$$
$$(\hat{\beta}_1 - \hat{\beta}_2 - t_\alpha\hat{\sigma}_{\hat{\phi}_3}, \hat{\beta}_1 - \hat{\beta}_2 + t_\alpha\hat{\sigma}_{\hat{\phi}_3}).$$

It is not correct to say that the true level of significance of tests based on these intervals is the stated one and we certainly cannot state that the probability is $1 - \alpha$ that simultaneously the three intervals above contain the true parameter point.

Remark A.15. To make the comparison between intervals given by the S-method and those yielded by the *t*-statistic concrete, let us use a specific example. Thus, take

$$\alpha = .05, \quad k = 2, \quad T - n - 1 = 30,$$

so that

$$F_\alpha = 3.32, \quad t_\alpha = 2.04.$$

Suppose that a sample yields

$$Q_* = \begin{bmatrix} .198 & .53 \\ .53 & 1.84 \end{bmatrix}, \quad \hat\sigma^2 = .05.$$

We can easily establish that any combination of estimates $\hat\beta_1, \hat\beta_2$ obeying

$$\frac{\hat\beta_*' Q_*^{-1} \hat\beta_*}{2\hat\sigma^2} = 220.57\hat\beta_1^2 + 23.74\hat\beta_2^2 - 127.10\hat\beta_1\hat\beta_2 > 3.32$$

will result in rejection of the null hypothesis

$$\beta_* = 0.$$

Further, we obtain

$$\hat\sigma_{\hat\phi_1}^2 = .0099, \quad \hat\sigma_{\hat\phi_2}^2 = .092, \quad \hat\sigma_{\hat\phi_3}^2 = .049.$$

The intervals based on the S-method are

$$(\hat\beta_1 - .256, \hat\beta_1 + .256), \quad (\hat\beta_2 - .782, \hat\beta_2 + .782),$$
$$(\hat\beta_1 - \hat\beta_2 - .570, \hat\beta_1 - \hat\beta_2 + .570).$$

The intervals based on (bilateral) *t*-tests, all at the nominal significance level of 5%, are

$$(\hat\beta_1 - .203, \hat\beta_1 + .203), \quad (\hat\beta_2 - .619, \hat\beta_2 + .619),$$
$$(\hat\beta_1 - \hat\beta_2 - .451, \hat\beta_1 - \hat\beta_2 + .451).$$

The reader should note that the invervals based on the S-method are appreciably wider than those based on bilateral *t*-tests. This is, indeed, one of the major arguments employed against the multiple comparisons test. In the general case, the comparison of the width of these two sets of intervals depends on the comparison of

$$(kF_{\alpha; k, T-n-1})^{1/2} \quad \text{and} \quad t_{\alpha; T-n-1}.$$

Since

$$t_{T-n-1}^2 = F_{1, T-n-1}$$

it follows that the comparison may also be said to rest on the difference

$$kF_{\alpha; k, T-n-1} - F_{\alpha; 1, T-n-1}.$$

Moreover, it may be verified (even by a casual look at tables of the F-distribution) that the difference above for $T - n - 1$ in the vicinity of 30 grows with k; hence, it follows that the more parameters we deal with the wider the intervals based on the S-method, relative to those implied by the (bilateral) t-tests.

Remark A.16. It is clear that in the context of the example in Remark A.15 and basing our conclusion on bilateral t-tests, any estimates obeying

$$\hat{\beta}_1 > .203, \qquad \hat{\beta}_2 > .619, \qquad \hat{\beta}_1 - \hat{\beta}_2 > .451$$

will lead us to accept

$$\beta_1 \neq 0, \qquad \beta_2 \neq 0, \qquad \beta_1 - \beta_2 \neq 0.$$

Any estimates obeying

$$\hat{\beta}_1 > .203, \qquad \hat{\beta}_2 < .619$$

will lead us to accept

$$\beta_1 \neq 0, \qquad \beta_2 = 0,$$

while any statistics obeying

$$\hat{\beta}_1 < .203, \qquad \hat{\beta}_2 > .619$$

will lead us to accept

$$\beta_1 = 0, \qquad \beta_2 \neq 0.$$

Using the S-method, however, would require for the cases enumerated above (respectively):

$$\hat{\beta}_1 > .256, \qquad \hat{\beta}_2 > .782, \qquad \hat{\beta}_1 - \hat{\beta}_2 > .570;$$
$$\hat{\beta}_1 > .256, \qquad \hat{\beta}_2 < .782;$$
$$\hat{\beta}_1 < .256, \qquad \hat{\beta}_2 > .782.$$

It is worth noting that if the parameter estimates were, in fact,

$$\hat{\beta}_1 = .22, \qquad \hat{\beta}_2 = .80$$

and $\hat{\sigma}^2$ and Q_* as in Remark A.15, the relevant F-statistic would have been

$$F_{2,30} = 3.50.$$

Since, for $\alpha = .05$, $F_{\alpha; 2, 30} = 3.32$ the hypothesis

$$\beta_* = 0$$

would have been rejected. If, subsequently, we were to use a series of bilateral t-tests each at the nominal level of significance of 5% we could not reject the hypothesis

$$\beta_1 \neq 0, \qquad \beta_2 \neq 0, \qquad \beta_1 - \beta_2 \neq 0$$

since the estimates

$$\hat{\beta}_1 = .22, \qquad \hat{\beta}_2 = .80, \qquad \hat{\beta}_1 - \hat{\beta}_2 = -.58$$

will define the intervals

$$(.017, .423), \qquad (.181, 1.419), \qquad (-1.031, -.129).$$

On the other hand, if we employ the S-method of multiple comparisons we could not reject

$$\beta_1 = 0, \qquad \beta_2 \neq 0, \qquad \beta_1 - \beta_2 \neq 0.$$

This is so since the estimates will define the intervals

$$(-0.036, .476), \qquad (.018, 1.582), \qquad (-1.150, -0.010),$$

and the conclusions reached by the two methods will differ.

In view of the fact that the nominal levels of significance of the t-test are incorrect, it might be better to rely more extensively, in empirical work, on the S-method for multiple comparisons.

Remark A.17. It is important to stress that the S-method is not to be interpreted as a sequential procedure, i.e., we should not think that the multiple tests procedure is to be undertaken *only* if the F-test rejects the null hypothesis, say

$$\beta_* = 0.$$

If we followed this practice we would obviously have a *conditional* test, just as in the case of the sequential t-tests. In such a context the multiple tests could not have the stated level of significance. Their correct significance level may be considerably lower and will generally depend on unknown parameters. In this connection see the exchange between H. Scheffè and R. A. Olshen [31].

The proper application of the S-method requires that the type of comparisons desired be formulated prior to estimation rather than be formulated and carried out as an afterthought following the rejection of the null hypothesis by the F-test.

3 The General Linear Model III

1 Generalities

In the two preceding chapters we have set forth, in some detail, the estimation of parameters and the properties of the resulting estimators in the context of the standard GLM. We recall that rather stringent assumptions were made relative to the error process and the explanatory variables. Now that the exposition has been completed it behooves us to inquire as to what happens when some, or all, of these assumptions are violated. The motivation is at least twofold. First, situations may, in fact, arise in which some nonstandard assumption may be appropriate. In such a case we would want to know how to handle the problem. Second, we would like to know what is the cost in terms of the properties of the resulting estimators if we operate under the standard assumptions that, as it turns out, are not valid. Thus, even though *we may not know* that the standard assumptions are, in part, violated we would like to know what is the cost in case *they are violated.*

It is relatively simple to give instances of the first type of motivation. Thus, consider the problem of estimating the household's demand functions for n commodities, the information deriving from a sample of N households. Each unit of observation (household) gives rise to n general linear models, one demand function for each commodity (or group of commodities). It would be quite plausible here to assume that even if one household's behavior is independent of another's, the error process in the demand for the ith commodity on the part of a given household would not be independent[1] of that in its demand for the jth commodity. Thus, if we consider the entire system of

---

[1] Actually more pervasive forms of dependence will materialize due to the budget restriction imposed on a household's consumption activities and the utility maximization hypothesis.

demand functions as one large GLM, its error process could not be expected to have a *scalar covariance matrix*. Consequently, we must develop techniques to handle such cases.

Another motivation, which is somewhat simpler to grasp, relates to the possibility—again in cross-sectional samples—that the error process, while it is one of independent random variables with zero mean, it is not one of identically distributed ones. For instance, if we are dealing with an aspect of a firm's operation and the sample consists of firms, we cannot, in general, claim that the error process is one of i.i.d. random variables, if only because of great differences *in the size of firms*. Thus, minimally the variance will vary from observation to observation. This phenomenon in which the error process is one of independent zero mean but nonidentically distributed random variables (minimally differing in their variance parameter) is termed *heteroskedasticity*. Also, it may very well be that the error process does not have zero mean. The zero mean assumption may be interpreted as stating that the forces impinging on the dependent variable—beyond the enumerated explanatory variables— are as likely to affect it positively as they are to affect it negatively. A nonzero (constant) mean would imply that such forces would, on balance, exert a positive or a negative impact on the dependent variable.

Finally, the error process may fail to be one of independent variables at all. It *may be an autoregression of the first order*, i.e., the error term u_t may obey

$$u_t = \rho u_{t-1} + \varepsilon_t$$

where $|\rho| < 1$ and $\{\varepsilon_t : t = 0, \pm 1, \pm 2, \ldots\}$ is a process consisting of i.i.d. random variables. For this case we need special techniques to ensure that we obtain efficient estimators; we would also need to know what would be the properties of the resulting estimators if we processed data ignoring the special character of the error process.

With regard to explanatory variables, a number of issues may be raised. One is simply the failure (or near failure) of the rank condition imposed on the matrix of explanatory variables. We shall deal with this in the next chapter. Another arises when *the explanatory variables are stochastic and correlated with the error process*. This encompasses a wide class of problems and will be examined later when we deal, e.g., with estimation in a simultaneous equations context.

2 Violation of Standard Error Process Assumptions

2.1 Nonzero Mean

In the discussions of the preceding two chapters we had always maintained the assumption that the error process exhibited zero mean. Here we shall examine the consequence of its violation. There are two possibilities; either the mean is constant or not. As we shall see, the latter case is best treated as a problem in missing variables—a subject that will be treated at a later stage. To be precise, let

$$\{u_t : t = 1, 2, \ldots\}$$

be the error process of a GLM and suppose

$$\mu_t = E(u_t) \quad \text{for all } t.$$

If the GLM is given by

$$y_t = \sum_{i=0}^{n} \beta_i x_{ti} + u_t$$

then

$$E(y_t | x_t) = \mu_t + \sum_{i=0}^{n} \beta_i x_{ti},$$

where μ_t is a nonstochastic quantity that varies with t.

Operating as if the error process exhibits zero mean is equivalent to omitting the "variable" μ_t. Presumably, the latter is unobservable. Consequently, we are omitting one of the determinants of the conditional mean of the dependent variable and the consequence of such an occurrence will be examined at a later stage. Thus, let us consider the case

$$\mu_t = \mu \quad \text{for all } t.$$

Defining

$$v_t = u_t - \mu$$

we see that we can write the GLM as

$$y_t = \mu + \sum_{i=0}^{n} \beta_i x_{ti} + v_t,$$

and if $x_{t0} = 1$ for all t the only consequence of a nonzero mean is to change the constant term of the equation; if the latter does not contain a constant term, then a consequence of the nonzero mean is that one (constant term) is introduced. Problem 1 at the end of this chapter completes what can, usefully, be said on this topic.

2.2 Nonscalar Covariance Matrix

For the standard GLM model we assume that the error vector obeys

$$E(u) = 0, \qquad \text{Cov}(u) = \sigma^2 I.$$

Let us formally examine the consequences of dealing with an error process such that

$$\text{Cov}(u) = \Sigma$$

where Σ is a positive definite matrix.

We can investigate this from two points of view. First, if it is known that the covariance matrix is as above, how do we estimate efficiently the parameters of the GLM? Second, what are the properties of the OLS estimator of the vector β, of the GLM, when its error process has a covariance matrix as above? To answer the first question, let the GLM be

$$y = X\beta + u \tag{1}$$

where, as is customary, y is $T \times 1$, X is $T \times (n + 1)$, β is $(n + 1) \times 1$, and u is $T \times 1$.

In the discussion to follow we shall operate with the following basic assumptions:

(A.1) the explanatory variables $\{x_{ti}: t = 1, 2, \ldots, T, i = 0, 1, 2, \ldots, n\}$ are nonstochastic and $\text{rank}(X) = n + 1$, where $X = (x_{ti})$ and, of course, $T \geq n + 1$;

(A.2) $\lim_{T \to \infty} X'\Sigma^{-1}X/T = Q$ exists and Q is a nonsingular matrix;

(A.3) the error process $\{u_t: t = 1, 2, \ldots, T\}$ has zero mean and for every T it has a nonsingular covariance matrix, i.e., $\text{Cov}(u) = \Sigma$.

Remark 1. It is important to realize that (A.1) *does not* imply (A.2). What it does imply (see also Problem 2) is that for finite $T > n$

$$X'\Sigma^{-1}X$$

is a nonsingular matrix.

We may now answer the question how to estimate the parameter β of (1) under (A.1), (A.2), and (A.3).

Since Σ is positive definite, there exists a nonsingular matrix B such that

$$\Sigma^{-1} = B'B.$$

Consider the transformed system

$$w = Z\beta + v \tag{2}$$

where

$$w = By, \qquad Z = BX, \qquad v = Bu.$$

Notice that the parameter vector of interest, β, was not affected by the transformation. We also see that if Σ is known the elements of Z are nonstochastic and

$$\text{rank}(Z) = n + 1$$

Moreover,

$$E(v) = 0, \qquad \text{Cov}(v) = I.$$

Thus, *the (relevant) basic conditions of Proposition 2 of Chapter 1 apply to the model in Equation (2) and consequently the OLS estimator in the context of that equation is BLUE.*

In fact, we have

Proposition 1. *Consider the GLM in (1) subject to (A.1), (A.2), and (A.3) and suppose Σ is known. Then*

$$\hat{\beta} = (X'\Sigma^{-1}X)^{-1}X'\Sigma^{-1}y$$

is

(i) *unbiased,*
(ii) *consistent,*
(iii) *best within the class of linear (in y) unbiased estimators, in the sense that if $\bar{\beta}$ is any other linear unbiased estimator, then $\mathrm{Cov}(\bar{\beta}) - \mathrm{Cov}(\hat{\beta})$ is positive semidefinite.*

PROOF. Upon substitution for y we find

$$\hat{\beta} = \beta + (X'\Sigma^{-1}X)^{-1}X'\Sigma^{-1}u,$$

and unbiasedness is immediate in view of (A.1) and (A.3). For consistency, it is sufficient to show that

$$\lim_{T \to \infty} \mathrm{Cov}(\hat{\beta}) = 0.$$

But

$$\mathrm{Cov}(\hat{\beta}) = \mathrm{E}[(X'\Sigma^{-1}X)^{-1}X'\Sigma^{-1}uu'\Sigma^{-1}X(X'\Sigma^{-1}X)^{-1}] = (X'\Sigma^{-1}X)^{-1} \quad (3)$$

and

$$\lim_{T \to \infty} \mathrm{Cov}(\hat{\beta}) = \lim_{T \to \infty} \frac{1}{T}\left(\frac{X'\Sigma^{-1}X}{T}\right)^{-1} = \lim_{T \to \infty} \frac{1}{T}Q^{-1} = 0,$$

the last equality following by (A.2).

To prove (iii) let Hy be any other linear unbiased estimator of β. Then we have that

$$\mathrm{E}(Hy) = HX\beta = \beta,$$

which implies

$$HX = I.$$

Define a matrix C by

$$H = (X'\Sigma^{-1}X)^{-1}X'\Sigma^{-1} + C$$

and observe that $CX = 0$. Thus

$$\mathrm{Cov}(Hy) = \mathrm{Cov}(\hat{\beta}) + C\Sigma C'.$$

Since $C\Sigma C'$ is positive semidefinite the proof is concluded.   q.e.d.

Remark 2. The estimator of Proposition 1 is said to be the *generalized least squares* (GLS) estimator or the *Aitken* estimator (in honor of the British mathematician A. C. Aitken, who first proposed it).

Corollary 1. *If in addition to the conditions of Proposition 1 we have that*

(A.4) *for every finite T, $u \sim N(0, \Sigma)$,*

then the Aitken estimator has the distribution

$$\hat{\beta} \sim N[\beta, (X'\Sigma^{-1}X)^{-1}].$$

PROOF. Obvious.

Frequently we deal with situations in which the covariance matrix of the errors may be represented as

$$\Sigma = \sigma^2 \Phi,$$

where Φ is a known (positive definite) matrix and σ^2 is a positive but unknown scalar. If the matrix of the decomposition above is now treated as decomposing Φ^{-1}, rather than Σ^{-1}, i.e., if

$$\Phi^{-1} = B'B,$$

then in the context of the transformed model in Equation (2) we have an exact analog of the GLM as examined in the two chapters immediately preceding. Thus, *all inference theory developed applies*, provided we estimate σ^2 by

$$\hat{\sigma}^2 = \frac{1}{T}(w - Z\hat{\beta})'(w - Z\hat{\beta}) = \frac{1}{T}\tilde{v}'\tilde{v}.$$

We can show that

$$\frac{\tilde{v}'\tilde{v}}{\sigma^2} = \frac{1}{\sigma}v'[I - Z(Z'Z)^{-1}Z']\frac{v}{\sigma} \sim \chi^2_{T-n-1},$$

that it is distributed independently of $\hat{\beta}$, and so on. There is only one aspect that differs appreciably from the situation considered earlier. This is the manner in which R^2 (the coefficient of determination of multiple regression) is to be defined. In the usual way it is given by

$$R^2 = 1 - \frac{\tilde{v}'\tilde{v}}{w' I\left(-\dfrac{ee'}{T}\right)w}.$$

In the context of the model examined in this section a number of results obtained in Chapter 1 do not hold. For example, we verified there that in the case where a constant term is included in the specification

$$\frac{1}{T}e'y = \frac{1}{T}e'\hat{y},$$

where $e = (1, 1, 1, \ldots, 1)'$ and $\hat{y}$ is the vector of the predicted values of the dependent variable. *This is not true in the present context*, i.e., if we put

$$\hat{w} = Z\hat{\beta}$$

it is *not* true that

$$\frac{1}{T} e'w = \frac{1}{T} e'\hat{w}.$$

Moreover, it is not true that the residuals sum to zero, i.e., *it is not true, in the present context, that*

$$e'\tilde{v} = e'(w - Z\hat{\beta}) = 0.$$

These two statements can be made whether the model, as originally specified, did or did not contain a constant term; consequently, if we define R^2 as above a number of its properties examined in Chapters 1 and 2 will fail to hold.

An important property established there is that

$$\frac{R^2}{1 - R^2} \frac{T - n - 1}{n}$$

is a test statistic for testing the hypothesis

$$H_0: \quad \beta_* = 0,$$

as against the alternative

$$H_1: \quad \beta_* \neq 0,$$

where β_* is the vector of coefficients of the bona fide explanatory variables, i.e.,

$$X = (e, X_1), \qquad \beta = (\beta_0, \beta_*')',$$

and thus

$$y = e\beta_0 + X_1\beta_* + u. \tag{4}$$

If R^2 is defined in the usual way, it cannot, in the present context, serve the purpose noted above. If we wish R^2 to continue serving this purpose we should define it differently. Let us see how this can be done. Applying the transformation to (4) we have

$$w = Be\beta_0 + BX_1\beta_* + v. \tag{5}$$

The interesting aspect to note is that the transformed model does not contain a "constant term," i.e., *it does not contain a parameter whose associated explanatory variable is a constant over all sample observations*—at least it would not be so in general. The new variables are (in vector or matrix form expressing all T observations)

$$z_{.0} = Be, \qquad Z_1 = BX_1.$$

The vector $z_{.0}$ would not, in general, contain only one distinct element (i.e., the variable would not assume the same value for all T observations), while Z_1 is simply the transform of the bona fide explanatory variables.

Normally, the operation of centering is carried out by the idempotent matrix

$$I - \frac{ee'}{T}.$$

This centers all variables about their respective sample means and, more importantly from the point of view of our immediate concern, *it eliminates the parameter* β_0. We are basically interested in the model as exhibited in (4) since this is the form embodying the relevant economic aspects of the phenomenon under study. The form exhibited in (2) is statistically convenient, but the variables in Z may not be intrinsically interesting. Thus, the hypothesis

$$\beta_* = 0$$

is still of considerable interest in judging the fit of the model to the data. Hence R^2 ought to be defined in such a way as to convey information directly on this issue.

Since the objective is to enable us to use it in testing the null hypothesis above it is only reasonable to look for a centering operation that eliminates β_0. Such an operation can be carried out by the idempotent matrix

$$I - \frac{Bee'B'}{\phi}$$

where $\phi = e'\Phi^{-1}e$. Consequently, we ought to define the coefficient of determination by

$$R^2 = 1 - \frac{\hat{v}'\hat{v}}{w'\left(I - \dfrac{Bee'B}{\phi}\right)w}. \tag{6}$$

Another way to approach this problem is to note that, to test the null hypothesis above, we may apply the results of Proposition 10 of Chapter 2. This entails regressing w on Be and w on Z. The sum of squared residuals from the first regression is

$$Q_0 = w'\left(I - \frac{Bee'B'}{\phi}\right)w.$$

The sum of squared residuals from the second regression is

$$Q_1 = \hat{v}'\hat{v} = w'(I - Z(Z'Z)^{-1}Z')w.$$

The test statistic is thus

$$\frac{Q_0 - Q_1}{Q_1} \frac{T - n - 1}{n} \sim F_{n, T-n-1}.$$

We observe that when R^2 is defined as in Equation (6)

$$\frac{R^2}{1 - R^2} = \frac{Q_0 - Q_1}{Q_1},$$

which is proportional to the desired test statistic. Unfortunately, however, when R^2 is defined as in Equation (6) *it is no longer the case that we can interpret it as the square of the correlation coefficient between the actual and predicted values of the dependent variable within the sample*—in the context of the representation in Equation (5). Nor is it the case that we can interpret it as the relative reduction in the variability of the dependent variable around its sample mean through the introduction of the (transformed) bona fide explanatory variables.

Let us now turn our attention to the associated problem: if the GLM of Equation (1) obeys (A.1) through (A.4), what are the properties of the OLS estimator $\tilde{\beta}$ of the parameter β and of the associated inference tests carried out in the "usual way?" The OLS estimator is given by

$$\tilde{\beta} = (X'X)^{-1}X'y = \beta + (X'X)^{-1}X'u. \tag{7}$$

We easily verify

$$E(\tilde{\beta}) = \beta$$

$$\text{Cov}(\tilde{\beta}) = \sigma^2(X'X)^{-1}X'\Phi X(X'X)^{-1},$$

where we have again written[2]

$$\text{Cov}(u) = \sigma^2\Phi.$$

We can show that

$$\lim_{T \to \infty} \text{Cov}(\tilde{\beta}) = \sigma^2 \lim_{T \to \infty} \frac{1}{T}\left(\frac{X'X}{T}\right)^{-1}\frac{X'\Phi X}{T}\left(\frac{X'X}{T}\right)^{-1} = 0$$

on the assumption that

$$\lim_{T \to \infty} \frac{X'\Phi X}{T}$$

exists as a finite element nonsingular matrix, thus ensuring that

$$\lim_{T \to \infty} \left(\frac{X'X}{T}\right)^{-1}$$

is a finite-element and well-defined matrix. Thus, the OLS estimator is still unbiased and consistent; it is, of course, less efficient than the Aitken estimator.

---

[2] Note that if Σ is a positive matrix we can always write it as $\Sigma = \sigma^2\Phi$ where $\sigma^2 > 0$ and Φ is positive definite. This involves no sacrifice of generality whatever. Only when we assert that Φ is known do we impose (significant) restrictions on the generality of the results.

The reader should also note a theoretical curiosity (see, in particular, Problem 7), viz., that an "Aitken-like" estimator of the form

$$\tilde{\beta} = (X'RX)^{-1}X'Ry$$

with arbitrary nonsingular matrix R will be unbiased and consistent provided

$$\lim_{T \to \infty} \frac{X'RX}{T}$$

exists as a finite-element well-defined nonstochastic nonsingular matrix.

The important issues, however, in terms of econometric practice are not to be found in the preceding. Rather, they are as follows: if we process the data on the assumption that the model obeys

$$\mathrm{Cov}(u) = \sigma^2 I.$$

while in fact it obeys

$$\mathrm{Cov}(u) = \sigma^2 \Phi,$$

what can we say about the properties of the various "test statistics," "goodness of fit," etc? It is to these issues that we now turn, having satisfied ourselves *that the estimators thus obtained are unbiased and consistent, but inefficient.*

Under assumptions (A.1) through (A.4) the estimator in (7) obeys

$$\tilde{\beta} \sim N[\beta, \sigma^2(X'X)^{-1}X'\Phi X(X'X)^{-1}].$$

Operating in the "usual way" we would obtain the "estimator"

$$\tilde{\sigma}^2 = \frac{(y - X\tilde{\beta})'(y - X\tilde{\beta})}{T - n - 1} = \frac{\tilde{u}'\tilde{u}}{T - n - 1}. \tag{8}$$

We observe that $\tilde{\sigma}^2$ is a *biased* and *inconsistent* estimator of σ^2. To see this note that

$$(T - n - 1)\mathrm{E}(\tilde{\sigma}^2) = \mathrm{E}[u'(I - M)u] = \mathrm{tr}\ \mathrm{E}(uu')(I - M) = \sigma^2\ \mathrm{tr}\ \Phi(I - M),$$

where

$$M = X(X'X)^{-1}X.$$

Since Φ is positive definite, we can write

$$\Phi = N\Lambda N'$$

where N is the (orthogonal) matrix of its characteristic vectors and Λ the diagonal matrix of the corresponding characteristic roots arranged in decreasing order of magnitude. Consequently,

$$\mathrm{tr}\ \Phi(I - M) = \mathrm{tr}\ \Lambda N'(I - M)N.$$

But $N'(I - M)N$ is positive semidefinite, so that its diagonal elements are nonnegative—in fact, they are less than or equal to unity since $N'MN$ is also positive semi definite. Putting

$$S = \frac{1}{T - n - 1} N'(I - M)N$$

we have

$$E(\tilde{\sigma}^2) = \sigma^2 \text{ tr } \Lambda S = \sigma^2 \sum_{i=1}^{T} \lambda_i s_{ii},$$

where

$$0 \le s_{ii} \le 1, \qquad \sum_{i=1}^{T} s_{ii} = 1.$$

Thus, unless all the roots are equal,

$$E(\tilde{\sigma}^2) \ne \sigma^2.$$

But if all the roots are equal, then

$$\Phi = \lambda I$$

so that we are, in effect, back to the standard case which we considered earlier.

To show inconsistency requires further specifications of the properties of the error process. Nonetheless, it would be sufficient for the present discussion to observe that

$$\plim_{T \to \infty} \tilde{\sigma}^2 = \plim_{T \to \infty} \frac{u'u}{T - n - 1},$$

which will not, in general, be σ^2! Observing that

$$\tilde{\beta} - \beta = (X'X)^{-1} X'u,$$
$$\tilde{u} = (I - M)u,$$

we see that, contrary to the situation in the standard case, $\tilde{\beta}$ and $\tilde{u}$ are not mutually independent. Consequently, estimating the "covariance matrix" of the estimator $\tilde{\beta}$ as

$$\text{“Cov}(\tilde{\beta})\text{”} = \tilde{\sigma}^2 (X'X)^{-1}$$

and *operating in the "usual way" in testing hypotheses on individual (or groups of) coefficients will not lead to statistics possessing the t- or F-distributions.*

Thus, e.g., if q_{ii} is the ith diagonal element of $(X'X)^{-1}$ and β_i is the ith element of β, it would not make any sense to operate with

$$\frac{\hat{\beta}_i - \beta_i^0}{\left(q_{ii} \dfrac{\tilde{u}'\tilde{u}}{T - n - 1} \right)^{1/2}}$$

for the purpose of testing the null hypothesis

$$H_0: \quad \beta_i = \beta_i^0,$$

where β_i^0 is a specific number.

The quantity above (commonly referred to as the t-ratio) *does not have the central t-distribution with $T - n - 1$ degrees of freedom*. Consequently, nothing can be concluded by contemplating such ratios. The same may be said of other customary test statistics such as the coefficient of determination of multiple regression R^2. It may, nevertheless, be interesting to ask: what is the connection between

$$\tilde{\sigma}^2 (X'X)^{-1}$$

and the appropriate covariance matrix of the OLS estimator

$$\sigma^2 (X'X)^{-1} X' \Phi X (X'X)^{-1}$$

Since the former is random while the latter is not, no direct comparison can be made. We can, however, compare the latter with the expected value of the former—or its probability limit. Specifically, let

$$k_T = \text{tr} \, \frac{\Phi(I - M)}{T - n - 1}$$

so that

$$E(\tilde{\sigma}^2) = \sigma^2 k_T$$

and consider

$$k_T (X'X)^{-1} - (X'X)^{-1} X' \Phi X (X'X)^{-1}$$
$$= (X'X)^{-1} X' N (k_T I - \Lambda) N' X (X'X)^{-1}$$

where we have again employed the decomposition $\Phi \doteq N\Lambda N'$. It is clear that if $k_T I - \Lambda$ is positive semidefinite or negative semidefinite so is the left member. Moreover, if $k_T I - \Lambda$ is indefinite then so is the left member as well. But the typical nonnull element of this matrix is

$$k_T - \lambda_j = \sum_{i=1}^{T} \lambda_i s_{ii} - \lambda_j,$$

which is the difference between the weighted sum of the roots and the jth root. In general, this difference will be positive for some indices j and negative for others. Hence $k_T I - \Lambda$ is, in general, indefinite and thus so is

$$E(\tilde{\sigma}^2)(X'X)^{-1} - \sigma^2 (X'X)^{-1} X' \Phi X (X'X)^{-1},$$

except in highly special cases.

The discussion in the last part of this section may be summarized in

Proposition 2. *Consider the GLM of Equation (1) subject to assumptions (A.1) through (A.4) and suppose further that*

$$\text{Cov}(u) = \sigma^2 \Phi,$$

where Φ is a known positive definite matrix. Let B be a nonsingular matrix such that $\Phi^{-1} = B'B$. Let $\hat{\beta} = (X'\Phi^{-1}X)^{-1}X'\Phi^{-1}y$ be the best linear unbiased consistent estimator of β. Let $\hat{u} = y - X\hat{\beta}$ and define

$$\hat{\sigma}^2 = \frac{1}{T - n - 1}\hat{u}'B'B\hat{u}.$$

Then the following statements are true.

(i) *$\hat{\beta}$ is also the ML estimator of β.*
(ii) *$\hat{\beta}$ and $\hat{u}$ are mutually independent.*
(iii) *$E(\hat{\sigma}^2) = \sigma^2$*
(iv) *$\hat{\beta} \sim N[\beta, \sigma^2(X'\Phi^{-1}X)^{-1}]$*
(v) *$(T - n - 1)(\hat{\sigma}^2/\sigma^2) \sim \chi^2_{T-n-1}$.*
(vi) *Tests of hypotheses on the vector β can be carried out as follows. If q_{ii} is the ith diagonal element of $(X'\Phi^{-1}X)^{-1}$, a test of the hypothesis*

$$H_0: \quad \beta_i = \beta_i^0$$

can be based on

$$\frac{\hat{\beta}_i - \beta_i^0}{(\sigma^2 q_{ii})^{1/2}} \sim t_{T-n-1}.$$

Similarly, tests of hypotheses on groups of coefficients may be based on the F-distribution. Thus, if we partition

$$\beta = \begin{pmatrix} \beta_{(1)} \\ \beta_{(2)} \end{pmatrix}$$

and partition conformably

$$(X'\Phi^{-1}X)^{-1} = \begin{bmatrix} S_{11} & S_{12} \\ S_{21} & S_{22} \end{bmatrix}$$

a test of the hypothesis

$$H_0: \quad \beta_{(2)} = \bar{\beta}_{(2)}$$

can be carried out by means of the test statistic

$$\frac{1}{k}[\hat{\beta}_{(2)} - \bar{\beta}_{(2)}]'\frac{S_{22}^{-1}}{\hat{\sigma}^2}[\hat{\beta}_{(2)} - \bar{\beta}_{(2)}] \sim F_{k, T-n-1},$$

where k is the number of elements in $\beta_{(2)}$. The preceding results can be obtained as applications of results in Chapter 2 if we operate with the transformed model $By = BX\beta + Bu$.

(vii) *If it is desired to define the coefficient of determination of multiple regression R^2 in such a way as to serve routinely for the test of the hypothesis*

$$H_0: \quad \beta_* = 0,$$

where $\beta = (\beta_0, \beta'_*)'$, β_0 being the constant term of the equation, we should define it as

$$R^2 = 1 - \frac{\hat{u}'B'B\hat{u}}{y'B'\left(I - \dfrac{Bee'B'}{\phi}\right)By}$$

where

$$e = (1, 1, 1, \ldots, 1)', \qquad \phi = e'B'Be.$$

(viii) *Let R be an arbitrary nonsingular matrix. Then $\tilde{\beta} = (X'RX)^{-1}X'Ry$ is a consistent and unbiased estimator of β provided $\lim_{T \to \infty} (1/T)X'RX$ exists as a nonsingular nonstochastic matrix.*

(ix) *The OLS estimator, a special case of the estimator in (viii) with $R = I$, obeys*

$$\tilde{\beta} \sim N[\beta, \sigma^2(X'X)^{-1}X'\Phi X(X'X)^{-1}].$$

(x) *Defining*

$$\tilde{u} = y - X\tilde{\beta} = (I - M)u, \qquad M = X(X'X)^{-1}X',$$

it is not true that $\tilde{u}$ and $\tilde{\beta}$ of (ix) are mutually independent. The "estimator" $\tilde{\sigma}^2 = (1/T)\tilde{u}'\tilde{u}$ is a biased and (generally) inconsistent estimator of σ^2.

(xi) *If we operate with $\tilde{\beta}$ of (ix) according to the development in Chapter 2 we are committing serious errors in that none of the theory developed there is applicable in the present context.*

(xii) *The "estimator" $\tilde{\sigma}^2 (X'X)^{-1}$ of the covariance matrix of the OLS estimator $\tilde{\beta}$ (of (ix)) represents, in the mean, neither an overestimate nor an underestimate, in the sense that*

$$E(\tilde{\sigma}^2)(X'X)^{-1} - \sigma^2(X'X)^{-1}X'\Phi X(X'X)^{-1}$$

is an indefinite matrix, except in highly special cases.

2.3 Heteroskedasticity

As we observed earlier, *heteroskedasticity* is the case for which the error process $\{u_t: t = 1, 2, \ldots\}$ of the general linear model is one of independent zero mean nonidentically distributed random variables such that

$$E(u_t^2) = \sigma_t^2, \qquad t = 1, 2, \ldots \quad .$$

Thus the covariance matrix Σ of the previous section is a diagonal matrix. The situation arises most commonly in cross-sectional studies. If, for example, we are interested in the dividend policies of firms and the observations we have are on individual firms' dividend disbursements (and other relevant variables), it would not be appropriate to think that, having employed the GLM representation, the error term for all firms in the sample would have the same variance. We may well be prepared to accept the assertion that the residual error in one firm's behavioral equation is *independent* of that in another's. *It would, however, be rather inappropriate to assert that the variance would be the same for all firms. This would be particularly so if firms differ appreciably in size.* Would, for example, the residual variation be of the same order for a firm which is 1000 times the size of another? Having posed this as a problem, the question arises: upon processing a body of data, how do we recognize it? A frequent practice is to examine the residuals of OLS and obtain the correlation of such residuals with "size" or some other variable that the investigator may suspect of determining the magnitude of the variance. In the example used above we may hypothesize that

$$\sigma_t^2 = \sigma^2 A_t,$$

where A_t is the capital stock—or some other "size" variable—for the tth firm. If this is, indeed, the case then we have exactly the same situation as in the previous section. Thus

$$\Sigma = \sigma^2 A, \qquad A = \text{diag}(A_1, A_2, \ldots, A_T),$$

with A known.

Consequently, the entire development there is applicable.

2.4 Autocorrelated Errors

It is often the case that the residuals from an OLS regression exhibit a behavior that is not intuitively compatible with the i.i.d. assumption. We would, for example, intuitively expect in such a case that the residuals would alternate in sign fairly regularly. We would not think it generally acceptable to have, say, the first third of the residuals all positive, the second third all negative, and so on. Such behavior would justifiably arouse suspicion. We may suspect either that some relevant variable has been omitted from the explanatory set or that the probability structure of the error process is not of the i.i.d. variety. We have, in part, dealt with the first eventuality earlier and will also deal with it in a subsequent chapter. Here, we shall examine the second eventuality, and we shall consider a rather special form of violation of the i.i.d. assumption.

Thus, suppose the error process is

$$u_t = \rho u_{t-1} + \varepsilon_t, \qquad |\rho| < 1,$$

such that

$$\{\varepsilon_t : t = 0, \pm 1, \pm 2, \ldots\}$$

is a sequence of i.i.d. random variables with zero mean and finite variance σ^2.

Remark 3. In the standard GLM the error process $\{u_t : t = 1, 2, \ldots\}$ was asserted to have the properties now ascribed to the ε-process. The u-process above is referred to as a *first-order autoregressive process* with parameter ρ. Sometimes ρ is also referred to as the *autocorrelation coefficient*. If we also asserted that the ε's were normally distributed then the u-process would have been termed a *first-order Markov process*.

The plan of discussion shall be as follows. First, we shall determine how the GLM ought to be estimated (efficiently) given that the error process is a first-order autoregression. Second, we shall ask: having estimated the model parameters by OLS methods, is there any test whereby we can determine whether the error process is one of i.i.d. random variables or is a first-order autoregression?

Our first task is to determine the second-moment properties of the u-process. Suppose that the latter "started up" at "time" α, and that the initial condition is thus u_α. Using the definition of the u-process successively we have

$$u_{\alpha+1} = \rho u_\alpha + \varepsilon_{\alpha+1}$$

$$u_{\alpha+2} = \rho^2 u_\alpha + \rho u_{\alpha+1} + \varepsilon_{\alpha+2}$$

$$\vdots$$

$$u_{\alpha+t'} = \rho^{t'} u_\alpha + \sum_{i=0}^{t'-1} \rho^i \varepsilon_{\alpha+t'-i}.$$

Effect the change in index $t' + \alpha = t$. Then, we have

$$u_t = \rho^{t-\alpha} u_\alpha + \sum_{i=0}^{t-\alpha-1} \rho^i \varepsilon_{t-1}.$$

Assuming the process to have started up indefinitely far in the past, i.e., letting $\alpha \to -\infty$, we conclude that

$$u_t = \sum_{i=0}^{\infty} \rho^i \varepsilon_{t-1}$$

which expresses the u-process as a function of the ε's without regard to "initial conditions." This representation is a "consequence" of the "stability" condition, $|\rho| < 1$.

It is clear, then, that

$$E(u_t) = 0$$

and that for $\tau > 0$

$$E(u_{t+\tau} u_t) = \text{Cov}(u_{t+\tau}, u_t)$$

$$= \sum_i \sum_j \rho^i \rho^j E(\varepsilon_{t+\tau-i} \varepsilon_{t-j})$$

$$= \sigma^2 \frac{\rho^\tau}{1 - \rho^2}.$$

Consequently, if we put $u = (u_1, u_2, \ldots, u_T)'$,

$$\mathrm{Cov}(u) = \frac{\sigma^2}{1 - \rho^2} \begin{bmatrix} 1 & \rho & \rho^2 & \cdots & \rho^{T-1} \\ \rho & 1 & \rho & \cdots & \rho^{T-2} \\ \rho^2 & \vdots & \vdots & & \vdots \\ \vdots & & & & \\ \rho^{T-1} & \rho^{T-2} & \rho^{T-3} & \cdots & 1 \end{bmatrix} = \sigma^2 V, \quad (9)$$

which provides an implicit definition of V. It may be verified that

$$V^{-1} = \begin{bmatrix} 1 & -\rho & 0 & & \cdots & & & 0 \\ -\rho & 1 + \rho^2 & -\rho & 0 & & & & \\ 0 & -\rho & 1 + \rho^2 & -\rho & & & & \vdots \\ \vdots & & & & \ddots & & \ddots & 0 \\ & & & & 0 & -\rho & 1 + \rho^2 & -\rho \\ 0 & & \cdots & & & 0 & -\rho & 1 \end{bmatrix} \quad (10)$$

and that

$$|V| = \frac{1}{1 - \rho^2}.$$

We gather these results in

Proposition 3. *Let*

$$u_t = \rho u_{t-1} + \varepsilon_t, \qquad |\rho| < 1,$$

so that

$$\{\varepsilon_t : t = 0, \pm 1, \pm 2, \ldots\}$$

is a sequence of independent identically distributed random variables with mean zero and finite variance $\sigma^2 > 0$, and suppose the first-order autoregressive process (the u-process above) has been in operation since the indefinite past. Then, the following statements are true:

(i) $u_t = \sum_{i=0}^{\infty} \rho^i \varepsilon_{t-i}$;
(ii) $\mathrm{Cov}(u) = \sigma^2 V$, $u = (u_1, u_2, \ldots, u_T)'$, *where V is as in Equation (9)*;
(iii) *The inverse of V is given by Equation (10)*,
(iv) $|V| = 1/(1 - \rho^2)$.

Let the model under consideration be that of Equation (1) subject to assumptions (A.1) through (A.3) with the additional specification that the error process be a first-order autoregression (as in Proposition 3 above). We note that we are dealing with a special case of the problem considered in Section 2.2, i.e., we have

$$\Phi = V$$

and V depends on only one unknown parameter, ρ. Thus, here it becomes possible to think of estimating β and Φ (i.e., ρ) simultaneously. In the discussion of Section 2.2, this issue was never raised, since if Φ is a general positive definite matrix of order T it contains at least T unknown parameters and we cannot possibly hope to estimate these unknown parameters (consistently) using a sample of size T. By contrast, in this case, we can not only investigate the Aitken estimator of β, i.e., the estimator examined in Section 2.2, when V is known, but the *feasible Aitken estimator as well.*

As may have been apparent in Section 2.2, the Aitken estimator is obtained by minimizing

$$(y - X\beta)'V^{-1}(y - X\beta)$$

with respect to β, for known V. The feasible Aitken estimator *minimizes the expression above with respect to both ρ and β.*

We notice a very important difference between the situation here and the situation encountered in earlier discussions. Earlier, the minimand (or maximand) was quadratic in the elements of the vector of interest, β. Thus, *the first-order conditions were linear in the parameter vector of interest.* Here, *however, the parameter vector of interest is $(\beta', \rho)'$ and the minimand is of higher order than quadratic in the elements of the vector of interest.* Thus, the feasible Aitken estimator is a *nonlinear least squares*[3] (NLLS) estimator and we must be prepared for difficulties not encountered in our earlier discussions.

Let us see exactly what, if any, additional difficulties are presented by this problem. The minimand is

$$S(\beta, \rho; y, X) = (y - X\beta)'B'B(y - X\beta).$$

Here

$$B = \begin{bmatrix} \sqrt{1-\rho^2} & 0 & \cdots & & 0 & 0 \\ -\rho & 1 & & & & \\ 0 & -\rho & 1 & & & \vdots \\ 0 & & & & & 0 \\ \vdots & & & -\rho & 1 & 0 \\ 0 & & \cdots & 0 & -\rho & 1 \end{bmatrix}, \qquad B'B = V^{-1}.$$

As in Section 2.2 for given ρ, we can minimize $S(\beta, \rho; y, X)$ to obtain

$$\hat{\beta}(\rho) = (Z'Z)^{-1}Z'w, \qquad Z = BX, \qquad w = By.$$

Inserting the solution in the minimand we obtain the concentrated minimand

$$S^*(\rho; y, X) = y'[V^{-1} - V^{-1}X(X'V^{-1}X)^{-1}X'V^{-1}]y,$$

which is now to be minimized with respect to ρ. The estimator we seek, say $\hat{\rho}$, must obey

$$S^*(\hat{\rho}; y, X) \le S^*(\rho; y, X) \tag{11}$$

[3] Such estimators are more appropriately called *minimum chi-square* (MCS) estimators.

for all admissible ρ. The concentrated minimand, however, is markedly nonlinear, and a solution cannot easily be found by the usual calculus methods. Thus, we shall adopt alternative approaches. One immediately suggests itself when we realize that, without appreciable loss of relevance, we can think of ρ as constrained to lie in the interval $[-1 + \delta_1, 1 - \delta_2]$ for arbitrarily small positive δ_1, δ_2, e.g., $\delta_1 = \delta_2 = .001$ or $\delta_1 = \delta_2 = .01$. The basic idea of the approach is to trace the curve $S^*(\rho; y, X)$ by evaluating it at a number of points $\{\rho_i : i = 1, 2, \ldots, m\}$ which adequately cover the interval $[-1 + \delta_1, 1 - \delta_2]$. When this is done we may select our estimator, say $\hat{\rho}$, by the condition

$$S^*(\hat{\rho}; y, X) \leq S^*(\rho_i; y, X), \qquad i = 1, 2, \ldots, m.$$

While the condition above does not do *exactly* what is dictated by (11) we can still approximate the solution required by (11) as closely as desired by taking the increment $\rho_i - \rho_{i-1}$ sufficiently small. The details of the computational strategy will be left to a later stage; for the moment we shall concentrate on exactly what is involved in obtaining this estimator.

We observe that the estimator above is simply a special case of the estimator examined in Proposition 2 and, as we noted then, it can be obtained as an OLS estimator in the context of a transformed system. The transformation, in the present case, is an extremely simple one. Thus, in view of the definition of B above,

$$w = \begin{pmatrix} \sqrt{1 - \rho^2}\, y_1 \\ y_2 - \rho y_1 \\ \vdots \\ y_T - \rho y_{T-1} \end{pmatrix}, \qquad Z = (z_{\cdot 0}, z_{\cdot 1}, \ldots, z_{\cdot n}), \qquad z_{\cdot i} = \begin{pmatrix} \sqrt{1 - \rho^2}\, x_{1i} \\ x_{2i} - \rho x_{1i} \\ x_{3i} - \rho x_{2i} \\ \vdots \\ x_{Ti} - \rho x_{T-1, i} \end{pmatrix}$$

Thus, nothing more is involved than taking quasi differences of successive observations—the exception being the first observation, whose transform is simply a multiple of the original observation. Once these facts are realized, however, it becomes quite obvious that the search for the global minimum of $S^*(\rho; y, X)$ can be carried out quite simply—given the existence of computer regression programs.

Thus, for each ρ_i carry out the transformations above, obtaining w and Z; regress w on Z, and *obtain the sum of squared residuals from this regression. This is simply* $S^*(\rho_i; y, X)$, which will be obtained for $\rho_i : i = 1, 2, \ldots, m$. Consequently, *select that regression for which the sum of squared residuals is smallest; the ρ corresponding to this regression*, say ρ_{i_0}, *is the estimate of ρ*, say $\hat{\rho}$. *The coefficient vector corresponding to this regression is the estimate of β*, say $\hat{\beta}(\hat{\rho})$.

Formally, we define $\hat{\rho}$ by

$$S^*(\hat{\rho}; y, X) \leq S^*(\rho_i; y, X), \qquad i = 1, 2, \ldots, m,$$

and

$$\hat{\beta}(\hat{\rho}) = (X'V_{\hat{\rho}}^{-1}X)^{-1}X'V_{\hat{\rho}}^{-1}y,$$

where $V_{\hat{\rho}}^{-1}$ means the matrix V^{-1} of Equation (10) evaluated at $\hat{\rho}$. Thus, the nonlinearity of the problem has been circumvented by repeated application of OLS. Moreover, this approach has definitely located—within the specified limits of numerical accuracy—the global minimum of the function and not merely a local stationary point. What this means is that, were we to attempt to solve

$$\frac{\partial S^*}{\partial \rho} = 0,$$

the resulting solution, say $\bar{\rho}$, might correspond either to a minimum, a maximum, or an inflection point; we can't know which until we examine the second-order derivative.

While the search procedure outlined above has certain advantages, it is computationally expensive if we insist on a single search operation using an interval of length, say, .01. This would involve approximately 200 regressions beginning with $\rho_1 = -.99$ and ending with $\rho_m = .99$ in steps of .01. Often, the following shortcut has been found quite accurate. First, use a coarse grid of ρ values, say $\rho_1 = -.99$, $\rho_2 = -.9$, $\rho_3 = -.8, \ldots, \rho_{21} = .99$. Once the region of apparent minimum is located, say for example $\rho_{14} = .3$, then we search the region (.2, .4) using a finer grid of points, the interval between adjacent points being, say, .01. We may also fit a quadratic in the neighborhood of the minimum of the second search and minimize this quadratic with respect to ρ.

What we hope to avoid is the situation depicted in Figure 1 below, where, even though the apparent minimum is at .3, the true global minimum is

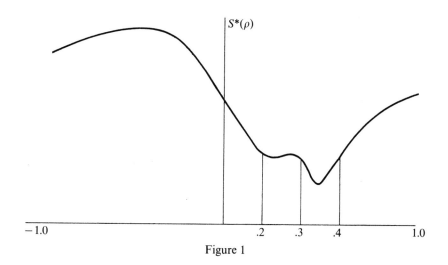

Figure 1

closer to .4. Of course a similar phenomenon may have occurred in a region far removed from the apparent minimum. Experience, however, indicates that this is not a frequent occurrence—and, at any rate, if one wants to avoid this problem one only has to shorten the interval between adjacent ρ's.

An alternative approach to the same problem is through the so-called Cochrane–Orcutt iteration, which we shall present in a slightly different way than originally presented by these authors [7]. The precedure is as follows. Let V_k^{-1} be the matrix V^{-1} of Equation (10) evaluated at the point ρ_k. For example, if $\rho_0 = 0$ then $V_0^{-1} = I$. Beginning with $\rho_0 = 0$ obtain the estimate

$$\hat{\beta}_{(0)} = (X'V_0^{-1}X)^{-1}X'V_0^{-1}y,$$

which is, of course, the OLS estimate. Compute the residuals $\hat{u}^{(0)} = y - X\hat{\beta}_{(0)}$ and obtain the estimate of ρ,

$$\hat{\rho}_{(1)} = \frac{\sum_{t=2}^{T} \hat{u}_t^{(0)}\hat{u}_{t-1}^{(0)}}{\sum_{t=2}^{T} \hat{u}_{t-1}^{(0)2}}.$$

Use $\hat{\rho}_1$ to evaluate V_1^{-1} and thus compute

$$\hat{\beta}_{(1)} = (X'V_1^{-1}X)^{-1}X'V_1^{-1}y$$

and the residual vector

$$\hat{u}^{(1)} = y - X\hat{\beta}_{(1)},$$

obtaining the second iterate

$$\hat{\rho}_{(2)} = \frac{\sum_{t=2}^{T} \hat{u}_t^{(1)}\hat{u}_{t-1}^{(1)}}{\sum_{t=2}^{T} \hat{u}_{t-1}^{(1)2}}.$$

We then use $\hat{\rho}_{(2)}$ to evaluate V_2^{-1} and so on until convergence is obtained, i.e., until

$$|\hat{\rho}_{(k-1)} - \hat{\rho}_{(k)}| < \delta,$$

where δ is a preassigned (small) positive quantity.

A proof that this procedure converges (in the case where the model is correctly specified) is given in Sargan [28]. It will not be repeated here since it is clearly beyond the scope of this book.

Notice that the Cochrane–Orcutt procedure also replaces a nonlinear problem by a sequence of linear problems. Thus, for given ρ it minimizes $S(\beta, \rho; y, X)$ to obtain β and, treating this β as fixed, determines ρ and so on until convergence is obtained.

The procedure actually suggested by the authors operates with

$$y_t - \rho y_{t-1} = \sum_{i=0}^{n} \beta_i(x_{ti} - \rho x_{t-1,i}) + (u_t - \rho u_{t-1}).$$

Beginning with $\rho_{(0)} = 0$ it computes the vector $\hat{\beta}_{(0)}$, whence it computes the residuals and the first iterate $\hat{\rho}_{(1)}$ and so on. The only difference between this and what we have examined above is that in the procedure as originally suggested by Cochrane and Orcutt one observation (the first) is unnecessarily eliminated from consideration.

Let us now summarize what is involved in feasible Aitken or minimum chi-square estimation of the parameters of a GLM whose error process is a first-order autoregression with parameter ρ.

Search procedure. Assuming that $\rho \in [-1 + \delta_1, 1 - \delta_2]$ for small $\delta_1, \delta_2 > 0$, subdivide the interval by points ρ_i, $i = 1, 2, \ldots, m$ so that $\rho_1 = -1 + \delta_1$, $\rho_m = 1 - \delta_2$. For each ρ_i form the matrix B and transform the data (y, X) to (w, Z); regress w on Z and obtain the sum of squared residuals from this regression. *The desired estimators of β, ρ, σ^2 are obtained from the regression exhibiting the smallest sum of squared residuals,* as follows. The estimator of ρ, say $\hat{\rho}$, is the ρ_i corresponding to this regression; the estimator of β, say $\hat{\beta}$, is the coefficient vector obtained from this regression; and the estimator of σ^2, say $\hat{\sigma}^2$, is the sum of the squared residuals of this regression divided by T—or by $T - n - 1$, it does not really matter. If $S(\beta, \rho; y, X)$ is the function to be minimized and if $\hat{\rho}$ is not equal to the end points $-1 + \delta_1$ or $1 - \delta_2$, then the procedure yields a numerical approximation to the solution of

$$\frac{\partial S}{\partial \beta} = 0, \qquad \frac{\partial S}{\partial \rho} = 0.$$

The accuracy of the approximation can be improved as much as is desired by making the interval between successive points sufficiently small, i.e., by making $|\rho_i - \rho_{i-1}| < \delta$ where δ is sufficiently small, and by finally fitting a quadratic at the end of the search operation.

Thus, the problem can be solved by a succession of ordinary regressions. A somewhat less expensive (if more inaccurate) procedure is to use, initially, a coarse grid of points in the search over ρ, and having located the region of the apparent minimum to search the region between the two adjacent points more intensively. In particular, if ρ_k represents the point of apparent minimum in the first search consider the region $[\rho_{k-1}, \rho_{k+1}]$ and search over a finer grid of points in this interval. The minimum over this interval will yield the desired estimators. If desired, a quadratic can be fitted and minimized between the points adjacent to the minimum.

Cochrane–Orcutt iteration. Obtain the OLS estimator of β and the OLS residuals; from the latter obtain the first iterate of ρ, say

$$\tilde{\rho}_{(1)} = \frac{\sum_{t=2}^{T} \tilde{u}_t^{(0)} \tilde{u}_{t-1}^{(0)}}{\sum_{t=2}^{T} \tilde{u}_{t-1}^{(0)2}}.$$

Using this $\tilde{\rho}_{(1)}$ form the matrix V_1^{-1}, which is the matrix V^{-1} of Equation (10) evaluated at $\tilde{\rho}_{(1)}$, and obtain the first iterate of β, say

$$\tilde{\beta}_{(1)} = (X'V_1^{-1}X)^{-1}X'V_1^{-1}y.$$

Using the residuals from this regression,

$$\tilde{u}^{(1)} = y - X\tilde{\beta}_{(1)},$$

compute the second iterate of ρ, say

$$\tilde{\rho}_{(2)} = \frac{\sum_{t=2}^{T} \tilde{u}_t^{(1)}\tilde{u}_{t-1}^{(1)}}{\sum_{t=2}^{T} \tilde{u}_{t-1}^{(1)2}}.$$

Form the matrix V_2^{-1}, which is again the matrix V^{-1} of Equation (10) evaluated at $\tilde{\rho}_{(2)}$, and obtain the second iterate of β, say

$$\tilde{\beta}_{(2)} = (X'V_2^{-1}X)^{-1}X'V_2^{-1}y.$$

Continue in this fashion until convergence is obtained, i.e., until, for a pre-assigned small quantity δ, at the kth iteration we find

$$|\tilde{\rho}_{(k)} - \tilde{\rho}_{(k-1)}| < \delta.$$

The converging iterates of this procedure also represent approximations to the solution of

$$\frac{\partial S}{\partial \beta} = 0, \qquad \frac{\partial S}{\partial \rho} = 0.$$

If the model is correctly specified the iteration will, normally, converge.

The relative merits of the two procedures are as follows. The search procedure—within limits of numerical accuracy—guarantees that we locate the global minimum. On the other hand, it is expensive. The iteration procedure may converge quite fast, in which case it would be less expansive. On the other hand, we have no guarantee that we have located the global minimum; nor do we have any guarantee that in *any particular application* the iteration will, indeed, converge. If the model is not correctly specified the iteration may cycle indefinitely.

Now that the computational aspects of these nonlinear estimators are settled, let us examine their properties. The first thing to note is that we do not have an explicit representation, for either estimator, that is a linear function of the observations on the dependent variable. Thus, the standard arguments employed in the preceding two chapters cannot be applied to show the consistency of the estimators for $\hat{\beta}$ and $\hat{\rho}$. Second, even if the error process were assumed to be normal we cannot hope to determine the density function of such estimators in the same straightforward way as we did earlier. These are two important casualties of the nonlinearities entailed by the autoregressive nature of the error terms. Nonetheless, consistency can be established by more involved arguments, and even without normality of the error process we can establish the asymptotic distribution of certain simple transforms of $\hat{\beta}$ and $\hat{\rho}$. We shall examine these issues, but not at the level of rigor or completeness exhibited in the preceding two chapters when dealing with the standard GLM. Observe, for example, that in the Cochrane–

Orcutt iteration the estimator of the autocorrelation coefficient is given, generally, by

$$\tilde{\rho} = \frac{\tilde{u}'\tilde{u}_{-1}}{\tilde{u}'_{-1}\tilde{u}_{-1}},$$

where

$$\tilde{u} = (\tilde{u}_2, \tilde{u}_3, \ldots, \tilde{u}_T)', \qquad \tilde{u}_{-1} = (\tilde{u}_1, \tilde{u}_2, \ldots, \tilde{u}_{T-1})$$

and

$$\tilde{u}_t = y_t - \sum_{i=0}^{n} \tilde{\beta}_i x_{ti}, \qquad t = 1, 2, \ldots, T.$$

This estimator obviously does depend on $\tilde{\beta}$ and, of course, is nonlinear in the dependent variable. Similarly,

$$\tilde{\beta} = (X'\tilde{V}^{-1}X)^{-1}X'\tilde{V}^{-1}y,$$

where $\tilde{V}^{-1}$ is the matrix of Equation (10) evaluated at $\tilde{\rho}$. This is also a non-linear function of $\tilde{\rho}$, and hence of the dependent variable. For the Cochrane–Orcutt iteration it is possible to prove consistency utilizing the results contained in Chapter 8. To this effect, we recall that we begin the process with the OLS estimator of β, which was shown to be consistent. We next observe that for the OLS estimator $\tilde{\beta}$,

$$\tilde{u} = y - X\tilde{\beta} = [I - X(X'X)^{-1}X']u,$$

$$\tilde{u}_{-1} = y_{-1} - X_{-1}\tilde{\beta} = u_{-1} - X_{-1}(X'X)^{-1}X'u,$$

where, in this context only,

$$y = (y_2, y_3, \ldots, y_T)' \qquad y_{-1} = (y_1, y_2, \ldots, y_{T-1})'$$

$$X = (x_{ti}), \qquad t = 2, 3, \ldots, T, i = 0, 1, 2, \ldots, n,$$

$$X_{-1} = (x_{ti}), \qquad t = 1, 2, \ldots, T - 1, i = 0, 1, 2, \ldots, n.$$

Thus

$$\tilde{u}'\tilde{u}_{-1} = u'u_{-1} - u'X_{-1}(X'X)^{-1}X'u - u'X(X'X)^{-1}X'u_{-1}$$
$$+ u'X(X'X)^{-1}X'X_{-1}(X'X)^{-1}X'u.$$

By the usual assumptions we know that

$$\lim_{T \to \infty} \frac{X'X}{T}$$

exists as a nonsingular nonstochastic matrix, and to this we may add that

$$\lim_{T \to \infty} \frac{X'X_{-1}}{T}$$

also exists as a well-defined nonstochastic matrix (but not necessarily a nonsingular one). With this in mind it is easy to see that

$$\underset{T \to \infty}{\text{plim}} \frac{1}{T} \tilde{u}'\tilde{u}_{-1} = \underset{T \to \infty}{\text{plim}} \frac{1}{T} u'u_{-1}.$$

Similar arguments will establish that

$$\underset{T \to \infty}{\text{plim}} \frac{1}{T} \tilde{u}'_{-1}\tilde{u}_{-1} = \underset{T \to \infty}{\text{plim}} \frac{1}{T} u'_{-1}u_{-1}.$$

Since

$$u = \rho u_{-1} + \varepsilon, \qquad \varepsilon = (\varepsilon_2, \varepsilon_3, \ldots, \varepsilon_T)',$$

we note that

$$u'u_{-1} = \rho u'_{-1}u_1 + \varepsilon'u_{-1}.$$

It may be shown (but we will not do so here since it is beyond the scope of this book) that

$$\underset{T \to \infty}{\text{plim}} \frac{u'_{-1}u_{-1}}{T} = \frac{\sigma^2}{1 - \rho^2}, \qquad \underset{T \to \infty}{\text{plim}} \frac{\varepsilon'u_{-1}}{T} = 0.$$

Consequently,

$$\underset{T \to \infty}{\text{plim}} \tilde{\rho} = \rho$$

and we see that $\tilde{\rho}$ is a consistent estimator of ρ. But then the first iterate

$$\tilde{\beta}_{(1)} = (X'\tilde{V}^{-1}X)^{-1}X'\tilde{V}^{-1}y$$

is also a consistent estimator of β. For this we recall again Proposition 20 of Chapter 8 and we thus conclude that

$$\underset{T \to \infty}{\text{plim}} \tilde{\beta}_{(1)} = \underset{T \to \infty}{\text{plim}} \left(\frac{X'V^{-1}X}{T} \right)^{-1} \left(\frac{X'V^{-1}y}{T} \right) = \beta,$$

where V^{-1} is now evaluated at the probability limit of $\tilde{\rho}$ above, which is, of course, the true parameter value. Incidentally, the last member of the equality above has been proved in Proposition 2 for a slightly more general case.

It is clear, then, that if we begin with a consistent estimator of β we get a consistent estimator of ρ, which yields again a consistent estimator of β, and so on until convergence. If the process converges, we have found $(\hat{\beta}, \hat{\rho})$ that are consistent estimators of β and ρ and at the same time are approximate solutions to the first order conditions

$$\frac{\partial S}{\partial \beta} = 0, \qquad \frac{\partial S}{\partial \rho} = 0.$$

We can also establish the consistency of the estimators yielded by the search procedure, but the arguments would be entirely too involved to present here. For this reason they are omitted. The interested reader, however, is referred to Dhrymes [8, Chapter 4]. If the model is correctly specified we would expect the two sets of estimators to coincide, i.e., the Cochrane–Orcutt and the search procedures would yield, roughly, the same estimates. Establishing the distributional aspects of the estimators (in both instances) involves the following. Let

$$\gamma = (\beta', \rho)'$$

and observe that, generally, the estimators obey

$$\frac{\partial S}{\partial \gamma}(\hat{\gamma}) = 0 \tag{12}$$

Expand this by the mean value theorem to obtain

$$\frac{\partial S}{\partial \gamma}(\hat{\gamma}) = \frac{\partial S}{\partial \gamma}(\gamma_0) + \frac{\partial^2 S}{\partial \gamma \partial \gamma}(\bar{\gamma})(\hat{\gamma} - \gamma_0), \tag{13}$$

where γ_0 is the true value of the unknown parameters and $\bar{\gamma}$ is such that

$$|\hat{\gamma} - \gamma_0| > |\bar{\gamma} - \gamma_0|.$$

Consequently, since $\hat{\gamma}$ converges in probability to γ_0 so does $\bar{\gamma}$. In view of (12) we can write (13) as

$$\sqrt{T}(\hat{\gamma} - \gamma_0) = -\left[\frac{1}{T}\frac{\partial^2 S}{\partial \gamma \partial \gamma}(\bar{\gamma})\right]^{-1}\frac{1}{\sqrt{T}}\frac{\partial S}{\partial \gamma}(\gamma_0).$$

In this case we can only obtain the distribution of the left side as $T \to \infty$. It is this that is referred to as the asymptotic distribution of the estimator $\hat{\gamma}$. All tests of significance carried out in empirical applications are based on this asymptotic distribution. It is clearly beyond the scope of this book to give the arguments establishing the asymptotic distribution above. What we can do is to explain what it means and to exhibit it for use in applications. First, we observe that

$$\sqrt{T}(\hat{\gamma} - \gamma_0)$$

is clearly a random variable indexed by the sample size T; its density function may be formally designated by

$$f_T(\cdot)$$

even though we may not know its form. Thus, to the estimator obtained above on the basis of a sample of size T there corresponds the function

$$f_T(\cdot).$$

If as T tends to infinity the sequence of functions $\{f_T\}$ converges pointwise (technically, at the points where they are continuous) to a function

$$f(\cdot)$$

then we say that

$$\sqrt{T}(\hat{\gamma} - \gamma_0)$$

converges in distribution to a random variable having density[4] $f(\cdot)$. In most cases encountered in econometrics the asymptotic distributions are normal and the arguments used to establish these conclusions are based in large measure on various forms of the central limit theorem. In the particular case under consideration one can show that, asymptotically,

$$\sqrt{T}(\hat{\gamma} - \gamma_0) \sim N(0, \sigma^2\Omega),$$

where

$$\Omega = \begin{bmatrix} \Omega_1 & 0 \\ 0 & \dfrac{1-\rho^2}{\sigma^2} \end{bmatrix}$$

and

$$\Omega_1 = \lim_{T \to \infty} \left(\frac{X'V^{-1}X}{T}\right)^{-1}.$$

Two things are clear from the preceding. First, asymptotically, $\hat{\beta}$ and $\tilde{\rho}$ are independent—and this occurs despite the fact that $\hat{\beta}$ is involved in the definition of $\hat{\rho}$ and vice versa. Second, we can estimate Ω consistently by

$$\hat{\sigma}^2\hat{\Omega} = \begin{bmatrix} \hat{\sigma}^2\hat{\Omega}_1 & 0 \\ 0 & 1-\hat{\rho}^2 \end{bmatrix}, \qquad \hat{\Omega}_1 = \left(\frac{X'\hat{V}^{-1}X}{T}\right)^{-1},$$

where $\hat{V}^{-1}$ is the matrix V^{-1} of Equation (10) evaluated at $\hat{\rho}$.

Tests of significance. It is obvious from the preceding that

$$\sqrt{T}(\hat{\beta} - \beta) \sim N(0, \sigma^2\Omega_1), \qquad \sqrt{T}(\hat{\rho} - \rho) \sim N[0, (1-\rho^2)]$$

Thus, if we wished to test the hypothesis

$$H_0: \quad \beta_{(2)} = \bar{\beta}_{(2)},$$

where $\beta_{(2)}$ consists of the last k-elements of β, we may proceed as follows. Partition Ω_1 conformably so that

$$\Omega_1 = \begin{bmatrix} R_{11} & R_{12} \\ R_{21} & R_{22} \end{bmatrix},$$

---

[4] Not all random variables have density functions. Strictly speaking, this statement should be phrased "... a random variable having distribution function ...." A distribution function need not be differentiable. Thus, the density function need not exist. But in this book all (continuous) random variables are assumed to have density functions.

and note that

$$\sqrt{T}(\hat{\beta}_{(2)} - \beta_{(2)}) \sim N(0, \sigma^2 R_{22}).$$

Consequently, asymptotically,

$$T(\hat{\beta}_{(2)}) - \beta_{(2)})' \frac{R_{22}^{-1}}{\sigma^2} (\hat{\beta}_{(2)} - \beta_{(2)}) \sim \chi_k^2.$$

Thus, if we denote by $\hat{\sigma}^2$ the consistent estimator of σ^2 and by $\hat{R}_2$ the consistent estimator of the appropriate submatrix of Ω_1, it may be shown that, asymptotically,

$$T(\hat{\beta}_{(2)} - \beta_{(2)})' \frac{\hat{R}_{22}^{-1}}{\hat{\sigma}^2} (\hat{\beta}_{(2)} - \beta_{(2)})$$

converges, in distribution, to a random variable, which is χ_k^2.

Observing that

$$T(\hat{\beta}_{(2)} - \beta_{(2)})' \frac{\hat{R}_{22}^{-1}}{\hat{\sigma}^2} (\hat{\beta}_{(2)} - \beta_{(2)}) = \frac{1}{\hat{\sigma}^2} (\hat{\beta}_{(2)} - \beta_{(2)})' \left(\frac{\hat{R}_{22}}{T}\right)^{-1} (\hat{\beta}_{(2)} - \beta_{(2)}),$$

we see that in carrying out tests of significance on β we need operate only with $\hat{\beta}$ and $(X'\hat{V}^{-1}X)^{-1}$; the factor T disappears. This means, of course, that we operate here exactly as we do in the context of the standard GLM *except that now these tests are exact only for large samples.* Moreover, *what was an F-test in the context of the standard GLM here becomes a chi-square test, and, of course, what was a t-test now becomes a normal test.* Incidentally, whether we use the search or the Cochrane–Orcutt iteration (provided it converges) *the last stage of the computation, i.e., the one from which we derive the estimators, yields a consistent estimate of the covariance matrix.* This is so since, in either case, to obtain the estimator of β we have to invert a matrix of the form

$$(X'V^{-1}X)$$

where: for the search procedure, V^{-1} is evaluated at $\hat{\rho}$, the estimator of ρ; and for the Cochrane–Orcutt iteration, it is evaluated at $\tilde{\rho}_{(k-1)}$, i.e., the estimate of ρ obtained from the residuals of the preceding iteration. If convergence is obtained at the kth iteration, $\tilde{\rho}_{(k)} \approx \tilde{\rho}_{(k-1)}$ and thus

$$(X'\tilde{V}_{k-1}^{-1}X)^{-1}$$

approximates very closely what is desired, viz.,

$$(X'\tilde{V}_k^{-1}X)^{-1}.$$

A test on ρ, say of the hypothesis

$$H_0: \quad \rho = \rho_0,$$

may be carried out using the statistic

$$\frac{\sqrt{T}(\hat{\rho} - \rho_0)}{\sqrt{(1 - \rho_0^2)}},$$

which is, asymptotically, $N(0, 1)$. Thus the inference problem in the context of the first order autoregressive errors model is completely solved.

Remark 4. The essence of the asymptotic distribution theory result, in the context of the model we operate with, is that knowledge of ρ is useful only in a small sample context. Asymptotically, whether we know ρ or we estimate it is immaterial. If we know ρ we have, for the Aitken estimator,

$$\hat{\beta} - \beta = (X'V^{-1}X)^{-1}X'V^{-1}u.$$

Thus

$$\sqrt{T}(\hat{\beta} - \beta) = \left(\frac{X'V^{-1}X}{T}\right)^{-1}\frac{X'V^{-1}u}{\sqrt{T}}$$

and it is clear that

$$\sqrt{T}(\hat{\beta} - \beta)$$

has mean zero and covariance matrix

$$\sigma^2\left(\frac{X'V^{-1}X}{T}\right).$$

A standard central limit theorem can be shown to yield the desired result. We hasten to add that *this is not a general phenomenon—if the matrix X contains lagged values of the dependent variable the result above will not hold.* Further development of such aspects may be found in Dhrymes [8].

Efficiency of OLS *with Nonscalar Covariance Matrix.*[5] We have shown earlier that, in general.

(a) if the GLM has error with nonscalar covariance matrix, the OLS estimator is inefficient;
(b) if, in the case of nonscalar covariance matrix, we estimate and carry out tests of hypotheses *as we would in the case of scalar covariance matrix*, we cannot, in general, say whether we under- or overestimate the variances of the coefficient estimators.

However, there are special circumstances under which the results above are not applicable.

In particular with respect to the result in (a) above consider again the GLM

$$y = X\beta + u, \qquad E(u) = 0, \qquad \text{Cov}(u) = \sigma^2\Phi,$$

_____
[5] This section may be omitted without essential loss of continuity.

where Φ is a suitable[6] positive definite matrix. Let X be[7] $T \times n$ and suppose that

$$X = Q_n A,$$

where A is $n \times n$ nonsingular and Q_n is $T \times n$ containing n of the (orthonormal) characteristic vectors of Φ. Let Λ be the matrix of characteristic roots and suppose we so arrange them and their corresponding characteristic vectors that

$$Q = (Q_n, Q_*), \quad \Lambda = \begin{bmatrix} \Lambda_n & 0 \\ 0 & \Lambda_* \end{bmatrix},$$

where Q is the $T \times T$ (orthogonal) matrix of the characteristic vectors of Φ. Evidently, Q_* is $T \times (T - n)$ and contains the characteristic vectors not corresponding to X; similarly, Λ_n contains the characteristic roots corresponding to Q_n. We have the obvious relations

$$\Phi = Q\Lambda Q', \quad Q'_n Q_n = I_n, \quad Q'_n Q_* = 0, \quad Q'_* Q_* = I_{T-n}.$$

Now, the OLS estimator of β is

$$\tilde{\beta} = (X'X)^{-1}X'y = (A'Q'_n Q_n A)^{-1}A'Q'_n y = A^{-1}Q'_n y. \tag{14}$$

The Aitken estimator is

$$\begin{aligned} \hat{\beta} &= (X'\Phi^{-1}X)^{-1}X'\Phi^{-1}y = (A'Q'_n Q\Lambda^{-1}Q'Q_n A)^{-1}A' \, Q'_n Q\Lambda^{-1}Q'y \\ &= A^{-1}\Lambda_n A'^{-1}A'\Lambda_n^{-1}Q'_n y \\ &= A^{-1}Q'_n y. \end{aligned} \tag{15}$$

Comparing (14) and (15) we conclude that the OLS and Aitken estimators coincide. Consequently, in this special case, the result in (a) above will not hold.

Let us now turn to the result under (b). We recall that at the end of Section 2.2 we had shown that, in the context of the GLM above, if we consider the covariance matrix of the OLS estimator to be

$$\sigma^2(X'X)^{-1}$$

and we estimate

$$\tilde{\sigma}^2 = \frac{1}{T - n} \tilde{u}'\tilde{u}, \quad \tilde{u} = y - X(X'X)^{-1}X'y = (I - M)y,$$

then we cannot tell, on the average, whether we over- or underestimate the variances of the elements of the estimator in (14).

---

[6] Clearly, Φ must be specified a bit more precisely. If left as a general positive definite matrix then it would, generally, contain more than T (independent) unknown parameters, and thus its elements could not be consistently estimated through a sample of T observations.

[7] In this context only, and for notational convenience, the dimension of X is set at $T \times n$; elsewhere, we shall continue to take X as $T \times (n + 1)$.

In fact, we had shown earlier that

$$E(\tilde{\sigma}^2) = \sigma^2 k_T, \qquad k_T = \frac{\text{tr } \Phi(I - M)}{T - n}.$$

In the context in which we operate, the covariance matrix of the estimator in (14) is

$$\text{Cov}(\tilde{\beta}) = \sigma^2(X'X)^{-1}X'\Phi X(X'X)^{-1} = \sigma^2 A^{-1}Q_n'\Phi Q_n A'^{-1},$$

while if we persist in operating on the assumption that the error process of the model has scalar covariance matrix we should take the covariance matrix of the estimator to be

$$``\text{Cov}(\tilde{\beta})" = \sigma^2(X'X)^{-1} = \sigma^2 A^{-1}A'^{-1}$$

and its estimator would be

$$``\text{Cov}(\tilde{\beta})" = \tilde{\sigma}^2 A^{-1}A'^{-1}, \qquad E(\tilde{\sigma}^2) = \sigma^2 k_T.$$

Thus, whether on the average we over- or underestimate depends on the difference

$$A^{-1}(k_T I - Q_n'\Phi Q_n)A'^{-1},$$

which is positive (semi) definite, negative (semi) definite or indefinite according as

$$k_T I - Q_n'\Phi_n Q_n$$

has the properties above.

Now suppose that (i) Q_n corresponds to the n smallest roots of Φ. In this case $Q_n'\Phi Q_n = \Lambda_n$ and we need to consider $k_T - \lambda_i, i = 1, 2, \ldots, n$. But in this case

$$\Phi(I - M) = Q\Lambda Q'[I - Q_n Q_n'] = Q_*\Lambda_* Q_*'$$

and

$$\text{tr}\left[\frac{\Phi(I - M)}{T - n}\right] = \frac{1}{T - n}\text{tr } \Lambda_* = \frac{1}{T - n}\sum_{i=1}^{T-n}\lambda_i,$$

it being understood that the $\lambda_i, i = 1, 2, \ldots, T$, have been arranged in order of decreasing magnitude. Thus, since

$$\lambda_{T-n+i} \le \lambda_j, \qquad j = 1, 2, \ldots, T - n, i = 1, 2, \ldots, n,$$

we conclude that, indeed,

$$k_T - \lambda_i \ge 0 \quad \text{and} \quad k_T I - Q_n'\Phi Q_n \ge 0,$$

and that, *in this case, we do, on the average, overestimate the variance of the estimator of the elements of β.*

Suppose, on the other hand, that (ii) Q_n corresponds to the n *largest* roots of Φ. Retracing our steps above we conclude that $Q_n'\Phi_n Q_n = \Lambda_n$ but now

$$\Lambda_* = \text{diag}(\lambda_{n+1}, \lambda_{n+2}, \ldots, \lambda_T)$$

and

$$\Lambda_n = \text{diag}(\lambda_1, \lambda_2, \ldots, \lambda_n),$$

i.e., Λ_* contains the $T - n$ smallest roots of Φ, while Λ_n contains the n largest! In this case

$$\text{tr}\left[\frac{\Phi(I - M)}{T - n}\right] = \frac{1}{T - n}\text{tr}\,\Lambda_* = \frac{1}{T - n}\sum_{i=1}^{T-n}\lambda_{i+n}.$$

The comparison now hinges on

$$k_T - \lambda_j = \frac{1}{T - n}\sum_{i=1}^{T-n}\lambda_{i+n} - \lambda_j, \quad j = 1, 2, \ldots, n.$$

Since

$$\lambda_{i+n} \le \lambda_j, \quad i = 1, 2, \ldots, T - n, j = 1, 2, \ldots, n,$$

we conclude that

$$k_T - \lambda_j \le 0 \quad \text{and} \quad k_T I - \Lambda_n \le 0$$

Thus, for this special case, the variances are, on the average underestimated.

Finally, suppose that (iii) Q_n corresponds to the n_2 smallest and the $n - n_2 = n_1$ largest roots of Φ. In this case

$$Q'\Phi Q = \Lambda, \qquad \text{tr}\left[\frac{\Phi(I - M)}{T - n}\right] = \frac{1}{T - n}\text{tr}\,\Lambda_*.$$

But now

$$\Lambda_n = \text{diag}(\lambda_1, \lambda_2, \ldots, \lambda_{n_1}, \lambda_{T-n_2+1}, \lambda_{T-n_2+2}, \ldots, \lambda_T),$$

$$\Lambda_* = \text{diag}(\lambda_{n_1+1}, \lambda_{n_1+2}, \ldots, \lambda_{T-n_2}),$$

and consequently

$$k_T - \lambda_j = \frac{1}{T - n}\left[\sum_{i=n_1+1}^{T-n_2}\lambda_i\right] - \lambda_j, \quad j = 1, 2, \ldots, n_1, T - n_2 + 1, \ldots, T,$$

may be either positive or negative.

Hence, in this—as well as in the general—case we have the result that some variances may be overestimated while others may be underestimated.

The preceding discussion, therefore reinforces the general result obtained earlier. *If we operate with the GLM as if its error process has a scalar covariance matrix, when in fact it does not, we cannot, in general, determine whether the estimated variances of the estimator of individual elements of β are, on the average, under- or overestimated and hence we cannot infer similar properties for the estimated t-ratios. We may be able to do so only in highly special cases.*

2.5 Tests for First-Order Autoregression: Durbin–Watson Theory

Here we are concerned with the following problem. In the context of the GLM

$$y = X\beta + u$$

we obtain the OLS estimator

$$\tilde{\beta} = (X'X)^{-1}X'y$$

and we wish to determine whether the evidence at hand supports the hypothesis that the error process is one of i.i.d. random variables. In the terminology introduced earlier, the statement that

$$\{u_t : t = 1, 2, \ldots\}$$

is a sequence of i.i.d. random variables is treated not as a maintained but as a testable hypothesis. How we carry out the test depends, of course, on what we consider to be the alternative. It has been customary in the analysis of economic time series data to consider as the alternative

$$u_t = \rho u_{t-1} + \varepsilon_t,$$

which is the specification examined extensively in the previous section. The test employed, almost exclusively, in empirical applications is the so-called Durbin–Watson test, named after its originators. We shall not recount here the origins of the test, but rather focus on what it is and how it may be applied. A full discussion may be found in the appendix to this chapter.

Defining the tridiagonal matrix

$$A = \begin{bmatrix} 1 & -1 & 0 & & \cdots & & & 0 \\ -1 & 2 & -1 & 0 & & & & \\ 0 & -1 & 2 & -1 & 0 & & & \vdots \\ & 0 & & & & & & \\ \vdots & & & & \vdots & & -1 & 0 \\ & & & & 0 & -1 & 2 & -1 \\ 0 & & \cdots & & & 0 & -1 & 1 \end{bmatrix} \tag{16}$$

and the vector of residuals

$$\tilde{u} = y - X\tilde{\beta},$$

the Durbin–Watson statistic is given by

$$d = \frac{\tilde{u}'A\tilde{u}}{\tilde{u}'\tilde{u}} = \frac{\sum_{t=2}^{T}(\tilde{u}_t - \tilde{u}_{t-1})^2}{\sum_{t=1}^{T}\tilde{u}_t^2}. \tag{17}$$

Remark 5. We recall that when dealing with the Cochrane–Orcutt iteration we defined the initial estimate of ρ as

$$\tilde{\rho} = \frac{\sum_{t=2}^{T}\tilde{u}_t\tilde{u}_{t-1}}{\sum_{t=2}^{T}\tilde{u}_{t-1}^2}.$$

Developing the statistic in (17) we see that

$$d = \frac{\sum_{t=2}^{T} \tilde{u}_t^2 + \sum_{t=2}^{T} \tilde{u}_{t-1}^2 - 2\sum_{t=2}^{T} \tilde{u}_t \tilde{u}_{t-1}}{\sum_{t=1}^{T} \tilde{u}_t^2} = 2(1 - \alpha\tilde{\rho}) - \frac{\tilde{u}_1^2 + \tilde{u}_T^2}{\tilde{u}'\tilde{u}},$$

where

$$\alpha = \left(\frac{\sum_{t=1}^{T-1} \tilde{u}_t^2}{\tilde{u}'\tilde{u}} \right)$$

Hence, approximately,

$$d \approx 2(1 - \tilde{\rho}).$$

Since we have shown earlier that $\tilde{\rho}$ is a consistent estimator of ρ, we see that

$$\plim_{T\to\infty} d = 2(1 - \rho),$$

and since $\rho \in (-1, 1)$ we conclude

$$\plim_{T\to\infty} d \in (0, 4).$$

Thus, intuitively it would seem that: if we obtain a d-value that is close to zero we would tend to conclude that $\rho \approx 1$, or, at any rate, that is positive; if the d value obtained from a given sample is near 2 we would tend to conclude that $\rho \approx 0$; finally, if the d value obtained is near 4 then we would tend to conclude that $\rho \approx -1$, or, at any rate, that it is negative.

Although the distribution of d is not tabulated, several approximations are available and the statements above can be made somewhat precise in terms of these approximations.

Return again to (17) and *add* the condition that

$$u \sim N(0, \sigma^2 I).$$

Recalling that

$$\tilde{u} = (I - M)u, \qquad M = X(X'X)^{-1}X',$$

we can write

$$d = \frac{u'(I - M)A(I - M)u}{u'(I - M)u}.$$

Unfortunately, the two quadratic forms of the ratio are not mutually independent. On the other hand, since their respective matrices commute we conclude by Proposition 53 of *Mathematics for Econometrics* that there exists an orthogonal matrix, say Q, that simultaneously diagonalizes them. In fact, it can be shown that Q can be chosen so that

$$Q'(I - M)A(I - M)Q = \Theta, \qquad Q'(I - M)Q = D,$$

where Θ and D are diagonal matrices containing the characteristic roots of $(I - M)A(I - M)$ and $I - M$ respectively. We observe that $I - M$ is an idempotent matrix of rank $T - n - 1$ and hence that

$$D = \begin{bmatrix} 0 & 0 \\ 0 & I \end{bmatrix}$$

the identity matrix being of order $T - n - 1$. We also note that A is of rank $T - 1$ (see Problem 10); moreover, its nonzero characteristic roots can be shown to be

$$\lambda_j = 2\left[1 - \cos\left(\frac{\pi j}{T}\right)\right], \qquad j = 1, 2, \ldots, T - 1,$$

so that they lie in the interval $(0, 4)$.

It is shown in the appendix to this chapter that if $k \, (\leq n + 1)$ of the columns of X are linear combinations of the k characteristic vectors corresponding to the k smallest characteristic roots of A then, *provided all roots are arranged in increasing order*, we have the inequalities

$$\lambda_{i+k} \leq \theta_{i+n+1} \leq \lambda_{i+n+1}, \qquad i = 1, 2, \ldots, T - n - 1,$$

where, as we noted above, the θ_i depend on the data matrix X while the λ_i do not. Consequently, we have

$$d = \frac{u'(I - M)A(I - M)u}{u'(I - M)u} = \frac{\sum_{i=1}^{T-n-1} \theta_{i+n+1}\, \xi_{i+n+1}^2}{\sum_{i=1}^{T-n-1} \xi_{i+n+1}^2}, \qquad (18)$$

where

$$\xi \sim N(0, I).$$

In view of the inequalities above we may define

$$d_{\mathrm{L}} = \frac{\sum_{i=1}^{T-n-1} \lambda_{i+k}\, \xi_{i+n+1}^2}{\xi' D \xi}, \qquad d_{\mathrm{U}} = \frac{\sum_{i=1}^{T-n-1} \lambda_{i+n+1}\, \xi_{i+n+1}^2}{\xi' D \xi}$$

where

$$d_{\mathrm{L}} \leq d \leq d_{\mathrm{U}}.$$

Finally, it is shown that if $F_{\mathrm{L}}(\cdot)$, $F(\cdot)$, $F_{\mathrm{U}}(\cdot)$ are the distribution functions of d_{L}, d, d_{U}, respectively, then for any r in their domain of definition these functions satisfy the inequality

$$F_{\mathrm{U}}(r) \leq F(r) \leq F_{\mathrm{L}}(r).$$

The distribution $F(\cdot)$ is, in principle, known; however, since it depends (because d depends) on the θ_i and the latter depend on the data matrix, X, it is difficult to tabulate. On the other hand, $F_{\mathrm{U}}(\cdot)$ and $F_{\mathrm{L}}(\cdot)$ *do not depend* on X and their significance points have, in fact, been tabulated. They appear in Tables I, Ia, and Ib, at the end of this volume.

Now suppose we wish to test the hypothesis

$$H_0: \quad \rho = 0,$$

as against

$$H_1: \quad \rho > 0.$$

Let r_L be a point such that $F_L(r_L) = \alpha$, where, say, $\alpha = .05$ or $\alpha = .025$ or $\alpha = .01$. We observe that $F(r_L) \le F_L(r_L)$. Thus, if the test statistic say $\bar{d}$, is such that

$$\bar{d} \le r_L,$$

we can be sure that

$$F(\bar{d}) < \alpha.$$

We also note that if r_U is a point such that

$$F_U(r_U) = \alpha$$

then

$$F(r_U) \ge F_U(r_U).$$

Consequently, if the test statistic, $\bar{d}$ is such that

$$\bar{d} \ge r_U$$

we can be sure that

$$F(\bar{d}) > \alpha.$$

Now, to test the hypothesis above we would normally wish to define a critical region, based on the distribution of d, such that: if the test statistic $\bar{d}$ obeys

$$\bar{d} \le d^*$$

we reject the null hypothesis; while if it obeys

$$\bar{d} > d^*$$

we accept the null hypothesis. Since tabulations for the significance points of $F(\cdot)$ do not exist we have to rely on tabulations for $F_L(\cdot)$ and $F_U(\cdot)$. But this means that the test must be of the form

$$\text{reject } H_0 \quad \text{if } \bar{d} \le r_L,$$

$$\text{accept } H_0 \quad \text{if } \bar{d} \ge r_U.$$

It is then obvious that if it so happens that

$$r_L < \bar{d} < r_U$$

we have no test, i.e., the test is inconclusive!

Clearly, if we wished to test

$$H_0: \quad \rho = 0,$$

as against to,

$$H_1: \quad \rho < 0,$$

we could operate with $4 - d$ and otherwise proceed exactly as before. *Again we should encounter a region of indeterminacy.*

Because the test is not always conclusive a number of approximations have arisen in an effort to remove the indeterminacy. It is not the intention here to survey the various approximations suggested. Durbin and Watson [13] report that for a number of cases of empirical relevance, the following approximation is quite accurate;

$$d \approx a + bd_U,$$

where a and b are chosen so that the mean and variance of d coincide with the mean and variance of $a + bd_U$. To see how this may be implemented we need to derive the first two moments of d.

It may be shown (but we shall not do so in this book) that in (18) d and $\sum_{i=1}^{T-n-1} \zeta_{i+n+1}^2$ are distributed independently. Hence in this case (see also Problem 12)

$$E(d) = \frac{E(d_1)}{E(d_2)}, \tag{19}$$

where

$$d_1 = \zeta' \Theta \zeta, \qquad d_2 = \zeta' D \zeta.$$

But

$$E(d_1) = \operatorname{tr} \Theta, \qquad E(d_2) = \operatorname{tr} D = T - n - 1.$$

Hence

$$E(d) = \frac{\sum_{i=n+2}^{T} \theta_i}{T - n - 1} = \bar{\theta}.$$

Similarly,

$$E(d^2) = \frac{E(d_1^2)}{E(d_2^2)}$$

and

$$E(d_1^2) = \sum_i \sum_j \theta_i \theta_j E(\zeta_i^2 \zeta_j^2) = \sum_{i \neq j} \theta_i \theta_j$$

$$+ 3 \sum_{i=n+2}^{T} \theta_i^2 = \left(\sum_{i=n+2}^{T} \theta_i \right)^2 + 2 \sum_{i=n+2}^{T} \theta_i^2$$

Similarly,

$$E(d_2^2) = \sum_i \sum_j (\xi_i^2 \xi_j^2) = (T - n + 1)(T - n - 1).$$

Since

$$Var(d) = E(d^2) - [E(d)]^2 = \frac{(\sum \theta_i)^2 + 2 \sum \theta_i^2}{(T - n + 1)(T - n - 1)} - \bar{\theta}^2$$

$$= \frac{2 \sum_{i=n+2}^{T} (\theta_i - \bar{\theta})^2}{(T - n + 1)(T - n - 1)},$$

we see that a and b are to be determined from

$$\bar{\theta} = a + b\bar{\lambda} = E(a + bd_U), \tag{20}$$

$$\sum_{i=1}^{T-n-1} (\theta_i - \bar{\theta})^2 = b^2 \sum_{i=1}^{T-n-1} (\lambda_{i+n+1} - \bar{\lambda})^2 = Var(a + bd_U),$$

where

$$\bar{\lambda} = \frac{1}{T - n - 1} \sum_{i=1}^{T-n-1} \lambda_{i+n+1}.$$

Let a and b, as determined from (20), be denoted by a^*, b^*; the distribution of

$$d^* = a^* + b^* d_U$$

is clearly induced by that of d_U. Denote it by $F_*(\cdot)$. Let r^* be a number such that

$$F_*(r^*) = \alpha.$$

The test of the hypothesis may now be specified as

$$\text{reject } \rho = 0 \quad \text{if } \bar{d} \leq r^*,$$

$$\text{accept } \rho = 0 \quad \text{if } \bar{d} > r^*,$$

where $\bar{d}$ is the test statistic. Since

$$\alpha = \Pr\{d^* \leq r^*\} = \Pr\left\{d_U \leq \frac{r^* - a^*}{b^*}\right\}$$

we see that, *using only the tabulated significance points of d_U, we can always resolve the ambiguities of the usual Durbin–Watson test.* Incidentally, the preceding shows that in solving (20) it is *convenient to choose b to be positive.*

To conclude this section let us recapitulate what is involved in the test of the i.i.d. assumption on the error process, relative to the first order autoregression alternative.

(i) Obtain the OLS estimator $\tilde{\beta} = (X'X)^{-1}X'y$ and the residual vector $\tilde{u} = y - X\tilde{\beta}$.

(ii) Compute the Durbin–Watson statistic $d = (\tilde{u}'A u)/(\tilde{u}'\tilde{u})$ where A is as given in (17). The Durbin–Watson statistic lies in the interval (0, 4). If it lies in (0, 2) and sufficiently far from 2 it indicates positive autocorrelation (i.e., $\rho > 0$). If it lies in (2, 4) and sufficiently far from 2 it indicates negative autocorrelation (i.e., $\rho < 0$).

(iii) For given level of significance α, the appropriate number of observations, T, and explanatory variables $n + 1$, find the significance points of the two bounding variables d_L and d_U that obey $d_L \le d \le d_U$. If the α-significance points are r_L and r_U respectively then: reject $\rho = 0$ (and thus accept $\rho > 0$) if the d-statistic, say $\bar{d}$, obeys $\bar{d} \le r_L$; accept $\rho = 0$ (and thus reject $\rho > 0$) if $\bar{d} \ge r_U$. If the $\bar{d}$-statistic obeys $r_L < \bar{d} < r_U$ then the test is inconclusive and no determination can be made regarding the hypothesis. In such a case consider the approximation $d \approx a + d_U$ and determine a and b in accordance with Equation (20), choosing $b > 0$. Let a^*, b^* be the parameters so chosen. Determine a number, r^*, such that $\alpha = \Pr\{d_U \le (r^* - a^*)/b^*.\}$ The test is then: reject $\rho = 0$ (and thus accept $\rho > 0$) if $\bar{d} \le r^*$; accept $\rho = 0$ (and thus reject $\rho > 0$) if $\bar{d} > r^*$.

(iv) If we are interested in the alternative $\rho < 0$ we consider $4 - d$ and apply to this statistic the procedure outlined above.

(v) Significance points for the bounding variables d_L and d_U, i.e., numbers r_L and r_U such that $\Pr\{d_L \le r_L\} = \alpha$ and $\Pr\{d_U \le r_U\} = \alpha$, are given at the end of this volume for $\alpha = .05$, $\alpha = .025$, and $\alpha = .01$.

Remark 6. It is important to stress that the significance tests above are not valid when the set of explanatory variables contains lagged endogenous variables. In many empirical applications it is common to have y_t depend on y_{t-1}. For such a case the significance points above do not apply. A fuller discussion of this aspect may be found in Dhrymes [8, Chapter 7].

2.6 Systems of GLM

It is often the case that we are confronted with a set of general linear models, say m in number. Thus, for example, if we deal with the demand for commodities on the part of a household, the demand for the ith commodity will depend on the household's (real) income and the (relative) prices of all (relevant) commodities. Under conditions of atomistic competition the household's activities in this market would not be expected to have an influence on its (household's) income or the relative prices of commodities. Hence, the household's demand for the ith commodity will, typically, constitute a GLM if the relation is expressed linearly in the parameters. Moreover the error term of the demand function for the ith commodity would be independent of the explanatory variables. Since we would be dealing with more than one commodity, we could thus have a set of general linear models, say m in number.

Similarly, in the theory of the firm if we consider the demand for the jth factor of production we would typically express it as a function of the relative factor prices. Again, under atomistic competition relative factor prices would not depend on the activities of the firm in the (factor) market so that we could assert that the relative factor prices are independent of the error term attaching to the factor demand functions. If the relations are formulated linearly in the parameters we are again confronted with a system of GLM's.

Consequently, let us examine what, if any, novel problems of estimation are presented by the system

$$y_{.j} = X_j \beta_{.j} + u_{.j}, \qquad 1 = 1, 2, \ldots, m.$$

In the above, $y_{.j}$ is a vector of T observations on the jth dependent variable, X_j is a $T \times k_j$ matrix of observations on the corresponding explanatory variables, $\beta_{.j}$ is the vector of unknown parameters in the jth GLM, and $u_{.j}$ is a T-element vector of "observations" on its error process.

We continue to assume that the standard conditions apply, i.e., that the explanatory variables are nonstochastic and

$$\text{rank}(X_j) = k_j, \qquad j = 1, 2, \ldots, m,$$

$$\lim_{T \to \infty} \frac{X_j' X_j}{T} = S_j, \qquad j = 1, 2, \ldots, m,$$

the S_j being nonsingular nonstochastic matrices. For each $j, j = 1, 2, \ldots, m$, the error process

$$\{u_{tj} : t = 1, 2, \ldots\}$$

is asserted to be a zero mean sequence of i.i.d. random variables with variance σ_{jj}.

At first the reader may think that the system above presents no problem since each GLM obeys the "standard assumptions" and thus the theory developed in Chapters 1 and 2 is fully applicable. This would, indeed, be the case if we were to estimate the parameters of these models *seriatim*, i.e., apply OLS techniques for $j = 1$ and thereby estimate $\beta_{.1}$; for $j = 2$ and thereby estimate $\beta_{.2}$, and so on. This, however, may not be an entirely satisfactory procedure. To see what are the issues involved, more precisely, let us employ a more revealing notation. Thus, let

$$y = (y_{.1}', y_{.2}', \ldots, y_{.m}')', \qquad X = \text{diag}(X_1, X_2, \ldots, X_m),$$
$$\beta = (\beta_{.1}', \beta_{.2}', \ldots, \beta_{.m}')', \qquad u = (u_{.1}', u_{.2}', \ldots, u_{.m}')',$$

and write the complete system as

$$y = X\beta + u. \tag{21}$$

In view of the assumptions made regarding each GLM we know that

$$\lim_{T \to \infty} \frac{X'X}{T} = \text{diag}(S_1, S_2, \ldots, S_m) = S$$

is a nonsingular nonstochastic matrix and that

$$\text{rank}(X) = k, \qquad k = \sum_{j=1}^{m} k_j.$$

We also know that

$$E(u) = 0.$$

We observe, however, that

$$\text{Cov}(u) = E(uu') = E\begin{bmatrix} u_{.1}u'_{.1} & u_{.1}u'_{.2} & \cdots & u_{.1}u'_{.m} \\ u_{.2}u'_{.2} & u'_{.2}u'_{.2} & \cdots & u_{.2}u'_{.m} \\ \vdots & \vdots & & \vdots \\ u_{.m}u'_{.1} & u_{.m}u'_{.2} & \cdots & u_{.m}u'_{.m} \end{bmatrix}.$$

The standard assumptions yield

$$E(u_{.i}u'_{.i}) = \sigma_{ii}I, \qquad i = 1, 2, \ldots, m,$$

where I is a $T \times T$ identity matrix; on the other hand, the standard assumptions convey no information on the off-diagonal blocks

$$E(u_{.i}u'_{.j}), \qquad i \neq j.$$

In line with the situation in a single GLM we may assert

$$E(u_{ti}u_{t'j}) = 0$$

for $t \neq t'$. However, there may well be a correlation between the ith and jth structural errors for the same index t. Hence, the natural extension of the assumptions under which we had operated in Chapters 1 and 2 is

$$E(u_{.i}u'_{.j}) = \sigma_{ij}I$$

where, as before, I is an identity matrix of dimension T. Consequently, we may write

$$\text{Cov}(u) = \begin{bmatrix} \sigma_{11}I & \sigma_{12}I & \sigma_{1m}I \\ \sigma_{21}I & \sigma_{22}I & \sigma_{2m}I \\ \vdots & & \\ \sigma_{m1}I & \sigma_{m2}I & \sigma_{mm}I \end{bmatrix} = \Sigma \otimes I \qquad (22)$$

where $\Sigma \otimes I$ denotes the Kronecker product of the matrices Σ and I (see *Mathematics for Econometrics*). Thus, in the model of (21) the standard assumptions regarding the explanatory variables are satisfied. For the error process, however, we have the situation described in Assumption (A.3) of Section 2.2, i.e., we have

$$\text{Cov}(u) = \Sigma \otimes I = \Phi.$$

But, by the discussion in that section we know that the efficient estimator is given by

$$\hat{\beta} = (X'\Phi^{-1}X)^{-1}X'\Phi^{-1}y.$$

If we add the assumption of normality for the error process, we then conclude that

$$\hat{\beta} \sim N[\beta, (X'\Phi^{-1}X)^{-1}].$$

The preceding is, of course, deduced as a special case of the results obtained in that section and need not be further elaborated. It should be repeated that *unless the elements of* Σ *are known,* $\hat{\beta}$ is *not* a feasible estimator. It is convenient to examine the nature of the estimator on the (unrealistic) assumption that Σ is known. This will allow us to concentrate on the essential features of the problem without the burden of accounting for the estimation of unknown parameters in the covariance matrix.

We first ask: how does the estimator above differ from the OLS estimator and when, if ever, would the two coincide? Now, the OLS estimator is given by

$$\tilde{\beta} = (X'X)^{-1}X'y.$$

A little reflection shows that

$$\tilde{\beta} = (\tilde{\beta}'._1, \tilde{\beta}'._2, \ldots, \tilde{\beta}'._m)',$$

where

$$\tilde{\beta}._i = (X_i'X_i)^{-1}X_i'y._i, \qquad i = 1, 2, \ldots, m.$$

Thus, the OLS estimator of the entire parameter set may be obtained *seriatim* by confining our attention to one equation (GLM) at a time. It is then apparent that this process of estimating the $\beta._i$, $i = 1, 2, \ldots, m$, utilizes information from the ith GLM only. Moreover, the parameters in Σ do not play any role whatever. By contrast, *the parameters in* Σ *do enter the Aitken procedure—which also utilizes information from all equations in estimating each* $\beta._i$, $i = 1, 2, \ldots, m$.

We shall now determine the conditions under which the OLS and Aitken procedures coincide. We have

Theorem 1. *Consider the system of* GLM

$$y._j = X_j \beta._j + u._j, \qquad j = 1, 2, \ldots, m,$$

where

$$X_j \quad is \quad T \times k_j, \qquad \beta._j \quad is \quad k_j \times 1,$$

and

$$\text{rank}(X_j) = k_j,$$

$$\lim_{T \to \infty} \frac{X_j'X_j}{T} = S_j,$$

$$E(u._j|X_i) = 0, \qquad \text{Cov}(u._i, u._j|X._k) = \sigma_{ij}I \quad for \quad i, j, k = 1, 2, \ldots, m.$$

Then the OLS *and Aitken estimators of the parameters* $\beta_{.j}$ *will concide if*

(a) $\sigma_{ij} = 0, i \neq j$,
(b) $X_j = X^*, j = 1, 2, \ldots, m$.

PROOF. Using the notation developed immediately above, we are dealing with the model

$$y = X\beta + u,$$

where

$$\text{Cov}(u) = \Sigma \otimes I = \Phi, \qquad X = \text{diag}(X_1, X_2, \ldots, X_m).$$

The Aitken estimator is

$$\hat{\beta} = (X'\Phi^{-1}X)^{-1}X'\Phi^{-1}y$$

while the OLS estimator is

$$\tilde{\beta} = (X'X)^{-1}X'y.$$

Suppose (a) holds. Then

$$X'\Phi^{-1}X = \text{diag}\left(\frac{X_1'X_1}{\sigma_{11}}, \frac{X_2'X_2}{\sigma_{22}}, \ldots, \frac{X_m'X_m}{\sigma_{mm}}\right).$$

Hence

$$(X'\Phi^{-1}X)^{-1}X'\Phi^{-1}y = (X'X)^{-1}X'y$$

which proves the first part of the theorem.

If

$$X_j = X^*$$

then

$$X = \text{diag}(X_1, \ldots, X_m) = (I \otimes X^*).$$

Consequently,

$$X'\Phi^{-1}X' = (I \otimes X^{*\prime})(\Sigma^{-1} \otimes I)(I \otimes X^*) = (\Sigma^{-1} \otimes X^{*\prime}X^*)$$

and

$$\hat{\beta} = (X'\Phi^{-1}X')^{-1}X'\Phi^{-1}y = (\Sigma^{-1} \otimes X^{*\prime}X^*)^{-1}(\Sigma^{-1} \otimes X^{*\prime})y$$
$$= [I \otimes (X^{*\prime}X^*)^{-1}X^{*\prime}]y.$$

The OLS estimator obeys

$$\tilde{\beta} = [(I \otimes X^{*\prime})(I \otimes X^*)]^{-1}[I \otimes X^{*\prime}]y = [I \otimes (X^{*\prime}X^*)^{-1}X^{*\prime}]y.$$

Comparing the two expressions above we conclude

$$\hat{\beta} = \tilde{\beta}. \quad \text{q.e.d.}$$

Remark 7. The intuitive content of the result above is as follows. When the error processes in the various GLM are correlated and the different GLM use different information sets, estimating the parameter set of each GLM one at a time is not a procedure that uses information efficiently. For, even though the m GLM appear to be structurally unrelated,[8] the fact that their error processes are correlated provides a link between them. Thus, if the ith GLM contains a different set of explanatory variables than the jth, then X_i will give some information on the unknown parameters $\beta_{\cdot j}$, $i \neq j$, and vice versa. This is so since X_i helps to determine $y_{\cdot i}$ and the latter (through $u_{\cdot i}$) is correlated with $y_{\cdot j}$, which obviously conveys information relevant for estimating $\beta_{\cdot j}$. Conversely, it is perfectly sensible that if all GLM contain the same set of explanatory variables, i.e., if

$$X_i = X^*, \qquad i = 1, 2, \ldots, m,$$

then there is no advantage to be gained by employing an Aitken procedure since each GLM contains exactly the same information that is relevant for the estimation of its coefficient vector. Similarly, if the error processes in the GLM are mutually independent (or minimally uncorrelated) then, even though the ith GLM may contain a different set of explanatory variables than the jth GLM, there is no way in which this information can be used in estimating the parameters of the latter. Consequently, in this case as well, Aitken is no improvement over OLS and again we have coincidence of the two procedures.

The question now arises as to whether a feasible Aitken estimator can be obtained in the present context. We recall that in section 2.2, when we examined the general problems posed by a nonscalar covariance matrix, we observed that unless the latter were further restricted we could not carry out the feasible Aitken procedure. In the preceding section we have examined at least one case where this is so, viz., the case of a first-order autoregressive error process. It is interesting that the system above offers another example of the existence of feasible Aitken estimators when the covariance parameters are unknown. It is also quite obvious how we may carry out such a scheme. By OLS methods obtain estimates of the coefficient vector in each GLM, thus obtaining the residual vectors

$$\tilde{u}_{\cdot i} = (I - M_i)y_{\cdot i} = (I - M_i)u_{\cdot i}, \qquad i = 1, 2, \ldots, m, \qquad (23)$$

where, of course,

$$M_i = X_i(X_i' X_i)^{-1} X_i' \qquad i = 1, 2, \ldots, m.$$

Consider

$$\tilde{\sigma}_{ij} = \frac{1}{T} \tilde{u}_{\cdot i}' \tilde{u}_{\cdot j} = \frac{1}{T} u_{\cdot i}'(I - M_i)(I - M_j)u_{\cdot j} \qquad (24)$$

---

[8] It is for this reason that A. Zellner who studied the problem of this section quite extensively termed it the problem of *seemingly unrelated regressions* [36].

and verify that

$$\plim_{T \to \infty} \tilde{\sigma}_{ij} = \plim_{T \to \infty} \frac{1}{T} u'_{\cdot i} u_{\cdot j} = \sigma_{ij}, \tag{25}$$

so that the quantity in (25) is a consistent estimator of the (i, j) element of the matrix Σ. Hence a consistent estimator of Φ has been produced, viz.,

$$\tilde{\Phi} = \tilde{\Sigma} \otimes I. \tag{26}$$

This is possible here since Φ has a fixed *finite number of unknown parameters no matter how large T is*. To be precise, Φ contains, at most, $m(m + 1)/2$ distinct unknown parameters, although its dimension is mT. The feasible Aitken estimator is thus

$$\tilde{\beta} = (X'\tilde{\Phi}^{-1}X)^{-1}X'\tilde{\Phi}^{-1}y. \tag{27}$$

This is clearly a consistent estimator of β and, indeed, it may be shown that if

$$\{u'_{t \cdot} : t = 1, 2, \ldots\}$$

is a sequence of independent identically distributed random vectors, where

$$u_{t \cdot} = (u_{t1}, u_{t2}, \ldots, u_{tm}),$$

then, asymptotically,

$$\sqrt{T}(\tilde{\beta} - \beta) \sim N(0, Q),$$

where

$$Q^{-1} = \lim_{T \to \infty} \frac{X'\Phi^{-1}X}{T}.$$

It is, then, interesting that asymptotically there is no cost to not knowing Σ, in the sense that the asymptotic distribution of the Aitken estimator

$$\hat{\beta} = (X'\Phi^{-1}X)^{-1}X'\Phi^{-1}y$$

(i.e., the estimator when Σ is known), and the feasible estimator

$$\tilde{\beta} = (X'\tilde{\Phi}^{-1}X)^{-1}X'\tilde{\Phi}^{-1}y$$

(i.e., the estimator when Σ is not known but is estimated consistently) are identical.

The salient results of this section may be recapitulated as follows. In the system of GLM

$$y_{\cdot i} = X_i \beta_{\cdot i} + u_{\cdot i}, \qquad i = 1, 2, \ldots, m,$$

suppose that

$$\mathrm{rank}(X_i) = k_i,$$

where X_i is $T \times k_i$ and $T > k_i$, and

$$\lim_{T \to \infty} \frac{X_i' X_i}{T} = S_i, \qquad i = 1, 2, \ldots, m,$$

the S_i being nonsingular nonstochastic matrices. Suppose further that

$$\lim_{T \to \infty} \frac{X' \Phi^{-1} X}{T} = Q^{-1}$$

exists as a nonsingular nonstochastic matrix, where

$$X = \text{diag}(X_1, X_2, \ldots, X_m), \qquad \Phi = \Sigma \otimes I,$$

and Σ is the (common) covariance matrix of

$$u_{t.}' = (u_{t1}, u_{t2}, \ldots, u_{tm})', \qquad t = 1, 2, \ldots, T.$$

Then, for any arbitrary (nonstochastic) nonsingular matrix Ψ such that

$$\frac{X' \Psi^{-1} X}{T}$$

is nonsingular and converges with T to a nonsingular matrix, the estimator

$$\beta^* = (X' \Psi^{-1} X)^{-1} X' \Psi^{-1} y$$

is *unbiased* and *consistent*. Within this class, the estimator with $\Psi = \Phi$ is the *efficient* (Aitken) estimator.

If Φ is *not known* the feasible Aitken estimator can be obtained by the following procedure. First, obtain the OLS estimator

$$\hat{\beta} = (X'X)^{-1} X'y$$

and the residual vectors

$$\tilde{u}_{.i} = [I - X_i(X_i' X_i)^{-1} X_i'] y_{.i}, \qquad i = 1, 2, \ldots, m.$$

Compute the consistent estimators

$$\tilde{\sigma}_{ij} = \frac{1}{T} \tilde{u}_{.i}' \tilde{u}_{.i}, \qquad i, j = 1, 2, \ldots, m,$$

and obtain the feasible Aitken estimator

$$\tilde{\beta} = (X' \tilde{\Phi}^{-1} X')^{-1} X' \tilde{\Phi}^{-1} y,$$

where

$$\tilde{\Phi} = \tilde{\Sigma} \otimes I, \qquad \tilde{\Sigma} = \tilde{\sigma}(_{ij}).$$

The asymptotic distribution of $\sqrt{T}(\tilde{\beta} - \beta)$ is given by

$$\sqrt{T}(\tilde{\beta} - \beta) \sim N(0, Q).$$

The covariance matrix Q can be consistently estimated by

$$\tilde{Q} = \left(\frac{X'\tilde{\Phi}^{-1}X}{T}\right)^{-1}.$$

Remark 8. Tests of significance on individual coefficients or sets of coefficients *may be based on the asymptotic distribution above, treating the appropriate element(s) of $\tilde{Q}$ as nonstochastic.* Thus, e.g., if $\beta_{(1)}$ consists of the first s elements of β and if Q_{11} is the $(s \times s)$ submatrix of Q corresponding to $\beta_{(1)}$ then, asymptotically,

$$T(\bar{\beta}_{(1)} - \beta_{(1)})'\tilde{Q}_{11}^{-1}(\bar{\beta}_{(1)} - \beta_{(1)}) \sim \chi_s^2.$$

Noting that

$$\tilde{Q} = T(X'\tilde{\Phi}^{-1}X)^{-1}$$

and putting

$$C = (X'\tilde{\Phi}^{-1}X)^{-1}$$

we observe that

$$\tilde{Q}_{11} = TC_{11},$$

where C_{11} is the submatrix of C corresponding to $\tilde{Q}_{11}$. Since

$$T\tilde{\Phi}_{11}^{-1} = C_{11}^{-1}$$

we conclude that

$$(\bar{\beta}_{(1)} - \beta_{(1)})'C_{11}^{-1}(\bar{\beta}_{(1)} - \beta_{(1)}) \sim \chi_s^2.$$

But this means that if, in the (computer) output of the estimation (regression) scheme *we treat the covariance matrix of the estimators as known in the formulation of significance tests, then such tests will be, asymptotically, exact.* In the case above, if $s = 1$ then such a test is asymptotically a normal test (or a χ_1^2-test since we are dealing with the square of the usual test statistic).

Somewhat imprecisely the result may be put as follows. If we continue operating with the feasible Aitken estimator above as we do with the GLM then, even though the procedure has no foundation for small samples, asymptotically we are completely justified—except that what was, in the GLM, a "t-test" is now a normal test, and what was an F-test is now a chi-square test.

QUESTIONS AND PROBLEMS

1. Consider the usual model

   $$y = X\beta + u, \qquad X = (e, X_1), \qquad e = (1, 1, \ldots, 1)', \qquad X_1 = (x_{.1}, x_{.2}, \ldots, x_{.n}),$$

   the $x_{.i}$ being T-element vectors on the (n) bona fide explanatory variables. Suppose $E(u) = \mu e$. Show that the OLS estimator of β_0 is biased while that of β_* is unbiased, where $\beta = (\beta_0, \beta_*')'$.

2. Let B be a $T \times T$ positive definite (symmetric) matrix. Let X be $T \times (n + 1)$ of rank $(n + 1)$. Show that $X'BX$ is a positive definite matrix. [*Hint*: for any $(n + 1)$-element column vector α show that $X\alpha = 0$ if and only if $\alpha = 0$.]

3. In Equation (2) show that if $\text{rank}(X) = n + 1$ then $\text{rank}(Z) = n + 1$ where $Z = BX$.

4. Give an alternative "proof" of Proposition 1, deducing it as a corollary to Proposition 2 of Chapter 1.

5. Verify that: (a)

$$I - \frac{Bee'B'}{\phi}$$

is an idempotent matrix of rank $T - 1$, where $\Phi^{-1} = B'B$, $e'\Phi^{-1}e = \phi$; (b) in the model as exhibited in Equation (5),

$$\left[I - \frac{Bee'B'}{\phi}\right]w = \left[I - \frac{Bee'B'}{\phi}\right]Z_1\beta_* + \left[I - \frac{Bee'B'}{\phi}\right]v,$$

and thus the conditional mean of w (given X) does not contain β_0; (c) in the expression for the Aitken estimator,

$$\hat{\beta}_0 = \frac{1}{\phi} e'B'(w - Z_1\beta_*),$$

$$\hat{\beta}_* = \left[Z_1'\left(I - \frac{Bee'B'}{\phi}\right)Z_1\right]^{-1} Z_1'\left(I - \frac{Bee'B'}{\phi}\right)w;$$

(d) $Z_1^*(Z_1^{*\prime}Z_1^*)^{-1}Z_1^{*\prime}(I - Z(Z' Z)^{-1}Z') = 0$, where

$$Z_1^* = \left[I - \frac{Bee'B'}{\phi}\right]Z_1.$$

[*Hint*: $Z_1^{*\prime}Z = (0, Z_1^{*\prime}Z_1^*)$, and use the partitioned inverse form for $(Z'Z)^{-1}$.]

6. In the model of Section 2.2 show that the coefficient of determination of multiple regession as defined in (6) can also be represented as

$$R^2 = \frac{\hat{\beta}_*' Z_1^{*\prime}Z_1^*\hat{\beta}_*}{w'\left[I - \dfrac{Bee'B'}{\phi}\right]w}.$$

[*Hint*: use Problem 5 of this chapter and Problems 6, 7, 8 of Chapter 2.]

7. In the context of the GLM of Equation (1) with $\text{Cov}(u) = \sigma^2\Phi$, show that for an arbitrary nonsingular matrix R, $\bar{\beta} = (X'RX)^{-1}X'Ry$ is an unbiased and consistent estimator of β. Show that questions of relative efficiency hinge on the proximity of R to the identity matrix (in the comparison to the OLS estimator) and on the proximity of R to Φ^{-1} (in the comparison to the Aitken estimator).

8. Verify (a) the inverse of V in Equation (9) is given by the matrix in Equation (10), (b) $|V| = 1/(1 - \rho^2)$. [*Hint*: consider V^{-1} and multiply its first row by ρ and add to the second row, thus verifying that $|V_T^{-1}| = |V_{T-1}^{-1}|$, where V_{T-1}^{-1} is a matrix of the same type as V^{-1} but of order $T - 1$.]

9. With regard to Equations (14) and (15) show that $Q_n' \Phi^{-1} Q_n = \Lambda_n^{-1}$, $Q_n' \Phi^{-1} = \Lambda_n^{-1} Q_n'$.

10. Show that the $T \times T$ matrix of Equation (16) is of rank $T - 1$. [*Hint*: add the first column to the second, the second to the third, etc.]

11. Following (18), why is $\xi \sim N(0, I)$ and *not* $\sim N(0, \sigma^2 I)$?

12. Verify, that if $d = d_1/d_2$, d_2 and d being mutually independent, then Equation (19) is valid. [*Hint*: $d_1 = d_2 d$.]

13. Why is it true that $E(d_1) = \text{tr} \, \Theta \, E(d_2) = T - n - 1$?

14. Verify that the first two moments of d_U have the same form as the first two moments of d.

Appendix

A.1 Durbin–Watson Theory

Here we explore, in somewhat greater detail, the issues involved in the derivation and use of the Durbin–Watson statistic.

Derivation. From Equation (18) of this chapter we have

$$d = \frac{u'(I - M)A(I - M)u}{u'(I - M)u}, \tag{A.1}$$

where it is assumed that

$$u \sim N(0, \sigma^2 I)$$

and

$$M = X(X'X)^{-1}X'.$$

For notational simplicity put

$$N = I - M$$

and note that, since N is idempotent,

$$N(NAN) = (NAN)N$$

i.e., the two matrices

$$N, \quad NAN$$

commute. We shall take advantage of this fact in order to greatly simplify the expression in (A.1). Because N is a symmetric idempotent matrix there exists an orthogonal matrix B such that

$$B'NB = D, \quad D = \text{diag}(0, I),$$

the identity matrix being of dimension equal to the rank of N, which is $T - n - 1$. Define

$$B'(NAN)B = C \tag{A.2}$$

and partition its rows and columns conformably with respect to D, i.e.,

$$C = \begin{bmatrix} C_{11} & C_{12} \\ C_{21} & C_{22} \end{bmatrix},$$

where C_{22} is $(T - n - 1) \times (T - n - 1)$ and C_{11} is $(n + 1) \times (n + 1)$. Observe that

$$
\begin{aligned}
CD &= B'(NAN)BB'NB \\
&= B'NANB \\
&= B'NNANB = B'NBB'NANB = DC.
\end{aligned}
$$

However,

$$CD = \begin{bmatrix} 0 & C_{12} \\ 0 & C_{22} \end{bmatrix}, \qquad DC = \begin{bmatrix} 0 & 0 \\ C_{21} & C_{22} \end{bmatrix}. \tag{A.3}$$

The relations above clearly imply

$$C_{12} = 0, \qquad C_{21} = 0,$$

so that consequently

$$C = \begin{bmatrix} C_{11} & 0 \\ 0 & C_{22} \end{bmatrix}.$$

Since A is a positive semidefinite (symmetric) matrix, C_{11} and C_{22} will have similar properties. Let E_i be the orthogonal matrix of characteristic vectors of C_{ii}, $i = 1, 2$, and Θ_i be the (diagonal) matrix of characteristic roots of C_{ii}, $i = 1, 2$, i.e.,

$$E_1' C_{11} E_1 = \Theta_1, \qquad E_2' C_{22} E_2 = \Theta_2.$$

It is clear that

$$E = \operatorname{diag}(E_1, E_2), \qquad \Theta = \operatorname{diag}(\Theta_1, \Theta_2)$$

are (respectively) the matrices of characteristic vectors and roots for the matrix C. Thus

$$E'CE = \Theta.$$

Bearing in mind what C is we have

$$E'B'(NAN)BE = \Theta. \tag{A.4}$$

From

$$B'NB = \begin{bmatrix} 0 & 0 \\ 0 & I \end{bmatrix} = D$$

we also see that

$$E'B'NBE = \begin{bmatrix} E'_1 & 0 \\ 0 & E'_2 \end{bmatrix} \begin{bmatrix} 0 & 0 \\ 0 & I \end{bmatrix} \begin{bmatrix} E_1 & 0 \\ 0 & E_2 \end{bmatrix} = \begin{bmatrix} 0 & 0 \\ 0 & I \end{bmatrix}. \tag{A.5}$$

Defining

$$Q = BE.$$

we note that

$$Q'Q = QQ' = (BE)'(BE) = (BE)(BE)' = I,$$

i.e., Q is orthogonal; moreover, (A.4) and (A.5) imply that

$$\begin{aligned} Q'NANQ &= \Theta, \\ Q'NQ &= D, \end{aligned} \tag{A.6}$$

where Θ is the diagonal matrix of the characteristic vectors of NAN.

If we put

$$\xi = Q'\left(\frac{u}{\sigma}\right)$$

we note that $\xi \sim N(0, I)$ and

$$d = \frac{\xi'\Theta\xi}{\xi'D\xi}, \tag{A.7}$$

Bounds on Characteristic Roots. The difficulty with the representation in (A.7) is that the numerator depends on the characteristic roots of NAN and thus, ultimately, on the data. A way out of this is found through the bounds, d_L and d_U, discussed earlier in the chapter. Let us now see how these bounds are established. We begin with

Lemma A.1. *The characteristic roots of the matrix A, as exhibited in Equation (16), are given by*

$$\lambda_j = 2\left\{1 - \cos\left[\frac{(j-1)\pi}{T}\right]\right\}, \qquad j = 1, 2, \ldots, T. \tag{A.8}$$

PROOF. We can write

$$A = 2I - 2A_*$$

where

$$A_* = \frac{1}{2} \begin{bmatrix} 1 & 1 & 0 & \cdots & & 0 \\ 1 & 0 & 1 & & & \\ 0 & 1 & 0 & & & \vdots \\ \vdots & & & & & 0 \\ & & & & 0 & 1 \\ 0 & \cdots & & 0 & 1 & 1 \end{bmatrix}, \tag{A.9}$$

Since

$$0 = |\lambda I - A| = |\lambda I - 2I + 2A_*| = (-2)^T |\mu I - A_*|,$$

where

$$\mu = \tfrac{1}{2}(2 - \lambda),$$

it follows that if we determine the characteristic roots of A_*, say

$$\{\mu_i : i = 1, 2, \ldots, T\},$$

then the (corresponding) characteristic roots of A are given by

$$\lambda_i = 2(1 - \mu_i), \qquad i = 1, 2, \ldots, T.$$

If μ and w are, respectively, a characteristic root and the corresponding characteristic vector of A_*, they satisfy the following set of equations:

$$\tfrac{1}{2}(w_1 + w_2) = \mu w_1; \tag{A.10}$$

$$\tfrac{1}{2}(w_{i-1} + w_{i+1}) = \mu w_i, \ i = 2, 3, \ldots, T - 1; \tag{A.11}$$

$$\tfrac{1}{2}(w_{T-1} + w_T) = \mu w_T. \tag{A.12}$$

In order to obtain an expression for μ and the elements of w, we note that the second set above may be rewritten as

$$w_{i+1} - 2\mu w_i + w_{i-1} = 0, \tag{A.13}$$

which is recognized as a second-order difference equation. The desired characteristic root and vector are related to the solution of the equation in (A.13). Its characteristic equation is

$$r^2 - 2\mu r + 1 = 0,$$

whose solutions are

$$r_1 = \mu + \sqrt{\mu^2 - 1}, \qquad r_2 = \mu - \sqrt{\mu^2 - 1}.$$

Since

$$r_1 + r_2 = 2\mu, \qquad r_1 r_2 = 1, \tag{A.14}$$

we conclude that

$$r_2 = \frac{1}{r_1}. \tag{A.15}$$

Thus for notational simplicity we shall denote the two roots by

$$r, \quad \frac{1}{r}.$$

From the general theory of solution of difference equations (see Section 5 of *Mathematics for Econometrics*), we know that the solution to (A.13) may be written as

$$w_t = c_1 r^t + c_2 r^{-t},$$

where c_1 and c_2 are constants to be determined by Equations (A.10) and (A.12). From (A.10) we find

$$(1 - r - r^{-1})(c_1 r + c_2 r^{-1}) + c_1 r^2 + c_2 r^{-2} = 0.$$

After considerable simplification this yields

$$(r - 1)(c_1 - c_2 r^{-1}) = 0, \tag{A.16}$$

which implies

$$c_2 = c_1 r.$$

Substituting in (A.12) and canceling c_1 yields

$$(1 - r)(r^T - r^{-T}) = 0, \tag{A.17}$$

which implies $r^{2T} = 1$, i.e., the solutions to (A.17) are the $2T$ roots of unity, plus the root $r = 1$. As is well known, the $2T$ roots of unity[9] are given by, say,

$$e^{i2\pi s/2T}.$$

The roots of the matrix are, thus,

$$\mu = \tfrac{1}{2}(r + r^{-1}) = \tfrac{1}{2}(e^{i2\pi s/2T} + e^{-i2\pi s/2T})$$

Since

$$e^{i2\pi s/2T} = e^{i2\pi(s - 2T)/2T}$$

it follows that the only distinct roots correspond to

$$r = e^{i\pi s/T}, \qquad s = 0, 1, 2, \ldots, T.$$

Moreover, the root

$$r_T = e^{i\pi} = -1$$

is inadmissible since the characteristic vector corresponding to it is

$$w_t = c_1 r_T^t + c_2 r_T^{-t} = c_1(-1)^t + c_1(-1)^{-(t-1)} = 0$$

is inadmissible. Consequently, the characteristic roots of the matrix A_* are given by

$$\mu_s = \tfrac{1}{2}(r_s + r_s^{-1}) = \tfrac{1}{2}(e^{i\pi s/T} + e^{-i\pi s/T}) = \cos(\pi s/T), \qquad s = 0, 1, \ldots, T - 1, \tag{A.18}$$

[9] Heuristically we may approach the problem as follows: since $e^{i2\pi s} = 1, s = 1, 2, \ldots$, we may write $r^{2T} = 1$ as $r = e^{i2s\pi/2T}$. In some sense this is a solution to the equation $r^{2T} = 1$, since if we raise both sides to the $2T$ power we get back the equation. Cancelling the factor 2 we get the solution $e^{is\pi/T}, s = 0, 1, 2, \ldots, T$. Extending the index s beyond T simply repeats the roots above.

and the corresponding characteristic roots of A by

$$\lambda_s = 2\left[1 - \cos\left(\frac{\pi s}{T}\right)\right], \qquad s = 0, 1, 2, \ldots, T - 1. \quad \text{q.e.d.} \quad (A.19)$$

Corollary A.1. *Let λ_s, as in (A.19), be the sth characteristic root of A. Then*

$$w_{ts} = \cos\left[\frac{(t - \frac{1}{2})\pi s}{T}\right], \qquad t = 1, 2, \ldots, T$$

is the corresponding characteristic vector.

PROOF. We first note that if μ_s, as in (A.18), is the sth characteristic root of A_* then

$$w_{ts} = c_1 r_s^t + c_1 r_s^{-t+1}, \qquad t = 1, 2, \ldots, T,$$

is the corresponding characteristic vector. If we choose

$$c_1 = \tfrac{1}{2} r_s^{-(1/2)}$$

then

$$w_{ts} = \tfrac{1}{2}[r_s^{t-(1/2)} + r_s^{-(t-(1/2))}] = \cos\left[\frac{(t - \frac{1}{2})\pi s}{T}\right], \qquad t = 1, 2, \ldots, T,$$

and it can be shown easily that

$$\sum_{t=1}^{T} w_{ts}^2 = \frac{T}{2}, \qquad \sum_{t=1}^{T} w_{ts} w_{ts'} = 0, \qquad s \neq s'.$$

Thus, we see that the vectors corresponding to the roots r_s are mutually orthogonal. If we wished we could have taken

$$c_1 = \frac{1}{2}\left(\sqrt{\frac{2}{T}}\right) r_s^{-(1/2)},$$

in which case we would have determined

$$w_{ts} = \sqrt{\frac{2}{T}} \cos\left[\frac{(t - \frac{1}{2})\pi s}{T}\right], \qquad t = 1, 2, \ldots, T, s = 0, 1, 2, \ldots, T - 1,$$

and thus ensured that

$$\sum_{t=1}^{T} w_{ts}^2 = 1.$$

Let

$$W = (w_{ts}), \qquad s = 0, 1, 2, \ldots, T - 1, t = 1, 2, \ldots, T,$$

the elements being as just defined above. We see that

$$AW = 2W - 2A_* W = W2(I - \overline{M}) = W\Lambda, \qquad (A.20)$$

where

$$\Lambda = \text{diag}(\lambda_1 \lambda_2, \ldots, \lambda_T), \qquad \overline{M} = \text{diag}(\mu_1, \mu_2, \ldots, \mu_T).$$

But (A.20) shows that W is the matrix of characteristic vectors of A.   q.e.d.

Remark A.1. Notice that since the roots above may be defined as

$$\lambda_j = 2\left[1 - \cos\left(\frac{\pi(j-1)}{T}\right)\right], \qquad j = 1, 2, \ldots, T,$$

and

$$\lambda_1 = 2[1 - \cos 0] = 0, \qquad \cos(\pi) = -1,$$

we can conclude that

$$\lambda_j \in [0, 4).$$

Notice further that they are arranged in increasing order of magnitude, i.e.,

$$\lambda_1 < \lambda_2 < \lambda_3 \cdots < \lambda_T.$$

Let us now turn to the relation between the roots of A, as established above, and those of

$$NAN \qquad N = I - X(X'X)^{-1}X'.$$

Since $(X'X)^{-1}$ is positive definite there exists a nonsingular matrix G such that

$$(X'X)^{-1} = GG' \tag{A.21}$$

Define

$$P = W'XG, \tag{A.22}$$

where W is the matrix of characteristic vectors of A. We have first

Lemma A.2. *The matrix*

$$P = W'XG$$

is a $T \times (n + 1)$ *matrix of rank* $(n + 1)$, *where G is as in* (A.21) *and W is the matrix of characteristic vectors of A. Moreover, its columns are mutually orthogonal.*

PROOF. The assertion that P is $T \times (n + 1)$ is evident. We further note

$$P'P = G'X'WW'XG = G'X'XG = I. \quad \text{q.e.d.}$$

Now consider the roots of NAN, i.e., consider

$$0 = |\theta I - NAN| = |\theta I - NW\Lambda W'N| = |W'W||\theta I - N^*\Lambda N^*|.$$

But

$$|W'W| = 1, \qquad N^* = W'NW = I - PP'.$$

Hence the roots of NAN are exactly those of

$$(I - PP')\Lambda(I - PP'),$$

where P is as defined in (A.22). It turns out that we can simplify this aspect considerably. Thus,

Lemma A.3. *Let $p_{\cdot i}$ be the ith column of P and let*

$$P_i = I - p_{\cdot i} p'_{\cdot i}.$$

Then

$$I - PP' = \prod_{i=1}^{n+1} P_i.$$

PROOF.

$$P_i P_j = (I - p_{\cdot i} p'_{\cdot i})(I - p_{\cdot j} p'_{\cdot j}) = I - p_{\cdot i} p'_{\cdot i} - p_{\cdot j} p'_{\cdot j}$$

since the columns of P are orthogonal. Moreover, the P_i are symmetric idempotent, i.e.,

$$P_i P'_i = P_i.$$

It follows therefore that

$$\prod_{i=1}^{n+1} P_i = I - \sum_{i=1}^{n+1} p_{\cdot i} p'_{\cdot i}.$$

Since

$$PP' = \sum_{i=1}^{n+1} p_{\cdot i} p'_{\cdot i}$$

we conclude

$$I - PP' = \prod_{i=1}^{n+1} P_i. \qquad \text{q.e.d.}$$

A very useful consequence of the lemma is that the problem may now be posed as follows: what is the relation of the roots of

$$\left| \theta I - \left(\prod_{i=1}^{n+1} P_i \right) \Lambda \left(\prod_{i=1}^{n+1} P_i \right) \right| = 0$$

to those of A, i.e., to the elements of Λ. Moreover, since

$$\left(\prod_{i=1}^{n+1} P_i \right) \Lambda \left(\prod_{i=1}^{n+1} P_i \right) = P_{n+1}(P_n \cdots P_2(P_1 \Lambda P_1)P_2 \cdots P_n)P_{n+1}$$

it follows that the problem may be approached recursively, by first asking: what are the relations between the elements of Λ and the roots of $P_1 \Lambda P_1$? If we answer that question then we have automatically answered the question: what are the relations between the roots of

$$P_2 P_1 \Lambda P_1 P_2$$

and those of

$$P_1 \Lambda P_1?$$

Hence, repeating the argument we can determine the relation between the roots of NAN and those of A. Before we take up these issues we state a very useful result.

Lemma A.4. *Let D be a nonsingular diagonal matrix of order m. Let α be a scalar and a, b be two m-element column vectors, and put*

$$H = D + \alpha ab'.$$

Then

$$|H| = \left[1 + \alpha \sum_{i=1}^{m} \left(\frac{a_i b_i}{d_{ii}}\right)\right] \prod_{j=1}^{m} d_{jj}.$$

PROOF. See Proposition 31 of *Mathematics for Econometrics*.

Lemma A.5. *The characteristic roots of*

$$P_1 \Lambda P_1,$$

arranged in increasing order, are the solution of

$$0 = \psi(\theta) = \theta f(\theta), \quad f(\theta) = \sum_{i=1}^{T} p_{i1}^2 \prod_{s \neq i} (\theta - \lambda_s),$$

and obey

$$\theta_i^{(1)} = 0, \quad \lambda_i \leq \theta_{i+1}^{(1)} \leq \lambda_{i+1}, \quad i = 1, 2, \ldots, T - 1.$$

PROOF. The characteristic roots of $P_1 \Lambda P_1$ are the solutions of

$$0 = |\theta I - P_1 \Lambda P_1| = |\theta I - \Lambda P_1| = |\theta I - \Lambda + \Lambda p_{\cdot 1} p_{\cdot 1}'|.$$

Taking

$$D = \theta I - \Lambda, \quad \alpha = 1, \quad \Lambda p_{\cdot 1} = a, \quad b = p_{\cdot 1},$$

applying Lemma A.4, and noting that

$$\sum_{i=1}^{T} p_{i1}^2 = 1,$$

we conclude

$$|\theta I - P_1 \Lambda P_1| = \theta f(\theta),\tag{A.23}$$

where

$$f(\theta) = \sum_{i=1}^{T} p_{i1}^2 \prod_{s \neq i} (\theta - \lambda_s).$$

We note that the characteristic equation of $P_1 \Lambda P_1$ as exhibited in the two equations above is a polynomial of degree T. Since $P_1 \Lambda P_1$ will, generally, be of rank $T - 1$, the polynomial equation

$$f(\theta) = 0$$

will not have a zero root. Indeed,

$$f(\lambda_1) = p_{11}^2 \prod_{s \neq 1} (\lambda_1 - \lambda_s) = (-1)^{T-1} p_{11}^2 \prod_{s=2}^{T} \lambda_s \neq 0$$

unless

$$p_{11} = 0$$

We remind the reader that in the preceding we employ notation

$$\lambda_j = 2\left\{1 - \cos\left[\frac{(j-1)\pi}{T}\right]\right\}, \qquad j = 1, 2, 3, \ldots, T,$$

so that the roots of A are arranged as

$$0 = \lambda_1 < \lambda_2 < \lambda_2 \cdots < \lambda_T < 4.$$

Now the roots of

$$\psi(\theta) = 0$$

are the one obvious zero root (associated with the factor θ) and the $T - 1$ (nonzero) roots of

$$f(\theta) = 0.$$

But for any $r \geq 2$,

$$f(\lambda_r) = (-1)^{T-r} p_{r1}^2 \prod_{i<r} (\lambda_r - \lambda_i) \prod_{i>r} (\lambda_i - \lambda_r) \neq 0$$

provided $p_{r1} \neq 0$. Assuming this to be so we have that if, say, $f(\lambda_r) > 0$, then $f(\lambda_{r+1}) < 0$. Thus, between the two roots of A, λ_r and λ_{r+1}, lies a root of $P_1 \Lambda P_1$. Denote such roots by

$$\theta_i^{(1)}, \qquad i = 1, 2, \ldots, T.$$

What the preceding states is that

$$\theta_1^{(1)} = 0, \qquad \lambda_i \leq \theta_{i+1}^{(1)} \leq \lambda_{i+1}, \qquad i = 1, 2, \ldots, T - 1. \quad \text{q.e.d.}$$

   The following Lemma is also important in the chain of our argumentation.

Lemma A.6. *Let*

$$Q_j = \left(\prod_{i=1}^{j} P_i\right) \Lambda \left(\prod_{i=1}^{j} P_i\right), \qquad j = 1, 2, \ldots, n+1.$$

Then Q_j has at least j zero roots.

PROOF. By definition

$$\prod_{i=1}^{j} P_i = I - \sum_{i=1}^{j} p_{\cdot i} p'_{\cdot i},$$

and, for $k \geq j$, we have

$$\text{rank}(Q_j) \leq T - j, \qquad j = 1, 2, 3, \ldots, N + 1,$$

which means that Q_j must have at least j zero roots.   q.e.d.

Lemma A.7. *Let Q_{j-1} be defined as in Lemma A.6. Let M_{j-1}, $\Theta^{(j-1)}$ be its associated matrices of characteristic vectors and roots respectively. Let Q_j and $\Theta^{(j)}$ be similarly defined. Then*

$$\theta_i^{(j)} = 0, \qquad i = 1, 2, \ldots, j,$$

$$\theta_i^{(j-1)} \leq \theta_{i+1}^{(j)} \leq \theta_{i+1}^{(j-1)}, \qquad i = j, j+1, \ldots, T.$$

PROOF. The first assertion is a restatement of the conclusion of Lemma A.6. For the second, consider

$$0 = |\theta I - Q_j| = |\theta I - P_j Q_{j-1} P_j| = |\theta I - \bar{P}_j \Theta^{(j-1)} \bar{P}_j|,$$

where

$$\bar{P}_j = M'_{j-1} P_j M_{j-1} = I - \bar{p}_{\cdot j} \bar{p}'_{\cdot j}, \qquad \bar{p}_{\cdot j} = M'_{j-1} p_{\cdot j}.$$

We note that

$$\bar{P}_j \bar{P}'_j = \bar{P}_j.$$

Thus

$$|\theta I - \bar{P}_j \Theta^{(j-1)} \bar{P}_j| = |\theta I - \Theta^{(j-1)} \bar{P}_j| = |\theta I - \Theta^{(j-1)} + \Theta^{(j-1)} \bar{p}_{\cdot j} \bar{p}'_{\cdot j}| = \psi(\theta),$$

where now

$$\psi(\theta) = \theta f(\theta), \qquad f(\theta) = \sum_{i=1}^{T} \bar{p}_{ij}^2 \prod_{s \neq i} (\theta - \theta_s^{(j-1)}).$$

Since we know that $\psi(\theta)$ has (at least) j zero roots we therefore know that $f(\theta)$ must have (at least) $j - 1$ zero roots. Hence

$$f(\theta_k^{(j-1)}) = \bar{p}_{kj}^2 \prod_{s \neq k} (\theta_k^{(j-1)} - \theta_s^{(j-1)}) = 0, \qquad k = 1, 2, \ldots, j-1,$$

which need not imply

$$\bar{p}_{kj} = 0, \qquad k = 1, 2, \ldots, j - 1.$$

Consequently, we can write

$$f(\theta) = \sum_{i=1}^{T} \bar{p}_{ij}^2 \prod_{s \neq i} (\theta - \theta_s^{(j-1)}).$$

and, for $k \geq j$, we have

$$f(\theta_k^{(j-1)}) = \bar{p}_{kj}^2 \prod_{s \neq k} (\theta_k^{(j-1)} - \theta_s^{(j-1)})$$

$$= (-1)^{T-k} \bar{p}_{kj}^2 \prod_{s<k} (\theta_k^{(j-1)} - \theta_s^{(j-1)}) \prod_{s>k} (\theta_s^{(j-1)} - \theta_k^{(j-1)}).$$

In general,

$$f(\theta_k^{(j-1)}) \neq 0, \qquad k \geq j,$$

provided $\bar{p}_{kj} \neq 0$. Thus if, e.g., $f(\theta_k^{(j-1)}) > 0$, then

$$f(\theta_{k+1}^{(j-1)}) < 0.$$

Consequently, a root of Q_j, say $\theta_{k+1}^{(j)}$, lies between $\theta_k^{(j-1)}$ and $\theta_{k+1}^{(j-1)}$. This is so since if

$$f(\theta_j^{(j-1)}) > 0,$$

then

$$f(\theta_{j+1}^{(j-1)}) < 0.$$

Thus, the first nonzero root of Q_j, viz., $\theta_{j+1}^{(j)}$ obeys

$$\theta_j^{(j-1)} \leq \theta_{j+1}^{(j)} \leq \theta_{j+1}^{(j-1)}.$$

Consequently, we have established

$$\theta_i^{(j)} = 0, \qquad i = 1, 2, \ldots, j,$$

$$\theta_i^{(j-1)} \leq \theta_{i+1}^{(j)} \leq \theta_{i+1}^{(j-1)}, \qquad i = j, j+1, \ldots, T-1. \quad \text{q.e.d.}$$

We may now prove

Theorem A.1. *Let*

$$\lambda_i, \qquad i = 1, 2, \ldots, T,$$

be the characteristic roots of Λ arranged as

$$0 = \lambda_1 < \lambda_2 < \cdots < \lambda_T < 4.$$

Let

$$\theta_i, \qquad i = 1, 2, \ldots, T,$$

be the roots of NAN similarly arranged in increasing order. Then, the following is true:

$$\theta_i = 0, \qquad i = 1, 2, \ldots, n + 1;$$

$$\lambda_j \leq \theta_{j+n+1} \leq \lambda_{j+n+1}, \qquad j = 1, 2, \ldots, T - n - 1.$$

PROOF. The first part of the theorem is evident. For the second part we note that from Lemma A.5 we have

$$\lambda_i \leq \theta^{(1)}_{i+1} \leq \lambda_{i+1}, \qquad i = 1, 2, \ldots, T - 1. \tag{A.24}$$

From Lemma A.7 we have that

$$\theta^{(1)}_{i+1} \leq \theta^{(2)}_{i+2} \cdots \leq \theta^{(n+1)}_{i+n+1} = \theta_{i+n+1}, \qquad i = 1, 2, \ldots, T - n - 1. \tag{A.25}$$

Thus (A.24) and (A.25) imply

$$\lambda_i \leq \theta_{i+n+1}, \qquad i = 1, 2, \ldots, T - n - 1. \tag{A.26}$$

Again using Lemma A.7,

$$\theta_{i+n+1} = \theta^{(n+1)}_{i+n+1} \leq \theta^{(n)}_{i+n+1} \cdots \leq \theta^{(1)}_{i+n+1}, \tag{A.27}$$

and (A.27) and (A.24) imply

$$\theta_{i+n+1} \leq \lambda_{i+n+1}, \qquad i = 1, 2, \ldots, T - n - 1. \tag{A.28}$$

Combining (A.26) and (A.28) we conclude

$$\lambda_i \leq \theta_{i+n+1} \leq \lambda_{i+n+1}, \qquad i = 1, 2, \ldots, T - n - 1. \quad \text{q.e.d.}$$

Let us now consider certain special cases. Thus, suppose k ($<n + 1$) columns of X are linear combinations of k characteristic vectors of A, say (for definiteness) those corresponding to the k smallest characteristic roots of A. We shall see that this type of X matrix will have an appreciable impact on the bounds determined in Theorem A.1.

The X matrix we deal with obeys

$$X = (X_1, X_2), \qquad X_1 = W_1 B,$$

where B is a nonsingular $k \times k$ matrix and W_1 is the matrix containing the characteristic vectors of A corresponding to its k smallest roots. Retracing our argument in the early part of this appendix we note that, in this case,

$$G = \begin{bmatrix} B^{-1} & -B^{-1}W'_1 X_2 C \\ 0 & C \end{bmatrix}, \tag{A.29}$$

where C is defined by

$$CC' = (X'_2 W_2 W'_2 X_2)^{-1}.$$

Remark A.2. The matrix C will always exist unless X_2 is such that

$$W'_2 X_2 = 0, \qquad W = (W_1, W_2),$$

or is not of full rank. This, however, is to be ruled out since X_2 is *not* a linear transformation of the first k characteristic vectors of A.

The matrix P of (A.22) is now

$$P = W'XG = \begin{bmatrix} I & 0 \\ 0 & P* \end{bmatrix},$$

where

$$P* = W'_2 X_2 C. \tag{A.30}$$

We verify that $P*$ is $(T - k) \times (n + 1 - k)$ and obeys

$$P*'P* = I.$$

Hence

$$I - PP' = \begin{bmatrix} 0 & 0 \\ 0 & I - P*P*' \end{bmatrix}. \tag{A.31}$$

We may now state

Theorem A.2. *Assume the conditions of Theorem A.1 and, in addition, that the data matrix of the GLM is*

$$X = (X_1, X_2), \qquad X_1 = W_1 B,$$

where B is $k \times k$ nonsingular and W_1 is the $T \times k$ matrix containing the characteristic vectors of A corresponding to, say, the k smallest characteristic roots. Then, the following is true regarding the relation between the roots $\theta_i, i = 1, 2, \ldots, T$, of NAN and $\lambda_i, i = 1, 2, \ldots, T$, of A:

$$\theta_i = 0, \qquad i = 1, 2, \ldots, n + 1;$$

$$\lambda_{j+k} \le \theta_{j+n+1} \le \lambda_{j+n+1}, \qquad j = 1, 2, \ldots, T - n - 1.$$

PROOF. The roots of NAN are exactly those of

$$(I - PP')\Lambda(I - PP').$$

For the special case under consideration,

$$P = \begin{bmatrix} I & 0 \\ 0 & P* \end{bmatrix},$$

where I is $k \times k$ and $P*$ is $(T - k) \times (n + 1 - k)$, its columns being orthogonal. Defining

$$\Lambda = \begin{bmatrix} \Lambda_1 & 0 \\ 0 & \Lambda* \end{bmatrix},$$

where

$$\Lambda* = \text{diag}(\lambda_{k+1}, \lambda_{k+2}, \ldots, \lambda_T),$$

we have that the characteristic equation of NAN is, in this special case,

$$\left| \theta I - \begin{bmatrix} 0 & 0 \\ 0 & (I - P*P*')\ \Lambda*(I - P*P*') \end{bmatrix} \right| = 0. \qquad (A.32)$$

But it is evident that (A.32) has k zero roots and that the remaining roots are those of

$$|\theta I - (I - P*P*')\Lambda*(I - P*P*')| = 0, \qquad (A.33)$$

the identity matrices in (A.33) being of order $T - k$. Let the roots of (A.33) be denoted by

$$\theta_i^*, \qquad i = 1, 2, \ldots, T - k.$$

But the problem in (A.33) is one for which Theorem A.1 applies. We thus conclude that

$$\theta_i^* = 0, \qquad i = 1, 2, \ldots, n + 1 - k. \qquad (A.34)$$

This is so since

$$I - P*P*'$$

is a $(T - k) \times (T - k)$ idempotent matrix of rank $T - n - 1$. Hence

$$(I - P*P*')\Lambda*(I - P*P*')$$

must have (at least)

$$T - k - (T - n - 1) = n + 1 - k$$

zero roots. Defining, further, the elements of $\Lambda*$ by

$$\lambda_i^* = \lambda_{i+k}, \qquad i = 1, 2, \ldots, T - k, \qquad (A.35)$$

and applying again Theorem A.1 we conclude

$$\lambda_i^* \leq \theta_{i+n+1-k}^* \leq \lambda_{i+n+1-k}^*, \qquad i = 1, 2, \ldots, T - n - 1 + k. \qquad (A.36)$$

Collecting results and translating to the unstarred notation we have

$$\theta_i = 0, \qquad i = 1, 2, \ldots, n + 1, \qquad (A.37)$$

$$\lambda_{j+k} \leq \theta_{j+n+1} \leq \lambda_{j+n+1}, \qquad j = 1, 2, \ldots, T - n - 1. \quad \text{q.e.d.}$$

Remark A.3. If k of the columns of X are linear transformations of k characteristic vectors of A—not necessarily those corresponding to the smallest or largest k characteristic roots—we proceed exactly as before, except that *now we should renumber the roots of A so that*

$$\lambda_i', \qquad i = 1, 2, \ldots, T,$$

and

$$\lambda_i', \qquad i = 1, 2, \ldots, k,$$

correspond to the specified characteristic vectors of A involved in the representation of the specified columns of X; without loss of generality we may take the latter to be the first k columns. The remaining roots we arrange in increasing order of magnitude, i.e.,

$$\lambda'_{k+1} < \lambda'_{k+2} < \lambda'_{k+3} < \cdots < \lambda'_T.$$

Proceeding as before we will obtain a relation just as in (A.32), from which we shall conclude that the equation there has k zero roots and that the remaining roots are those of the analog of (A.33). The elements of the matrix Λ_* will be, in the present case,

$$\lambda'_{i+k}, \qquad i = 1, 2, \ldots, T - k.$$

Putting

$$\lambda^*_i = \lambda'_{i+k}$$

we will thus conclude that the roots of NAN, say θ_i, obey

$$\begin{aligned} \theta_i &= 0, \qquad i = 1, 2, \ldots, n + 1 \\ \lambda^*_i \le \theta^*_i &\le \lambda^*_{i+n+1-k}, \qquad i = 1, 2, \ldots, T - n - 1 + k, \end{aligned} \qquad \text{(A.38)}$$

where, of course,

$$\theta^*_i = \theta_{i+n+1}.$$

Evidently, if we have as in the case of Theorem A.2 that the k roots are the k smallest characteristic roots then

$$\lambda^{*\prime}_i = \lambda_{i+k}, \qquad \lambda^{*\prime}_{i+n+1-k} = \lambda_{i+n+1},$$

and the bounds are exactly as before. On the other hand, if the characteristic vectors in question are those corresponding to the *first and last* $k - 1$ *roots* of A, i.e., to λ_1 and $\lambda_{T-k+2}, \lambda_{T-k+3}, \ldots, \lambda_T$, then

$$\lambda^{*\prime}_i = \lambda_{i+1}, \qquad i = 1, 2, \ldots, T - k.$$

Thus the bounds become

$$\begin{aligned} \theta_i &= 0, \qquad i = 1, 2, \ldots, n + 1, \\ \lambda_{i+1} &\le \theta_{i+n+1} \le \lambda_{i+n+2-k}, \qquad i = 1, 2, \ldots, T - n - 1. \end{aligned} \qquad \text{(A.39)}$$

For the special case

$$k = n + 1$$

(A.37) yields

$$\lambda_{i+n+1} \le \theta_{i+n+1} \le \lambda_{i+n+1}, \qquad \text{(A.40)}$$

which implies

$$\lambda_{i+n+1} = \theta_{i+n+1}, \qquad i = 1, 2, \ldots, T - n - 1.$$

On the other hand, (A.39) yields

$$\lambda_{i+1} \leq \theta_{i+n+1} \leq \lambda_{i+1}, \qquad i = 1, 2, \ldots, T - n - 1, \qquad (A.41)$$

which implies

$$\lambda_{i+1} = \theta_{i+n+1}, \qquad i = 1, 2, \ldots, T - n - 1. \qquad (A.42)$$

Remark A.4. The effect of having k vectors of X expressible as linear combinations of k characteristic vectors of A is to make the bounds on the roots of NAN tighter.

Remark A.5. Referring to Theorem A.1 we note that the smallest characteristic root of A is

$$\lambda_1 = 0.$$

One can easily verify that

$$e = (1, 1, \ldots, 1)'$$

is a characteristic vector corresponding to this root. But in most cases the GLM will contain a constant term, and hence the appropriate bounds can be determined quite easily from (A.37) to be

$$\lambda_{i+1} \leq \theta_{i+n+1} \leq \lambda_{i+n+1}, \qquad i = 1, 2, \ldots, T - n - 1. \qquad (A.43)$$

Remark A.6. In the previous chapter we considered the special case where X is a linear transformation of $n + 1$ of the characteristic vectors of V^{-1}. It will be recalled that in such a case the OLS and Aitken estimators of the parameter vector β will be the same. It might also be noted in passing that if u is a T-element random vector having the joint density

$$K \exp\{-\tfrac{1}{2}\alpha[u'(D + \gamma A)u]\}, \qquad (A.44)$$

where K is a suitable constant, $\alpha > 0$, D is positive definite, A is symmetric, and γ is scalar such that $D + \gamma A$ is a positive definite matrix, then the uniformly most powerful test of the hypothesis

$$\gamma = 0,$$

as against

$$\gamma < 0,$$

is provided by $r < r_*$ where

$$r = \frac{\tilde{u}'A\tilde{u}}{\tilde{u}'D\tilde{u}}$$

and r_* is a suitable constant, determined by the desired level of significance. Here it is understood that we deal with the model

$$y = X\beta + u$$

and that the columns of X are linear combinations of $n + 1$ *of the characteristic vectors of A.* In the definition above we have

$$\tilde{u} = [I - X(X'X)^{-1}X']u = Nu.$$

This result is due to Anderson [1] and [2].

It would appear that for the cases we have considered in this book (i.e., when the errors are normally distributed), taking

$$D = I, \qquad \alpha = \frac{(1 - \rho)^2}{\sigma^2}, \qquad \gamma = \frac{\rho}{(1 - \rho)^2}, \qquad \text{(A.45)}$$

and A as in Equation (16) of the chapter we are dealing with autoregressive errors obeying

$$u_t = \rho u_{t-1} + \varepsilon_t. \qquad \text{(A.46)}$$

Thus, testing

$$H_0: \quad \gamma = 0,$$

as against

$$H_1: \quad \gamma < 0,$$

is equivalent, in the context of (A.45) and (A.46), to testing

$$H_0: \quad \rho = 0, \qquad \text{(A.47a)}$$

as against

$$H_1: \quad \rho < 0. \qquad \text{(A.47b)}$$

Thus, to paraphrase the Anderson result: *if we are dealing with a GLM whose error structure obeys* (A.46) *then a uniformly most powerful* (UMP) *test for the hypothesis* (A.47) *will exist when the density function of the error obeys* (A.44) *and, moreover, the data matrix X is a linear transformation of an appropriate submatrix of the matrix of characteristic vectors of A. Furthermore, when these conditions hold the* UMP *test is the Durbin–Watson test, which uses the d-statistic computed from* OLS *residuals, as defined in Equation* (17) *of this chapter.* Examining these conditions a bit more closely, however, shows a slight discrepancy. If we make the assignments in (A.45) then

$$\alpha[D + \gamma A] = \frac{1}{\sigma^2}
\begin{bmatrix}
1 + \rho^2 - \rho & -\rho & 0 & \cdots & 0 \\
-\rho & 1 + \rho^2 & -\rho & & \vdots \\
0 & \vdots & \vdots & & 0 \\
\vdots & -\rho & 1 + \rho^2 & \cdots & -\rho \\
0 & 0 & -\rho & \cdots & 1 + \rho^2 - \rho
\end{bmatrix}.$$

But the (inverse of the) covariance matrix of the error terms obeying (A.46) with the ε's i.i.d. and normal with variance σ^2 would be

$$\frac{1}{\sigma^2} V^{-1} = \frac{1}{\sigma^2} \begin{bmatrix} 1 & -\rho & 0 & \cdots & & 0 \\ -\rho & 1+\rho^2 & -\rho & & & \vdots \\ 0 & -\rho & 1+\rho^2 & -\rho & & \\ & & & & & 0 \\ \vdots & & & -\rho & 1+\rho^2 & -\rho \\ 0 & & \cdots & & -\rho & 1 \end{bmatrix}.$$

A comparison of the two matrices shows that they differ, although rather slightly, in the upper left- and lower right-hand corner elements. They would coincide, of course, when

$$\rho = 0 \quad \text{or} \quad \rho = 1.$$

We note, further, that

$$\sigma^2 \alpha[D + \gamma A] = (1 - \rho)^2 I + \rho A. \tag{A.48}$$

Hence, if W is the matrix of characteristic vectors of A then

$$[(1 - \rho)^2 I + \rho A]W = [(1 - \rho)^2 I + \rho \Lambda]W.$$

This shows:

(a) the characteristic roots of the matrix in (A.48) are given by $\psi_i = (1 - \rho)^2 + \rho \lambda_i$, $i = 1, 2, \ldots, T$, where λ_i are the corresponding characteristic roots of A;

(b) if W is the matrix of characteristic vectors of A then it is also that of the matrix in (A.48).

Remark A.7. What implications emerge from the lengthy remark regarding tests for autocorrelation? We have, in particular, the following:

(i) in a very strict sense, we can never have UMP tests for the case we have considered in this chapter since the matrix of the quadratic form in (A.44) subject to the parametric assignments given in (A.45) is *never the same* as the (inverse of the) covariance matrix of the error process in (A.46) when the ε's are i.i.d. and normal with mean zero and variance σ^2. The difference, however, is small—the (1, 1) and (T, T) elements differ by $-\rho(1 - \rho)$. Thus, the difference is positive when $\rho < 0$ and negative when $\rho > 0$. It vanishes when $\rho = 0$ or $\rho = 1$;

(ii) *if we are prepared to ignore the differences* in (i) and thus consider the roots and vectors of $(1 - \rho)^2 I + \rho A$ and V^{-1} as the same, *then a* UMP *test will exist and will be the Durbin–Watson test only in the special case where the data matrix X is a linear transformation of n + 1 of the characteristic vectors of A—and hence of V^{-1}*;

(iii) When X is a linear transformation of $n + 1$ of the characteristic vectors of V^{-1}, as we established in the current chapter, the OLS and Aitken estimators of β coincide.

Remark A.8. The preceding discussion reveals a very interesting aspect of the problem. Presumably, we are interested in testing for autocorrelation in the error process, because if such is the case OLS will not be an efficient procedure and another (efficient) estimator is called for. The test utilized is the Durbin–Watson test. Yet when this test is optimal, i.e., a UMP test, OLS is an efficient estimator—hence the result of the test would not matter. On the other hand, when the results of the test would matter, i.e., when in the presence of autocorrelation OLS is inefficient, *the Durbin–Watson test is not UMP.*

Bounds on the Durbin–Watson Statistic. Let us now return to the problem that has motivated much of the discussion above. We recall that for testing the null hypothesis

$$H_0: \quad \rho = 0,$$

as against the alternative

$$H_1: \quad \rho < 0,$$

where it is understood that the error terms of the GLM obey (A.46) and the ε's are i.i.d. normal random variables with mean zero and variance σ^2, we use the test statistic

$$d = \frac{\xi' \Theta \xi}{\xi' D \xi},$$

where ξ is a $T \times 1$ vector obeying

$$\xi \sim N(0, \sigma^2 I),$$

$$D = \begin{bmatrix} 0 & 0 \\ 0 & I_{T-n-1} \end{bmatrix},$$

$$\Theta = \text{diag}(\theta_1, \theta_2, \ldots, \theta_T),$$

and the θ_i are the characteristic roots of NAN arranged in increasing order. Noting that

$$\theta_i = 0, \qquad i = 1, 2, \ldots, n + 1,$$

we may thus rewrite the statistic more usefully as

$$d = \frac{\sum_{i=1}^{T-n-1} \theta_{i+n+1} \xi_{i+n+1}^2}{\sum_{i=1}^{T-n-1} \xi_{i+n+1}^2}. \tag{A.49}$$

Considering now the bounds as given by Theorem A.2 let us define

$$d_L = \frac{\sum_{i=1}^{T-n-1} \lambda_{i+k} \xi_{i+n+1}^2}{\sum_{i=1}^{T-n-1} \xi_{i+n+1}^2}. \tag{A.50}$$

$$d_U = \frac{\sum_{i=1}^{T-n-1} \lambda_{i+n+1} \xi_{i+n+1}^2}{\sum_{i=1}^{T-n-1} \xi_{i+n+1}^2}, \tag{A.51}$$

and thus conclude

$$d_L \leq d \leq d_U. \tag{A.52}$$

Remark A.9. An important byproduct of the derivations above is that *the bounds d_L and d_U do not depend on the data matrix X.* It is the tabulation of the significance points of d_L and d_U that one actually uses in carrying out tests for the presence of autocorrelation.

Remark A.10. Consider the special cases examined in Remark A.3. If X is a linear transformation of the $n + 1$ characteristic vectors corresponding to the $n + 1$ smallest characteristic roots of A, then

$$\lambda_{i+n+1} = \theta_{i+n+1}$$

and hence, in this case,

$$d = d_U.$$

On the other hand, when the $n + 1$ characteristic vectors above correspond to the smallest (zero) and the n largest characteristic roots of A, then the condition in (A.42) holds. Hence, in this special case, given the bounds in (A.39) we have

$$d = d_L.$$

In these two special cases the test for autocorrelation may be based on the exact distribution of the test (Durbin–Watson) statistic, since the relevant parts of the distribution of d_L and d_U have been tabulated.

Use of the Durbin–Watson Statistic. Let $F_L(\cdot)$, $F(\cdot)$, and $F_U(\cdot)$ be (respectively) the distribution functions of d_L, d, and d_U and let r be a point in the range of these random variables. Then by definition

$$\begin{aligned} Pr\{d_L \leq r\} &= F_L(r), \\ Pr\{d \leq r\} &= F(r), \\ Pr\{d_U \leq r\} &= F_U(r). \end{aligned} \tag{A.53}$$

Now, it is clear that

$$Pr\{d_U \leq r\} \leq Pr\{d \leq r\} \tag{A.54}$$

since

$$d_U \leq r \quad \text{implies} \quad d \leq r.$$

But the converse is not true. Similarly note that

$$d_L > r \quad \text{implies} \quad d > r$$

but that the converse is not true.

This means that

$$1 - Pr\{d_L \leq r\} = Pr\{d_L > r\} \leq Pr\{d > r\} = 1 - Pr\{d \leq r\},$$

which in turn implies

$$Pr\{d \leq r\} \leq Pr\{d_L \leq r\}. \tag{A.55}$$

Combining (A.53), (A.54), and (A.55) we have

$$F_U(r) \leq F(r) \leq F_L(r). \tag{A.56}$$

But this immediately suggests a way for testing the autocorrelation hypothesis. Let r_L be a number such that

$$F_L(r_L) = 1 - \alpha,$$

where α is the chosen level of significance, say

$$\alpha = .10 \quad \text{or} \quad \alpha = .05 \quad \text{or} \quad \alpha = .025.$$

If $F_L(\cdot)$ were the appropriate distribution function and d were the Durbin–Watson (hereafter abbreviated D.W.) statistic, in a given instance, the acceptance region would be

$$d \leq r_L \tag{A.57a}$$

and the rejection region

$$d > r_L. \tag{A.57b}$$

The level of significance of the test would be α. What is the consequence, for the properties of the test, of the inequalities in (A.56)? Well, since

$$F(r_L) \leq F_L(r_L),$$

it follows that the number r^* such that

$$F(r^*) = 1 - \alpha$$

obeys

$$r^* \geq r_L.$$

Thus, in using the acceptance region in (A.57a) we are being too conservative, in the sense that we could have

$$d > r_L \quad \text{and at the same time} \quad d \leq r^*.$$

Conversely, let r_U be a number such that

$$F_U(r_U) = 1 - \alpha.$$

Arguing as before we establish

$$r^* \leq r_U.$$

If we define the rejection region as

$$d > r_U, \tag{A.58}$$

we again see that we reject conservatively, in the sense that we could have

$$d \leq r_U \quad \text{but at the same time} \quad d > r^*.$$

The application of the D.W. test in practice makes use of both conditions (A.57) and (A.58), i.e., we accept

$$H_0: \quad \rho = 0$$

if, for a given statistic d, (A.57a) is satisfied, and we accept

$$H_1: \quad \rho < 0$$

only if (A.58) is satisfied. In so doing we are being very conservative, in the sense that if

$$d \leq r_L \quad \text{then surely} \quad d \leq r^*,$$

and if

$$d > r_U \quad \text{then surely} \quad d > r^*.$$

A consequence of this conservatism, however, is that we are left with a region of indeterminacy. Thus, if

$$r_L < d < r_U,$$

then we have no rigorous basis of accepting either

$$H_0: \quad \rho = 0 \quad \text{or} \quad H_1: \quad \rho < 0.$$

If the desired test is

$$H_0: \quad \rho = 0,$$

as against

$$H_1: \quad \rho > 0,$$

we proceed somewhat differently. Let α again be the level of significance and choose two numbers say r_L and r_U, such that

$$F_L(r_L) = \alpha \quad \text{and} \quad F_U(r_U) = \alpha.$$

In view of (A.56), the number r^* such that

$$F(r^*) = \alpha$$

obeys

$$r_L \leq r^* \leq r_U.$$

Now the *acceptance region* is defined as

$$d \geq r_U \qquad\qquad\qquad\qquad \text{(A.59)}$$

while the *rejection region* is defined as

$$d \leq r_L. \qquad\qquad\qquad\qquad \text{(A.60)}$$

Just as in the preceding case we are being conservative, and the consequence is that we have, again, a region of indeterminacy

$$r_L < d < r_U.$$

Let us now recapitulate the procedure for carrying out a test of the hypothesis that the error terms in a GLM are a first order autoregressive process.

(i) Obtain the residuals

$$\tilde{u} = y - X\tilde{\beta}, \qquad \tilde{\beta} = (X'X)^{-1}X'y.$$

(ii) Compute the D.W. statistic

$$d = \frac{\tilde{u}'A\tilde{u}}{\tilde{u}'\tilde{u}},$$

where A is as defined in Equation (16) of the chapter.

(iii) Choose the level of significance, say α.

(a) If it is desired to test

$$H_0: \quad \rho = 0,$$

as against

$$H_1: \quad \rho < 0,$$

determine, from the tabulated distributions two numbers r_L, r_U such that $F_L(r_L) = 1 - \alpha$, $F_U(r_U) = 1 - \alpha$. If $d \leq r_L$, accept $\rho = 0$. If $d \geq r_U$, accept $\rho < 0$. If $r_L < a < r_U$, the result of the test is inconclusive and other means must be found for determining whether $\rho = 0$ or $\rho < 0$ is to be accepted as true.

(b) If it is desired to test

$$H_0: \quad \rho = 0,$$

as against

$$H_1: \quad \rho > 0,$$

with level of significance α, determine from the tabulated distributions two numbers, say r_L and r_U, such that $F_L(r_L) = \alpha$, $F_U(r_U) = \alpha$. If $d \geq r_U$, accept the hypothesis $\rho = 0$. If $d \leq r_L$, accept the hypothesis $\rho > 0$. If $r_L < d < r_U$, the result of the test is indeterminate.

Remark A.11. Tabulations of $F_L(\cdot)$ and $F_U(\cdot)$ exist typically in the form of 5% significance points (i.e., values of r_L and r_U) for varying numbers of observations and explanatory variables (exclusive of the constant term). Such tabulations are constructed from the point of view of the test

$$H_0: \quad \rho = 0,$$

as against

$$H_1: \quad \rho > 0.$$

It is suggested that when we are interested in the hypothesis

$$H_0: \quad \rho = 0,$$

as against

$$H_1: \quad \rho < 0,$$

we use 4-d as the test statistic and the r_L, r_U significance points from the tabulated distributions.

Remark A.12. The existing tabulations assume that the data matrix X contains one column that is (a multiple of) a characteristic vector of A. As we have remarked earlier the vector

$$e = (1, 1, 1, \ldots, 1)'$$

is the characteristic vector corresponding to the smallest (zero) characteristic root of A. Consequently, the bounds for the roots of NAN for this case are

$$\lambda_{i+1} \leq \theta_{i+n+1} \leq \lambda_{i+n+1}, \qquad i = 1, 2, \ldots, T - n - 1,$$

and the tabulations are based on

$$d_L = \frac{\sum_{s=1}^{T-n-1} \lambda_{s+1} \zeta_{s+n+1}^2}{\sum_{s=1}^{T-n-1} \zeta_{s+n+1}^2},$$

$$d_U = \frac{\sum_{s=1}^{T-n-1} \lambda_{s+n+1} \zeta_{s+n+1}^2}{\sum_{s=1}^{T-n-1} \zeta_{s+n+1}^2}.$$

The reader should note, therefore, that if the **GLM** *under consideration does not contain a constant term, then the tabulated percentage points of the* **D.W.** *statistic are not applicable. This is so since in this case we are dealing with* $k = 0$ *and the lower bound should be defined as*

$$d_L' = \frac{\sum_{s=1}^{T-n-1} \lambda_s \zeta_{s+n+1}^2}{\sum_{s=1}^{T-n-1} \zeta_{s+n+1}^2}.$$

We observe that, since $\lambda_1 = 0$,

$$d_L - d_L' = \frac{\sum_{s=1}^{T-n-1} (\lambda_{s+1} - \lambda_s)\zeta_{s+n+1}^2}{\sum_{s=1}^{T-n-1} \zeta_{s+n+1}^2} \geq 0.$$

Thus, the tabulated distribution is inappropriate for the case of the excluded constant term. This can be remedied by running a regression with a constant term and carrying out the test in the usual way. At the cost of being redundant *let us stress again that there is nothing peculiar with the* **D.W.** *statistic when the* **GLM** *does not contain a constant term. It is merely that the existing tabulations are inappropriate for this case in so far as the lower bound is concerned; the tabulations are quite appropriate, however, for the upper bound.*

Remark A.13. At the end of this volume we present more recent tabulations of the bounds of the D.W. statistics giving 1%, 2.5%, and 5% significance points. *As in the earlier tabulations it is assumed that the* **GLM** *model does contain a constant term; thus, these tabulations are inappropriate when there is no constant term.*

Remark A.14. Two aspects of the use of D.W. tabulations deserve comment. First, it is conceivable that the test statistic will fall in the region of indeterminancy and hence that the test will be inconclusive. A number of suggestions have been made for this eventuality, the most useful of which is the use of the approximation

$$d \approx a + bd_U,$$

where a and b are fitted by the first two moments of d. The virtue of this is that the test is based on existing tabulations of d_U. The reader interested in exploring the details of this approach is referred to Durbin and Watson [13].

Remark A.15. An alternative to the D.W. statistic for testing the hypothesis that the error terms of a GLM constitute a first-order autoregression may be based on the asymptotic distribution of the natural estimator of ρ obtained from the residuals. Thus, e.g., in the GLM

$$y = X\beta + u$$

let $\tilde{u}_t$, $t = 1, 2, \ldots, T$, be the OLS residuals. An obvious estimator of ρ is

$$\tilde{\rho} = \frac{\sum_{t=2}^{T} \tilde{u}_t \tilde{u}_{t-1}}{\sum_{t=1}^{T-1} \tilde{u}_t^2},$$

obtained by regressing $\tilde{u}_t$ on $\tilde{u}_{t-1}$ and suppressing the constant term. It may be shown that if the GLM does not contain lagged dependant variables then, asymptotically,

$$\sqrt{T}(\tilde{\rho} - \rho) \sim N(0, 1 - \rho^2).$$

Consequently, if the sample is reasonably large a test for the presence of autoregression (of the first order) in the errors may be carried out on the basis of the statistic

$$\sqrt{T}\tilde{\rho},$$

which, under the null hypothesis (of no autoregression) will have the distribution

$$\sqrt{T}\tilde{\rho} \sim N(0, 1).$$

A.2 Gaps in Data

It frequently happens with time series data that observations are noncontiguous. A typical example is the exclusion of a certain period from the sample as representing nonstandard behavior. Thus, in time series studies of consumption behavior one usually excludes observations for the period 1941–1944 or 1945; this is justified by noting the shortages due to price controls during the war years. In such a case we are presented with the following problem: if we have a model with autoregressive errors, what is the appropriate form of the autoregressive transformation and of the D.W. statistic when there are gaps in the sample?

We shall examine the following problem. Suppose in the GLM

$$y = X\beta + u$$

observations are available for

$$t = 1, 2, \ldots, r, r + k + 1, \ldots, T,$$

so that at "time" r there is a gap of k observations. How do we obtain efficient estimators (a) when the error process obeys

$$u_t = \rho u_{t-1} + \varepsilon_t?$$

(b) when the error process obeys

$$u_t = \rho_1 u_{t-1} + \rho_2 u_{t-2} + \varepsilon_t?$$

Moreover, how in the case (a) do we test the hypothesis $\rho = 0$? To provide an answer to these problems, we recall that in the standard first-order autoregression the estimation proceeds, conceptually, by determining a matrix M such that Mu consists of uncorrelated—and in the case of normality, of independent—elements, it being understood that u is the vector of errors of the GLM.

What is the analog of M for the process

$$u_t = \rho u_{t-1} + \varepsilon_t, \qquad t = 1, 2, \ldots, T,$$

when observations for $t = r + 1, r + 2, \ldots, r + k$ are missing? The usual transformation, through the matrix M referred to above, yields

$$\sqrt{1 - \rho^2} u_1,$$
$$u_2 - \rho u_1,$$
$$\vdots$$
$$u_T - \rho u_{T-1}.$$

This is not feasible in the present case since the observations u_t for $r + 1 \leq t \leq r + k$ are missing; in particular, the observation following u_r is u_{r+k+1}. Remembering that the goal is to replace u by a vector of uncorrelated elements, we note that for $1 \leq j \leq k + 1$

$$u_{r+j} = \sum_{s=0}^{\infty} \rho^s \varepsilon_{r+j-s}$$

$$= \sum_{s=0}^{j-1} \rho^s \varepsilon_{r+j-s} + \sum_{s=j}^{\infty} \rho^s \varepsilon_{r+j-s} \qquad (A.61)$$

$$= \rho^j u_r + \sum_{s=0}^{j-1} \rho^s \varepsilon_{r+j-s}.$$

Thus

$$u_{r+j} - \rho^j u_r = \sum_{s=0}^{j-1} \rho^s \varepsilon_{r+j-s}.$$

We observe that

$$\mathrm{Var}\!\left(\sum_{s=0}^{j-1}\rho^{s}\varepsilon_{r+j-s}\right)=\sigma^{2}\!\left(\frac{1-\rho^{2j}}{1-\rho^{2}}\right)=\sigma^{2}\phi^{2},$$

and that

$$u_{t}-\rho u_{t-1}$$

has variance σ^{2} for $t\le r$ and $t\ge r+k+2$; thus, we conclude

$$\frac{1}{s}\,(u_{r+k+1}-\rho^{k}u_{r})$$

also has variance σ^{2} and is independent of the other terms. Hence the matrix

$$M_{1}=\begin{bmatrix}
\sqrt{1-\rho^{2}} & 0 & & & \cdots & & & & 0 \\
-\rho & 1 & & & & & & & \\
0 & \ddots & \ddots & & & & & & \\
 & & -\rho & 1 & \ddots & & & & \vdots \\
 & & & -(\rho^{k}/s) & (1/s) & & & & \\
 & \vdots & & & \ddots & & & & \\
 & & & & & -\rho & 1 & & \\
 & & & & & & \ddots & \ddots & 0 \\
0 & & & \cdots & & & 0 & -\rho & 1
\end{bmatrix}$$

where $1/s = [(1-\rho^{2})/(1-\rho^{2k})]^{1/2}$, implies that $M_{1}u$ has covariance matrix $\sigma^{2}I$ and mean zero.

Hence the autoregressive transformation is of the form

$$\sqrt{1-\rho^{2}}\,u_{1},$$

$$u_{2}-\rho u_{1},$$

$$u_{3}-\rho u_{2},$$

$$\vdots$$

$$u_{r}-\rho u_{r-1},$$

$$\frac{1}{s}\,(u_{r+k+1}-\rho^{k+1}u_{r}),$$

$$u_{r+k+2}-\rho u_{r+k+1},$$

$$\vdots$$

$$u_{T}-\rho u_{T-1},$$

with the corresponding transformation on the dependent and explanatory variables. Estimation by the *search method* proceeds as before, i.e., for given ρ we compute the transformed variables and carry out the OLS procedure. We do this for a set of ρ-values that is sufficiently dense in the interval $(-1, 1)$. The estimator corresponds to the coefficients obtained in the regression exhibiting the smallest sum of squared residuals. Aside from this more complicated transformation, the situation is entirely analogous to the standard case where no observations are missing.

In their original paper Cochrane and Orcutt (C.O.) did not provide a procedure for handling the missing observations case (see [7]). Carrying on in their framework, however, one might suggest the following. Regress the dependent on the explanatory variables, neglecting the fact that some observations are missing. Obtain the residuals

$$\tilde{u}_t, \qquad t = 1, 2, \ldots, r, r + k + 1, r + k + 2, \ldots, T.$$

Obtain

$$\tilde{\rho} = \frac{\sum_{t=2}^{r} \tilde{u}_t \tilde{u}_{t-1} + \sum_{i=2}^{T-r-k} \tilde{u}_{r+k+i} \tilde{u}_{r+k+i-1}}{\sum_{t=1}^{'T-1} \tilde{u}_t^2}$$

where

$$\sum_{t=1}^{T-1}{}' \tilde{u}_t^2$$

indicates that the terms for $t = r + 1, r + 2, \ldots, r + k$ have been omitted. Given this $\tilde{\rho}$ compute

$$y_t - \tilde{\rho} y_{t-1}, \qquad t = 1, 2, \ldots, r,$$

$$y_t - \tilde{\rho} y_{t-1}, \qquad t = r + k + 2, \ldots, T,$$

and similarly for the explanatory variables. Carry out the regression with the transformed variables. Obtain the residuals and, thus, another estimate of ρ. Continue in this fashion until convergence is obtained.

If one wishes to test for the presence of first-order autoregression in the residuals then, following the initial regression above, compute the "adjusted" D.W. statistic

$$d^* = \frac{\sum_{t=2}^{r} (\tilde{u}_t - \tilde{u}_{t-1})^2 + \sum_{t=r+k+2}^{r} (\tilde{u}_t - \tilde{u}_{t-1})^2}{\sum_{t=1}^{'T-1} \tilde{u}_t^2}. \tag{A.62}$$

Clearly, the procedures outlined above apply for a data gap at any point $r \geq 1$. Obviously if $r = 1$ or if $r = T - k$ then the "gap" occasions no problems whatever since all sample observations are contiguous.

Remark A.16. It should be pointed out that the statistic in (A.62) cannot be used in conjunction with the usual D.W. tables. This is so since

$$d^* = \frac{\tilde{u}' A^* \tilde{u}}{\tilde{u}' \tilde{u}},$$

just as in the standard D.W. case. However, the matrix A^* is not of the same form as the matrix A in the usual case; in fact, if we put A_r, A_{T-k-r} for two matrices of order r, $T - k - r$ respectively and of the same form as in the standard case, then

$$A^* = \text{diag}(A_r, A_{T-k-r}).$$

This makes it clear, however, that A^* *has two zero roots while the matrix A in the usual case has only one zero root.* Thus, to test for first-order autoregression in the "gap" case it is better to rely on asymptotic theory if the sample is moderately large. The estimator $\tilde{\rho}$, as given in the discussion of the C.O. procedure, may be easily obtained from the OLS residuals and, asymptotically, behaves as

$$\sqrt{T}(\tilde{\rho} - \rho) \sim N(0, 1 - \rho^2).$$

To close this section we derive the appropriate autoregressive transformation when the error process is a second-order autoregression and there is a gap in the data. Such an error specification is often required for quarterly data.

As in the earlier discussion, what we wish is to represent the observation immediately following the gap in terms of the adjacent observations(s). To be concrete, suppose

$$u_t = \rho_1 u_{t-1} + \rho_2 u_{t-2} + \varepsilon_t, \tag{A.63}$$

where

$$\{\varepsilon_t : t = 0, \pm 1, \pm 2, \ldots\}$$

is a sequence of i.i.d. random variables with mean zero and variance σ^2. We require the process above to be stable, i.e., we require that for $\sigma^2 < \infty$ the variance of the u_t's also be finite. Introduce the lag operator L such that, for any function x_t,

$$Lx_t = x_{t-1}$$

and, in general,

$$L^s x_t = x_{t-s}, \qquad s \geq 0.$$

For $s = 0$ we set $L^0 = I$, the identity operator. Polynomials in the lag operator L are isomorphic to polynomials in a real (or complex) indeterminate, i.e., ordinary polynomials like

$$P(t) = a_0 + a_1 t + a_2 t^2 + \cdots + a_n t^n,$$

where the a_i, $i = 0, 1, \ldots, n$ are real numbers and t is the real (or complex) indeterminate (i.e., the "unknown"). This isomorphism means that whatever operations may be performed on ordinary polynomials can also be performed in the same manner on polynomials in the lag operator L. The reader desiring greater detail is referred to Dhrymes [8, Chapters 1 and 2]. Noting that

$$u_{t-1} = Lu_t, \qquad u_{t-2} = L^2 u_t,$$

we can write

$$(I - \rho_1 L - \rho_2 L^2)u_t = \varepsilon_t.$$

Note also that

$$(I - \rho_1 L - \rho_2 L^2) = (I - \lambda_1 L)(I - \lambda_2 L)$$

for

$$\lambda_1 + \lambda_2 = \rho_1, \qquad -\lambda_1\lambda_2 = \rho_2. \tag{A.64}$$

Thus, the process in (A.63) can also be represented as

$$u_t = \frac{I}{(I - \lambda_1 L)(I - \lambda_2 L)} \varepsilon_t.$$

But $I/(I - \lambda_1 L)$ behaves like

$$\frac{1}{1 - \lambda_1 t} = \sum_{i=0}^{\infty} \lambda_1^i t^i.$$

Hence

$$u_t = \sum_{j=0}^{\infty} \lambda_1^j \sum_{i=0}^{\infty} \lambda_2^i \varepsilon_{t-i-j}.$$

Assuming that $|\lambda_2| \le |\lambda_1| < 1$, we can rewrite the double sum above as

$$u_t = \sum_{i=0}^{\infty} c_i \varepsilon_{t-i} \tag{A.65}$$

where

$$c_i = \frac{\lambda_1^{i+1} - \lambda_2^{i+1}}{\lambda_1 - \lambda_2}.$$

If in the sequence

$$u_t, \qquad t = 1, 2, \ldots, T,$$

there is a gap of length k at $t = r$, it means that u_r is followed by u_{r+k+1}, u_{r+k+2}, and so on. For observations u_t, $2 < t \le r$, we know that the transformation

$$u_t - \rho_1 u_{t-1} - \rho_2 u_{t-2}$$

yields i.i.d. random variables, viz., the ε's. For $t > r$, however, we encounter a difficulty. The observation following u_r is u_{r+k+1}. Thus, blindly applying the transformation yields

$$u_{r+k+1} - \rho_1 u_r - \rho_2 u_{r-1}.$$

But the expression above is *not* ε_{r+k+1}. So the problem may be formulated as: what coefficients should we attach to u_r and u_{r-1} in order to render the difference a function only of $\{\varepsilon_t : t = r + 1, r + 2, \ldots, r + k + 1\}$?

It is for this purpose that the expression in (A.65) is required. We note

$$u_{r+k+1} = \sum_{i=0}^{\infty} c_i \varepsilon_{r+k+1-i} = \sum_{i=0}^{k+1} c_i \varepsilon_{r+k+1-i} + \sum_{i=k+2}^{\infty} c_i \varepsilon_{r+k+1-i}.$$

But, putting $j = i - (k+2)$ yields

$$\sum_{i=k+2}^{\infty} c_i \varepsilon_{r+k+1-i} = \sum_{j=0}^{\infty} c_{k+2+j} \varepsilon_{r-1-j}$$

Similarly,

$$u_r = \varepsilon_r + \sum_{j=0}^{\infty} c_{j+1} \varepsilon_{r-1-j}, \qquad u_{r-1} = \sum_{j=0}^{\infty} c_j \varepsilon_{r-1-j}.$$

Thus,

$$u_{r+k+1} - \alpha u_r - \beta u_{r-1} = \sum_{i=0}^{k+1} c_i \varepsilon_{r+k+1-i} - \alpha \varepsilon_r$$

$$+ \sum_{j=0}^{\infty} (c_{k+2+j} - \alpha c_{j+1} - \beta c_j) \varepsilon_{r-1-j}$$

and we require

$$c_{k+2+j} - \alpha c_{j+1} - \beta c_j = 0.$$

This is satisfied by the choice

$$\alpha = \frac{\lambda_1^{k+2} - \lambda_2^{k+2}}{\lambda_1 - \lambda_2}, \qquad \beta = -\lambda_1 \lambda_2 \frac{\lambda_1^{k+1} - \lambda_2^{k+1}}{\lambda_1 - \lambda_2}. \qquad \text{(A.66)}$$

It is, of course, apparent that

$$\alpha = c_{k+1}, \qquad \rho_2 = -\lambda_1 \lambda_2,$$

and thus

$$u_{r+k+1} - \alpha u_r - \beta u_{r-1} = u_{r+k+1} - c_{k+1} u_r - \rho_2 c_k u_{r-1} = \sum_{i=0}^{k} c_i \varepsilon_{r+k+1-i}.$$

Similarly, if we wish to find quantities γ and δ such that[10]

$$u_{r+k+2} - \gamma u_r - \delta u_{r-1}$$

is a function of at most only $\{\varepsilon_t: t = r+1, r+2, \ldots, r+k+2\}$, we conclude, following the same procedure as above, that

$$\gamma = c_{k+2}, \qquad \delta = \rho_2 c_{k+1}. \qquad \text{(A.67)}$$

---

[10] If one wished, one could seek to determine coefficients γ and δ such that $u_{r+k+2} - \gamma u_{r+k+1} - \delta u_r$ satisfies conditions similar to those above.

Thus,

$$u_{r+k+2} - \gamma u_r - \delta u_{r-1} = u_{r+k+2} - c_{k+2}u_r - \rho_2 c_{k+1}u_{r-1} = \sum_{i=0}^{k+1} c_i \varepsilon_{r+k+2-i}.$$

Put

$$
\begin{aligned}
v_t &= u_t, & t &= 1, 2, \ldots, r \\
&= u_t - c_{k+1}u_r - \rho_2 c_k u_{r-1}, & t &= r + k + 1 \\
&= u_t - c_{k+2}u_r - \rho_2 c_{k+1}u_{r-1}, & t &= r + k + 2 \\
&= u_t, & t &= r + k + 3, r + k + 4, \ldots, T.
\end{aligned}
$$

To produce the transformation desired we need only derive an expression for the variances and covariances of v_1, v_2, those of v_{r+k+1}, v_{r+k+2}, as well as an expression for the coefficients c_i, appearing immediately below (A.65), involving only ρ_1 and ρ_2. Now,

$$\text{Var}(u_t) = \sigma_{00} = \rho_1^2 \sigma_{00} + \rho_2^2 \sigma_{00} + 2\rho_1\rho_2\sigma_{01} + \sigma^2,$$

where

$$\sigma_{01} = \text{Cov}(u_t, u_{t-1}).$$

Moreover,

$$\sigma_{01} = \rho_1 \sigma_{00} + \rho_2 \sigma_{01},$$

which yields

$$\sigma_{00} = \left\{\frac{1 - \rho_2}{(1 - \rho_2)(1 - \rho_2^2) - \rho_1^2(1 + \rho_2)}\right\}\sigma^2, \qquad \sigma_{01} = \left[\frac{\rho_1}{1 - \rho_2}\right]\sigma_{00}.$$

Thus

$$\text{Var}(v_t) = \sigma_{00}, \qquad t = 1, 2,$$
$$\text{Cov}(v_t, v_{t-1}) = \sigma_{01}, \qquad t = 2.$$

In addition,

$$\text{Var}(v_{r+k+1}) = \sigma^2\left[\sum_{i=0}^{k} c_i^2\right], \qquad \text{Var}(v_{r+k+2}) = \sigma^2\left[\sum_{i=0}^{k+1} c_i^2\right],$$

$$\text{Cov}(v_{r+k+2}, v_{r+k+1}) = \sigma^2\left[\sum_{i=0}^{k} c_i c_{i+1}\right].$$

Now, if x and y are two random variables, it is well known that

$$-\left(\frac{\sigma_{xy}}{\sigma_{xx}}\right)x + y$$

is uncorrelated with x, where, obviously,

$$\sigma_{xx} = \text{Var}(x), \qquad \sigma_{xy} = \text{Cov}(x, y).$$

Consequently,

$$-\left(\frac{\rho_1}{1 - \rho_2}\right) u_1 + u_2$$

is uncorrelated with u_1. Similarly,

$$-\left(\frac{\sum_{i=0}^{k} c_{i+1} c_i}{\sum_{i=0}^{k} c_i^2}\right) v_{r+k+1} + v_{r+k+2}$$

is uncorrelated with v_{r+k+1}.

Define the lower triangular matrix

$$S = \begin{bmatrix}
s_{11} & 0 & & & \cdots & & & & 0 \\
s_{21} & s_{22} & 0 & & & & & & \\
-\rho_2 & -\rho_1 & 1 & 0 & & & & & \\
0 & -\rho_2 & -\rho_1 & 1 & & & & & \\
\vdots & & & & & & & & \vdots \\
0 & \cdots & s_{r+1,r-1} & s_{r+1,r} & s_{r+1,r+1} & & & & \\
0 & \cdots & s_{r+2,r-1} & s_{r+2,r} & s_{r+2,r+1} & s_{r+2,r+2} & & & \\
& & & & -\rho_2 & -\rho_1 & 1 & 0 & \\
\vdots & & & & & & & 0 & \\
& & & & & & & 0 & \\
0 & & & \cdots & & & -\rho_2 & -\rho_1 & 1
\end{bmatrix},$$

$$(A.68)$$

where

$$s_{11} = \left\{\frac{1 - \rho_2}{(1 - \rho_2)(1 - \rho_2^2) - (1 + \rho_2)\rho_1^2}\right\}^{-1/2}$$

$$s_{22} = \left\{\frac{1 - \rho_2}{[(1 - \rho_2)^2 - \rho_1^2]^{1/2}}\right\} s_{11}$$

$$s_{21} = -\left\{\frac{\rho_1}{[(1 - \rho_2)^2 - \rho_1^2]^{1/2}}\right\} s_{11}$$

$$s_{r+1,r-1} = -\rho_2 c_k s_{r+1,r+1}, \quad s_{r+1,r} = -c_{k+1} s_{r+1,r+1},$$

$$s_{r+1,r+1} = \left[\sum_{i=0}^{k} c_i^2\right]^{-1/2},$$

$$s_{r+2,r-1} = -\rho_2 c_{k+1} s_{r+2,r+2} - \rho_2 c_k s_{r+2,r+1},$$

$$s_{r+2,r} = -c_{k+2} s_{r+2,r+2} - c_{k+1} s_{r+2,r+1},$$

$$s_{r+2,r+1} = -\left(\frac{\sum_{i=0}^{k} c_{i+1} c_i}{\sum_{i=0}^{k} c_i^2}\right) s_{r+2,r+2},$$

$$s_{r+2,r+2} = \left\{\frac{(\sum_{i=0}^{k} c_i^2)(\sum_{i=0}^{k+1} c_i^2) - (\sum_{i=0}^{k} c_{i+1} c_i)^2}{\sum_{i=0}^{k} c_i^2}\right\}^{-1/2}.$$

To express the coefficients c_i in terms of ρ_1, ρ_2 (instead of λ_1, λ_2, as we did earlier) we proceed as follows. Consider the recursive relation (neglecting the ε's)

$$u_{r+1} = \rho_1 u_r + \rho_2 u_{r-1},$$

$$u_{r+2} = \rho_1 u_{r+1} + \rho_2 u_r = (\rho_1^2 + \rho_2) u_r + \rho_1 \rho_2 u_{r-1}.$$

If we put, in general,

$$u_{r+s} = c_s^* u_r + d_s u_{r-1}$$

we obtain the recursions

$$c_s^* = \rho_1 c_{s-1}^* + \rho_2 c_{s-2}^*, \qquad d_s = \rho_1 d_{s-1} + \rho_2 d_{s-2}$$

with the "initial conditions"

$$c_0^* = 1, \qquad c_{-1}^* = 0, \qquad d_0 = 0, \qquad d_{-1} = 1. \tag{A.69}$$

But from (A.66) and (A.67) we easily see that

$$c_s^* = c_s, \qquad d_s = \rho_2 c_{s-1}, \tag{A.70}$$

where the c_s's are exactly the quantities defined just below in, (A.65). Computing, recursively, a few of the coefficients c_s and taking the initial conditions (A.69) into account, we find

$$c_1 = \rho_1, \qquad\qquad\qquad c_5 = \rho_1^5 + 4\rho_1^3 \rho_2 + 3\rho_1 \rho_2^2,$$

$$c_2 = \rho_1^2 + \rho_2, \qquad\qquad c_6 = \rho_1^6 + 5\rho_1^4 \rho_2 + 6\rho_1^2 \rho_2^2 + \rho_2^3,$$

$$c_3 = \rho_1^3 + 2\rho_1 \rho_2, \qquad\quad c_7 = \rho_1^7 + 6\rho_1^5 \rho_2 + 10\rho_1^3 \rho_2^2 + 4\rho_1 \rho_2^3,$$

$$c_4 = \rho_1^4 + 3\rho_1^2 \rho_2 + \rho_2^2, \quad\; c_8 = \rho_1^8 + 7\rho_1^6 \rho_2 + 15\rho_1^4 \rho_2^2 + 10\rho_1^2 \rho_2^3 + \rho_2^4,$$

or, in general,

$$c_s = \sum_{i=0}^{[s/2]} a_{si} \rho_1^{s-2i} \rho_2^i, \qquad s \geq 1, \tag{A.71}$$

where $[s/2]$ is the integral part of $s/2$;

$$a_{si} = a_{s-1,i} + a_{s-2,i-1}, \qquad i \geq 1, s \geq 2, \tag{A.72}$$

and for *all* s

$$a_{s0} = 1, \qquad a_{sj} = 0, \qquad j > \left[\frac{s}{2}\right], s \geq 1,$$

while for *even* s

$$a_{s,[s/2]} = 1;$$

and the "initial" conditions are

$$a_{00} = 0.$$

The recursion in (A.51) together with the conditions just enumerated completely describes the coefficients

$$\{c_i: i = 0, 1, 2, \ldots\}$$

and thus completely determines the elements of the matrix S in terms of the parameters ρ_1, ρ_2. What this accomplishes is the following. Suppose that

$$u_t = \rho_1 u_{t-1} + \rho_2 u_{t-2} + \varepsilon_t$$

and that the sequence

$$\{\varepsilon_t: t = 0, \pm 1, \pm 2, \ldots\}$$

is one of i.i.d. random variables with mean zero and variance σ^2.
   Let

$$u = (u_1, u_2, \ldots, u_r, u_{r+k+1}, u_{r+k+2}, \ldots, u_T)'.$$

Then Su is a vector of uncorrelated random elements whose mean is zero and whose (common) variance is σ^2. If we assert that the elements of the ε-sequence obey, in addition,

$$\varepsilon_t \sim N(0, \sigma^2) \quad \text{for all } t$$

then

$$Su \sim N(0, \sigma^2 I),$$

where I is the identity matrix of order $T - k$.
   One could, clearly, estimate the parameters of the GLM, exhibiting a gap of length k, and whose errors are a second-order autoregression, by a search procedure applied to the transformed data Sy, SX, where S is as defined in (A.68). We may summarize the discussion of this section in

Theorem A.3. *Consider the* GLM

$$y = X\beta + u,$$

where y is $(T - k) \times 1, X$ is $(T - k) \times (n + 1), u$ is $(T - k) \times 1$, and there is a gap of length k in the observations as follows:

$$\{y_t, x_{ti}: t = 1, 2, \ldots, r, r + k + 1, r + k + 2, \ldots, T, i = 0, 1, 2, \ldots, n\}.$$

Provided

(a) $\text{rank}(X) = n + 1$,
(b) $(\text{p})\lim_{T \to \infty}(X'X/T)$ *is positive definite*,
(c) $E(u|X) = 0$,

the following is true.

(i) *If, in addition,* $\text{Cov}(u \mid X) = \sigma^2 I$, *then the OLS estimator of* β *is consistent, unbiased, and efficient.*
(ii) *If* $u_t = \rho u_{t-1} + \varepsilon_t$ *and* $\{\varepsilon_t : t = 0, \pm 1, \pm 2, \ldots\}$ *is a sequence of i.i.d. random variables with mean zero and variance* σ^2, *the OLS estimator* $\tilde{\beta} = (X'X)^{-1}X'y$ *is unbiased, consistent, but inefficient. The (feasible) efficient estimator is obtained as*

$$\hat{\beta} = (X'\tilde{M}_1'\tilde{M}_1 X)^{-1} X'\tilde{M}_1'\tilde{M}_1 y,$$

where M_1 *is a* $(T - k) \times (T - k)$ *matrix with elements that are all zero except:*

$$m_{jj} \begin{cases} = \sqrt{1 - \rho^2} & \text{for } j = 1 \\[2mm] = \left[\dfrac{1 - \rho^2}{1 - \rho^{2k}}\right]^{1/2} & \text{for } j = r + 1 \\[2mm] = 1 & \text{otherwise}; \end{cases}$$

$$m_{j+1, j} \begin{cases} = -\rho^k \left[\dfrac{1 - \rho^2}{1 - \rho^{2k}}\right]^{1/2} & \text{for } j = r \\[2mm] = -\rho \cdot & \text{otherwise}. \end{cases}$$

The matrix $\tilde{M}_1$ *is obtained by substituting for* ρ *its estimator* $\tilde{\rho}$. *The latter may be obtained by the usual search method applied to the transformed model*

$$M_1 y = M_1 X \beta + M_1 u$$

or by a suitable extension of the C.O. method. In the observation gap model above the OLS residuals, say $\tilde{u}_t$, *may be used to compute a D.W.-like statistic*

$$\frac{\sum_{t=2}^{r} (\tilde{u}_t - \tilde{u}_{t-1})^2 + \sum_{t=r+k+2}^{T} (\tilde{u}_t - \tilde{u}_{t-1})^2}{\sum_{t=1}^{r} \tilde{u}_t^2 + \sum_{t=r+k+1}^{T} \tilde{u}_t^2}.$$

The usual tabulations, however, for the D.W. statistic are not appropriate in this instance and one should, if the sample is moderately large, apply asymptotic theory.

(iii) *If* $u_t = \rho_1 u_{t-1} + \rho_2 u_{t-2} + \varepsilon_t$, *the sequence* $\{\varepsilon_t : t = 0, \pm 1, \pm 2, \ldots\}$, *being one of i.i.d. random variables with mean zero and variance* σ^2, *and if the process is stable, i.e., the roots of* $z^2 - \rho_1 z - \rho_2 = 0$ *are less than unity in absolute value, then the OLS estimator is unbiased and consistent but is is inefficient. The (feasible) efficient estimator is obtained by the search method, which minimizes*

$$(Sy - SX\beta)'(Sy - SX\beta)$$

over the range $\rho_1 \in (-2, 2)$, $\rho_2 \in (-1, 1)$. The estimator is

$$\hat{\beta} = (X'\tilde{S}'\tilde{S}X)^{-1}X'\tilde{S}'\tilde{S}y,$$

where S is a $(T - k) \times (T - k)$ matrix all of whose elements are zero except:

$$S_{jj} \begin{cases} = \left\{ \dfrac{1 - \rho_2}{(1 - \rho_2)(1 - \rho_2^2) - (1 + \rho_2)\rho_1^2} \right\}^{-1/2}, & j = 1 \\[3mm] = \left\{ \dfrac{1 - \rho_2}{[(1 - \rho_2)^2 - \rho_1^2]^{1/2}} \right\} S_{11}, & j = 2 \\[3mm] = 1, & j = 3, \ldots, r \\[3mm] \left[\displaystyle\sum_{i=0}^{k} c_i^2 \right]^{-1/2}, & j = r + 1 \\[3mm] = \left[\dfrac{(\sum_{i=0}^{k} c_i^2)(\sum_{i=0}^{k+1} c_i^2) - (\sum_{i=0}^{k} c_i c_{i+1})}{\sum_{i=0}^{k} c_i^2} \right]^{-1/2}, & j = r + 2 \\[3mm] = 1, & j = r + 3, \ldots, T - k; \end{cases}$$

$$S_{j+1,j} \begin{cases} = \left\{ -\dfrac{\rho_1}{[(1 - \rho_2)^2 - \rho_1^2]^{1/2}} \right\} S_{11}, & j = 1 \\[3mm] = -\rho_1, & j = 2, 3, \ldots, r - 1 \\[3mm] = -c_{k+1} S_{r+1, r+1}, & j = r \\[3mm] = -\left(\dfrac{\sum_{i=0}^{k} c_i c_{i+1}}{\sum_{i=0}^{k} c_i^2} \right) S_{r+2, r+2}, & j = r + 1 \\[3mm] = -\rho_1, & j = r + 2, r + 3, \ldots, T - k; \end{cases}$$

$$S_{j+2,j} \begin{cases} = -\rho_2. & j = 1, 2, \ldots, r - 2 \\[2mm] = -\rho_2 c_k S_{r+1, r+1}, & j = r - 1 \\[2mm] = -c_{k+2} S_{r+2, r+2} - c_{k+1} S_{r+2, r+1}, & j = r \\[2mm] = -\rho_2, & j = r + 1, r + 2, \ldots, T - k; \end{cases}$$

$$S_{j+3,j} \begin{cases} = 0, & j = 1, 2, \ldots, r - 2 \\[2mm] = -\rho_2(c_{k+1} S_{r+2, r+2} + c_k S_{r+2, r+1}), & j = r - 1 \\[2mm] = 0, & j = r, r + 1, \ldots, T - k. \end{cases}$$

The coefficients c_s above are given by

$$c_s = \sum_{i=0}^{[s/2]} a_{si} \rho_1^{s - 2i} \rho_2^i,$$

where $[s/2]$ denotes the integral part of $s/2$, and

$$a_{si} = a_{s-1,i} + a_{s-2,i-1}, \qquad i \geq 1, s \geq 2,$$

where

$$a_{00} = 0$$

$$a_{s0} = 1, \qquad a_{sj} = 0, \qquad j > \left[\frac{s}{2}\right], s \geq 1,$$

and for even s

$$a_{s,[s/2]} = 1.$$

The General Linear Model IV

4

In this chapter we take up the problems occasioned by the failure of the rank condition (for the matrix of explanatory variables). This problem arises as a matter of course in analysis of variance (or covariance) models where some of the variables are classificatory. In this case, we are led to the construction of "dummy" variables representing the classificatory schemes. Since all such classificatory schemes are exhaustive, it is not surprising that the "dummy" variables are linearly dependent and, thus, the rank condition for the data matrix fails.

Another instance in which the problem arises is in (aggregate) economic time series which exhibit a high degree of intercorrelation although not exact linear dependence. For example, if we take value added, as well as employment and capital stock, in manufacturing for the United States over the period 1949–1972 we would find a very high degree of correlation between employment (labor) and capital stock (capital). If we regress value added on capital and labor, we shall, typically, obtain regression coefficients which are theoretically unacceptable—for example we may obtain a *negative* coefficient for capital! Whatever the reason for this strange result—and we are not discussing this issue here—empirical researchers tend to attribute this phenomenon to the high degree of correlation among the explanatory variables. Here, the rank condition for the data matrix is not, strictly, violated but the moment matrix $X'X$ is so ill conditioned that doubts may be expressed about the (computational) validity of the ensuing regression results. This latter case is commonly referred to as *the* problem of multicollinearity,[1] although

---

[1] The term is actually a misnomer. Two linearly dependent variables are said to be *collinear* since they lie on the same line. Thus if $\alpha_1 x_1 + \alpha_2 x_2 = 0$ then we have $x_2 = -(\alpha_1/\alpha_2)x_1$ for $\alpha_2 \neq 0$. Three or more linearly dependent variables, however, lie on the *same plane* and this case is more properly referred to as *coplanarity*.

the term, strictly speaking, refers to the failure of the rank condition. In the following sections we shall examine, separately, the case of exact and near collinearity for continuous variables as well as the case of exact collinearity arising in the context of analysis of variance or analysis of covariance problems.

1 Multicollinearity: Failure of the Rank Condition

1.1 Definition of the Problem

When dealing with the GLM

$$y = X\beta + u, \tag{1}$$

discussion of multicollinearity in the literature of econometrics exhibits two aspects. One is quite precise and deals exactly with the problem as conveyed by the term, i.e., the case where the columns of X exhibit linear dependencies. The other is quite fuzzy and deals with the case of "near dependence" of such columns. We shall discuss the first aspect with some precision, touch on the second aspect, and examine some of the proposed "remedies."[2]

Thus, in the GLM above, suppose

$$\text{rank}(X) < n + 1. \tag{2}$$

We remind the reader that, by convention, the GLM has a constant term and contains n bona fide explanatory variables. The normal equations are

$$X'X\beta = X'y. \tag{3}$$

Since, however, X is *not* of full (column) rank the matrix $X'X$ is singular and *consequently a unique solution does not exist.* When this is so, of course, it is not possible to estimate the impact exerted, individually, by each of the explanatory variables on the dependent variable i.e., we cannot identify, separately, the elements of the vector β.

We are, nonetheless, quite able to say *something* about the vector of co-efficients, β. First, referring to Section 3 of *Mathematics for Econometrics*, we see, in the terminology of that discussion, that we are dealing with the inconsistent system

$$u = y - X\beta.$$

A least squares solution is given by

$$\hat{\beta} = X_s y,$$

---

[2] For full comprehension of the discussion in the first part of this chapter the reader is well advised to master Section 3 of *Mathematics for Econometrics*.

where X_s is an s-inverse (least squares inverse) of X. The latter is given by

$$X_s = (X'X)_c X'$$

where $(X'X)_c$ is a c-inverse (conditional inverse) of $X'X$. Thus, a least squares solution is

$$\hat{\beta} = (X'X)_c X'y. \tag{4}$$

Of course, the same result would have been reached had we operated directly with (3). Now the difficulty with (4) is that this is not the only solution to the problem, and it is interesting to inquire as to whether there are any functions of the coefficient vector that are invariant with respect to the particular choice of the c-inverse. This is, in fact, answered by Proposition 82 of *Mathematics for Econometrics*. Precisely, let

$$G = CX.$$

Then

$$G\hat{\beta}$$

is invariant to the choice of the c-inverse. One interesting choice is

$$G = X.$$

Then,

$$X\hat{\beta} = X(X'X)_c X'y$$

is invariant to the choice of the c-inverse. For completeness we shall sketch a proof of this fact. By definition of a c-inverse, we have

$$(X'X)(X'X)_c(X'X) = X'X.$$

Premultiply by the g-inverse (generalized inverse) of X', which is unique, and postmultiply by X_g—which is the g-inverse of X—to obtain

$$X_g' X'X(X'X)_c X'XX_g = X_g' X'XX_g. \tag{5}$$

But from the properties of g-inverses we have

$$X_g' X'X = XX_g X = X, \qquad X'XX_g = X'X_g' X' = X'$$

and the relation in (5) can be written

$$X(X'X)_c X' = XX_g. \tag{6}$$

But the right side in (6) is uniquely determined. Hence, the left is uniquely determined as well. Consequently, the "within the sample" predictions of the dependent variable

$$\hat{y} = X\hat{\beta}$$

are uniquely defined, even though the rank condition fails and the individual components of the coefficient vector β cannot be identified.

Now, compute the vector of residuals

$$\hat{u} = y - X\hat{\beta} = [I - X(X'X)_c X]y$$

and recall that

$$\frac{\hat{u}'\hat{u}}{T - n - 1}$$

has served as an unbiased estimator of σ^2—the common variance of the error terms. Is this a quantity that is invariant to the choice of the c-inverse? We note that

$$I - X(X'X)_c X' = I - XX_g.$$

Moreover, Proposition 76 of *Mathematics for Econometrics* shows that $I - XX_g$ is an idempotent matrix. Consequently,

$$\hat{u}'\hat{u} = y'[I - XX_g]y. \tag{7}$$

Since the g-inverse is *unique* it follows that $I - XX_g$ is unique and, thus, that $\hat{u}'\hat{u}$ is invariant to the choice of the c-inverse.

Further discussion of this aspect will be facilitated if we introduce

Definition 1. In the context of the GLM

$$y = X\beta + u,$$

a set of parameters or a function of parameters is said to be *estimable* if there exists a function of the observations which is an unbiased estimator of the set or function of the parameters in question.

The set or function is said to be *linearly estimable* if there exists a function of the observations, linear in the dependent variable, which is an unbiased estimator of the set or function of the parameters in question.

In light of this definition and the definition preceding it, we have

Proposition 1. *Consider the GLM*

$$y = X\beta + u$$

where

$$\text{rank}(X) < n + 1$$

and otherwise subject to the standard assumptions. Then:

(i) *$G\beta$ is a linearly estimable function if G lies in the row space of X, i.e., if these exists a matrix C such that*

$$G = CX.$$

(ii) *σ^2 is an estimable parameter.*

PROOF. Let G be in the row space of X and $\hat\beta$ as in (4). We have, using (6) and the properties of g-inverses,

$$\begin{aligned} G\hat\beta &= CX(X'X)_c X'y \\ &= CXX_g X\beta + CXX_g u \\ &= CX\beta + CXX_g u = G\beta + GX_g u. \end{aligned}$$

Thus

$$E(G\hat\beta \mid X) = G\beta,$$

which proves the first part of the proposition. For the second part, note that from Equation (7)

$$\hat u'\hat u = (\beta'X' + u')(I - XX_g)(X\beta + u) = u'(I - XX_g)u.$$

Consequently

$$E(\hat u'\hat u) = \operatorname{tr}[(I - XX_g)E(uu')] = \sigma^2 \operatorname{tr}(I - XX_g) = \sigma^2(T - r),$$

where T is the number of observations (also the dimension of the identity matrix) and r is the rank of X and hence of XX_g. (See Proposition 72 in *Mathematics for Econometrics*). Thus,

$$\hat\sigma^2 = \frac{\hat u'\hat u}{T - r}$$

is an unbiased estimator of σ^2.   q.e.d.

Corollary 1. *If, in addition, we assume that the elements of the error process $\{u_t : t = 1, 2, \ldots\}$ are normally distributed then*

$$G\hat\beta \sim N(G\beta, \sigma^2 GX_g X'_g G').$$

PROOF. Obvious from the fact that

$$G\hat\beta = G\beta + GX_g u$$

and

$$u \sim N(0, \sigma^2 I).   \text{q.e.d.}$$

Remark 1. In the interesting case $G = X$ we find

$$X\hat\beta \sim N(X\beta, \sigma^2 XX_g X'_g X').$$

But

$$XX_g X'_g X' = XX_g XX_g = XX_g$$

and XX_g is a $T \times T$ matrix of rank r. Hence, the density

$$X\hat\beta \sim N(X\beta, \sigma^2 XX_g)$$

has a singular covariance matrix.

The other aspect of the problem alluded to above refers to the case where $X'X$ is not, strictly speaking, singular, but the investigator feels it is so close to singularity as to vitiate the results of the estimation process. Since this is a somewhat vague statement it is difficult to see exactly what is to be made of it. First, we note that, provided the inverse $(X'X)^{-1}$ is *accurately obtained* (in the sense that the inversion process is free of round off error of appreciable magnitude), there is no "problem." The OLS estimator of β is still a BLUE and, on the assumption of normality for the errors,

$$\hat{\beta} \sim N[\beta, \sigma^2(X'X)^{-1}].$$

We remind the reader that all inference and distributional results are conditional on the data matrix X. Given X, the OLS procedure has extracted from the data the best linear unbiased estimator. *Nothing more can be done, given X, and there is no problem in the sense that some aspect of the procedure cannot be carried out or that some property of the estimator fails to hold.* The only conceivable problem that may arise relates to the accuracy of the inversion procedure for $(X'X)^{-1}$ and this is a problem in computation, not theoretical inference. To recapitulate this aspect, if $X'X$ is not strictly singular and the inverse $(X'X)^{-1}$ is accurately obtained, the OLS procedure still gets us exactly what we "bargained for," i.e., the BLUE of β. Hence, *given the sample*, there is no problem. This is not to say that the investigator facing a nearly multicollinear system will be content with the results. His uneasiness may arise because the diagonal elements of $\hat{\sigma}^2(X'X)^{-1}$ are "too large" and hence the bounds on the estimated coefficients may be "too wide." This, however, *is not, strictly speaking, an indication of failure of the procedure; it is rather a shortcoming of the sample in that it does not permit sharp inferences to be made.*

Thus, a proper remedy is to obtain another sample or, if the data are inherently nearly multicollinear, to reformulate the investigation so as to deal with linear combinations of coefficients whose estimator may be free of multicollinearity.

1.2 Recognition of Multicollinearity and Proposed Remedies

Recognition. It is frequently proposed that one way of recognizing multicollinearity as a problem is to look at the off diagonal elements of the sample correlation matrix. These give the intercorrelation among the explanatory variables. The farther away they are from zero (in absolute value) the more pressing the problem. While there is a measure of truth in this statement, it is false as a general proposition. This is perhaps best illustrated by an example.

EXAMPLE 1. Consider a GLM whose data matrix, X, upon centering of observations and division by the appropriate (sample) standard errors yields the *sample correlation matrix*

$$\begin{bmatrix} 1 & .4 & .4 \\ .4 & 1 & -.68 \\ .4 & -.68 & 1 \end{bmatrix}$$

Here, the intercorrelations are modest: .4 between x_1 and x_2; .4 between x_1 and x_3; $-.68$ between x_2 and x_3. Yet, the determinant of this matrix is

$$\det \begin{bmatrix} 1 & .4 & .4 \\ .4 & 1 & -.68 \\ .4 & -.68 & 1 \end{bmatrix} = 1 - (.68)^2 - .8(.672) = 0.$$

Thus, here we have strict multicollinearity among the three variables x_1, x_2, x_3 even though no simple intercorrelation exceeds, in absolute value, .7!

On the other hand suppose another sample yields the (sample) correlation matrix

$$\begin{bmatrix} 1 & .9 & .9 \\ .9 & 1 & .9 \\ .9 & .9 & 1 \end{bmatrix}$$

The determinant of this matrix is

$$\det \begin{bmatrix} 1 & .9 & .9 \\ .9 & 1 & .9 \\ .9 & .9 & 1 \end{bmatrix} = (1 - .81) - .9(.9 - .81) + .9(.81 - .9) = .028.$$

Thus, here all intercorrelations are .9, yet the three variables x_1, x_2, x_3 are linearly independent and the correlation matrix is clearly nonsingular.

Remark 2. The import of the example is that high intercorrelations constitute neither a necessary nor a sufficient condition for the presence of multicollinearity. The grain of truth, referred to earlier, is that as any one or more off diagonal elements of the correlation matrix converge to 1 (or -1) we obviously *do have multicollinearity*. Short of that event, however, it is not clear how "near" or exact multicollinearity can be characterized simply so that it can become evident by inspection of the sample intercorrelations of the explanatory variables.[3]

Another phenomenon that is often alleged to disclose the presence of multicollinearity is the discovery, upon estimation of the parameters of a GLM, that tests of significance show the coefficients of all bona fide variables to be insignificantly different from zero (by the application of the t-test to individual coefficients) while R^2 is shown to be "significant" by the F-test!

---

[3] A more satisfactory approach may be to consider the ratio between the largest and smallest characteristic root of the correlation matrix; if this ratio exceeds, say, 500 we may conclude that the coefficient estimates obtained by the regression are, numerically, unreliable.

Again, while this also contains a measure of truth, such a phenomenon does not necessarily disclose the presence of multicollinearity. It is true that "near multicollinearity" might eventuate in this configuration of results, but such a configuration might also come about without multicollinearity, "near" or exact.

The argument to be given, is due to Geary and Leser [17], who also provide examples of other, seemingly paradoxical, results. Suppose we have the usual GLM and parameters have been estimated; the data have been expressed as deviations from sample means and the explanatory variables have been divided by their (sample) standard deviations. The result may be written as

$$y_t = \sum_{i=1}^{n} b_i x_{ti} + \hat{u}_t,$$

where the b_i are the estimated coefficients and the $\hat{u}_t$ are the regression residuals. It involves no loss in relevance to suppose that the $b_i > 0$. This is so since if some $b_i < 0$ we can redefine the corresponding explanatory variable to be $-x_i$.

We note that, in view of the normalizations,

$$\sum_{t=1}^{T} x_{ti}^2 = T, \qquad \sum_{t=1}^{T} x_{ti} x_{tj} = T r_{ij},$$

where r_{ij} is the sample correlation coefficient between the ith and jth explanatory variables. We recall that the F-statistic for testing the "significance of the regression" or the "significance of the coefficient of determination, R^2" or for testing that the vector of coefficients of the bona fide variables is significantly different from zero (all the preceding statements are equivalent) is given by

$$F_{n,\,T-n-1} = \frac{\sum_{t=1}^{T} (\sum_{i=1}^{n} b_i x_{ti})^2}{\sum_{t=1}^{T} \hat{u}_t^2} \frac{T-n-1}{n} = \frac{T}{n\hat{\sigma}^2} \sum_{j=1}^{n} \sum_{i=1}^{n} b_i r_{ij} b_j,$$

where

$$\hat{\sigma}^2 = \frac{\sum_{t=1}^{T} \hat{u}_t^2}{T-n-1}.$$

Now, the t-ratio for the ith coefficient is

$$t_i = \frac{b_i}{\sqrt{\hat{\sigma}^2 q_{ii}}}, \qquad i = 1, 2, \ldots, n,$$

where

$$R = (r_{ij}), \qquad i, j = 1, 2, \ldots, n,$$

and q_{ii} is the ith diagonal element of (R^{-1}/T). Thus,

$$T q_{ii} = \frac{R_{ii}}{|R|} = r^{ii},$$

where here it is understood that the elements of R^{-1} are denoted by

$$R^{-1} = (r^{ij}), \qquad i, j = 1, 2, \ldots, n,$$

and that R_{ii} is the cofactor of the ith diagonal element of R. Consequently, we can write

$$b_i = \frac{t_i \sqrt{\hat{\sigma}^2 r^{ii}}}{\sqrt{T}}. \tag{9}$$

Substituting in (8) we obtain

$$F_{n, T-n-1} = \frac{1}{n} \sum_{i=1}^{n} \sum_{j=1}^{n} t_i t_j r_{ij} \sqrt{r^{ii} r^{jj}}, \tag{10}$$

which establishes the precise connection between the F-statistic and the t-ratios of individual coefficients—as they are obtained after a regression is carried out. We remind the reader that, by construction, $t_i \geq 0$ for all i. It is quite easy to demonstrate that if all $r_{ij} \geq 0$ it is possible for individual coefficients to be pronounced "insignificant" by the t-test based on the statistics t_i, while the F-test will yield a "significant" pronouncement for the regression. This will be so, quite evidently, if all the r_{ij} are close to unity, which, of course, corresponds to the case where the explanatory variables are, pairwise, nearly collinear. Such an event is, obviously, not paradoxical *since in such a case we know from the preceding discussion that the individual coefficients cannot be identified, although $X\beta$ is an estimable function.* This is the kernel of truth referred to earlier. Unfortunately, however, the individual coefficients *could be pronounced "insignificant" by the t-test* (applied *seriatim*) while the overall regression could be pronounced "significant" by the F-test *even though the explanatory variables are mutually uncorrelated.* This is truly a paradoxical situation. The following example will illustrate the circumstances under which it may occur.

EXAMPLE 2. First recall that if t has the central t-distribution with r degrees of freedom, then t^2 is F-distributed with 1 and r degrees of freedom. Since most t-tests are bilateral, i.e., when we test the hypothesis $\beta_i = 0$ the alternative is $\beta_i \neq 0$, we may build up from tables of the F-distribution the significance points of both the F- and t-tests. We thus find, for the somewhat typical case $n = 5$, $T - n - 1 = 20$, the data set forth in Table 1, below.

Table 1 Significance Points for Various F-distributed Variables

Level of Significance	$F_{5, 20}$	$F_{1, 20}$
0.10	2.16	2.97
0.05	2.71	4.35
0.025	3.29	5.87
0.01	4.10	8.10

Now suppose that in Equation (10), $r_{ij} = 0\ i \neq j$. It follows that $|R| = 1$, $R_{ii} = 1, r^{ii} = 1$. Thus, ·

$$F_{n,\,T-n-1} = \frac{1}{n} \sum_{i=1}^{n} t_i^2$$

so that, here, the F-statistic is the average of the t^2-statistics.

If, for example, we carry out tests of significance at the 10% level and the t^2-statistics are, individually, less than 2.97 we shall pronounce each coefficient "insignificant" (at the 10% level). If their average, however, is greater than 2.16 we shall pronounce the entire regression "significant." Similarly, if the t^2-statistics are, individually, less than 4.35 we shall pronounce each coefficient "insignificant" at the 5% level; if their average, on the other hand, is more than 2.71 we shall pronounce the entire regression to be "significant."

Notice that here all explanatory variables are mutually uncorrelated so that the question of near collinearity does not arise. It is merely a phenomenon in which a series of t-tests and an F-test give contradictory answers, and is meant to illustrate the proposition that "insignificance" by a series of t-tests and "significance" by means of an F-test does not necessarily imply the presence of collinearity.

The reader may recall, in this connection, the discussion in the Appendix of Chapter 2 regarding the relationship between the F-test, the S-method of multiple comparisons, and a series of t-tests.

Proposed Remedies. While in the preceding discussion we identified multicollinearity as a problem of the sample data whose remedy properly lies with the procurement of a different sample or reformulation of the objectives of the investigation, still we find in the literature of econometrics a number of other proposed "remedies." The principal suggestions are:

(i) dropping one or more variables;
(ii) use of the generalized inverse;
(iii) use of (some of) the principal components of the explanatory variables;
(iv) use of Ridge regression and Bayesian methods.

Perhaps the most commonly implemented procedure is the first. Indeed, we shall employ it at a later point when we discuss the analysis of variance model. *When employed, however, the user should be aware that a different set of parameters is estimated depending on which variables are eliminated.*

(i) *Dropping one or more variables.* We discuss this procedure for the case of strict collinearity in some detail. For the case of near collinearity we shall reexamine the problem when we discuss aspects of misspecification theory. Thus, in the GLM of Equation (1) suppose

$$\text{rank}(X) = r < n + 1, \tag{11}$$

and otherwise let the standard assumptions hold. By renumbering variables, if necessary, partition

$$X = (X_1, X_2)$$

in such a way that

$$\text{rank}(X_1) = r$$

and

$$X_2 = X_1 D \tag{12}$$

for a suitable matrix D. This is possible since the condition in (11) means that $n + 1 - r$ of the columns of X can be expressed as linear combinations of the remaining r columns. Partition, conformably with X,

$$\beta = (\beta^{(1)\prime}, \beta^{(2)\prime})',$$

and rewrite (1) as

$$y = X_1 \beta^{(1)} + X_2 \beta^{(2)} + u.$$

Given the relation in (12) we have

$$y = X_1 \beta^{(1)} + X_1 D \beta^{(2)} + u = X_1 \bar{\beta}^{(1)} + u,$$

were

$$\bar{\beta}^{(1)} = \beta^{(1)} + D \beta^{(2)}. \tag{13}$$

Thus, dropping the variables $x_{.r+1}, x_{.r+2}, \ldots, x_{.n+1}$ *will not result in "better" estimates of the coefficients of the variables* $x_{.1}, x_{.2}, \ldots, x_{.r}$, *i.e., of the vector* $\beta^{(1)}$, *but rather in estimates of the vector* $\bar{\beta}^{(1)}$, *which is an entirely different matter!*

Moreover, just which variables we drop is a matter of choice. The fact that X is of rank r means that $n + 1 - r$ of its columns can be expressed as a linear combination of the remaining r. But, typically, we can choose these r columns in a number of ways—at most $\binom{n+1}{r}$. Thus, if instead of dropping the last $n + 1 - r$ variables we had dropped the first $n + 1 - r$ variables we would have estimated an entirely different set of parameters, even in the case where the sets retained have several variables in common. This aspect is perhaps made clearer by means of a simple example.

EXAMPLE 3. Consider the GLM

$$y_t = \sum_{i=1}^{n} x_{ti} \beta_i + u_t$$

where variables are stated as deviations from sample means. In particular, the model above does not contain a constant term. Now suppose the explanatory variables are multicollinear. Precisely, suppose that the matrix

$$X = (x_{ti}), \qquad t = 1, 2, \ldots, T, \qquad i = 1, 2, \ldots, n,$$

is of rank $r < n$. This means, in terms of the discussion above, that there exists a matrix D such that, say,

$$X_2 = X_1 D,$$

where

$$X_1 = (x_{.1}, x_{.2}, \ldots, x_{.r}), \qquad X_2 = (x_{.r+1}, \ldots, x_{.n}).$$

Another way of expressing this relation is to write

$$X \begin{pmatrix} D \\ -I \end{pmatrix} = 0.$$

Now suppose that $r = n - 1$ so that D is simply an $(n - 1)$-element vector. This yields

$$x_{\cdot n} = X_1 d, \tag{14}$$

where now

$$X_1 = (x_{\cdot 1}, x_{\cdot 2}, \ldots, x_{\cdot n-1}).$$

The coefficient vector estimated by the regression of y on X_1 is

$$\bar{\beta}^{(1)} = \beta^{(1)} + d\beta_n, \qquad \beta^{(1)} = (\beta_1, \beta_2, \ldots, \beta_{n-1})'. \tag{15}$$

But since d cannot have all elements null suppose $d_{n-1} \neq 0$. From (14) we can also write

$$x_{\cdot n-1} = X_1^* d^*,$$

where now

$$X_1^* = (x_{\cdot 1}, x_{\cdot 2}, \ldots, x_{\cdot n-2}, x_{\cdot n}),$$

$$d^* = \left(-\frac{d_1}{d_{n-1}}, \ldots, -\frac{d_{n-2}}{d_{n-1}}, \frac{1}{d_{n-1}} \right).$$

Regressing y on X_1^* will estimate the vector

$$\bar{\beta}^{*(1)} = \beta^{*(1)} + d^*\beta_{n-1}, \qquad \beta^{*(1)} = (\beta_1, \beta_2, \ldots, \beta_{n-2}, \beta_n)'.$$

The sets of regressors in X_1 and X_1^* have all elements in common except for their last elements. However, the coefficients estimated in the two regressions for their common elements may be quite different. Thus, the coefficient of $x_{\cdot i}$, in the first regression is

$$\beta_i + d_i\beta_n, \qquad i = 1, 2, \ldots, n - 2,$$

while in the second regression it is

$$\beta_i - \frac{d_i}{d_{n-1}} \beta_{n-1}, \qquad i = 1, 2, \ldots, n - 2,$$

and these two quantities may be very different.

Remark 3. Frequently, empirical investigators find that upon dropping different sets of variables from a regression the estimates of the coefficients of retained variables change, often appreciably, thus occasioning considerable consternation. On the other hand, if such coefficient estimates change only imperceptibly this is treated as a source of satisfaction regarding the "stability" of the coefficients in question. Neither reaction is fully warranted. As the preceding discussion suggests, dropping different sets of variables means that one is estimating different parameters, even if they pertain to the same explanatory variable! Getting the same estimates for the coefficient of a given (retained) variable even though different sets have been dropped need only mean that the variable in question is orthogonal to the set of included variables in both instances. Again, while there is a grain of truth in such rules of thumb, they do not constitute a rigorous basis for detecting or remedying the problem of multicollinearity.

The discussion above has concentrated on the case of exact collinearity but, evidently, similar considerations apply to the case of near collinearity as well.

(ii) *Use of the g-inverse.* The second proposed remedy consists of the use of the g-inverse. For the standard GLM, the normal equations of least squares are

$$X'X\beta = X'y.$$

But

$$\mathrm{rank}(X) < n + 1,$$

so that the inverse $(X'X)^{-1}$ does not exist. The use of the g-inverse yields

$$\bar{\beta} = (X'X)_{\mathrm{g}} X'y. \tag{16}$$

Since the g-inverse is unique we have the illusion that we have solved the problem. Let us see, exactly, what we have done. From Proposition 72 of *Mathematics for Econometrics* we have that

$$(X'X)_{\mathrm{g}} = X_{\mathrm{g}} X'_{\mathrm{g}}.$$

Thus,

$$(X'X)_{\mathrm{g}} X'y = X_{\mathrm{g}} X'_{\mathrm{g}} X'y = X_{\mathrm{g}} X X_{\mathrm{g}} y = X_{\mathrm{g}} y$$

and (16) may be rewritten

$$\bar{\beta} = X_{\mathrm{g}} y, \tag{17}$$

which is a uniquely determined $(n + 1)$-element vector. What is its relation to the true parameter vector β? Since

$$y = X\beta + u,$$

substituting we find

$$\bar{\beta} = X_{\mathrm{g}} X\beta + X_{\mathrm{g}} u.$$

Consequently,

$$\bar{\beta} \sim N(X_{\mathrm{g}} X\beta, \sigma^2 X_{\mathrm{g}} X'_{\mathrm{g}}).$$

Proposition 72 similarly shows that $\bar{\beta}$ has a *singular* distribution since

$$\mathrm{rank}(X_{\mathrm{g}} X'_{\mathrm{g}}) = \mathrm{rank}(X_{\mathrm{g}}) = \mathrm{rank}(X) < n + 1,$$

while $X_{g} X'_{\mathrm{g}}$ is of dimension $n + 1$. Thus, $\bar{\beta}$ is a biased estimator of β. This is so since Corollary 16 of *Mathematics for Econometrics* shows that

$$X_{\mathrm{g}} X = I$$

only if

$$\mathrm{rank}(X) = n + 1.$$

But perhaps the most revealing aspect of what the g-inverse approach entails in this context is provided by Proposition 85 of *Mathematics for Econometrics*. There, if we look upon our problem as one in which *we* seek a solution to the inconsistent system of equations

$$u = y - X\beta$$

it is shown that

$$\bar{\beta} = X_g y$$

is the minimum norm least squares (MNLS) solution, i.e., of the many possible solutions to the normal equations

$$X'X\beta = X'y$$

the one exhibited in (17) *has minimum norm.* Precisely, *if* $\underline{\beta}$ *is any other vector satisfying the normal equations, then*

$$\underline{\beta}'\underline{\beta} > \bar{\beta}'\bar{\beta}. \tag{18}$$

Thus, uniqueness is purchased at the price of the condition in (18). But minimality of norm has, typically, no justification in the economics of the model, nor does it appear to have any intuitively appealing interpretation.

(iii) *Use of principal components.* The reader will not miss any important aspect of the procedure if, for the moment, he thinks of principal components simply as a data transformation procedure. For clarity of exposition we deal again with the case of exact collinearity, reserving discussion of the case of near collinearity until we introduce the theory of misspecification. As before, the problem is that

$$\text{rank}(X) = r < n.$$

Let D and P be, respectively, the matrices of characteristic roots and vectors of the matrix $X'X$. Thus

$$D = \text{diag}(d_1, d_2, \ldots, d_n),$$

$$P'P = I.$$

It is understood here that all explanatory variables have been centered around their respective sample mean and have been divided by the square root of their respective sample variance. Thus we have

$$X'XP = PD. \tag{19}$$

In the model

$$y = X\beta + u$$

we note that

$$X\beta = XPP'\beta = Z\gamma, \qquad Z = XP, \quad \gamma = P'\beta.$$

Hence, simply as a matter of data transformation and reparametrization we can write the model as

$$y = Z\gamma + u. \tag{20}$$

If X were of full rank we could estimate γ as

$$\hat{\gamma} = (Z'Z)^{-1}Z'y$$

and thereby estimate β as

$$\hat{\beta} = P\hat{\gamma}.$$

This is so since β and γ are equivalent parametric representations connected by

$$P\gamma = PP'\beta = \beta. \tag{21}$$

If X is not of full rank it is not possible to do this; we could however, do something appropriately analogous. In view of the rank condition we can write

$$D = \begin{bmatrix} D_r & 0 \\ 0 & 0 \end{bmatrix},$$

where D_r is the diagonal matrix containing the nonzero characteristic roots. Partition P conformably; thus

$$P = (P_r, P_*),$$

where P_r contains the first r columns of P, corresponding to the nonzero roots. From (19) we observe the following:

$$P'X'XP = \begin{bmatrix} P'_r X'XP_r & P'_r X'XP_* \\ P'_* X'XP_r & P'_* X'XP_* \end{bmatrix} = \begin{bmatrix} D_r & 0 \\ 0 & 0 \end{bmatrix}.$$

From this we conclude that

$$XP_* = 0.$$

Consequently,

$$XPP'\beta = (XP_r, 0) \begin{pmatrix} P'_r \\ P'_* \end{pmatrix} \beta = XP_r P'_r \beta.$$

Since this must be true for all vectors β we conclude

$$X = XP_r P'_r. \tag{22}$$

The "principal components" procedure is as follows. Define

$$Z_r = XP_r$$

and regress y on Z_r, obtaining

$$\hat{\gamma}_{(r)} = (Z'_r Z_r)^{-1}Z'_r y = D_r^{-1}P'_r X'y. \tag{23}$$

By analogy with (21) obtain the pseudoestimate of β,

$$\hat{\beta} = P_r \hat{\gamma}_{(r)} = P_r D_r^{-1} P_r' X' y. \tag{24}$$

We shall now show that (24) is exactly the estimate obtained by the use of the g-inverse. We recall that this estimator is given in (17) as

$$\bar{\beta} = X_g y.$$

To show its equivalence with $\hat{\beta}$ of (24) *it will be sufficient to show that*

$$X_g = P_r D_r^{-1} P_r' X'. \tag{25}$$

But note:

(i) $XX_g = XP_r D_r^{-1} P_r' X'$, which is symmetric;
(ii) $X_g X = P_r D_r^{-1} P_r' X' X = P_r P_r'$, which is symmetric;
(iii) $XX_g X = XP_r P_r' = X$ by Equation (22);
(iv) $X_g XX_g = P_r D_r^{-1} P_r' X' X P_r D_r^{-1} P_r' X' = P_r D_r^{-1} P_r' X' = X_g$.

This, in terms of the properties defining the g-inverse, shows that X_g is, indeed, given by (25), and thus demonstrates that (17) and (24) are identical estimators.

Remark 4. The procedure above was termed an application of principal components theory for the following reasons. If x is an m-element random vector with mean μ and covariance matrix Σ, let

$$\Lambda = \operatorname{diag}(\lambda_1, \lambda_2, \ldots, \lambda_m)$$

be the diagonal matrix of the characteristic roots of Σ (arranged in decreasing order), and let

$$A = (\alpha_{\cdot 1}, \alpha_{\cdot 2}, \ldots, \alpha_{\cdot m})$$

be the orthogonal matrix of the associated characteristic vectors. The variables

$$\zeta_i = \alpha_{\cdot i}' x, \qquad i = 1, 2, \ldots, m,$$

are said to be *the principal components of the vector* x. We observe that

$$\operatorname{Var}(\zeta_i) = \alpha_{\cdot i}' \Sigma \alpha_{\cdot i} = \lambda_i, \qquad i = 1, 2, \ldots, m,$$

and, moreover,

$$\operatorname{Cov}(\zeta_i, \zeta_j) = \alpha_{\cdot i}' \Sigma \alpha_{\cdot j} = 0, \qquad i \neq j$$

Finally, note that principal components preserve the generalized variance and variability of the random vector, defined as the determinant and the trace, respectively, of the covariance matrix, Σ. This is so, since

$$\Lambda = A\Sigma A'$$

and

$$|\Lambda| = |\Sigma|, \quad \operatorname{tr} \Lambda = \operatorname{tr} \Sigma.$$

In the preceding discussion, if the explanatory variables are first centered about their respective sample means and if we think of such variables as being random, then

$$\frac{1}{T} X'X$$

is a consistent estimate of their "covariance matrix." Consequently, D and P are consistent estimators, respectively, of the matrices of characteristic roots and vectors. As noted earlier the procedure need not be understood in the context of principal components theory. It may be viewed simply as a trans-formation of data operation. In empirical applications one usually defines characteristic roots and vectors in terms of the sample correlation matrix (rather than the sample covariance matrix). This is exactly what we have done earlier and is done in order to avoid problems connected with widely differing units in which the variables may be measured.

Remark 5. Although it was shown that the "principal components" approach is equivalent to the g-inverse approach, the former is a far more fruitful way of looking at the problem. In dealing with "near multicollinear-ity" the g-inverse approach can only produce the *standard OLS estimator since, strictly speaking X'X is, in this case, nonsingular.* However, in the principal components approach we have the added flexibility of using a more limited number of principal components. It is very often the case with economic time series that a limited number of principal components will account for much of the variability in the data. Particularly, suppose that the last $n - r$ roots of $X'X$ are quite small and are considered to be too in-accurately obtained (i.e., they are thought to suffer excessively from com-putational roundoff errors). This would mean that their associated characteristic vectors are also very inaccurately obtained. Thus, the investigator may decide that their computational inaccuracy far outweighs their informational content and for this reason operate only with the first r components. The estimator thus obtained is

$$\hat{\gamma}_{(r)} = (Z_r'Z_r)^{-1}Z_r'y, \qquad Z_r = XP_r,$$

P_r being the matrix consisting of the first r columns of P, which is the matrix of characteristic vectors of $X'X$. We may still compute

$$\hat{\beta} = P_r\hat{\gamma}_{(r)},$$

but here we should bear in mind that this estimator of β no longer corresponds to the g-inverse estimator.

(iv) *Use of Ridge regression* (RR). This is an estimation procedure introduced by the chemical engineer Hoerl [19], [20]. In its simplest form it consists of adding a constant multiple of the identity matrix to $X'X$ before solving the normal equations of OLS. Although the main usefulness of this method, if any, could be in the case of near collinearity we shall present and motivate it for the case of exact multicollinearity. As in other cases we shall reserve discussion of near collinearity aspects until we deal with misspecification analysis.

A number of motivations may be provided for RR, e.g., as an estimator that minimizes the sum of squared residuals subject to a norm constraint on the vector of estimates, as an estimator that yields in some sense a smaller mean squared error (MSE)[4] matrix than the OLS estimator, or, finally, as a Bayesian estimator.

Thus, suppose in the context of the usual GLM we wish to obtain an estimator subject to

$$\beta'\beta = c,$$

where c is a specified constant. The problem is, thus, to minimize

$$(y - X\beta)'(y - X\beta) + k(\beta'\beta - c)$$

and the first-order conditions are

$$-X'(y - X\beta) + k\beta = 0, \tag{26}$$

$$\beta'\beta = c.$$

The solution will obey

$$\hat\beta(k) = (X'X + kI)^{-1}X'y. \tag{27}$$

In general, the estimator in (27) will be defined even if

$$\text{rank}(X) < n. \tag{28}$$

Remark 6. Recall that in all discussions involving collinearity "remedies" we assume that data have been centered about the appropriate sample mean. Hence the rank condition in (28).

Remark 7. The motivation above leads to a *nonlinear estimator for* β. This is so since k is a Lagrangian multiplier and as such it will have to satisfy Equation (26). But this means that, in general, it would depend on y!

Another motivation that would be quite appropriate in the case of "near collinearity" is to determine the factor k in such a way that it "minimizes" the MSE matrix of the estimator in (27) relative to that of the OLS estimator.

By definition, the MSE matrix of (27) is

$$E[\hat\beta(k) - \beta][\hat\beta(k) - \beta]' = \sigma^2 A^{-1}X'XA^{-1} + k^2 A^{-1}\beta\beta'A^{-1}, \tag{29}$$

where

$$A = X'X + kI.$$

---

[4] Let $\hat\theta$ be an estimator of a parameter vector θ. The MSE matrix of $\hat\theta$ is defined by

$$\text{MSE}(\hat\theta) = E(\hat\theta - \theta)(\hat\theta - \theta)'$$

If $E(\hat\theta) = \bar\theta$ we note that $\text{MSE}(\hat\theta) = \text{Cov}(\hat\theta) + (\bar\theta - \theta)(\bar\theta - \theta)'$, so that the MSE matrix is the sum of the covariance and bias matrices of an estimator.

The first term on the right side of (29) is the variance component, while the second is the bias component. For notational simplicity put

$$MSE(k) = E[\hat{\beta}(k) - \beta][\hat{\beta}(k) - \beta]',$$

and note that, provided $(X'X)^{-1}$ exists,

$$MSE(0) = \sigma^2(X'X)^{-1}$$

is simply the covariance matrix of the OLS estimator. Thus, in the case of "near collinearity" it becomes possible to pose the problem: find a k such that

$$tr[MSE(0) - MSE(k)] > 0.$$

Letting P and D be, respectively, the matrices of characteristic vectors and roots of $X'X$ we can show that

$$tr[MSE(0) - MSE(k)] = \sigma^2 \sum_{i=1}^{n} \frac{1}{d_i} - \sigma^2 \sum_{i=1}^{n} \left[\frac{d_i}{(d_i + k)^2} \right] - k^2 \sum_{i=1}^{n} \left(\frac{s_i}{d_i + k} \right)^2,$$

where

$$s = P'\beta, \qquad s = (s_1, s_2, \dots, s_n)'.$$

Collecting terms we find

$$tr[MSE(0) - MSE(k)] = \sum_{i=1}^{n} \left[\frac{(2\sigma^2 d_i)k + (\sigma^2 - s_i^2 d_i)k^2}{d_i(d_i + k)^2} \right]. \qquad (30)$$

Since the d_i are all positive, if

$$\sigma^2 - s_i^2 d_i \geq 0$$

then any $k > 0$ will make the right side of (30) positive. On the other hand, if

$$\sigma^2 - s_i^2 d_i < 0$$

then any k obeying

$$0 < k < \frac{2\sigma^2 d_i}{s_i^2 d_i - \sigma^2}$$

will have the effect of making the right side of (30) positive. Hence, there would always exist a k such that

$$tr[MSE(0) - MSE(k)] > 0,$$

although such k is not unique and does depend on the data matrix X as well as the unknown parameter vector β.

An alternative motivation is to take a Bayesian approach. Thus, recall the discussion at the end of Chapter 2, where in dealing with the GLM we determined that the mean of the posterior distribution of the parameter vector β is given by

$$b_1 = (X'X + \sigma^2 h^2 I)^{-1}[X'Xb + \sigma^2(h^2 I)b_0],$$

where $h^{-2}I$ and b_0 are, respectively, the covariance matrix and mean vector of the prior distribution, and b is the OLS estimator of β. Thus, the mean vector of the posterior distribution of β, given the sample information y and X, is

$$b_1 = (X'X + \sigma^2 h^2 I)^{-1}[X'y + (\sigma^2 h^2)b_0] \tag{31}$$

If in (31) we take

$$b_0 = 0$$

and put

$$k = \sigma^2 h^2$$

we have the RR estimator!

Thus, if in a Bayesian context, we agree that, given the data y and X, we are to "estimate" the parameter β by the mean of its posterior distribution, and if we further take the prior mean to be zero, we can obtain the RR estimator!

Once the procedures above are understood, it is easy to generalize the RR estimator as follows. From the principal components approach we know that we can write the GLM as

$$y = Z\gamma + u, \qquad Z = XP, \qquad \gamma = P'\beta,$$

where P is the matrix of characteristic vectors of $X'X$. If we have an estimator of γ, say $\hat{\gamma}$, we can always obtain an estimator of β, viz.,

$$\hat{\beta} = P\hat{\gamma}.$$

Thus, let the generalization of the RR estimator, defined above, be

$$\hat{\beta}(K) = P(Z'Z + K)^{-1}Z'y,$$

where

$$K = \mathrm{diag}(k_1, k_2, \ldots, k_n).$$

Noting that

$$Z'Z = P'X'XP = D,$$

where

$$D = \mathrm{diag}(d_1, d_2, \ldots, d_n)$$

is the matrix of characteristic roots of $X'X$, we find

$$\hat{\beta}(k) - \beta = P(D + K)^{-1}Z'Z\gamma - P\gamma + P(D + K)^{-1}Z'u.$$

Hence

$$\mathrm{MSE}[\hat{\beta}(K)] = P[(D + K)^{-1}Z'Z - I]\gamma\gamma'[Z'Z(D + K)^{-1} - I]P'$$
$$+ \sigma^2 P(D + K)^{-1}Z'Z(D + K)^{-1}P'.$$

But

$$(D + K)^{-1}Z'Z - I = (D + K)^{-1}[Z'Z + K - K] - I = -(D + K)^{-1}K,$$

and consequently

$$\text{MSE}[\hat{\beta}(K)] = P(D + K)^{-1} K \gamma \gamma' K (D + K)^{-1} P'$$
$$+ \sigma^2 P(D + K)^{-1} D(D + K)^{-1} P'. \tag{32}$$

Moreover,

$$\text{tr MSE}[\hat{\beta}(K)] = \sum_{i=1}^{n} \left[\frac{k_i^2 \gamma_i^2 + \sigma^2 d_i}{(d_i + k_i)^2} \right]. \tag{33}$$

We can now explicitly minimize (33) with respect to the k_i to find

$$\frac{\partial \, \text{tr MSE}[\hat{\beta}(K)]}{\partial k_i} = \frac{2(d_i \gamma_i^2 k_i - \sigma^2 d_i)}{(d_i + k_i)^3} = 0, \qquad i, 1, 2, \ldots, n.$$

Solving yields the minimizing values

$$k_i = \frac{\sigma^2}{\gamma_i^2}, \qquad i = 1, 2, \ldots, n. \tag{34}$$

To see the relationship of the generalized RR estimator to Bayesian methods we express the estimator as

$$\hat{\beta}(K) = P(Z'Z + K)^{-1} Z'y = (X'X + PKP')^{-1} X'y, \tag{35}$$

which is recognized as the mean of the posterior distribution of β, given the data y and X, and the prior density, which is

$$N(0, \sigma^{-2} PKP').$$

Admittedly, this is a contrived interpretation, but it is a Bayesian interpretation nonetheless.

Remark 8. What RR methods have established is that if we are willing to depart from the class of unbiased estimators there exist procedures that minimize some function of the MSE matrix. These procedures, however, are nonoperational, in the sense that they involve knowledge of the (generally) unknown parameter vector β, as well as σ^2. In the determination of K it will not do to replace β and σ^2 by estimates thereof since it is not clear at all whether doing so will preserve the optimality properties of the RR estimator. Thus, we have an existence result but we generally lack the means of implementing it; moreover, it is not known what are the properties of estimators that substitute, in the determination of K, estimates of β and σ^2.

Remark 9. It is useful for the reader to bear in mind that all methods for "remedying" the multicollinearity problem, save for the one involving the more or less arbitrary elimination of variables, are closely related. Indeed, for the case of strict multicollinearity the g-inverse and principal components approaches are formally identical if in the latter we use all components corresponding to nonzero roots. Bayesian and RR methods are similar in that they utilize extra sample information.

Remark 10. Having cataloged the main proposed "solutions" to the multicollinearity problem we must again stress that in the case of exact multicollinearity we cannot escape the fact that *all individual parameters cannot be identified.* Thus, we must reformulate our objectives so that we can make do with the estimable functions that may be available in such a situation, or else abandon the search for unbiased estimators and be content with estimators that are biased but may have other desirable properties. The case of "near collinearity" will be examined in greater detail in the context of misspecification analysis. It bears reiterating that, given the data matrix X, the OLS estimator gives the best linear unbiased estimator (BLUE) *so long as $X'X$ is nonsingular.* Under "near" multicollinearity, $X'X$ *is still nonsingular and if its inverse is correctly obtainable, that is all we are entitled to, or can, expect from the data.* It may be unfortunate if the estimated covariance matrix has large diagonal elements, but this is merely a reflection of the fact that the sample is not very informative on the parameters of interest. This is not failure of the method; it is a "deficiency" of the data. No amount of statistical manipulation can enrich the data!

2 Analysis of Variance: Categorical Explanatory Variables

It is not unusual in empirical investigations to deal with a model in which some of the explanatory variables are continuous while others are discontinuous or categorical. Thus, for example, if we are dealing with an investment model and the sample is cross sectional, i.e., the basic unit of observation is the firm, it would not be unreasonable to formulate the dependency as follows: investment expenditures may depend on current and past capital rental, but also on the type of activity the firm is engaged in. This is most usefully taken into account by having the constant term vary from one group of firms to another according to the standard industrial classification code of each firm. (The preceding, of course, assumes that rentals vary sufficiently from firm to firm to permit identification of the relation.)

Or suppose we have a cross-sectional sample of households and we are interested in studying their demand for transportation services. Again, it would not be unreasonable to hypothesize that the latter depends on the distance between residence and locus of work of the employed members of the household and their number, on the income of the household, perhaps on the race and sex of the head of household, etc. In both cases, it is clearly seen that some variables are continuous—for example, rental of capital, income of households, distance from locus of work; while others are discontinuous or categorical—for example, the industrial classification code for firms, the race and sex of the head of households, etc.

Let us now examine what, if any, additional problems are created by the introduction of such categorical variables. To approach the matter gradually we first consider the case where *all* explanatory variables are categorical. Thus, consider a variable of interest y which may be characterized by two attributes: attribute I has m classifications, while attribute II has k classifications. An observation, then, is characterized by these two atttributes and is well-denoted by

$$y_{ij}$$

which means that y belongs to the ith classification of attribute I and the jth classification of attribute II. The model may be formulated as

$$y_{ij} = \alpha_i^* + \beta_j^* + u_{ij}, \tag{36}$$

whose interpretation is: the variable of interest, y, has a constant mean, which is made up of the additive impacts of being in the ith classification of attribute I and the jth classification of attribute II.

One limitation of the formulation above is that it does not permit interaction between classifications in attribute I and attribute II. This means that knowing the sequences

$$\{\alpha_i^* : i = 1, 2, \ldots, m\}, \qquad \{\beta_j^* : j = 1, 2, \ldots, k\}$$

would enable us to compute the mean of y_{ij} for all i and j. The prototype of this model is of agricultural origins: y_{ij} may be the yield of wheat, and α_i^* describes the effect of cultivation in the ith type of soil while β_j^* describes the effect of using the jth type of fertilizer. The term u_{ij} represents the cumulative effect of all other variables—such as, for example, weather, variations in seed quality—and is taken to be a random variable. Absence of interaction, in this context, means that the soil effect (when the ith type of soil is used) is the same no matter what fertilizer is utilized and is, thus, a limitation of some severity. In most econometric applications, however, absence of interaction is not an appreciable limitation. Most commonly the model in (36) is written as

$$y_{ij} = \mu + \alpha_i + \beta_j + u_{ij}. \tag{37}$$

In (37), μ is called the general mean, α_i ($i = 1, 2, \ldots, m$) the row effects, and β_j ($j = 1, 2, \ldots, k$) the column effects. The parameters, typically, obey the restriction

$$\sum_{i=1}^{m} \alpha_i = 0, \qquad \sum_{j=1}^{k} \beta_j = 0, \tag{38}$$

although the restriction in (38) is not always imposed.[5]

---

[5] It is necessary to impose the restriction in (38) only if we wish to reconcile the models in (36) and (37). On the other hand the model in (37) could be postulated *ab initio*, in which case such restriction need not be imposed.

The connection between the parameters in (36) and (37) is as follows:

$$\mu = \alpha + \beta, \qquad \alpha_i = \alpha_i^* - \alpha, \qquad \beta_j = \beta_j^* - \beta,$$

$$\alpha = \frac{\sum_{i=1}^{m} \alpha_i^*}{m}, \qquad \beta = \frac{\sum_{j=1}^{k} \beta_j^*}{k}. \tag{39}$$

We now consider the problem of estimating the parameters of (37). As is usual, we shall assert that the error components (the u_{ij}) are independent identically distributed random variables with mean zero and variance σ^2. For the moment, let us assume that we have only one observation per cell, i.e., there is only one observation having the ith classification of attribute I and the jth classification of attribute II. To handle the estimation problem, in the context of the model in (37), we define the *dummy variables*

$$W_i \begin{cases} = 1 & \text{if observation belongs to } i\text{th classification of attribute I} \\ = 0 & \text{otherwise,} \end{cases}$$

$$\tag{40}$$

$$Z_j \begin{cases} = 1 & \text{if observation belongs to the } j\text{th classification of attribute II} \\ = 0 & \text{otherwise.} \end{cases}$$

The model may, thus, be written compactly as

$$y_{ij} = \mu + \sum_{r=1}^{m} \alpha_r W_r + \sum_{s=1}^{k} \beta_s Z_s + u_{ij}. \tag{41}$$

If we arrange observations lexicographically, i.e.,

$$y = (y_{11}, y_{12}, y_{13}, \ldots, y_{1k}, y_{21}, y_{22}, \ldots, y_{2k}, \ldots, y_{m1}, y_{m2}, y_{m3}, \ldots, y_{mk})',$$

we can write

$$y = X^* \gamma^* + u, \tag{42}$$

where u is arranged in the same order as y,

$$\gamma^* = (\mu, \alpha_1, \alpha_2, \ldots, \alpha_m, \beta_1, \beta_2, \ldots, \beta_k)'$$

and

$$X^* = \begin{bmatrix} 1 & 1 & 0 & 0 & 0 & 1 & 0 & 0 \\ 1 & 1 & 0 & 0 & 0 & 0 & 1 & 0 \\ 1 & 1 & 0 & 0 & 0 & 0 & 0 & 1 \\ 1 & 0 & 1 & 0 & 0 & 1 & 0 & 0 \\ 1 & 0 & 1 & 0 & 0 & 0 & 1 & 0 \\ 1 & 0 & 1 & 0 & 0 & 0 & 0 & 1 \\ 1 & 0 & 0 & 1 & 0 & 1 & 0 & 0 \\ 1 & 0 & 0 & 1 & 0 & 0 & 1 & 0 \\ 1 & 0 & 0 & 1 & 0 & 0 & 0 & 1 \\ 1 & 0 & 0 & 0 & 1 & 1 & 0 & 0 \\ 1 & 0 & 0 & 0 & 1 & 0 & 1 & 0 \\ 1 & 0 & 0 & 0 & 1 & 0 & 0 & 1 \end{bmatrix} = (e, W^*, Z^*).$$

Here, obviously, e is the first column of X^* and corresponds to the general mean, W^* consists of the next four columns of X^* and contains the "observations" on the row "dummy variables," while Z^* consists of the last three columns of X^* and contains the "observations" on the column "dummy variables." In the representation above we assume $m = 4, k = 3$.

We see, therefore, that the case where the explanatory variables are all categorical has a representation as an ordinary GLM.

Remark 11. The model we are dealing with is also termed the analysis of variance (ANOVA) model since one of its uses may be to elucidate how much of the variance of the dependent variable is accounted by row effects and how much by column effects. This particular aspect is not of great significance in econometric work.

Returning to the model of (42), we may attempt to estimate it by OLS methods. We shall soon discover, however, that OLS methods, in this context, are not without problems. Even a cursory examination will disclose that *the columns of X^* are linearly dependent.* Thus, adding the second, third, fourth, and fifth columns gives the first column, as does adding the last three columns. Of course, it is clear how these linear dependencies arise, since every observation has attached to it a classification for attribute I *and* a classification for attribute II—the two classification schemes being exhaustive. It is intuitively evident from the preceding that *we cannot estimate separately all the row and column effects as well as the general mean.* We can, however, estimate differences in effects, the so-called *contrasts.* Examples of contrasts are

$$\alpha_i - \alpha_m, \qquad \beta_j - \beta_k, \qquad i = 1, 2, \ldots, m - 1, j = 1, 2, \ldots, k - 1.$$

These are quite meaningful parameters, in this context, since we *are* typically interested in whether all row and column effects are the same. Thus, in the argicultural example above a natural question to ask is: are all fertilizers equally effective? This has a natural formulation as a test of significance on the contrasts.

Now, it turns out that there is an exceedingly simple way in which contrasts can be estimated. In particular, suppose we are interested in the contrasts with respect to the mth row and kth column effects. *This may be done by simply suppressing, in the matrix X^*, the columns corresponding to W_m and Z_k;* the resulting matrix is of full rank and, thus, the theory of the standard GLM developed in earlier chapters applies completely. This is made clear in

Proposition 2. *Consider the analysis of variance model in which there are two attributes, attribute I with m classifications and attribute II with k classifications, i.e.,*

$$y_{ij} = \mu + \sum_{r=1}^{m} \alpha_r W_r + \sum_{s=1}^{k} \beta_s Z_s + u_{ij}, \qquad i = 1, 2, \ldots, m, j = 1, 2, \ldots, k,$$

where the W_r, Z_s are "dummy" variables. The following statements are true:
(i) The parameter vector

$$\gamma^* = (\mu, \alpha_1, \alpha_2, \ldots, \alpha_m, \beta_1, \beta_2, \ldots, \beta_k)'$$

is not estimable.
(ii) The parameter vector

$$\gamma = (\mu + \alpha_m + \beta_k, \alpha_1 - \alpha_m, \alpha_2 - \alpha_m, \ldots,$$
$$\alpha_{m-1} - \alpha_m, \beta_1 - \beta_k, \beta_2 - \beta_k, \ldots, \beta_{k-1} - \beta_k)'$$

is estimable and is obtained by suppressing in the model W_m and Z_k, i.e., by regressing y_{ij} on all the dummy variables with the exception of W_m and Z_k.

PROOF. Construct the matrix of dummy variables

$$X^* = (e, W^*, Z^*),$$

where: e is a column vector of mk elements consisting entirely of 1's; W^* and Z^* are the matrices of "observations" on the dummy variables corresponding, respectively, to the attribute I and II classifications. It is clear that the model can be written as

$$y = X^*\gamma^* + u,$$

the elements in y (and u) being the observations arranged lexicographically in a column. The matrix X^* is of rank $m + k - 1$. This is so since, if we define e_m to be an m-element column consisting entirely of 1's and e_k to be a k-element column vector consisting entirely of 1's, we see that

$$W^*e_m = e, \qquad Z^*e_k = e, \tag{43}$$

which shows that the columns of X^* exhibit two linear dependencies. Denoting by $W_{\cdot i}$ the ith column of W^* and by $Z_{\cdot j}$ the jth column of Z^* we have from (43)

$$W_{\cdot m} = e - \sum_{i=1}^{m-1} W_{\cdot i}, \qquad Z_{\cdot k} = e - \sum_{j=1}^{k-1} Z_{\cdot j}. \tag{44}$$

Substituting, we note

$$X^*\gamma^* = e\mu + \sum_{i=1}^{m-1} \alpha_i W_{\cdot i} + \alpha_m \left(e - \sum_{i=1}^{m-1} W_{\cdot i} \right) + \sum_{i=1}^{k-1} \beta_j Z_{\cdot j} + \beta_k \left(e - \sum_{j=1}^{k-1} Z_{\cdot j} \right),$$

and on regrouping we find

$$X^*\gamma^* = (\mu + \alpha_m + \beta_k)e + \sum_{i=1}^{m-1} (\alpha_i - \alpha_m)W_{\cdot i} + \sum_{j=1}^{k-1} (\beta_j - \beta_k)Z_{\cdot j} = X\gamma,$$

where X is the submatrix of X^* obtained by eliminating from the latter $W_{\cdot m}$ and $Z_{\cdot k}$. Consequently, the model may be written, equivalently, as

$$y = X\gamma + u, \tag{46}$$

where now X is of full rank. Thus, the OLS estimator of γ is BLUE, which completes the proof.

Corollary 2. *If the error process of* (42) *obeys*

$$u \sim N(0, \sigma^2 I)$$

then

$$\hat{\gamma} \sim N[\gamma, \sigma^2(X'X)^{-1}].$$

Remark 12. It is apparent from the argument in the proof of the proposition above that if we desire to estimate, say, the contrasts $\alpha_i - \alpha_4$ or $\beta_j - \beta_8$, then we may do so by suppressing the two columns corresponding to W_4 and Z_8. The constant term in the resulting regression is an estimate of $\mu + \alpha_4 + \beta_8$, the coefficient of W_i is an estimate of $\alpha_i - \alpha_4$, while the coefficient of Z_j is an estimate of $\beta_j - \beta_8$.

Remark 13. The method above is easily extended to three-way classification problems—or, more generally, to cases where there are more than two attributes. It involves, essentially, a lexicographic ordering of the observations, definition of the dummy variables, and construction of the X^* matrix. If, for example, there are three attributes with m_1, m_2, m_3 classifications, respectively, the X^* matrix will have $(m_1 m_2 m_3)$ rows and $m_1 + m_2 + m_3 + 1$ columns. Its rank will be $m_1 + m_2 + m_3 - 2$, and estimating the contrasts can again be obtained by suppressing an appropriate set of three dummy variables. Precisely, suppose there are three attributes with m_i classifications in the ith attribute, $i = 1, 2, 3$. Define the dummies: W_{j_1} for the first attribute, $j_1 = 1, 2, \ldots, m_1$; Z_{j_2} for the second attribute, $j_2 = 1, 2, \ldots, m_2$; and Q_{j_3} for the third attribute, $j_3 = 1, 2, \ldots, m_3$. Let their corresponding effects be $\alpha_{j_1}, \beta_{j_2}, \gamma_{j_3}$. By suppressing $W_{m_1}, Z_{m_2}, Q_{m_3}$ and performing an ordinary regression we are estimating

$$\mu + \alpha_{m_1} + \beta_{m_2} + \gamma_{m_3}, \qquad \alpha_{j_1} - \alpha_{m_1}, \qquad \beta_{j_2} - \beta_{m_2}, \qquad \gamma_{j_3} - \gamma_{m_3},$$
$$j_i = 1, 2, \ldots, m_i - 1, \quad i = 1, 2, 3.$$

Remark 14. If more than one observation is available per cell, essentially the same considerations prevail. The only difference is that some rows are repeated. For example, suppose that for the (i, j) cell we have n_{ij} observations. Thus, we can write

$$y_{ijt} = \mu + \alpha_i + \beta_j + u_{ijt},$$
$$t = 1, 2, \ldots, n_{ij}, \qquad i = 1, 2, \ldots, m, \qquad j = 1, 2, \ldots, k.$$

It is seen that the row of the matrix X^* corresponding to y_{ijt} *is the same for all* t. The rank of the matrix X^* is still $m + k - 1$ and the same procedure as outlined above will yield estimators for the contrasts.

EXAMPLE 4. Suppose in the two-way classification model we have $m = 2$ and $k = 2$, for y_{11} we have 3 observations, while for y_{12} we have 2 and for the remaining, one. The matrix X^* is given by

$$X^* = \begin{bmatrix} 1 & 1 & 0 & 1 & 0 \\ 1 & 1 & 0 & 1 & 0 \\ 1 & 1 & 0 & 1 & 0 \\ 1 & 1 & 0 & 0 & 1 \\ 1 & 1 & 0 & 0 & 1 \\ 1 & 0 & 1 & 1 & 0 \\ 1 & 0 & 1 & 0 & 1 \end{bmatrix}.$$

The rank of X^* is still 3; if we suppress the fifth and third columns—thus suppressing the dummy variables that correspond to the second classification in the two attributes—we have

$$X = \begin{bmatrix} 1 & 1 & 1 \\ 1 & 1 & 1 \\ 1 & 1 & 1 \\ 1 & 1 & 0 \\ 1 & 1 & 0 \\ 1 & 0 & 1 \\ 1 & 0 & 0 \end{bmatrix}, \qquad X'X = \begin{bmatrix} 7 & 5 & 4 \\ 5 & 5 & 3 \\ 4 & 3 & 4 \end{bmatrix},$$

$$(X'X)^{-1} = \frac{1}{17} \begin{bmatrix} 11 & -8 & -5 \\ -8 & 12 & -1 \\ -5 & -1 & 10 \end{bmatrix}.$$

If we regress the dependent variable y on the variables contained in X, the estimator is

$$\hat{\gamma} = (X'X)^{-1} X'y,$$

and its expected value is

$$E(\hat{\gamma}) = (X'X)^{-1} X'X^*\gamma^* = \frac{1}{17} \begin{bmatrix} 11 & -8 & -5 \\ -8 & 12 & -1 \\ -5 & -1 & 10 \end{bmatrix} \begin{bmatrix} 7 & 5 & 2 & 4 & 3 \\ 5 & 5 & 0 & 3 & 2 \\ 4 & 3 & 1 & 4 & 0 \end{bmatrix} \begin{pmatrix} \mu \\ \alpha_1 \\ \alpha_2 \\ \beta_1 \\ \beta_2 \end{pmatrix}.$$

Carrying out the matrix multiplication, we find

$$E(\hat{\gamma}) = \frac{1}{17} \begin{bmatrix} 17 & 0 & 17 & 0 & 17 \\ 0 & 17 & -17 & 0 & 0 \\ 0 & 0 & 0 & 17 & -17 \end{bmatrix} \begin{pmatrix} \mu \\ \alpha_1 \\ \alpha_2 \\ \beta_1 \\ \beta_2 \end{pmatrix} = \begin{bmatrix} \mu + \alpha_2 + \beta_2 \\ \alpha_1 - \alpha_2 \\ \beta_1 - \beta_2 \end{bmatrix},$$

which, in part, illustrates Proposition 2 and the ensuing discussion.

In the remainder of this section we shall deal with the formal aspects of determining what functions of the vector γ^* are estimable, and the reader for whom this is of no interest may omit the discussion to follow without loss of continuity. We are still operating with the two-way classification model.

Definition 2. Let $h_{\cdot i}$ be an $(m + k + 1)$-element column vector; the function $h'_{\cdot i} \gamma^*$ is said to be *linearly estimable* if and only if there exists a linear (in y) unbiased estimator for it. A set of r $(< m + k + 1)$ estimable functions $h'_{\cdot i} \gamma^*$, $i = 1, 2, \ldots, r$, is said to be *linearly independent* if the matrix

$$H = (h_{\cdot 1}, h_{\cdot 2}, \ldots, h_{\cdot r})$$

is of rank r.

In the context of our discussion it is of interest to know how many linearly independent estimable functions there are. If we determine them, then we can determine the totality of estimable functions.

We shall lead to this result by means of some auxiliary discussion.

Proposition 3. *Consider the model of (42). There exists an orthogonal reparametrization such that we are dealing with a model of full rank. In particular, the reparametrization is*

$$y = Q\delta + u,$$

where

$$Q = X^* P_r, \qquad \delta = P'_r \gamma^*, \qquad P = (P_r, P_*),$$

and P is the orthogonal matrix of characteristic vectors of $X^{*\prime} X^*$, P_r being the submatrix corresponding to the r $(= m + k - 1)$ nonzero (positive) characteristic roots.

PROOF. Since $X^{*\prime} X^*$ is a positive semidefinite matrix, there exists an orthogonal matrix P such that

$$P' X^{*\prime} X^* P = \begin{bmatrix} D_r & 0 \\ 0 & 0 \end{bmatrix},$$

where D_r is a diagonal matrix (of order $m + k - 1$) containing the nonzero roots of $X^{*\prime} X^*$. Partition

$$P = (P_r, P_*)$$

and observe that

$$X^* P_* = 0$$

Thus,

$$y = X^* \gamma^* + u = X^* P P' \gamma^* + u = X^* (P_r, P_*) \begin{pmatrix} P'_r \\ P'_* \end{pmatrix} \gamma^* + u = Q\delta + u,$$

where

$$Q = X^*P_r, \qquad \delta = P_r'\gamma^*.$$

We note that, since D_r is nonsingular and

$$P_r'X^{*\prime}X^*P_r = D_r,$$

$Q = X^*P_r$ is of full rank, which completes the proof.

Corollary 3. *The vector*

$$\delta = P_r'\gamma^*$$

represents a set of linearly independent estimable functions.

PROOF. We note that

$$\hat\delta = (Q'Q)^{-1}Q'y = (Q'Q)^{-1}Q'X^*\gamma + (Q'Q)^{-1}Q'u.$$

Thus,

$$E(\hat\delta) = (Q'Q)^{-1}Q'X^*\gamma^*.$$

But

$$Q'Q = P_r'X^{*\prime}X^*P_r = D_r, \qquad P_r'X^{*\prime}X^* = D_rP_r',$$

which shows

$$E(\hat\delta) = P_r'\gamma^*.$$

The proof is completed by noting that P_r is of dimension $(m + k + 1) \times (m + k - 1)$ and its columns are linearly independent (because they are mutually orthogonal).   q.e.d.

Proposition 4. *In the case under study, there exist at most $m + k - 1$ linearly independent estimable functions.*

PROOF. If we attempt to minimize

$$(y - X^*\gamma^*)'(y - X^*\gamma^*)$$

we obtain the normal equations

$$X^{*\prime}X^*\gamma^* = X^{*\prime}y,$$

which admit of an infinite number of solutions, the general form of which is (according Corollary 17 of *Mathematics for Econometrics*)

$$\hat\gamma^* = (X^{*\prime}X^*)_g X^{*\prime}y + [I - (X^{*\prime}X^*)_g(X^{*\prime}X^*)]h.$$

In the above, $(X^{*\prime}X^*)_g$ is the generalized inverse of $X^{*\prime}X^*$, and h is an arbitrary vector. Let G be $r \times (m + k + 1)$ and consider

$$G\gamma^*.$$

The quantity above is estimable if

$$G[I - (X^{*\prime}X^*)_{\mathrm{g}}(X^{*\prime}X^*)] = 0.$$

But this implies that there exists a matrix C such that

$$G = CX^*,$$

which in turn implies

$$\mathrm{rank}(G) \leq \mathrm{rank}(X^*) = m + k - 1.$$

Definition 2 then completes the proof of the proposition.   q.e.d.

Corollary 4. *There exist exactly $m + k - 1$ linearly independent (linearly) estimable functions.*

PROOF. Obvious from Propositions 3 and 4.

Remark 15. In Proposition 3, we exhibited r $(= m + k - 1)$ linearly independent (linearly) estimable functions, viz.,

$$P_r^\prime \gamma^*.$$

It is clear that all estimable functions can be generated from $P_r^\prime \gamma^*$. Thus, let M be any matrix such that

$$M\gamma^*$$

is estimable. This means that if X_{g}^* is the g-inverse of X^* the estimator of $M\gamma^*$ is

$$MX_{\mathrm{g}}^* y$$

and, moreover, that

$$\mathrm{E}[MX_{\mathrm{g}}^* y] = MX_{\mathrm{g}}^* X^* \gamma^* = M\gamma^*, \qquad M = CX^*.$$

We must be also able to express

$$M\gamma^* = AP_r^\prime \gamma^*$$

for a suitable matrix A. Since γ^* is arbitrary this entails the condition

$$CX^* = AP_r^\prime.$$

Recalling from Equation (22) that, in this context,

$$X^* = X^* P_r P_r^\prime,$$

we have

$$CX^* = CX^* P_r P_r^\prime = AP_r^\prime,$$

and thus we need only choose

$$A = CX^* P_r$$

Consequently, all estimable functions can be generated by the linearly independent set

$$P'_r \gamma*$$

as asserted, implicitly, by the corollary above.

3 Analysis of Covariance: Some Categorical and Some Continuous Explanatory Variables

The analysis of covariance model arises far more frequently in econometric applications than does the analysis of variance model. Suppose, for example, we wish to study the dividend practices of firms. We may well hypothesize that the dividend payout ratio (i.e., the proportion of dividends to profits) for a given firm depends on some of the latter's continuous characteristics, such as the recent rate of profit growth, as well as on some discontinuous (classificatory) ones, such as the industrial classification of the firm (which is, of course, a discrete, categorical variable). Often, categorical variables arise also because the investigator feels that the dependence is neither linear nor smooth—being more like a step function. In the example just mentioned we may feel that the payout ratio depends also on the firm's size, measured, say, in terms of the size of its capital stock. But the dependence may not be conceived as being linear or smooth. To express this dependence one may define a set of exhaustive size classifications—for example, 0 to 1,000,000, 1 to 10,000,000, etc.—and define appropriate size dummies. These considerations will lead to a model in which the intercept of the relation (if the latter is linear) is stated in terms of the "general mean" (μ of the earlier section), the industrial activity effect α_i (i.e., the effect of being in the ith industrial group classification such as, for example, being a textile or an aircraft producing firm, etc.), and a size effect β_j (i.e., being in the jth size class). The data matrix of such a model will be

$$X* = (X_1^*, X_2),$$

where X_1^* contains the categorical variables, while X_2 contains the observations on the continuous variables. This model does not present any novel problems. Just as in the preceding section, X_1^* *will not be of full rank.* Hence $X*$ will not be of full rank. If the matrix of continuous variables X_2 is of full rank *then by suppressing the columns in X_1^* corresponding to one classification per attribute we shall obtain a model of full rank in which the coefficients of retained dummy variables are simply the contrasts* (with respect to the omitted ones). The way we can prove the truth of this assertion exactly parallels the proof of Proposition 2 and for this reason further discussion on this point will be omitted.

Remark 16. Suppose we deal with the analysis of covariance model

$$y_{ijt} = \mu + \sum_{r=1}^{m} \alpha_r W_r + \sum_{s=1}^{k} \beta_s Z_s + \sum_{g=1}^{n} \delta_g x_{ijgt} + u_{ijt}$$

where the dummy variables W_r, Z_s are as defined earlier and x_{ijgt} denotes the tth observation on the gth continuous variable corresponding to the (i, j) cell. Thus, we are dealing, in principle, with mk GLMs, i.e., one for each cell (corresponding to classification i of attribute I and classification j of attribute II). *What enables us to treat them as one single model is the assertion that the coefficients of the continuous variables are the same across cells and that the error process has the same distribution over all elements of the sample.* Consequently, the analysis of covariance model is simply just a specialized form of the GLM in which the constant term of the model is decomposed into the contributions made by the various classifications of two attributional characteristics—for this two-way classification scheme. In particular, $\mu + \alpha_i + \beta_j$ can be interpreted as the constant term of the equation if we confine our attention to the (i, j) cell. As before, if we suppress the dummy variables W_m and Z_k and apply OLS procedures, the constant term of the estimated equation is an estimate of $\mu + \alpha_m + \beta_k$; if we add to this estimate the estimated coefficients of W_i and $Z_j, i = 1, 2, \ldots, m - 1, j = 1, 2, \ldots, k - 1$, we obtain an estimate of $\mu + \alpha_i + \beta_j$. If we wish to test the hypothesis that the attribute classification effects are all equal we simply test the hypothesis

$$H_0: \quad \gamma = 0,$$

as against the alternative

$$H_1: \quad \gamma \neq 0,$$

where

$$\gamma = (\gamma'_{(1)}, \gamma_{(2)})'$$

and

$$\gamma_{(1)} = (\alpha_1 - \alpha_m, \alpha_2 - \alpha_m, \ldots, \alpha_{m-1} - \alpha_m)',$$
$$\gamma_{(2)} = (\beta_1 - \beta_k, \beta_2 - \beta_k, \ldots, \beta_{k-1} - \beta_k)'.$$

The procedures by which such tests can be carried out have been discussed extensively in Chapter 2.

5 Misspecification Analysis and Errors in Variables

1 Introduction

In previous discussion we have examined the standard form of the GLM and the consequences of violating assumptions pertaining to the i.i.d. properties of the error term and the rank of the data matrix X.

Here we take up an important and recurring problem in empirical research involving the use of an inappropriate data matrix. This may eventuate either because the appropriate data matrix is inherently unobservable or because, even though (in principle) observable, the pertinent information is not available.

An example of the first instance occurs in the context of the "permanent income" hypothesis. More precisely, the hypothesis specifies the relationship

$$C = \alpha + \beta Y + U \tag{1}$$

where C and Y are, respectively, permanent consumption and income, and U is an error term independent of Y. Unfortunately, such quantities are inherently unobservable and one further theorizes that they are related to observed consumption and income by

$$c = C + u_1, \qquad y = Y + u_2, \tag{2}$$

where c and y are, respectively, observed consumption and income, and u_i, $i = 1, 2$, are error terms independent of C, Y.

If any analysis is to be carried out in the context of (1) it is clear that the requisite data is lacking—and, indeed, in principle unobtainable. However, it may be possible to use the relations in (2) to rewrite (1) as

$$c = \alpha + \beta y + u, \tag{3}$$

218

where

$$u = U - u_1 - \beta u_2.$$

Ostensibly in (3) we deal with the same parametric configuration as in (1) except that the data matrix to be employed in estimation, if a sample of size T were available, is

$$(y, X_*), \qquad y = (y_1, y_2, \ldots, y_T)', \qquad x = (x_1, x_2, \ldots, x_T)', \qquad X_* = (e, x),$$

e being a T-element vector of 1's, while the appropriate data matrix would have been

$$(Y, X^*), \qquad Y = (Y_1, Y_2, \ldots, Y_T)',$$

$$X = (X_1, X_2, \ldots, X_T)', \qquad X^* = (e, X).$$

In this particular case we have the additional complication that x and u are correlated.

The situation above is said to constitute an *errors in variables* (EIV) *model* and represents the consumption function suggested by Milton Friedman. More fully, u_1 represents transitory consumption, which is taken to be a zero mean i.i.d. random variable with variance σ_1^2, while u_2 represents transitory income, which is also taken to be a zero mean i.i.d. random variable with variance σ_2^2. Although it does not appear to have any special significance, the assumption is also made that u_1 and u_2 are independent of the structural error U.

There are many instances, however, in which the appropriate data matrix is, in principle, observable, but observations on one or more relevant variables are lacking. Thus, e.g., in the Cobb–Douglas production function context we have

$$Q = AK^\alpha L^\beta e^u,$$

where Q, K, L, u are, respectively, output, capital, labor, and a random variable. Thus, we would require data on output, capital, and labor services. Data on output is available through the various value added series published (in the United States) by the Department of Commerce. Similarly, data on labor services is available through the various manhours worked series. Unfortunately, capital services data is, typically, not available. *What we do have is data on the capital stock.* But since the extent of capital utilization varies from period to period depending on the state of industry demand, frictional factors in labor employment, etc., the dependence of capital services on the capital stock may not be of the simple form

$$K_t = K_t^{*\gamma} e^{u_{t1}}, \qquad t = 1, 2, \ldots,$$

where K_t^* is capital stock, γ is a constant, and u_{t1} is a zero mean i.i.d. random variable.

Thus, if we rewrite the production function as

$$\ln Q_t = \ln A + \alpha \ln K_t + \beta \ln L_t + u_t, \qquad t = 1, 2, \ldots, T,$$

the appropriate data matrix will be

$$(y, X), \qquad y = (\ln Q_1, \ln Q_2, \dots, \ln Q_T)', \qquad X = (e, x_{.1}, x_{.2}),$$

$$x_{.1} = (\ln K_1, \ln K_2, \dots, \ln K_T)', \qquad x_{.2} = (\ln L_1, \ln L_2, \dots, \ln L_T)'.$$

In the absence of information on capital services we would use

$$X^* = (e, x_{.1}^*, x_{.2})$$

where

$$x_{.1}^* = (\ln K_1^*, \ln K_2^*, \dots, \ln K_T^*).$$

The investigation of the consequences of using the data matrix

$$(y, X^*)$$

instead of

$$(y, X)$$

to estimate the parameters of the model, in the absence of an explicitly stated relationship between X and X^*, is referred to as *misspecification analysis*. Properly speaking, the term ought to be applied to the general problem of determining the consequences of an incorrect or improper premise in the argument leading to a conclusion regarding the properties of an estimator.

Be that as it may, we see that the formal differences between the EIV model and misspecification analysis, as the term is commonly used in the literature of econometrics, are rather slight.

Out of deference to tradition we shall treat these two topics separately, and in the context of misspecification analysis we shall further examine the problem of near collinearity.

2 Misspecification Analysis

2.1 General Theory in the Context of the GLM

The typical problem considered is the following. Suppose we have the GLM

$$y = X\beta + u \tag{4}$$

under the standard assumptions. Suppose further that, either through ignorance or nonavailability, we use the data matrix X^*, instead of X, in estimating the vector β. What properties can we claim for this "estimator"? Precisely, the "estimator" employed is

$$\hat{\beta}^* = (X^{*\prime}X^*)^{-1}X^{*\prime}y, \tag{5}$$

and we wish to examine its mean and covariance characteristics and their relation to β, the parameter we seek to estimate. In order to analyse these issues we need to know the correct specification for y, which is, of course, given in (4). Thus, substituting from (4) in (5) we have

$$\hat{\beta}^* = (X^{*\prime}X^*)^{-1}X^{*\prime}X\beta + (X^{*\prime}X^*)^{-1}X^{*\prime}u.$$

Put

$$P = (X^{*\prime}X^*)^{-1}X^{*\prime}X$$

and notice that P is the matrix of the regression coefficients in the regression of the "correct" on the "misspecified" variables, i.e., the regression of the variables contained in X on those contained in X^*.

To proceed further with establishing the properties of this "estimator" we need to specify the relationship of the "misspecified" variables to the structural error. If we assert the analog of the appropriate standard condition we have

$$E(u\,|\,X^*) = 0, \qquad \mathrm{Cov}(u\,|\,X^*) = \sigma^2 I \tag{6}$$

or, in general, that the elements of u and X^* are uncorrelated or mutually independent. Given (6) we easily conclude

$$E(\hat{\beta}^*) = P\beta, \qquad \mathrm{Cov}(\hat{\beta}^*) = \sigma^2 (X^{*\prime}X^*)^{-1}. \tag{7}$$

It is clear that unless $P\beta = \beta$ the estimator above is biased. Thus, it is useful to obtain its mean squared error (MSE) matrix,

$$\mathrm{MSE}(\hat{\beta}^*) = E(\hat{\beta}^* - \beta)(\hat{\beta}^* - \beta)' = (P - I)\beta\beta'(P - I) + \sigma^2 (X^{*\prime}X^*)^{-1}. \tag{8}$$

We have therefore the following fundamental proposition:

Proposition 1. *Let*

$$y = X\beta + u$$

be a GLM *subject to the standard assumptions. Suppose that in estimating β the data matrix X is not available, but that instead we use the (available) data matrix X^*. Suppose, further, that*[1]

$$\mathrm{rank}(X^*) = n + 1, \qquad (\text{p})\lim_{T \to \infty} \left(\frac{X^{*\prime}X^*}{T} \right) = M_{x^*x^*},$$

$$E(u\,|\,X^*) = 0, \qquad \mathrm{Cov}(u\,|\,X^*) = \sigma^2 I,$$

---

[1] We remind the reader that (p)lim means the probability limit or the ordinary limit, whichever is relevant in the context.

where $M_{x^*x^*}$ is a positive definite matrix. Then the OLS "estimator" using the misspecified data matrix

$$\hat{\beta}^* = (X^{*\prime}X^*)^{-1}X^{*\prime}y$$

has the following properties:

  (i) it is biased and its bias is given by $(P - I)\beta$;
 (ii) it is inconsistent and its inconsistency is given by $(\bar{P} - I)\beta$, where

$$\bar{P} = \underset{T \to \infty}{(\text{p})\lim} (X^{*\prime}X^*)^{-1}X^{*\prime}X;$$

(iii) its MSE error matrix is

$$\text{MSE}(\hat{\beta}^*) = (P - I)\beta\beta'(P - I) + \sigma^2(X^{*\prime}X^*)^{-1}$$

and, moreover,

$$\underset{T \to \infty}{(\text{p})\lim} \text{MSE}(\hat{\beta}^*) = (\bar{P} - I)\beta\beta'(P - I).$$

PROOF. Since the expectation of $\hat{\beta}^*$ was given in (7), (i) is proved by noting that the bias of the estimator is

$$E(\hat{\beta}^*) - \beta = (P - I)\beta.$$

For (ii) we note that

$$\underset{T \to \infty}{\text{p}\lim} \hat{\beta}^* = \underset{T \to \infty}{(\text{p})\lim} P\beta + \underset{T \to \infty}{\text{p}\lim} \left(\frac{X^{*\prime}X^*}{T}\right)^{-1}\left(\frac{X^{*\prime}u}{T}\right) = \bar{P}\beta.$$

This is so since, under the standard assumptions,

$$\underset{T \to \infty}{(\text{p})\lim}\left(\frac{X^{*\prime}X^*}{T}\right) = M_{x^*x^*}$$

while, under the premises of the proposition,

$$\underset{T \to \infty}{\text{p}\lim}\left(\frac{X^{*\prime}u}{T}\right) = 0.$$

In (ii) it is implicitly assumed that

$$\underset{T \to \infty}{(\text{p})\lim}\left(\frac{X^{*\prime}X}{T}\right) = M_{x^*x}$$

exists, so that

$$\bar{P} = M_{x^*x^*}^{-1}M_{x^*x}.$$

The *inconsistency* of an estimator is defined by

$$\underset{T \to \infty}{\text{p}\lim} \hat{\beta}^* - \beta = (\bar{P} - I)\beta.$$

Finally, the MSE matrix of the "estimator" $\hat{\beta}*$ has been established in (8), and we note that since

$$(\text{p})\lim_{T \to \infty} \left(\frac{X^{*\prime}X^*}{T} \right) = M_{x^*x^*}$$

exists as a well defined positive definite matrix,

$$(\text{p})\lim_{T \to \infty} (X^{*\prime}X^*)^{-1} = 0,$$

which completes the proof.   q.e.d.

Although the result above is quite general and comprehensive its abstraction may prevent the reader from fully appreciating its implications. In the remainder of this section we shall examine a few of the more common instances in which the proposition may have relevance.

Case of one misspecified variable. Suppose we have a situation in which only one of the variables, say the last, is misspecified. Thus,

$$X = (X_1, x_{\cdot n}), \qquad X^* = (X_1, x^*_{\cdot n}),$$

so that X and X^* have the matrix X_1 in common. The bias of the estimator $\hat{\beta}*$ now becomes

$$(P - I)\beta = (p_{\cdot n} - e_{\cdot n})\beta_n,$$

where

$$p_{\cdot n} = (X^{*\prime}X^*)^{-1}X^{*\prime}x_{\cdot n}, \qquad e_{\cdot n} = (0, 0, \ldots, 1)',$$

so that $p_{\cdot n}$ is the vector of regression coefficients in the regression of the correct (omitted) variable on all the included variables and $e_{\cdot n}$ is an $(n + 1)$-element vector all of whose elements are zero save the last, which is unity. In the preceding, we remind the reader that

$$X_1 = (x_{\cdot 0}, x_{\cdot 1}, \ldots, x_{\cdot n-1}), \qquad \beta_{(1)} = (\beta_0, \beta_1, \ldots, \beta_{n-1})'$$
$$\beta = (\beta'_{(1)}, \beta_n)',$$

and that

$$x_{\cdot 0} = (1, 1, \ldots, 1)'$$

is the vector of "observations" on the "variable" corresponding to the constant term of the GLM.

Consequently, we see that provided the variables in X_1 are *not* orthogonal to $x_{\cdot n}$, and $x^*_{\cdot n}$, so that

$$X'_1 x_{\cdot n} \neq 0, \qquad X'_1 x^*_{\cdot n} \neq 0$$

misspecifying just one variable induces a "misspecification bias" on the estimators of the coefficients of all variables and not merely on the coefficient of the misspecified variable. The extent of this bias is proportional to the true coefficient of the misspecified variable.

The mean squared error matrix becomes

$$\text{MSE}(\hat{\beta}^*) = \beta_n^2 (p_{\cdot n} - e_{\cdot n})(p_{\cdot n} - e_{\cdot n})' + \sigma^2 (X^{*\prime} X^*)^{-1}.$$

Partitioning $(X^{*\prime} X^*)^{-1}$ conformably with

$$\beta = \begin{pmatrix} \beta_{(1)} \\ \beta_n \end{pmatrix},$$

we have

$$(X^{*\prime} X^*)^{-1} = Q = \begin{bmatrix} Q_{11} & Q_{12} \\ Q_{21} & Q_{22} \end{bmatrix},$$

where

$$Q_{11} = [X_1'(I - s_n x_{\cdot n}^* x_{\cdot n}^{*\prime}) X_1]^{-1}, \qquad s_n = (x_{\cdot n}^{*\prime} x_{\cdot n}^*)^{-1},$$

$$Q_{22} = [x_{\cdot n}^{*\prime}(I - X_1(X_1' X_1)^{-1} X_1') x_{\cdot n}^*]^{-1} = s_{nn}^{-1}$$

$$Q_{12} = Q_{21}' = -Q_{11} X_1' x_{\cdot n}^* s_n.$$

We remind the reader that s_{nn} is simply the sum of squared residuals in the regression of the misspecified variable employed, i.e., $x_{\cdot n}^*$, on the correctly specified variables, i.e., those in X_1. The result of the preceding discussion can be summarized in

Corollary 1. *Consider the situation of Proposition 1 and suppose that only one variable is misspecified, i.e.,*

$$X = (X_1, x_{\cdot n}), \qquad X^* = (X_1, x_{\cdot n}^*).$$

Then, provided

$$X_1' x_{\cdot n} \neq 0, \quad \text{or} \quad X_1' x_{\cdot n}^* \neq 0$$

the coefficients of all variables are biased and inconsistent. The bias is given by

$$b(\hat{\beta}^*) = (p_{\cdot n} - e_{\cdot n}) \beta_n$$

and the inconsistency is given by

$$i(\hat{\beta}^*) = (\bar{p}_{\cdot n} - e_{\cdot n}) \beta_n,$$

where

$$p_{\cdot n} = (X^{*\prime} X^*)^{-1} X^{*\prime} x_{\cdot n}, \qquad \bar{p}_{\cdot n} = (p)\lim_{T \to \infty} p_{\cdot n}.$$

In the special case where

$$X_1' x_{\cdot n} = 0, \qquad X_1' x_{\cdot n}^* = 0$$

then

$$p_{\cdot n} = \begin{pmatrix} 0 \\ \vdots \\ 0 \\ p_{nn} \end{pmatrix},$$

and consequently the coefficients of all variables other than the misspecified one are unbiasedly and consistently estimated.

Case of one omitted variable. This case is formally the same as the one considered above for the special case

$$x_{tn}^* = 0 \quad \text{for all } t.$$

Because, however, certain rank conditions are violated it is best treated separately.

Using the same notation as above, we have now

$$X = (X_1, x_{\cdot n}), \qquad X^* = X_1,$$

so that the nth variable is suppressed in carrying out the regression. Needless to say, the vector β^* now "estimates" only the elements

$$\beta_{(1)} = (\beta_0, \beta_1, \ldots, \beta_{n-1})',$$

the coefficient β_n being, implicitly, estimated by zero. Thus, now

$$\begin{aligned}
\hat{\beta}_{(1)}^* &= (X^{*\prime}X^*)^{-1}X^{*\prime}y \\
&= (X_1'X_1)^{-1}X_1'[X_1\beta_{(1)} + x_{\cdot n}\beta_n + u] \\
&= \beta_{(1)} + r_{\cdot n}\beta_n + (X_1'X_1)^{-1}X_1'u,
\end{aligned}$$

where

$$r_{\cdot n} = (X_1'X_1)^{-1}X_1'x_{\cdot n},$$

i.e., it is the vector of regression coefficients in the regression of the excluded on the included variables. Without any additional assumptions we conclude

$$E(\hat{\beta}_{(1)}^*) = \beta_{(1)} + r_{\cdot n}\beta_n \tag{9}$$

and

$$\text{MSE}(\hat{\beta}_{(1)}^*) = \beta_n^2 r_{\cdot n}r_{\cdot n}' + \sigma^2(X_1'X_1)^{-1}. \tag{10}$$

As before, omitting one variable has bias and inconsistency implications for the estimates of the coefficients of all variables of the GLM.

However, in the special case where

$$X_1'x_{\cdot n} = 0,$$

i.e., the excluded variable is orthogonal to the included variables, we note

$$r_{\cdot n} = 0,$$

so that

$$\text{MSE}(\hat{\beta}^*) = \sigma^2(X_1'X_1)^{-1}.$$

Consequently, for this case $\hat{\beta}_{(1)}^*$ coincides with the OLS estimator of $\beta_{(1)}$, and for this set of parameters $\hat{\beta}_{(1)}^*$ is the BLUE. Needless to say this is *not* so if we consider β (rather than $\beta_{(1)}$) as the vector of interest. The results of the discussion above may be summarized in

Corollary 2. *Consider the situation in Proposition* 1 *and suppose further that*

$$X = (X_1, x_{\cdot n}), \qquad X^* = X_1.$$

Then, provided

$$X_1' x_{\cdot n} = 0,$$

the estimator $\hat{\beta}^*_{(1)}$ *is the same as the OLS estimator of* $\beta_{(1)}$ *and as such is a* BLUE.
If, on the other hand,

$$X_1' x_{\cdot n} \neq 0,$$

then the "estimator" $\hat{\beta}^*_{(1)}$ *is biased and inconsistent as an estimator of* $\beta_{(1)}$. *Its bias is given by*

$$b(\hat{\beta}^*_{(1)}) = r_{\cdot n} \beta_n, \qquad r_{\cdot n} = (X_1' X_1)^{-1} X_1' x_{\cdot n},$$

while its inconsistency is given by

$$i(\hat{\beta}^*_{(1)}) = \bar{r}_{\cdot n} \beta_n, \qquad \bar{r}_{\cdot n} = \underset{T \to \infty}{(p)\lim} r_{\cdot n}.$$

In any event, the "estimator" of β_n, *which is implicitly taken to be zero, is generally biased and inconsistent—except in the obvious special case where it so happens that*

$$\beta_n = 0.$$

Goodness of fit and misspecification. Occasionally, the situation arises in empirical work where the correct specification is not known. One has, say, two candidates

$$y = X_1 \beta + u_1, \qquad y = X_2 \beta + u_2,$$

where X_1, X_2 are two matrices each of dimension $T \times (n + 1)$.

It is frequently asserted, in this context, that the regression exhibiting the highest R^2 identifies the correct model. While in some circumstances this is a true statement it has to be interpreted with considerable caution. Clearly, given *any* two specifications one would invariably exhibit a higher R^2 than the other. Clearly, one model is not a subset of the other since X_1 and X_2 are of the same dimension but contain different variables (although they may have some in common). In this context the standard tests for the "significance" of the difference in the coefficients of determination for the two regressions, $R_1^2 - R_2^2$, would not be applicable. Hence we should be reduced to the position of stating that if $R_1^2 > R_2^2$ then

$$y = X_1 \beta + u_1$$

is the correct specification and this would appear a singularly facile and simpleminded way of arriving at the "truth." What we may say with some formal justification is the following. If of the two models

$$y_1 = X_1\beta + u_1, \qquad y_2 = X_2\beta + u_2$$

one is known to be the true model, then if upon regression we find that the coefficient of determination of multiple regression for the first model, say R_1^2, is larger than the corresponding coefficient for the second model, say R_2^2, then we can "conclude" that

$$y = X_1\beta + u_1$$

is the "true" model.

The formal justification rests on the following large sample argument. By definition

$$R_1^2 = 1 - \frac{\hat{u}_1'\hat{u}_1}{y^{*\prime}y^*}, \qquad R_2^2 = 1 - \frac{\hat{u}_2'\hat{u}_2}{y^{*\prime}y^*},$$

where

$$y^* = \left(I - \frac{ee'}{T}\right)y, \qquad e = (1, 1, \ldots, 1)',$$

$$\hat{u}_i = (I - M_i)y, \qquad M_i = X_i(X_i'X_i)^{-1}X_i', \qquad i = 1, 2.$$

If the first model is the true one then

$$\hat{u}_1 = (I - M_1)u_1, \qquad \hat{u}_2 = (I - M_2)X_1\beta + (I - M_2)u_1.$$

Thus, if the standard assumptions apply to the two data matrices X_1, X_2, then

$$\underset{T\to\infty}{\text{plim}} R_1^2 = \underset{T\to\infty}{\text{plim}} R_2^2 + \underset{T\to\infty}{\text{plim}} \frac{\beta'X_1'(I - M_2)X_1\beta}{y^{*\prime}y^*} \geq \underset{T\to\infty}{\text{plim}} R_2^2.$$

This shows that the probability limit of the coefficient of determination of the true model is not less than the corresponding limit for the incorrectly specified model. The same of course would apply to the sum of squares of the residuals with the obvious reversal of signs. Precisely, we have that

$$\underset{T\to\infty}{\text{plim}} \left(\frac{\hat{u}_2'\hat{u}_2}{T}\right) = \underset{T\to\infty}{\text{plim}} \left(\frac{\hat{u}_1'\hat{u}_1}{T}\right) + (\text{p})\underset{T\to\infty}{\text{lim}} \left[\frac{\beta'X_1(I - M_2)X_1\beta}{T}\right]$$

$$\geq \underset{T\to\infty}{\text{plim}} \left(\frac{\hat{u}_1'\hat{u}_1}{T}\right).$$

Consequently, we can argue that, for sufficiently large samples, the probability is high that

$$\hat{u}_2'\hat{u}_2 \geq \hat{u}_1'\hat{u}_1, \qquad R_1^2 \geq R_2^2,$$

if, in fact, the first model is the true one.

Evidently, if it is not known that one of the models is the true one no inference can be derived about the "truth" by simply contemplating the residuals or the R^2's of the two regressions.

2.2 Proxy Variables and Their Use

It is a frequent occurrence in applied econometric research that one (or more) of the variables of a hypothesized GLM is not available—either because data has not been collected even though it is possible to collect them or because the variable is inherently unobservable. In such circumstances it is a common practice to resort to so-called "proxy variables," i.e., variables that substitute for the missing variable. It is not always clear what criteria are to be satisfied by proxy variables. More frequently than not the criterion is that the investigator believes the proxy to be of the same genre as the (missing) variable of interest and, presumably, to exhibit substantial correlation with the latter.

The question then is posed: should we omit the missing variable from the GLM, thus committing a misspecification error, or should we use a proxy? Unfortunately, *in either case we are committing a misspecification error, something that cannot be avoided short of obtaining information on the missing variable.* Then, the proper definition of the problem is: would we do better by committing the (misspecification) error of omitting the missing variable or by using in its stead a proxy variable?

Much of the groundwork for answering this question has been laid in the preceding section; the proxy variable alternative evidently corresponds to the case of one misspecified variable treated therein.

We recall that the MSE matrix in the case of the proxy variable is

$$\mathrm{MSE}(\hat{\beta}^*) = \beta_n^2 (p_{\cdot n} - e_{\cdot n})(p_{\cdot n} - e_{\cdot n})' + \sigma^2 (X^{*\prime}X^*)^{-1}, \tag{11}$$

where $p_{\cdot n} = (X^{*\prime}X^*)^{-1}X^{*\prime}x_{\cdot n}$ and $e_{\cdot n}$ is an $(n+1)$-element vector all of whose elements are zero save the last, which is unity.

It will be convenient to express the various quantities in the mean squared error matrix above in partitional inverse form; this will greatly facilitate discussion later on. It may be shown (see Problem 3) that

$$(X^{*\prime}X^*)^{-1} = \begin{bmatrix} (X_1'X_1)^{-1} + \left(\dfrac{1}{s_{nn}}\right) r_{\cdot n}^* r_{\cdot n}^{*\prime} & -\left(\dfrac{1}{s_{nn}}\right) r_{\cdot n}^* \\[2ex] -\left(\dfrac{1}{s_{nn}}\right) r_{\cdot n}^{*\prime} & \dfrac{1}{s_{nn}} \end{bmatrix} \tag{12}$$

where

$$r_{\cdot n}^* = (X_1'X_1)^{-1}X_1'x_{\cdot n}^*, \qquad s_{nn} = x_{\cdot n}^{*\prime}M_1 x_{\cdot n}^*, \qquad M_1 = I - X_1(X_1'X_1)^{-1}X_1'. \tag{13}$$

Thus, (see Problem 4)

$$p_{\cdot n} - e_{\cdot n} = \begin{pmatrix} r_{\cdot n} - \phi r_{\cdot n}^* \\ \phi - 1 \end{pmatrix} \tag{14}$$

where

$$r_{\cdot n} = (X_1' X_1)^{-1} X_1' x_{\cdot n}, \qquad \phi = \frac{x_{\cdot n}^{*'} M_1 x_{\cdot n}}{x_{\cdot n}^{*'} M_1 x_{\cdot n}^{*}} \qquad (15)$$

Now, in the case where the missing variable is ignored, in effect we estimate

$$\hat{\beta} = \begin{pmatrix} (X_1' X_1)^{-1} X_1' y \\ 0 \end{pmatrix}, \qquad (16)$$

and the bias and covariance matrix of this estimator are given by

$$b(\hat{\beta}) = \begin{pmatrix} r_{\cdot n} \\ -1 \end{pmatrix} \beta_n, \qquad \mathrm{Cov}(\hat{\beta}) = \sigma^2 \begin{bmatrix} (X_1' X_1)^{-1} & 0 \\ 0 & 0 \end{bmatrix}. \qquad (17)$$

Hence, we can write quite easily

$$\mathrm{MSE}(\hat{\beta}^*) = b(\hat{\beta}^*) b(\hat{\beta}^*)' + \sigma^2 (X^{*'} X^*)^{-1}, \qquad (18)$$

where

$b(\hat{\beta}^*) b(\hat{\beta}^*)'$

$$= \beta_n^2 \begin{bmatrix} r_{\cdot n} r_{\cdot n}' + \phi^2 r_{\cdot n}^* r_{\cdot n}^{*'} - \phi r_{\cdot n} r_{\cdot n}^{*'} - \phi r_{\cdot n}^* r_{\cdot n}' & (\phi - 1) r_{\cdot n} - \phi(\phi - 1) r_{\cdot n}^* \\ (\phi - 1) r_{\cdot n}' - \phi(\phi - 1) r_{\cdot n}^{*'} & (\phi - 1)^2 \end{bmatrix}, \qquad (19)$$

and

$$\mathrm{MSE}(\hat{\beta}) = b(\hat{\beta}) b(\hat{\beta})' + \mathrm{Cov}(\hat{\beta}), \qquad (20)$$

where

$$b(\hat{\beta}) b(\hat{\beta})' = \beta_n^2 \begin{bmatrix} r_{\cdot n} r_{\cdot n}' & -r_{\cdot n} \\ -r_{\cdot n}' & 1 \end{bmatrix}, \qquad \mathrm{Cov}(\hat{\beta}) = \sigma^2 \begin{bmatrix} (X_1' X_1)^{-1} & 0 \\ 0 & 0 \end{bmatrix}. \qquad (21)$$

We are now in a position to answer the question posed earlier in this section, viz, whether it is preferable to suppress (from consideration) the missing variable or to use a proxy in its stead. This may be answered in terms of mean squared error efficiency, in which case we are led to consider

$$\mathrm{MSE}(\hat{\beta}^*) - \mathrm{MSE}(\hat{\beta}) = \beta_n^2 \begin{bmatrix} \phi^2 r_{\cdot n}^* r_{\cdot n}^{*'} - \phi r_{\cdot n} r_{\cdot n}^{*'} - \phi r_{\cdot n}^* r_{\cdot n}' & \phi r_{\cdot n} - \phi(\phi - 1) r_{\cdot n}^* \\ \phi r_{\cdot n}' - \phi(\phi - 1) r_{\cdot n}^{*'} & (\phi - 1)^2 - 1 \end{bmatrix}$$

$$+ \frac{\sigma^2}{s_{nn}} \begin{bmatrix} r_{\cdot n}^* r_{\cdot n}^{*'} & -r_{\cdot n}^* \\ -r_{\cdot n}^{*'} & 1 \end{bmatrix}. \qquad (22)$$

Again, in answering the question we may do so for the case where we are only concerned with the first n elements of the vector β or for the case where we are concerned with all its elements.

In the former case the matrix of interest is

$$A = \beta_n^2(\phi^2 r_{\cdot n}^* r_{\cdot n}^{*\prime} - \phi r_{\cdot n}^* r_{\cdot n}^{\prime} - \phi r_{\cdot n} r_{\cdot n}^{*\prime}) + \frac{\sigma^2}{s_{nn}} r_{\cdot n}^* r_{\cdot n}^{*\prime}, \tag{23}$$

while in the latter case the matrix of interest is the one exhibited in (22). Clearly, we cannot give an unambiguous answer for all proxies. Indeed, for some proxies the matrix in (22) and/or (23) may be positive (semi)definite, for others negative (semi)definite, while for yet others it may be indefinite. Thus, some proxies may be quite unsuitable, others quite suitable, while for yet others the indications for their use are not clearcut. So let us consider the case frequently invoked by practicing econometricians, i.e., the case where it is asserted that

$$x_{\cdot n} = \theta x_{\cdot n}^* + \varepsilon \tag{24}$$

where ε is a T-element vector of random variables having the i.i.d. property with

$$E(\varepsilon | X^*) = 0, \qquad \text{Cov}(\varepsilon | X^*) = \sigma_\varepsilon^2 I. \tag{25}$$

Under these circumstances we note that if we take expectations in (15) with respect to ε we find

$$E(\phi | X^*) = \theta, \qquad E(r_{\cdot n} | X^*) = \theta r_{\cdot n}^*. \tag{26}$$

Consequently, when the missing variable is given by (24) the difference in (22), after taking probability limits, simplifies to

$$\text{MSE}(\hat{\beta}^*) - \text{MSE}(\hat{\beta}) = \begin{bmatrix} \left(\dfrac{\sigma^2}{s_{nn}} - \beta_n^2 \theta^2\right) r_{\cdot n}^* r_{\cdot n}^{*\prime} & -\left(\dfrac{\sigma^2}{s_{nn}} - \beta_n^2 \theta\right) r_{\cdot n}^* \\[2ex] -\left(\dfrac{\sigma^2}{s_{nn}} - \beta_n^2 \theta\right) r_{\cdot n}^{*\prime} & \left(\dfrac{\sigma^2}{s_{nn}} - \beta_n^2\right) + \beta_n^2(\theta - 1)^2 \end{bmatrix}. \tag{27}$$

We see immediately that if we are only interested in the coefficients of the available variables use of the proxy may make matters worse, to the extent that

$$\frac{\sigma^2}{s_{nn}} - \beta_n^2 \theta^2 \geq 0.$$

This could be so, in particular, when the proxy variable is sufficiently collinear with the available variables; this is so since

$$s_{nn} = x_{\cdot n}^{*\prime}(I - X_1(X_1' X_1)^{-1} X_1') x_{\cdot n}^*$$

and, thus, is nothing more than the sum of the squared residuals in the regression of $x_{\cdot n}^*$ on X_1!

Remark 1. The result just stated rests, in part, on our preoccupation with the coefficients of the available variables. It is quite conceivable that

$$\frac{\sigma^2}{S_{nn}} - \beta_n^2 \theta^2 \geq 0,$$

while

$$\left(\frac{\sigma^2}{S_{nn}} - \beta_n^2\right) + \beta_n^2(\theta - 1)^2 \leq 0.$$

This would mean that the matrix in (27) is indefinite and that, despite the fact that the coefficients of the available variables are "better" estimated when the missing variable is suppressed, the coefficient of the missing variable is "better" estimated through the use of the proxy.

The situation described would eventuate, for example, if

$$0 \leq \frac{\sigma^2}{S_{nn}} - \beta_n^2 \theta^2 \leq \beta_n^2[(1 + \theta) - (1 - \theta)](1 - \theta).$$

In particular, the reader will verify that

$$\frac{\sigma^2}{S_{nn}} = .56, \qquad \beta_n^2 = 1, \qquad \theta = .6$$

produces

$$\frac{\sigma^2}{S_{nn}} - \beta_n^2 \theta^2 = .2, \qquad \frac{\sigma^2}{S_{nn}} - \beta_n^2 + \beta_n^2(\theta - 1)^2 = -.28.$$

Needless to say, there are many parametric configurations for which the choice between proxy and no proxy is not clearcut and the matrix in (27) is indefinite in a more complicated way.

Remark 2. Consider the case now where $x_{\cdot n}$ is actually available but one chooses not to use it in the regression. In this particular case $\theta = 1$ and the matrix in (27) reduces to

$$\left(\frac{\sigma^2}{S_{nn}} - \beta_n^2\right)\begin{bmatrix} r_{\cdot n} r'_{\cdot n} & -r_{\cdot n} \\ -r'_{\cdot n} & 1 \end{bmatrix} = \left(\frac{\sigma^2}{S_{nn}} - \beta_n^2\right)\begin{pmatrix} r_{\cdot n} \\ -1 \end{pmatrix}(r'_{\cdot n}, -1),$$

which is evidently either positive or negative semidefinite according as

$$\frac{\sigma^2}{S_{nn}} - \beta_n^2 \gtrless 0.$$

But this would indicate that if $x_{\cdot n}$ is "sufficiently collinear" with the variables in X_1 we may gain in MSE efficiency by suppressing it! We shall return to this issue at a later section.

Remark 3. Clearly, for some cases, the use of a proxy is to be preferred to simply ignoring the missing variable. This would be so for the type of proxy exhibited in (24) if (σ^2/s_{nn}) is not large while β_n^2 is of appreciable size, i.e., if there is no marked collinearity between the proxy and the available variables and if the missing variable plays an important role (as indexed, say, by the size of its coefficient) in determining the behavior of the dependent variable.

Remark 4. Notice, referring to Equation (19), that for proxies of the type exhibited in (24) the bias component of the MSE matrix reduces (in the limit) to

$$b(\hat{\beta}*)b(\hat{\beta}*)' = \beta_n^2 \begin{bmatrix} 0 & 0 \\ 0 & (\theta - 1)^2 \end{bmatrix},$$

so that the coefficients of the available variables are estimated without bias!

2.3 Near Collinearity

In this section we shall examine various proposed remedies to the problem of "near collinearity" in the context of misspecification analysis. Indeed, in applied work one rarely encounters "strict collinearity." Thus, the issues considered in Chapter 4 are not likely to be very relevant in empirical research. In particular, one of the "remedies" considered, viz., that of the generalized inverse estimator, will, in the case of "near collinearity," produce exactly the OLS estimator! This is so since $X'X$ is not, strictly speaking, singular. Of the other remedies mentioned there we shall consider the practice of dropping variables, the principal components estimator, as well as the ridge regression/Bayesian solution.

Principal Components Estimator. We first review, briefly, the mechanics of this estimator. Let

$$y = X\beta + u \tag{28}$$

be the standard GLM and suppose that all data have been centered about the appropriate sample means.[2]

Obtain the matrices of characteristic roots and vectors of the matrix $X'X$. Thus

$$X'XA = AR, \tag{29}$$

where

$$A = (a_{.1}, a_{.2}, \ldots, a_{.n}), \qquad R = \mathrm{diag}(r_1, r_2, \ldots, r_n),$$

$$r_1 \geq r_2 \geq \cdots \geq r_n,$$

the $a_{.i}$ and r_i being, respectively, the characteristic vectors and corresponding characteristic roots, $i = 1, 2, \ldots, n$.

---

[2] In the discussion of all near collinearity remedies we shall assume that data have been centered about the appropriate sample means.

The "observations" on the principal components are defined by

$$Z = XA. \tag{30}$$

Note that the ith column of Z is given by

$$z_{\cdot i} = \sum_{j=1}^{n} x_{\cdot j} a_{ji} = Xa_{\cdot i}, \qquad i = 1, 2, \ldots, n,$$

so that it is a linear combination of all of the columns of the matrix X. If we are faced with the problem of near collinearity this means that one or more of the roots r_i, $i = 1, 2, \ldots, n$, are "close" to zero and so small as to be indistinguishable. For definiteness, suppose that the last $n - k$ roots fall in that category and that the first k roots account for, say, 95% or more of the variability of the data, i.e.,

$$\frac{\sum_{i=1}^{k} r_i}{\sum_{i=1}^{n} r_i} \geq .95.$$

Noting that A is an orthogonal matrix we have that

$$X = ZA'$$

and substituting in (28) we find

$$y = Z\gamma + u, \tag{31}$$

where

$$\gamma = A'\beta. \tag{32}$$

It is clear that if we use (31) as the basic model we can estimate by OLS

$$\hat{\gamma} = (Z'Z)^{-1}Z'y \tag{33}$$

and then obtain, using (32),

$$\hat{\beta} = A\hat{\gamma} = A(A'X'XA)^{-1}A'Xy = (X'X)^{-1}X'y. \tag{34}$$

But this means that the OLS estimator of β can be obtained in two steps:

(i) obtain the principal components of the explanatory variables and obtain the regression coefficients in the regression of y on Z;

(ii) use the relation in (32) and the regression coefficients in (33) to obtain the estimator of β.

Of course this is a wasteful way of obtaining the OLS estimator of β! However, the exercise is instructive in that it suggests a way of coping with the problem of "near collinearity."

If, as suggested above, we are confronted with a situation in which the last $n - k$ roots are very near zero and hence, numerically very unreliable, we may use in (i) above only the first k components.

From (33) we see that the elements of $\hat{\gamma}$ can be obtained one at a time, i.e., they are uncorrelated. Thus, for example,

$$\hat{\gamma}_i = \frac{1}{r_i} z'_{\cdot i} y, \qquad i = 1, 2, \ldots, n.$$

This is so since, using (29), we have

$$Z'Z = A'X'XA = R$$

Hence, using only the first k components means estimating

$$\hat{\gamma}_{(k)} = (Z'_k Z_k)^{-1} Z'_k y, \tag{35}$$

where

$$Z_k = X A_k, \qquad A = (A_k, A_*), \qquad \gamma = (\gamma'_{(k)}, \gamma'_*)'.$$

A_k is the submatrix of A containing the characteristic vectors corresponding to the k largest roots and γ has been partitioned conformably with A.

We then estimate β by

$$\hat{\beta} = A_k \hat{\gamma}_{(k)}.$$

What are the properties of this estimator? From (32) we have that

$$\beta = A_k \gamma_{(k)} + A_* \gamma_*.$$

Since

$$E(\hat{\gamma}_{(k)}) = \gamma_{(k)}, \qquad Cov(\hat{\gamma}_{(k)}) = \sigma^2 R_k^{-1},$$

where

$$R = \begin{bmatrix} R_k & 0 \\ 0 & R_* \end{bmatrix}, \qquad R_k = \operatorname{diag}(r_1, r_2, \ldots, r_k),$$

it follows that

$$E(\hat{\beta}) = A_k \gamma_{(k)}, \qquad Cov(\hat{\beta}) = \sigma^2 A_k R_k^{-1} A'_k, \tag{36}$$

and consequently, for the principal components estimator we have

$$MSE(\hat{\beta}_{PC}) = A_* \gamma_* \gamma'_* A'_* + \sigma^2 A_k R_k^{-1} A'_k. \tag{37}$$

The question then may naturally arise: when are we justified in choosing this biased estimator over the OLS estimator? For the latter we have

$$MSE(\hat{\beta}_{OLS}) = \sigma^2 (X'X)^{-1} = \sigma^2 (ARA')^{-1} = \sigma^2 [A_k R_k^{-1} A'_k + A_* R_*^{-1} A'_*]. \tag{38}$$

Thus

$$MSE(\hat{\beta}_{OLS}) - MSE(\hat{\beta}_{PC}) = \sigma^2 A_* R_*^{-1} A'_* - A_* \gamma_* \gamma'_* A'_*, \tag{39}$$

and the choice depends crucially on the tradeoff between the bias of the principal components estimator and the ill conditioning of $X'X$—or equivalently, on the magnitude of its smallest characteristic roots.

If we further simplify the criterion for choice, by considering the trace of the matrix in (39), we find

$$\psi = \sigma^2 \sum_{i=k+1}^{n} \left(\frac{1}{r_i}\right) - \gamma'_* \gamma_* = \sigma^2 \sum_{i=k+1}^{n} \left(\frac{1}{r_i}\right) - \beta'\beta + \gamma'_{(k)}\gamma_{(k)}. \qquad (40)$$

Remark 5. The import of the preceding discussion is as follows. If, in an empirical context, we are faced with a "nearly collinear" sample we may obtain $\hat{\beta}_{OLS}$, $\hat{\sigma}^2_{OLS}$ as well as the matrices A, R. We may be satisfied that the last $n - k$ roots are very inaccurately obtained and thus coping with "near collinearity" requires us to suppress the last $n - k$ components.

Proceeding now as in the discussion above we may obtain $\hat{\beta}_{PC}$, i.e., the estimator of β implied by the principal components approach, which uses only the first k components. Having done so we may check the criterion in (40) by computing

$$\hat{\sigma}^2_{OLS} \sum_{i=k+1}^{n} \left(\frac{1}{r_i}\right) - \hat{\beta}'\hat{\beta} + \hat{\gamma}'_{(k)}\hat{\gamma}_{(k)}$$

and determining whether it is positive.

Similarly, if we are in doubt as to whether we should or should not use the kth component we may compute

$$\frac{\hat{\sigma}^2_{OLS}}{r_k} - \hat{\gamma}^2_k.$$

If it is positive we conclude that the estimator of β based on k components is "inefficient" relative to the estimator of β based on $(k - 1)$ components— according to the trace of the MSE matrix criterion.

We stress that the procedure above is based on judgemental considerations and is not to be construed as a rigorous method for determining the number of principal components to be used in any given empirical context. In order for this to be a rigorous procedure *it is necessary for us to know the parameters* β and σ^2, in which case, of course, the problem would be obviated.

Dropping "Collinear Variables." The main features of this procedure have been discussed earlier in somewhat different contexts. Here we shall provide for completeness an outline. Suppose it is determined that $n - k$ of the explanatory variables are "highly collinear." This would mean that if, say, we partition

$$X = (X_k, X_*)$$

so that X_k contains the first k explanatory variables, then

$$X_* \approx X_k D$$

for some appropriate nonstochastic matrix D. We hasten to caution the reader that, in most cases, what variables we put in X_k (so long as there are k of them) and what variables we put in X_* (so long as there are $n - k$ of them) is completely arbitrary.

Comparing with the principal components approach we note that instead of using the first k components of X this procedure simply discards X_*, i.e., *it uses the first k variables*, and thus estimates, implicitly,

$$\hat{\beta} = \begin{pmatrix} (X_k' X_k)^{-1} X_k' y \\ 0 \end{pmatrix}.$$

We observe

$$\mathrm{E}(\hat{\beta}) = \begin{pmatrix} P_k \beta_* + \beta_{(k)} \\ 0 \end{pmatrix}, \qquad P_k = (X_k' X_k)^{-1} X_k' X_*, \tag{41}$$

$$\mathrm{Cov}(\hat{\beta}) = \sigma^2 \begin{bmatrix} (X_k' X_k)^{-1} & 0 \\ 0 & 0 \end{bmatrix},$$

where the partitioning is the natural one induced by our using only the variables in X_k. It follows that

$$\mathrm{MSE}(\hat{\beta}) = P \beta_* \beta_*' P' + \mathrm{Cov}(\hat{\beta}), \tag{42}$$

where

$$P = \begin{pmatrix} P_k \\ -I \end{pmatrix}$$

In order to compare the estimator above with the OLS estimator we need an expression for the MSE matrix of the latter that is partitioned conformably with the partition implicit in (41).

We recall from earlier discussion that

$$(X'X)^{-1} = \begin{bmatrix} (X_k' M_* X_k)^{-1} & -P_k (X_*' M_k X_*)^{-1} \\ -(X_*' M_k X_*)^{-1} P_k' & (X_*' M_k X_*)^{-1} \end{bmatrix},$$

where, evidently,

$$\begin{aligned} M_k &= I - X_k (X_k' X_k)^{-1} X_k', \\ M_* &= I - X_* (X_*' X_*)^{-1} X_*'. \end{aligned} \tag{43}$$

It may be established (see Problem 7) that

$$(X_k' M_* X_k)^{-1} = (X_k' X_k)^{-1} + P_k (X_*' M_k X_*)^{-1} P_k'. \tag{44}$$

Hence

$$\mathrm{MSE}(\hat{\beta}_{\mathrm{OLS}}) = \sigma^2 \begin{bmatrix} (X_k' X_k)^{-1} + P_k (X_*' M_k X_*)^{-1} P_k' & -P_k (X_*' M_k X_*)^{-1} \\ -(X_*' M_k X_*)^{-1} P_k' & (X_*' M_k X_*)^{-1} \end{bmatrix}. \tag{45}$$

Consequently, from (42) we find

$$\mathrm{MSE}(\hat{\beta}_{\mathrm{OLS}}) - \mathrm{MSE}(\hat{\beta}) = P[\sigma^2 (X_*' M_k X_*)^{-1} - \beta_* \beta_*'] P', \tag{46}$$

which is the multivariate generalization of the situation discussed in Remark 2.

Remark 6. What the preceding discussion establishes is the following: If, in the context of a GLM whose (centered) data matrix is X, the investigator feels that "near collinearity" is present, so that $n - k$ explanatory variables can be expressed, approximately, as a linear transformation of the remaining k explanatory variables, then dropping $n - k$ variables and thus implicitly "estimating" their coefficients to be zero (with zero covariance matrix) will produce an estimator which is efficient relative to the OLS estimator in the mean squared error sense if and only if

$$\sigma^2 (X'_* M_k X_*)^{-1} - \beta_* \beta'_*$$

is a positive semi definite matrix. In the expression above X_* is the matrix containing the observations on the variables that are dropped and β_* is the vector containing their true coefficients. The matrix $X'_* M_k X_*$ is recognized as the second moment matrix of the residuals of the regression of the suppressed variables, X_*, on the retained variables, X_k.

Remark 7. The result above is exactly analogous to that obtained earlier when we considered proxy variables and briefly entertained the notion that the missing variable be simply dropped from consideration. The matrix $X'_* M_k X_*$ is the multivariate generalization of s_{nn}, defined in Equation (13), which is simply the sum of squared residuals in the regression of $x_{.n}$ on the variables $x_{.1}, x_{.2}, \ldots, x_{.n-1}$.

We may summarize the discussion regarding the principal components and suppression of variables approaches to the problem of "near collinearity" in

Theorem 1. *Consider the GLM*

$$y = X\beta + u$$

and suppose that the data have been centered about their respective sample means. Suppose further that the data matrix is felt by the investigator to be "nearly collinear" so that only k of the n principal components need be used and, alternatively, in the suppression of variables option only k explanatory variables need be used. Accordingly, partition

$$X = (X_k, X_*)$$

so that X_* contains the suppressed variables.
Similarly, let A be the matrix of characteristic vectors of $X'X$ and partition

$$A = (A_k, A_*)$$

so that A_* corresponds to the $(n - k)$ suppressed principal components.
Let

$$\tilde{\beta} = \begin{pmatrix} (X'_k X_k)^{-1} X'_k y \\ 0 \end{pmatrix}$$

be the estimator of β implied by the suppression of variables option. Let

$$\hat{\beta}_{OLS} = (X'X)^{-1}X'y$$

be the OLS estimator and let

$$\hat{\beta}_{PC} = A_k(Z'_k z_k)^{-1}Z'_k y$$

be the estimator of β implied by the principal components option (that uses only the first k components). Then the following statements are true:

(i)
$$E(\hat{\beta}_{OLS}) = \beta, \qquad E(\hat{\beta}_{PC}) = A_k \gamma_{(k)}, \qquad E(\tilde{\beta}) = \begin{pmatrix} \beta_{(k)} + P_k \beta_* \\ 0 \end{pmatrix},$$

where $\beta_{(k)}$ is the subvector of β containing its first k elements (corresponding to the variables in X_k), $\gamma_{(k)}$ is the vector containing the first k elements of $\gamma = A'\beta$, i.e., $\gamma = (\gamma'_{(k)}, \gamma'_*)'$, and P_k is given by

$$P_k = (X'_k X_k)^{-1}X'_k X_*;$$

(ii) $$\text{bias}(\hat{\beta}_{OLS}) \equiv b(\hat{\beta}_{OLS}) = 0, \qquad b(\tilde{\beta}) = P\beta_*, \qquad b(\hat{\beta}_{PC}) = A_* \gamma_*,$$

where

$$P = \begin{pmatrix} P_k \\ -I \end{pmatrix};$$

(iii)
$$\text{Cov}(\hat{\beta}_{OLS}) = \sigma^2(X'X)^{-1}, \qquad \text{Cov}(\tilde{\beta}) = \sigma^2 \begin{bmatrix} (X'_k X_k)^{-1} & 0 \\ 0 & 0 \end{bmatrix},$$

$$\text{Cov}(\hat{\beta}_{PC}) = \sigma^2 A_k R_k^{-1} A'_k,$$

where R is the diagonal matrix of characteristic roots of $X'X$ with r_i, $i = 1, 2, \ldots, n$, such that $r_1 \geq r_2 \leq r_3 \geq \cdots \geq r_n$ and

$$R = \begin{bmatrix} R_k & 0 \\ 0 & R_* \end{bmatrix};$$

(iv) $$\text{MSE}(\hat{\beta}_{OLS}) - \text{MSE}(\tilde{\beta}) = P[\sigma^2(X'_* M_k X_*)^{-1} - \beta_* \beta'_*]P',$$

$$\text{MSE}(\hat{\beta}_{OLS}) - \text{MSE}(\hat{\beta}_{PC}) = A_*[\sigma^2 R_*^{-1} - \gamma_* \gamma'_*]A'_*,$$

where

$$\gamma_* = A'_* \beta.$$

Remark 8. The results in (iv) above provide criteria under which dropping variables or principal components may produce an estimator that dominates OLS in the mean squared error matrix sense.

The reader should note, however, that these criteria depend on the unknown parameters β and σ^2. Hence, they are not strictly speaking operational. Substituting the OLS estimates $\hat{\sigma}^2$ or $\hat{\beta}$ in (iv) above cannot be guaranteed, even in a large sample sense, to produce the same conclusions as stated in the theorem. Hence, these are best understood as judgmental criteria, to be employed by the investigator in the face of what he may consider to be a severe case of "near collinearity."

Despite the substantial insights afforded us by the results of the theorem in dealing with the problem of "near collinearity" and the option of dropping variables or dropping principal components we have so far no indication as to which of these two options may be preferable. This may be answered by considering directly

$$\text{MSE}(\tilde{\beta}) - \text{MSE}(\hat{\beta}_{\text{PC}}) = P\beta_* \beta'_* P' - A_* A'_* \beta \beta' A_* A'_* + \sigma^2 \begin{bmatrix} (X'_k X_k)^{-1} & 0 \\ 0 & 0 \end{bmatrix}$$
$$- \sigma^2 A_k R_k^{-1} A'_k.$$

It is difficult to say anything regarding the positive or negative semidefiniteness of the matrix in the equation above. However, if we alter our criterion to the trace, i.e., if we consider relative efficiency in the trace of the mean squared error matrix sense, we find

$$\text{tr}[\text{MSE}(\tilde{\beta}) - \text{MSE}(\hat{\beta}_{\text{PC}})] = \beta'_* P' P \beta_* - \beta' A_* A'_* \beta + \sigma^2 \sum_{i=1}^{k} \left(\frac{1}{r_i^*} - \frac{1}{r_i} \right), \quad (47)$$

where $r_1^* \geq r_2^* \geq \cdots \geq r_k^*$ are the characteristic roots of $X'_k X_k$. It may be shown[3] that

$$r_i \geq r_i^*, \quad i = 1, 2, \ldots, k, \quad (48)$$

and thus we conclude that

$$\sigma^2 \sum_{i=1}^{k} \left(\frac{1}{r_i^*} - \frac{1}{r_i} \right) \geq 0.$$

Indeed, in the typical case the strict inequality will hold. Unfortunately, it does not appear possible to say very much about the difference of the two quadratic forms

$$\beta'_* P' P \beta_* - \beta' A_* A'_* \beta.$$

It remains only a presumption that between these two options it will more often be the case that principal components will dominate.

Remark 9. Although it does not appear possible in a formal way to establish the desirability of dropping (an equivalent number of) principal components instead of variables, there is a practical sense in which the principal components version is more desirable. Frequently, some variables are collinear because of certain policy restrictions. If a policy reversal is contemplated and it is desired to forecast its implications, it would be preferable to do so in the context of the principal components option. At least in this

---

[3] Demonstrating this fact is clearly beyond the scope of this book. The interested reader, however, may find a discussion of this aspect in Bellman [4, pp. 114–115].

instance we would have an "estimate" for the coefficients of all relevant variables. In the dropping of variables option we would, generally, have an estimate for only a subset of such coefficients. Hence significant aspects of such implications will be obscured or escape our notice entirely.

Ridge Regression. In the preceding chapter when we discussed ridge regression in the context of exact multicollinearity we paired it with Bayesian methods; by contrast, we paired the principal components and generalized inverse approaches.

In the context of "near collinearity" the generalized inverse solution is evidently exactly the same as OLS, since the matrix $X'X$ is, in principle, invertible. The Bayesian solution exhibits no special features when we deal with "near" as distinct from "exact" collinearity. Hence, it need not be discussed here.

Consequently, in our discussion of the ridge regression (RR) option we will dwell on its similarities with the principal components (PC) and OLS estimators. We recall that the generalized ridge regression (GRR) estimator is obtained as

$$\hat{\beta}_{GRR} = A(R + K)^{-1}A'X'y, \tag{49}$$

where

$$K = \text{diag}(k_1, k_2, \ldots, k_n)$$

contains n quantities to be determined, while A and R are, respectively, the matrices of characteristic vectors and roots of $X'X$.

The standard RR estimator is the special case where

$$K = kI, \tag{50}$$

so that

$$\hat{\beta}_{RR} = A(R + kI)^{-1}A'X'y. \tag{51}$$

In both (49) and (51) it is assumed that

$$k_i > 0, \qquad k > 0, \qquad i = 1, 2, \ldots, n.$$

In order to see the basic similarity between RR, GRR, and PC estimation it is convenient to introduce the diagonal matrix

$$D = \text{diag}(d_1, d_2, \ldots, d_n)$$

and to write the estimators in (49) and (51) more revealingly as

$$\hat{\beta}_D = ADR^{-1}D'A'X'y. \tag{52}$$

A little reflection will show that the OLS estimator is the special case with

$$D = I. \tag{53}$$

The PC estimator (using only k components) is the special case

$$D = \begin{bmatrix} I_k & 0 \\ 0 & 0 \end{bmatrix}. \tag{54}$$

The simple RR estimator is the case with

$$D = (R + kI)^{-1/2}R^{1/2}, \tag{55}$$

where, e.g., the expression R^{α} means the diagonal matrix whose ith diagonal element is

$$r_i^{\alpha}, \qquad i = 1, 2, \ldots, n,$$

and α is a suitable scalar.

Finally, the GRR estimator is given by the choice

$$D = (R + K)^{-1/2}R^{1/2}. \tag{56}$$

Thus, given the extensive discussion devoted to the PC option it does not appear advisable to dwell on the merits of RR or GRR as a means of coping with "near collinearity."

It would appear that what we have here is a situation of attaching varying weights to the characteristic roots of $X'X$ in forming the estimator of the parameter vector β. In the OLS case we attach the same weight (unity) to all roots—even those that appear to be obtained with substantial computational error. In PC we attach unitary weights to the "large" roots (i.e., the first k roots) while we attach zero weights to the "small" ones, i.e., the last $n - k$ roots.

In the PR or GRR version we attach weights which decline as the magnitude of the roots declines. Thus the weight attached to the ith root is

$$\left(\frac{r_i}{r_i + k_i} \right)^{1/2}, \qquad i = 1, 2, \ldots, n,$$

in the case of GRR, and

$$\left(\frac{r_i}{r_i + k} \right)^{1/2}, \qquad i = 1, 2, \ldots, n,$$

in the case of the standard RR.

As a practical matter one would not want to employ RR or GRR until substantial new results on the properties of these estimators become available. The reason for this cautionary approach is that since in ridge regression we are afforded considerably more latitude in dealing with the data than in the case of dropping variables or dropping principal components, the temptation would be to manipulate our choice of the k_i until our preconceptions regarding the parameters are embodied in the estimates. With considerably more degrees of freedom it will be almost certain that with any given sample we shall be able to do so. Such procedures, however, would be self-defeating in the long run since it is not clear that we would have learned very much from the sample observations regarding the elements of the vector β, although we would have the illusion that we had.

The reader interested in RR or GRR as a data analytic technique may consult the works of Hoerl and Kennard [21], [22]; Vinod [33]; Marquardt [24]; as well as the Monte Carlo studies of McDonald and Galarneau [25], and Newhouse and Oman [26].

3 Errors in Variables (EIV): Bivariate Model

3.1 Inconsistency of the OLS Estimator

The formal aspects of this problem may be described, in the simplest possible case, as follows. Suppose it is known that two variables of interest, Y, X, are connected by the relation

$$Y = \alpha_0 + \alpha_1 X + u, \tag{57}$$

where u is a structural error (random variable). Unfortunately, Y and X cannot be observed; what *is* observed (and observable) at time t is

$$x_t = X_t + u_{t1}, \qquad y_t = Y_t + u_{t2}, \qquad t = 1, 2, \ldots \quad . \tag{58}$$

It is clear that (57) and (58) can be combined to yield

$$y_t = \alpha_0 + \alpha_1 x_t + v_t, \qquad v_t = s_{t\cdot}.(1, -\alpha_1, 1)', \qquad s_{t\cdot} = (u_t, u_{t1}, u_{t2}). \tag{59}$$

The feature of (59) that distinguishes it from the standard general linear model is that, even if we assume

$$\{s'_{t\cdot} : t = 1, 2, \ldots\}$$

is a sequence of i.i.d. random variables such that

$$E(s'_{t\cdot}) = 0, \qquad \mathrm{Cov}(s'_{t\cdot}) = \mathrm{diag}(\sigma_{00}, \sigma_{11}, \sigma_{22}),$$

we still have

$$\mathrm{Cov}(x_t, v_t) = -\alpha_1 \sigma_{11} \neq 0, \qquad E(v_t | x_t) = -\alpha_1 u_{t1} \neq 0,$$

so that the explanatory variable and the error term are *correlated*. From a certain point of view, the situation described above is a special case of the misspecification problem dealt with in earlier sections. To see this, consider the special case where

$$u_{t2} \equiv 0.$$

In the specification error context we use instead of X a variable X^* *without precisely stating what is the connection between them.* In the EIV context the connection between X, X^* is specified (partially) as in (58). It is customary, and often quite plausible in the context, additionally to require in (58) that u_{t2} and X_t be mutually independent and that u_{t1} and Y_t also be mutually independent. Thus, in the EIV model we know, typically, from the specification, that

$$(\text{p})\lim_{T \to \infty} \frac{1}{T} \sum_{t=1}^{T} X_t^* X_t = (\text{p})\lim_{T \to \infty} \frac{1}{T} \sum_{t=1}^{T} X_t^2 > 0,$$

assuming the limits to be well defined. This is the sense in which the EIV model is a special case of the situation considered when examining the misspecification problem.

We now return to (59) and consider two issues. First, if using the observables $(y_t, x_t), t = 1, 2, \ldots, T$, we carry out a regression, what are the properties of the resulting regression coefficients or estimators of α_0 and α_1? Second, are there procedures that yield "better" estimators than the regression procedure (OLS), and if so what (if any) additional information is needed to implement them?

To answer the first question, centering observations about the corresponding sample means, we find

$$\tilde{\alpha}_1 = \frac{\sum (x_t - \bar{x})(y_t - \bar{y})}{\sum (x_t - \bar{x})^2}, \qquad \tilde{\alpha}_0 = \bar{y} - \tilde{\alpha}_1 \bar{x},$$

$$\bar{x} = \frac{1}{T} \sum x_t, \qquad \bar{y} = \frac{1}{T} \sum y_t,$$

the summation being over $t = 1, 2, \ldots, T$.

But we have

$$y_t - \bar{y} = \alpha_1(x_t - \bar{x}) + v_t - \bar{v}, \qquad x_t - \bar{x} = (X_t - \bar{X}) + u_{t1} - \bar{u}_1.$$

After substitution in (60) we find

$$\tilde{\alpha}_1 = \alpha_1 + \frac{\sum (x_t - \bar{x})(v_t - \bar{v})}{\sum (x_t - \bar{x})^2}. \tag{61}$$

Without a more precise specification of the distribution function of u_t, u_{t1}, and u_{t2}, it is not possible to evaluate the expectation of $\tilde{\alpha}_1$. In some cases of interest the expectation may even fail to exist. It is, however, rather simple to evaluate the probability limit in (61). To this effect we require the additonal assumption

$$(\text{p})\lim_{T \to \infty} \frac{1}{T} X' \left(I - \frac{ee'}{T} \right) X = \sigma_x^2, \qquad X = (X_1, X_2, \ldots, X_T)',$$

exists as a positive quantity, where e is a column vector all of whose (T) elements are unity. Now

$$\text{plim}_{T \to \infty} \frac{1}{T} \sum_{t=1}^{T} (x_t - \bar{x})^2 = \sigma_x^2 + \sigma_{11}.$$

Moreover,

$$\sum (x_t - \bar{x})(v_t - \bar{v}) = \sum (X_t - \bar{X})(v_t - \bar{v})$$
$$+ \sum (u_{t1} - \bar{u}_1)[u_t - \bar{u} - \alpha_1(u_{t1} - \bar{u}_1) + u_{t2} - \bar{u}_2].$$

Dividing by T and taking probability limits above yields $-\alpha_1\sigma_{11}$. Thus,

$$\underset{T \to \infty}{\text{plim}} \, \tilde{\alpha}_1 = \alpha_1 - \frac{\alpha_1\sigma_{11}}{\sigma_x^2 + \sigma_{11}} = \alpha_1\left(\frac{\sigma_x^2}{\sigma_x^2 + \sigma_{11}}\right) < \alpha_1,$$

and we see that, provided σ_x^2 and σ_{11} are *positive*, $\tilde{\alpha}_1$ "underestimates" α_1, in the sense that its probability limit is *always* less than α_1.

In general, the larger is σ_x^2 relative to σ_{11} the "better" is α_1 estimated by $\tilde{\alpha}_1$. Intuitively, what this means is that if x_t is "dominated" by its "systematic component" X_t, the error committed by using the former instead of the latter in the regression is small, and conversely.

The discussion above is summarized in

Theorem 2. *Consider the model*

$$Y_t = \alpha_0 + \alpha_1 X_t + u_t$$

and suppose observations are available only on

$$x_t = X_t + u_{t1}, \qquad y_t = Y_t + u_{t2}.$$

If one considers

$$\tilde{\alpha}_1 = \frac{\sum (x_t - \bar{x})(y_t - \bar{y})}{\sum (x_t - \bar{x})^2}$$

as an estimator of α_1 then, provided

$$\underset{T \to \infty}{(\text{p})\lim} \, \frac{1}{T} X'\left(I - \frac{ee'}{T}\right)X = \sigma_x^2 > 0,$$

$\tilde{\alpha}_1$ is an inconsistent estimator of α_1; more precisely

$$\bar{\alpha}_1 = \underset{T \to \infty}{\text{plim}} \, \tilde{\alpha}_1 = \alpha_1 = \alpha_1\left(\frac{\sigma_x^2}{\sigma_x^2 + \sigma_{11}}\right) < \alpha_1.$$

The expectation of $\tilde{\alpha}_1$ cannot be determined without further specification.

Remark 10. Note that

$$\frac{\sigma_x^2}{\sigma_x^2 + \sigma_{11}} = \frac{1}{1 + (\sigma_{11}/\sigma_x^2)}$$

and, thus, what matters regarding the properties of $\tilde{\alpha}_1$ is the behavior of the ratio $\sigma_{11}/\sigma_x^2 = \lambda$. We note that

$$\lim_{\lambda \to 0} \tilde{\alpha}_1 = \alpha_1$$

while

$$\lim_{\lambda \to \infty} \tilde{\alpha}_1 = 0$$

3.2 Wald and ML Estimators

It is apparent from the preceding that given only the information (y_t, x_t), the parameters α_0 and α_1 cannot be estimated consistently by OLS methods. A rather simple procedure was proposed by Wald [35] that would yield consistent estimators provided some additional information were available. Wald's method rests on the simple observation that in order to determine a line we only require two distinct points lying on it. The line in question is

$$E(Y_t|X_t) = \alpha_0 + \alpha_1 X_t.$$

If we could determine two points lying on this line we could certainly determine α_0 and α_1. *Unfortunately, we only observe y_t and x_t, which need not lie on that line.* Now suppose we *could order observations*

$$(y_1, x_1), (y_2, x_2), \ldots, (y_T, x_T)$$

corresponding to the X component of x. Thus, to x_t in the order above corresponds X_t, where $X_1 \le X_2 \le X_3 \le \cdots \le X_T$. In the following it involves no loss of relevance to suppose T is even and to compute

$$\bar{y}^{(1)} = \frac{1}{T_1}\sum_{t=1}^{T_1} y_t, \quad \bar{y}^{(2)} = \frac{1}{T_2}\sum_{t=T_1+1}^{T} y_t, \quad \bar{x}^{(1)} = \frac{1}{T_1}\sum_{t=1}^{T_1} x_t,$$

$$\bar{x}^{(2)} = \frac{1}{T_2}\sum_{t=T_1+1}^{T} x_t, \quad T_2 = T - T_1, \quad T_1 = \frac{T}{2}.$$

We observe that

$$\operatorname*{plim}_{T \to \infty}[\bar{y}^{(2)} - \bar{y}^{(1)}] = \alpha_1 \operatorname*{(p)lim}_{T \to \infty}[\bar{X}^{(2)} - \bar{X}^{(1)}],$$

$$\operatorname*{plim}_{T \to \infty}[\bar{x}^{(2)} - \bar{x}^{(1)}] = \operatorname*{(p)lim}_{T \to \infty}[\bar{X}^{(2)} - \bar{X}^{(1)}],$$

(62)

where $\bar{X}^{(2)}, \bar{X}^{(1)}$ correspond in the obvious way to $\bar{x}^{(2)}, \bar{x}^{(1)}$. In order for (62) to be used in subsequent arguments we must have that

$$\operatorname*{(p)lim}_{T \to \infty} \frac{1}{T}\sum_{t=1}^{T} X_t$$

exists as a well-defined finite quantity. If this condition is satisfied, it is clear from (62) that the points $\bar{y}^{(1)}$, $\bar{y}^{(2)}$, $\bar{x}^{(1)}$, $\bar{x}^{(2)}$ converge in probability to points lying on the line

$$E(Y_t \mid X_t) = \alpha_0 + \alpha_1 X_t$$

and, consequently, they could be used to determine this line. In particular, putting

$$\tilde{\alpha}_1 = \frac{\bar{y}^{(2)} - \bar{y}^{(1)}}{\bar{x}^{(2)} - \bar{x}^{(1)}}, \qquad \tilde{\alpha}_0 = \bar{y} - \tilde{\alpha}_1 \bar{x},$$

where $\bar{y}$ and $\bar{x}$ are, respectively, the sample means of y and x, we easily conclude that

$$\plim_{T \to \infty} \tilde{\alpha}_1 = \alpha_1, \qquad \plim_{T \to \infty} \tilde{\alpha}_0 = \alpha_0.$$

Remark 11. Notice that here it is the additional information regarding the order of the x_t according to the size of the X_t that permits consistent estimation of the unknown parameters. In general, of course, such information will not be available since the problem is that the $\{X_t : t = 1, 2, \ldots\}$ are *not observed or observable.* Thus, the practical significance of this solution is rather limited.

Remark 12. Notice, also, that the Wald estimator is an instrumental variables estimator. Thus, let e_1 be a T-element row vector whose elements are either 0 or $1/T_1$; similarly, let e_2 be a T-element row vector whose elements are either zero or $1/T_2$ ($T_1 = T_2 = T/2$). Suppose information on the ranking of X_t is available, so that

$$(e_2 - e_1)'y = \bar{y}^{(2)} - \bar{y}^{(1)}, \qquad (e_2 - e_1)'x = \bar{x}^{(2)} - \bar{x}^{(1)},$$

$$y = (y_1, y_2, \ldots, y_T)', \qquad x = (x_1, x_2, \ldots, x_T)'.$$

It is clear, then, that $e_2 - e_1$ is the instrumental variable that yields Wald's estimator. Notice that $e_2 - e_1$ satisfies the requirements for an instrumental variable, i.e.,

$$\plim_{T \to \infty}(e_2 - e_1)'x \neq 0, \qquad \plim_{T \to \infty}(e_2 - e_1)'v = 0, \qquad v = (v_1, v_2, \ldots, v_T)'.$$

The consistency of the Wald estimator would then follow from the general proposition regarding the consistency of all instrumental variables estimators.

Let us now see how the EIV problem will manifest itself in a maximum likelihood context. Thus consider again the model[4]

$$y_t = \alpha + \beta X_t + u_t, \qquad x_t = X_t + u_{t1},$$

---

[4] Notice that here we are implicitly assuming that the dependent variable is not observed with error.

and assume in addition that

$$\{(u_t, u_{t1}): t = 1, 2, \ldots\}$$

is a sequence of i.i.d. random vectors with mean zero and nonsingular covariance matrix Σ. We still assume that X_t and (u_t, u_{t1}) are mutually independent and that

$$(\text{p})\lim_{T \to \infty} \frac{1}{T} X'\left(I - \frac{ee'}{T}\right)X = \sigma_x^2 > 0$$

exists as a well-defined nonstochastic quantity. Now, the joint (log) likelihood function of the observations is given by

$$L = -T \ln(2\pi) - \frac{T}{2} \ln |\Sigma|$$

$$-\frac{1}{2}\sum_{t=1}^{T} (y_t - \alpha - \beta X_t, x_t - X_t)\Sigma^{-1}\begin{pmatrix} y_t - \alpha - \beta X_t \\ x_t - X_t \end{pmatrix}. \tag{63}$$

We have to maximize (63) with respect to $X_t, t = 1, 2, \ldots, T, \alpha, \beta, \sigma_{11}, \sigma_{12}, \sigma_{22}$. To accomplish this we employ stepwise maximization, i.e., we first maximize with respect to X_t and substitute in (63); we maximize then with respect to α, then β, and so on.[5]

Differentiating (63) with respect to X_t, and setting the result equal to zero, yields

$$X_t = \frac{(\beta, 1)\Sigma^{-1}(y_t - \alpha, x_t)'}{c}, \qquad c = (\beta, 1)\Sigma^{-1}(\beta, 1)'.$$

Consequently, we have

$$y_t - \alpha - \beta X_t = \left(\frac{1}{c}\right)(\beta, 1)\Sigma^{-1}(0, y_t - \alpha - \beta x_t)',$$

$$x_t - X_t = \left(-\frac{1}{c}\right)(\beta, 1)\Sigma^{-1}(y_t - \alpha - \beta x_t, 0)'.$$

Noting that

$$|\Sigma|c = (\beta, 1)|\Sigma|\Sigma^{-1}(\beta, 1) = (\beta, 1)\begin{bmatrix} \sigma_{22} & -\sigma_{12} \\ -\sigma_{21} & \sigma_{11} \end{bmatrix}(\beta, 1) = (1, -\beta)\Sigma(1, -\beta)',$$

[5] It can be shown that this is equivalent to maximizing simultaneously with respect to all parameters.

we have, substituting in the last term of (63),

$$
-\frac{1}{2}\frac{(\beta, 1)\Sigma^{-1}\begin{pmatrix} \sigma^{22} & -\sigma^{12} \\ -\sigma^{21} & \sigma^{11} \end{pmatrix}\Sigma^{-1}(\beta, 1)'}{c^2}\sum_{t=1}^{T}(y_t - \alpha - \beta x_t)^2
$$

$$
= -\frac{1}{2}\frac{(\beta, 1)\Sigma^{-1}(\beta, 1)'}{|\Sigma|c^2}\sum_{t=1}^{T}(y_t - \alpha - \beta x_t)^2
$$

$$
= -\frac{1}{2}\frac{\sum_{t=1}^{T}(y_t - \alpha - \beta x_t)^2}{(1, -\beta)\Sigma(1, -\beta)'}. \tag{64}
$$

Maximizing (64) with respect to α we obtain

$$
\tilde{\alpha} = \bar{y} - \beta\bar{x}, \qquad \bar{y} = \frac{1}{T}e'y, \qquad \bar{x} = \frac{1}{T}e'x,
$$

and consequently the concentrated likelihood function, expressed now solely in terms of β and Σ, becomes

$$
L^* = -T\ln(2\pi) - \frac{T}{2}\ln|\Sigma| - \frac{T}{2}\frac{(1, -\beta)A(1, -\beta)'}{(1, -\beta)\Sigma(1, -\beta)'} \tag{65}
$$

where

$$
A = \frac{1}{T}(y, x)'[I - e(e'e)e'](y, x),
$$

e being a T-element column vector all of whose elements are unity. Since the matrix in brackets above is a symmetric idempotent matrix, it is clear that A is at least positive semidefinite. Moreover, Σ is positive definite. Thus, (see *Mathematics for Econometrics*) there exists a nonsingular matrix W such that

$$
\Sigma = W'W, \qquad A = W'\Lambda W,
$$

where Λ is a diagonal matrix, the diagonal elements of which are the solutions to

$$
|\lambda\Sigma - A| = 0. \tag{66}
$$

Putting $\xi = W(1, -\beta)'$, we can write the last term in (65) as

$$
-\frac{T}{2}\frac{(1, -\beta)A(1, -\beta)'}{(1, -\beta)\Sigma(1, -\beta)'} = -\frac{T}{2}\frac{\xi'\Lambda\xi}{\xi'\xi}.
$$

Since

$$
\frac{\xi'\Lambda\xi}{\xi'\xi} = \sum_{i=1}^{r}\lambda_i\frac{\xi_i^2}{\xi'\xi}, \qquad r = 2,
$$

it is clear that

$$\min_i \lambda_i \le \frac{\xi'\Lambda\xi}{\xi'\xi} \le \max_i \lambda_i.$$

Consequently, (65) will be maximized with respect to $(1, -\beta)$ if we choose $\zeta = (1, -\tilde{\beta})'$ as the characteristic vector (of A in the metric of Σ) corresponding to the smallest characteristic root, say $\hat{\lambda}$, in (66). We have, then,

$$A\zeta = \hat{\lambda}\Sigma\zeta.$$

Consequently,

$$\frac{\zeta'A\zeta}{\zeta'\Sigma\zeta} = \hat{\lambda}$$

and ζ thus maximizes the last term of (65). Hence,

$$\zeta' = (1, -\tilde{\beta})$$

is the desired estimator. Unfortunately, however, ζ *cannot be computed unless* Σ *is known, at least up to a scalar multiple.* Thus, in this general case, we cannot obtain maximum likelihood estimators of the unknown parameters. Indeed, suppose we tried *first* to maximize (65) with respect to the elements of Σ, and subsequently maximize with respect to β. We would obtain

$$\frac{\partial L^*}{\partial \Sigma} = -\frac{T}{2} \Sigma^{-1} + \frac{T}{2} \frac{c_1}{c_2} (1, -\beta)'(1, -\beta) = 0,$$

where

$$c_1 = (1, -\beta)A(1, -\beta)', \qquad c_2 = [(1, -\beta)\Sigma(1, -\beta)]^2,$$

which implies that the "estimator" of Σ^{-1}, whatever it might be, is always a *singular* matrix; but this is contrary to the assumption underlying the estimation procedure. Thus, we conclude that *in this general case maximum likelihood estimators for the parameters of the model cannot be obtained.*

Suppose, however, that Σ is known up to a scalar multiple, i.e., suppose

$$\Sigma = \sigma^2 \Sigma_0$$

with Σ_0 *known.* Then the concentrated likelihood function of (65) can be written as

$$L^* = -T \ln(2\pi) - \frac{T}{2} \ln |\Sigma_0| - T \ln \sigma^2 - \frac{T}{2\sigma^2} \frac{(1, -\beta)A(1, -\beta)'}{(1, -\beta)\Sigma_0(1, -\beta)'}. \tag{67}$$

Maximizing (67) with respect to σ^2 we find

$$\hat{\sigma}^2 = \frac{1}{2} \frac{(1, -\beta)A(1, -\beta)'}{(1, -\beta)\Sigma_0(1, -\beta)'}$$

Inserting in (67) we have

$$L^{**} = -T[\ln(2\pi) + 1] - \frac{T}{2} \ln|\Sigma_0| + T \ln 2 - T \ln\left[\frac{(1, -\beta)A(1, -\beta)'}{(1, -\beta)\Sigma_0(1, -\beta)'}\right],$$

and maximizing L^{**} with respect to β is equivalent to *minimizing*

$$\frac{(1, -\beta)A(1, -\beta)'}{(1, -\beta)\Sigma_0(1, -\beta)'}$$

with respect to β. But this is the problem we have already solved and the solution determined was to choose $(1, -\tilde{\beta})$ as the characteristic vector corresponding to the smallest characteristic root of

$$|\lambda\Sigma_0 - A| = 0.$$

Since Σ_0 is now a *known* matrix there is no problem in determining the smallest characteristic root of the equation above and its associated characteristic vector. Characteristic vectors, of course, are unique only up to a scalar multiple, but in the present case uniqueness of $\tilde{\beta}$ is ensured by the fact that *in the characteristic vector the first element is constrained to be unity.*

Remark 13. In the tradition of the econometrics literature it is assumed that $\sigma_{12} = 0$; thus Σ_0 is a diagonal matrix and the assumptions under which we operate require us to know the *ratio*, say, σ_{22}/σ_{11}. While this is a somewhat less restrictive assumption than the one required to implement Wald's estimator, still there is no doubt that the requirement that we know the ratio of the variances σ_{22}/σ_{11} is quite a restrictive one.

4 Errors in Variables (EIV): General Model

4.1 Derivation of the Estimator

In this section we shall examine the general model in which some of the explanatory variables are observed with error while others are observed without error. Since whether the dependent variable is observed with or without error is irrelevant, we shall not take this aspect into consideration. Thus, we shall be dealing with the model

$$y_t = x_{t.}\alpha + w_{t.}\beta + u_{t0} = z_{t.}\delta + u_{t0},$$
$$w_{t.}^* = w_{t.} + u_{t.}^*,$$
(68)

where, evidently,

$$z_{t.} = (x_{t.}, w_{t.}), \qquad \delta = (\alpha', \beta')'.$$

$w_{t.}, x_{t.}$ are, respectively, r- and s-element row vectors containing the variables observed with and without error at time t. The second equation in (68) indicates that the observations available on $w_{t\cdot}$, i.e., $w_{t\cdot}^*$, are related to the variable of interest through an additive random component. As before we assume that $w_{t\cdot}$ and u_t^* are mutually independent. Define the vector

$$u_{t\cdot} = (u_{t0}, u_{t\cdot}^*)$$

and assume that $\{u_{t\cdot}': t = 1, 2, \ldots\}$ is a sequence of i.i.d. random variables with

$$E(u_{t\cdot}'|X) = 0, \qquad \mathrm{Cov}(u_{t\cdot}'|X) = \Sigma.$$

We partition

$$\Sigma = \begin{bmatrix} \sigma_{00} & \Sigma_{12} \\ \Sigma_{21} & \Sigma_{22} \end{bmatrix}, \qquad \Sigma^{-1} = \begin{bmatrix} \sigma^{11} & \Sigma^{12} \\ \Sigma^{21} & \Sigma^{22} \end{bmatrix}$$

conformably with the composition of the vector $u_{t\cdot}$.

The problem is to estimate α, β by maximum likelihood methods. The joint density of the sample in terms of the u's is

$$(2\pi)^{-T(r+1)/2} |\Sigma|^{-T/2} \exp\left(-\frac{1}{2} \sum_{t=1}^{T} u_{t\cdot} \Sigma^{-1} u_{t\cdot}'\right).$$

Since the Jacobian of the transformation from $u_{t\cdot}$ to $(y_t, w_{t\cdot}^*)$ is unity, we have that the (log) likelihood function in terms of $(y_t, w_{t\cdot}^*)$ is

$$L = -\frac{T(r+1)}{2} \ln(2\pi) - \frac{T}{2} \ln|\Sigma|$$

$$-\frac{1}{2} \sum_{t=1}^{T} (y_t - x_{t\cdot}\alpha - w_{t\cdot}\beta, w_{t\cdot}^* - w_{t\cdot})\Sigma^{-1}(y_t - x_{t\cdot}\alpha - w_{t\cdot}\beta, w_{t\cdot}^* - w_{t\cdot})'.$$

$$(69)$$

Differentiating (69) with respect to $w_{t\cdot}$ and setting the result equal to zero, we find

$$\sigma^{11}(y_t - x_{t\cdot}\alpha)\beta' + w_{t\cdot}^*\Sigma^{21}\beta' + (y_t - x_{t\cdot}\alpha)\Sigma^{12} + w_{t\cdot}^*\Sigma^{22}$$

$$= w_{t\cdot}(\sigma^{11}\beta\beta' + \Sigma^{21}\beta' + \beta\Sigma^{12} + \Sigma^{22}).$$

We note that

$$\sigma^{11}\beta\beta' + \Sigma^{21}\beta' + \beta\Sigma^{12} + \Sigma^{22} = (\beta, I)\Sigma^{-1}(\beta, I)'$$

and

$$[\sigma^{11}(y_t - x_{t\cdot}\alpha) + w_{t\cdot}^*\Sigma^{21}]\beta' + (y_t - x_{t\cdot}\alpha)\Sigma^{12} + w_{t\cdot}^*\Sigma^{22}$$

$$= (y_t - x_{t\cdot}\alpha, w_{t\cdot}^*)\Sigma^{-1}(\beta, I)'.$$

Consequently we find

$$w_{t\cdot} = (y_t - x_{t\cdot}\alpha, w_{t\cdot}^*)\Sigma^{-1}(\beta, I)'[(\beta, I)\Sigma^{-1}(\beta, I)']^{-1},$$

$$w_{t\cdot}^* - w_{t\cdot} = -(y_t - x_{t\cdot}\alpha - w_{t\cdot}^*\beta, 0)\Sigma^{-1}(\beta, I)'[(\beta, I)\Sigma^{-1}(\beta, I)']^{-1}.$$

In order to proceed in the same fashion as we did for the bivariate model, it is necessary to substitute for w_t. in $y_t - x_t.\alpha - w_t.\beta$. It may easily be verified that when this is done we arrive at some very complicated expression so that nothing appears very transparent. The problem is, essentially, that no useful representation for the inverse of $(\beta, I)\Sigma^{-1}(\beta, I)'$ exists. *Partly for this reason, let us impose, at this stage, the condition, common in the tradition of the econometrics literature, that the errors of observation are independent of (minimally uncorrelated with) the structural error.* This means that Σ is of the form

$$\Sigma = \begin{bmatrix} \sigma_{00} & 0 \\ 0 & \Sigma_{22} \end{bmatrix}, \qquad \sigma_{00} = E(u_{t0}^2), \qquad \Sigma_{22} = \text{Cov}(u_{t.}^{*\prime}). \tag{70}$$

We see, then, that

$$(\beta, I)\Sigma^{-1}(\beta, I)' = \frac{\beta\beta'}{\sigma_{00}} + \Sigma_{22}^{-1}.$$

It is easily shown (see Proposition 33 of *Mathematics for Econometrics*) that

$$[(\beta, I)\Sigma^{-1}(\beta, I)']^{-1} = \Sigma_{22} - \frac{\Sigma_{22}\beta\beta'\Sigma_{22}}{\sigma_{00} + \beta'\Sigma_{22}\beta}$$

and, moreover,

$$\Sigma^{-1}(\beta, I)'[(\beta, I)\Sigma^{-1}(\beta, I)']^{-1} = \begin{pmatrix} \dfrac{\beta'\Sigma_{22}}{\mu} \\ I - \dfrac{\beta\beta'\Sigma_{22}}{\mu} \end{pmatrix},$$

$$\mu = \sigma_{00} + \beta'\Sigma_{22}\beta.$$

Thus,

$$w_t. = \left(\frac{1}{\mu}\right)(y_t - x_t.\alpha - w_{t.}^*\beta)\beta'\Sigma_{22} + w_{t.}^*,$$

and consequently

$$w_t.\beta = \frac{\beta'\Sigma_{22}\beta}{\mu}(y_t - x_t.\alpha - w_{t.}^*\beta) + w_{t.}^*\beta.$$

Moreover,

$$w_{t.}^* - w_t. = -\left(\frac{1}{\mu}\right)(y_t - x_t.\alpha - w_{t.}^*\beta)\beta'\Sigma_{22},$$

$$y_t - x_t.\alpha - w_t.\beta = \left(\frac{\sigma_{00}}{\mu}\right)(y_t - x_t.\alpha - w_{t.}^*\beta),$$

and

$$(y_t - x_t.\alpha - w_t.\beta, w_{t.}^* - w_t.) = \left(\frac{y_t - x_t.\alpha - w_{t.}^*\beta}{\mu}\right)(\sigma_{00}, -\beta'\Sigma_{22}). \tag{71}$$

Taking into account the restrictions in (70) and substituting (71) in (69) we obtain the concentrated likelihood

$$L^* = -\frac{T(r+1)}{2}\ln(2\pi) - \frac{T}{2}\ln|\Sigma_{22}| - \frac{T}{2}\ln\sigma_{00} - \frac{1}{2\mu}\sum_{t=1}^{T}(y_t - x_{t.}\alpha - w_{t.}^*\beta)^2.$$

(72)

Maximizing (72) with respect to α we obtain

$$\hat{\alpha} = (X'X)^{-1}X'(y - W^*\beta),$$

(73)

where X is the $T \times s$ matrix whose tth row is $x_{t.}$ and W^* the $T \times r$ matrix whose tth row is $w_{t.}^*$. Upon substitution of (73) in (72) we note that the last term there becomes proportional to

$$-\frac{\gamma'A\gamma}{\gamma'\Sigma\gamma},$$

where

$$A = \frac{1}{T}(y, W^*)'(I - X(X'X)^{-1}X')(y, W^*), \qquad \gamma = (1, -\beta')'.$$

(74)

Formally, this is exactly the problem we had encountered in the bivariate model. As in that case it is also apparent here that maximum likelihood estimators cannot be obtained unless more information is available regarding the matrix Σ. This becomes obvious if we proceed to maximize (72) with respect to the elements of Σ_{22}. Differentiation yields

$$-\frac{T}{2}\Sigma_{22}^{-1} + \frac{T}{2}\frac{\beta\beta'}{\mu^2}\frac{1}{T}\sum_{t=1}^{T}(y_t - x_{t.}\alpha - w_{t.}^*\beta)^2 = 0,$$

which implies that the estimator of Σ_{22} is *not* defined. *Since $\beta\beta'$ is of rank one, we are again forced to conclude that unless we know Σ_{22} up to a scalar multiple we cannot proceed.* To this effect, suppose that

$$\Sigma = \sigma_{00}\begin{bmatrix} 1 & 0 \\ 0 & \Sigma_0 \end{bmatrix} = \sigma_{00}\Sigma_0^*, \qquad \Sigma_0^* = \begin{bmatrix} 1 & 0 \\ 0 & \Sigma_0 \end{bmatrix},$$

(75)

where Σ_0 is now a *known* matrix. The concentrated likelihood function, making use of (73) and (74), becomes

$$L^* = -\frac{T(r+1)}{2}\ln(2\pi) - \frac{T}{2}\ln|\Sigma_0| - \frac{T(r+1)}{2}\ln\sigma_{00} - \frac{T}{2\sigma_{00}}\frac{\gamma'A\gamma}{\gamma'\Sigma_0^*\gamma}.$$

(76)

Maximizing with respect to σ_{00} yields

$$\hat{\sigma}_{00} = \frac{1}{r+1}\frac{\gamma'A\gamma}{\gamma'\Sigma_0^*\gamma}$$

and substituting in (76) we have

$$
L^{**} = -\frac{T(r+1)}{2}[\ln(2\pi)+1] - \frac{T}{2}\ln|\Sigma_0| + \frac{T(r+1)}{2}\ln(r+1)
$$
$$
-\frac{T(r+1)}{2}\ln\left(\frac{\gamma'A\gamma}{\gamma'\Sigma_0^*\gamma}\right), \tag{77}
$$

which now has to be maximized with respect to β.

As we saw in the preceding section,

$$
\min_i \lambda_i \le \frac{\gamma'A\gamma}{\gamma'\Sigma_0^*\gamma} \le \max_i \lambda_i,
$$

where the λ_i $(i = 1, 2, \ldots, r+1)$ are the solutions of $|\lambda\Sigma_0^* - A| = 0$. The vector γ is, thus, to be chosen as the characteristic vector corresponding to $\hat{\lambda} = \min_i \lambda_i$, i.e., the vector satisfying

$$
A\gamma = \hat{\lambda}\Sigma_0^*\gamma. \tag{78}
$$

Even though characteristic vectors are unique only up to a scalar multiple, β is *uniquely* determined, since $\gamma = (1, -\beta')'$, i.e. *we normalize the characteristic vector in (78) by requiring that its first element be unity.* The remaining elements yield the ML estimator of $-\beta$. In view of the preceding we also conclude

$$
\hat{\sigma}_{00} = \frac{\hat{\lambda}}{r+1}. \tag{79}
$$

From (78) substituting in (73) yields the estimator of α.

Let us now recapitulate the steps involved in obtaining ML estimators of the parameters of the general EIV model as exhibited in (68) subject to the restriction on the covariance matrix as given in (75).

(i) Form the matrix

$$
A = \frac{1}{T}(y, W^*)'(I - X(X'X)^{-1}X')(y, W^*).
$$

(ii) Find the smallest characteristic root of A in the metric of Σ_0^*, as the latter is given in (75), say $\hat{\lambda}$, and its associated characteristic vector, say $\hat{\gamma}$.

(iii) Normalize $\hat{\gamma}$ so that its first element is unity; the remaining elements constitute the estimator of $-\beta$, say $-\hat{\beta}$.

(iv) Substitute $\hat{\beta}$ in (73) to obtain the estimator of α; thus,

$$
\hat{\alpha} = (X'X)^{-1}X'(y - W^*\hat{\beta}).
$$

(v) Estimate the scale factor

$$
\hat{\sigma}_{00} = \frac{\hat{\lambda}}{r+1}.
$$

Remark 14. If all explanatory variables are subject to errors of observation then under step (i) the matrix A becomes $A = (1/T)(y, W^*)'(y, W^*)$.

Remark 15. It is clear that even if normality is not assumed the estimators developed above have an interpretation as *Aitken or minimum chi-square (MCS) estimators*; this is true for $\hat{\alpha}$, $\hat{\beta}$ but not for σ_{00}. Thus, consider again the model in (68) and substitute from the second into the first equation to obtain

$$y_t = x_{t.}\alpha + w_{t.}^*\beta + u_t - u_{t.}^*\beta.$$

The error term $v_t = (u_t, u_{t.}^*)(1, -\beta')'$ constitutes a sequence of i.i.d. random variables with mean zero and variance $\sigma_{00}(1, -\beta')\Sigma_0^*(1, -\beta')'$. The quantity to be minimized in Aitken or MCS estimation is

$$\frac{1}{\sigma_{00}} \frac{\sum_{t=1}^T (y_t - x_{t.}\alpha - w_{t.}^*\beta)^2}{\gamma'\Sigma_0^*\gamma}, \qquad \gamma = (1, -\beta')'.$$

Minimizing first with respect to α yields

$$\tilde{\alpha} = (X'X)^{-1}X'(y - W^*\beta).$$

Inserting into the minimand we find

$$\frac{T}{\sigma_{00}} \frac{\gamma'A\gamma}{\gamma'\Sigma_0^*\gamma}, \qquad A = \frac{1}{T}(y, W^*)'[I - X(X'X)^{-1}X'](y, W^*).$$

Minimizing with respect to γ yields exactly the same solution as determined above. Thus, the normality assumption is not at all essential in deriving estimators of parameters in the EIV model.

4.2 Asymptotic Properties[6]

In this section we shall demonstrate the consistency of the EIV estimator of the structural parameters and give its asymptotic distribution. The derivation of the latter, however, represents a task well beyond the scope of this book and for this reason it will not be given.

We begin by showing the consistency of $\hat{\lambda}$ as an estimator of σ_{00}, and more particularly as an estimator of the smallest characteristic root of

$$|\lambda\Sigma_0 - \bar{A}| = 0, \tag{80}$$

where

$$\bar{A} = \operatorname*{plim}_{T \to \infty} A = \bar{A}_1 + \bar{A}_2 + \bar{A}_3. \tag{81}$$

[6] The material in this section is somewhat more involved than usual. The reader will not experience a loss of continuity if he merely familiarizes himself with the conclusions and omits the derivations entirely.

The meaning of $\bar{A}_1, \bar{A}_2, \bar{A}_3$ is established as follows. By definition

$$A = \frac{1}{T}(y, W^*)'N(y, W^*), \qquad N = I - X(X'X)^{-1}X', \qquad W^* = W + U^*.$$

Define

$$A_1 = \frac{1}{T}\begin{bmatrix} \delta'Z'NZ\delta & \delta'Z'NW \\ W'NZ\delta & W'NW \end{bmatrix}, \qquad A_2 = \frac{1}{T}\begin{bmatrix} u'Nu & 0 \\ 0 & U^{*'}NU^* \end{bmatrix},$$

$$A_3 = \frac{1}{T}\begin{bmatrix} 2u'NZ\delta & \delta'Z'NU^* + U^{*'}NZ\delta + u'NW^* \\ U^{*'}NZ\delta + \delta'Z'NU^* + W^{*'}Nu & U^{*'}NW + W'NU^* \end{bmatrix}.$$

$$(82)$$

where $Z = (X,W)$, $u = (u_{10}, u_{20}, \ldots, u_{T0})'$, $U^* = (u_{ti}^*)$.

In view of the assumptions regarding the variables of the model we easily conclude that

$$\operatorname*{plim}_{T\to\infty} A_3 = \bar{A}_3 = 0, \qquad \operatorname*{plim}_{T\to\infty} A_2 = \bar{A}_2 = \sigma_{00}\Sigma_0^*,$$

$$(83)$$

$$\operatorname*{plim}_{T\to\infty} A_1 = \bar{A}_1 = (\beta, I)'V(\beta, I), \qquad V = (\mathrm{p})\operatorname*{lim}_{T\to\infty} \frac{1}{T} W'NW.$$

It would be useful, before we proceed with the problem at hand, to elucidate the assumptions implicit in our discussions so far regarding the explanatory variables of the model. In particular, we have required that

$$(\mathrm{p})\operatorname*{lim}_{T\to\infty} \frac{Z'Z}{T} = Q = \begin{bmatrix} Q_{11} & Q_{12} \\ Q_{21} & Q_{22} \end{bmatrix}$$

$$(84)$$

exists as a positive definite matrix, where evidently

$$(\mathrm{p})\operatorname*{lim}_{T\to\infty} \frac{1}{T} X'X = Q_{11}, \qquad (\mathrm{p})\operatorname*{lim}_{T\to\infty} \frac{1}{T} W'W = Q_{22}, \qquad (\mathrm{p})\operatorname*{lim}_{T\to\infty} \frac{X'W}{T} = Q_{12}.$$

The assumption in (84) is the exact analog of the standard assumption regarding explanatory variables in the context of the typical GLM. In view of the assumption in (84) it is clear that

$$V = Q_{22} - Q_{21}Q_{11}^{-1}Q_{12}$$

$$(85)$$

and as such it is an $r \times r$ nonsingular (positive definite) matrix.

Now that the notation and necessary background have been established we recall that the characteristic roots of a matrix A are a continuous function of its elements. Hence, by the results in Chapter 8 we conclude that the roots of

$$|\lambda\Sigma_0^* - A| = 0$$

converge in probability to the roots of

$$|\lambda\Sigma_0^* - \bar{A}| = 0.$$

$$(86)$$

But

$$\bar{A} = \sigma_{00}\Sigma_0^* + (\beta, I)'V(\beta, I), \tag{87}$$

and we further observe that

$$0 = |\lambda\Sigma_0^* - \bar{A}| = |(\lambda - \sigma_{00})\Sigma_0^* - (\beta, I)'V(\beta, I)|. \tag{88}$$

Consequently, by Proposition 52 of *Mathematics for Econometrics* the roots of (86), say λ_i, obey

$$\lambda_i \geq \sigma_{00}, \qquad i = 1, 2, \ldots, r + 1. \tag{89}$$

Moreover, because

$$(\beta, I)'V(\beta, I)$$

is a singular matrix one of the roots of (88) is zero.

Thus, we conclude that the smallest of the roots in (86) obeys $\lambda_{\min} = \sigma_{00}$. What the preceding has established is that if $\hat{\lambda}$ is the smallest root of

$$|\lambda\Sigma_0^* - A| = 0$$

then

$$\plim_{T \to \infty} \hat{\lambda} = \sigma_{00}. \tag{90}$$

Remark 16. The result in Equation (90) shows that the ML estimator for σ_{00} is not consistent in the context of the EIV model as examined in this section. While it may appear odd to have a ML estimator be inconsistent we observe that this is an easily correctible problem. Instead of estimating σ_{00} by

$$\hat{\sigma}_{00} = \frac{\hat{\lambda}}{r + 1}$$

we simply estimate it by

$$\tilde{\sigma}_{00} = (r + 1)\hat{\sigma}_{00} = \hat{\lambda},$$

which is a consistent estimator.

To establish the consistency and other asymptotic properties of the EIV estimator of $\hat{\delta}$ we note that it obeys

$$\begin{bmatrix} y'Ny & y'NW^* \\ W^{*'}Ny & W^{*'}NW^* \end{bmatrix}\begin{pmatrix} 1 \\ -\hat{\beta} \end{pmatrix} = T\hat{\lambda}\begin{bmatrix} 1 & 0 \\ 0 & \Sigma_0 \end{bmatrix}\begin{pmatrix} 1 \\ -\hat{\beta} \end{pmatrix},$$
$$\hat{\alpha} = (X'X)^{-1}X'(y - W^*\hat{\beta}).$$

The relevant parts of these two equations may be written more conveniently as

$$\begin{bmatrix} X'X & X'W^* \\ 0 & W^{*'}NW^* - T\hat{\lambda}\Sigma_0 \end{bmatrix}\hat{\delta} = \begin{pmatrix} X' \\ W^{*'}N \end{pmatrix}y. \tag{91}$$

Dividing through by T, substituting for y, and taking probability limits we find

$$\begin{bmatrix} Q_{11} & Q_{12} \\ 0 & Q_{22} - Q_{21}Q_{11}^{-1}Q_{12} \end{bmatrix} \bar{\delta} = \begin{bmatrix} Q_{11} & Q_{12} \\ 0 & Q_{22} - Q_{21}Q_{11}^{-1}Q_{12} \end{bmatrix} \delta, \quad (92)$$

where

$$\bar{\delta} = \operatorname*{plim}_{T \to \infty} \hat{\delta}.$$

In view of the assumption in (84) we conclude

$$\operatorname*{plim}_{T \to \infty} \hat{\delta} = \delta, \quad (93)$$

which shows the consistency of the EIV estimator of the structural coefficients.

To examine the asymptotic distribution aspects of this estimator we again rely on (91) where, upon substituting for y and rearranging terms, we find

$$\sqrt{T}(\hat{\delta} - \delta) = \frac{1}{\sqrt{T}} \left[\begin{pmatrix} \left(\dfrac{X'X}{T}\right)^{-1} X' \\ 0 \end{pmatrix} + SW'N \right] (u - U^*\beta) + S \frac{1}{\sqrt{T}} U^{*\prime}Nu$$

$$+ \sqrt{T}(\hat{\lambda} - \lambda)S\Sigma_0 \beta - S \frac{1}{\sqrt{T}}(U^{*\prime}NU^* - \sqrt{T}\Sigma_{22})\beta$$

where

$$S = \begin{pmatrix} -(X'X)^{-1}X'W^* \\ I \end{pmatrix} \left(\frac{W^{*\prime}NW^*}{T} - \hat{\lambda}\Sigma_0 \right)^{-1}. \quad (94)$$

In view of the fact that

$$\frac{1}{\sqrt{T}}(U^{*\prime}NU^* - \sqrt{T}\Sigma_{22}) \sim \frac{1}{\sqrt{T}}(U^{*\prime}U^* - \sqrt{T}\Sigma_{22})$$

because

$$\frac{U^{*\prime}X}{T}\left(\frac{X'X}{T}\right)^{-1}\frac{U^{*\prime}X}{\sqrt{T}} \sim 0 \cdot \frac{U^{*\prime}X}{\sqrt{T}},$$

and similarly that

$$\frac{U^{*\prime}Nu}{\sqrt{T}} \sim \frac{U^{*\prime}u}{\sqrt{T}},$$

we can simplify the expression above to

$$\sqrt{T}(\hat{\delta} - \delta) \sim \frac{1}{\sqrt{T}} \left[\left(\begin{pmatrix} \left(\frac{X'X}{T}\right)^{-1} X' \\ 0 \end{pmatrix} \right) + SW'N \right] (u - U^*\beta) + S \frac{1}{\sqrt{T}} U^{*\prime} u$$

$$+ \sqrt{T}(\hat{\lambda} - \lambda) S\Sigma_0 \beta - S \frac{1}{\sqrt{T}} (U^{*\prime}U^* - \sqrt{T}\Sigma_{22})\beta.$$

(95)

From (95), by the application of the appropriate central limit theorems, we can establish that

$$\sqrt{T}(\hat{\delta} - \delta) \sim N(0, \mu\Psi),$$

(96)

where

$$\Psi = Q^{-1} + Q^{-1} \begin{bmatrix} 0 & 0 \\ 0 & R \end{bmatrix} Q^{-1},$$

$$R = \sigma_{00} \left(\Sigma_0 + \frac{2\sigma_{00}}{\mu} \Sigma_0 \beta\beta'\Sigma_0 \right).$$

(97)

While (97) gives the covariance matrix of the asymptotic distribution of the EIV estimator, for purposes of inference we require at least a consistent estimator for it. This is easily obtained as

$$\tilde{\Psi} = \tilde{Q}^{-1} + \tilde{Q}^{-1} \begin{bmatrix} 0 & 0 \\ 0 & \tilde{R} \end{bmatrix} \tilde{Q}^{-1},$$

$$\tilde{Q} = \frac{1}{T} \begin{bmatrix} X'X & X'W^* \\ W^{*\prime}X & W^{*\prime}W^* - T\tilde{\sigma}_{00}\Sigma_0 \end{bmatrix},$$

(98)

$$\tilde{R} = \tilde{\sigma}_{00} \left(\Sigma_0 + \frac{2\tilde{\sigma}_{00}}{\tilde{\mu}} \Sigma_0 \hat{\beta}\hat{\beta}'\Sigma_0 \right),$$

$$\tilde{\mu} = \tilde{\sigma}_{00}(1 + \hat{\beta}'\Sigma_0\hat{\beta}), \qquad \tilde{\sigma}_{00} = \hat{\lambda}.$$

The discussion of the EIV model estimator may be summarized in

Theorem 3. *Consider the model*

$$y = Z\delta + u, \qquad Z = (X, W), \qquad \delta = \begin{pmatrix} \alpha \\ \beta \end{pmatrix},$$

and suppose that W is not observable, but instead we observe

$$W^* = U^* + W.$$

Suppose further that

$$\underset{T \to \infty}{(p)\lim} \frac{Z'Z}{T} = Q$$

exists as a positive definite matrix, that

$$\begin{pmatrix} u_{t0} \\ u_{t\cdot}^{*\prime} \end{pmatrix} \sim N(0, \sigma_{00}\Sigma_0^*), \qquad \Sigma_0^* = \frac{1}{\sigma_{00}} \begin{bmatrix} \sigma_{00} & 0 \\ 0 & \Sigma_{22} \end{bmatrix},$$

and that

$$\frac{1}{\sigma_{00}} \Sigma_{22} = \Sigma_0$$

is known. Let

$$\left\{ \begin{pmatrix} u_{t0} \\ u_{t\cdot}^{*\prime} \end{pmatrix} : t = 1, 2, \ldots \right\}$$

be a sequence of i.i.d. *random variables, where evidently* u_{t0} *is the structural error at time* t—*the* tth *element of* u—*and* $u_{t\cdot}^{*\prime}$ *is the vector of observational errors, i.e., the* tth *column of the matrix* $U^{*\prime}$. *Then the following statements are true.*

(i) *The* ML *estimator of* δ *is given by the solution of*

$$\begin{bmatrix} X'X & X'W^* \\ 0 & W^{*\prime}NW^* - T\hat{\lambda}\Sigma_0 \end{bmatrix} \hat{\delta} = \begin{pmatrix} X' \\ W^{*\prime}N \end{pmatrix} y,$$

where $\hat{\lambda}$ *is the smallest characteristic root of* A *in the metric of* Σ_0, *and* $A = (1/T)(y, W^*)'[I - X(X'X)^{-1}X'](y, W^*)$.

(ii) $\operatorname*{plim}_{T \to \infty} \hat{\lambda} = \sigma_{00}$, *and* $\operatorname*{plim}_{T \to \infty} \hat{\delta} = \delta$.

(iii) *Asymptotically,* $\sqrt{T}(\hat{\delta} - \delta) \sim N(0, \mu\Psi)$, *where*

$$\Psi = Q^{-1} + Q^{-1} \begin{bmatrix} 0 & 0 \\ 0 & R \end{bmatrix} Q^{-1},$$

$$R = \sigma_{00}\left(\Sigma_0 + \frac{2\sigma_{00}}{\mu}\Sigma_0\beta\beta'\Sigma_0\right), \qquad \mu = \sigma_{00}(1 + \beta'\Sigma_0\beta).$$

(iv) *The covariance matrix of the asymptotic distribution may be estimated consistently by the quantities in Equation (98).*

(v) *Tests of significance may be based, in the usual way, on the asymptotic distribution. Thus, for example, putting* $\Psi = (\Psi_{ij})$, $\sqrt{T}(\hat{\delta}_i - \delta_i)/\sqrt{\tilde{\mu}\tilde{\Psi}_{ii}} \sim N(0, 1)$, *and so on, where* δ_i *is the* ith *element of* δ.

5 Misspecification Error Analysis for EIV Models

5.1 The general case

In the previous two sections we examined the problem of estimating the parameters of an EIV model by ML methods under a certain set of restrictive assumptions on the structural and observational "errors." We also briefly examined the asymptotic distribution of the resulting parameter estimators, and thus deduced how tests of significance and other inference problems should be handled. In even earlier sections we have employed misspecification error analysis to determine the nature of the "bias" or "inconsistency" involved in various contexts. In this section we shall employ misspecification error analysis to examine certain other aspects of the inference problem when EIV models are estimated improperly—by OLS.

The model under consideration is that in Theorem 3. If its (coefficient) parameters are estimated by OLS we have

$$\tilde{\delta} = (Z^{*\prime}Z^{*})^{-1}Z^{*\prime}y, \qquad Z^{*} = (X, W^{*}). \tag{99}$$

Since we can write

$$y = Z^{*}\delta + v, \qquad v = u - U^{*}\beta,$$

we have, upon substitution,

$$\tilde{\delta} - \delta = \left(\frac{Z^{*\prime}Z^{*}}{T}\right)^{-1}\frac{1}{T}Z^{*\prime}(u - U^{*}\beta).$$

Upon taking probability limits, we obtain

$$\operatorname*{plim}_{T \to \infty}(\tilde{\delta} - \delta) = -\left\{Q + \begin{bmatrix} 0 & 0 \\ 0 & \Sigma_{22} \end{bmatrix}\right\}^{-1}\begin{pmatrix} 0 \\ \Sigma_{22}\beta \end{pmatrix} \tag{100}$$

In contrast to the bivariate model of Section 3.1, where the direction of inconsistency was calculable, it is apparent from (100) that here it is not possible to do so except under very special circumstances.

Even though the direction of inconsistency cannot, in general, be unambiguously determined, there may still be other aspects of the inference problem to which we can address ourselves. Thus, for example, what is the impact of treating an EIV model as if its variables are free from errors of observation, on the usual t-ratios or on the coefficient of determination of multiple regression R^2? Frequently, judgments are made regarding whether a model is to be preferred over another based on R^2 and whether a hypothesis regarding the economics of some issues is to be accepted, based on the "significance" of a regression coefficient.

Here, let us denote, by

$$\tilde{\delta}_{*} = (Z^{*\prime}Z^{*})^{-1}Z^{*\prime}y \quad \text{and} \quad \tilde{\delta} = (Z^{\prime}Z)^{-1}Z^{\prime}y$$

the estimator of δ using (respectively) Z^*, the data matrix containing errors of observation, and Z, the data matrix not containing errors of observation. Evidently, the last-mentioned "estimator" cannot be obtained. If it could, then, of course, there would be no problem to deal with in the present context. The coefficient of determination obtained in the first case is

$$R^{*2} = \frac{\tilde{\delta}'_* Z^{*'} Z^* \tilde{\delta}_*}{y'y}, \qquad \frac{R^{*2}}{1 - R^{*2}} = \frac{\tilde{\delta}'_* Z^{*'} Z^* \tilde{\delta}_*}{\tilde{u}^{*'} \tilde{u}^*}, \qquad (101)$$

where it is assumed that the elements of Z^* and y have been centered about their respective sample means.[7]

In the above,

$$\tilde{u}^* = y - Z^*(Z^{*'}Z^*)^{-1}Z^{*'}y.$$

Except for a scalar multiple, the statistic $R^{*2}/(1 - R^{*2})$ is the F-statistic one uses to test the "goodness of fit" of the model, i.e., the "significance" (collectively) of the coefficients of the variables (other than the constant term) contained in the vector δ.

It is not possible, in the general case, to determine the distribution of the quantities in (101). We could, however, determine the probability limit of $R^{*2}/(1 - R^{*2})$ and compare it with the probability limit of the F-statistic as it would have been computed had we access to the error free variables.[8] If an unambiguous conclusion can be derived from this comparison we can then argue that for sufficiently large samples operating with variables not free from error would tend to either reject models too frequently, or not frequently enough. The numerator of (101) yields, upon substitution,

$$\frac{1}{T} y'Z^*(Z^{*'}Z^*)^{-1}Z^{*'}y.$$

Expanding, we have

$$\frac{1}{T} y'Z^*(Z^{*'}Z^*)^{-1}Z^{*'}y = \frac{1}{T} [u'Z^*(Z^{*'}Z^*)^{-1}Z^{*'}u + 2u'Z^*(Z^{*'}Z^*)^{-1}Z^{*'}Z\delta$$

$$+ \delta'Z'Z^*(Z^{*'}Z^*)^{-1}Z^{*'}Z\delta].$$

In view of the conditions under which we operate we immediately conclude, upon taking probability limits, that the limit above is

$$\delta'Q(Q + P)^{-1}Q\delta, \qquad P = \begin{bmatrix} 0 & 0 \\ 0 & \Sigma_{22} \end{bmatrix}.$$

For the denominator we observe that

$$\frac{1}{T} \tilde{u}^{*'} \tilde{u}^* = \frac{1}{T} [u'M_* u + 2u'M_* Z\delta + \delta'Z'M_* Z\delta],$$

[7] This is done for convenience only; notice, then, that $\tilde{\delta}_*$ does *not* contain the constant term of the model.

[8] Notice that the problem posed here has been discussed in a very general way in Section 2.1.

and upon taking probability limits we find

$$\sigma_{00} + \delta'Q\delta - \delta'Q(Q + P)^{-1}Q\delta.$$

In the case where error-free variables are employed we have

$$\frac{R^2}{1 - R^2} = \frac{\tilde{\delta}'Z'Z\tilde{\delta}}{\tilde{u}'\tilde{u}}, \qquad \tilde{u} = [I - Z(Z'Z)^{-1}Z']u,$$

and upon taking probability limits we have

$$\frac{\delta'Q\delta}{\sigma_{00}}.$$

Consequently,

$$\plim_{T \to \infty} \left(\frac{R^2}{1 - R^2} - \frac{R^{*2}}{1 - R^{*2}} \right) = \frac{\sigma_{00}(\phi - \omega) + \phi(\phi - \omega)}{\sigma_{00}(\sigma_{00} + \phi - \omega)}, \qquad (102)$$

where

$$\phi = \delta'Q\delta, \qquad \omega = \delta'Q(Q + P)^{-1}Q\delta.$$

We shall now show that for any δ, $\phi - \omega \geq 0$. But this is so if and only if $Q - Q(Q + P)^{-1}Q$ is positive semidefinite; this, in turn is so if and only if $Q + P - Q$ is positive semidefinite, which it obviously is (refer to the definition of P above). We conclude therefore, that in general we would have

$$\plim_{T \to \infty} \frac{R^2}{1 - R^2} > \plim_{T \to \infty} \frac{R^{*2}}{1 - R^{*2}}. \qquad (103)$$

Remark 17. We used the strict inequality sign in (103) since, in general, we would not expect $\phi = \omega$ except in highly special cases, as for example when $\Sigma_{22} = 0$, a trivial case, or when $\beta = 0$, also a trivial case. The reader should bear in mind, however, that this has not been formally demonstrated.

Remark 18. The result in (103) shows that, in the limit, the F-statistic is unambiguously understated when operating by the usual regression methods with an EIV model. Consequently, we would expect to reject models "too frequently."

Remark 19. If a model is "accepted" by a "significant" F-statistic as above, it is rather unlikely that the conclusion would be reversed. More precisely, if a model is accepted when variables are suspected to be observed with error, it is the implication of (103) that the decision is not likely to be reversed in the case where error-free observations become available; we still, however, have to operate with (potentially) inconsistent estimators for the model's parameters.

We now turn to the question of whether the impact on t-ratios can be unambiguously determined. We note that when operating with an EIV model by regression methods, we shall estimate the error variance by

$$\tilde{\sigma}_{00}^* = \frac{1}{T}\, \tilde{u}^{*\prime}\tilde{u}^*$$

while in the error-free case we should have

$$\tilde{\sigma}_{00} = \frac{1}{T}\, \tilde{u}'\tilde{u}$$

We have already shown in the preceding that

$$\plim_{T\to\infty}(\tilde{\sigma}_{00}^* - \tilde{\sigma}_{00}) = \phi - \omega \geq 0, \tag{104}$$

and that, typically, strict inequality in (104) will hold. It bears repeating that what is meant by operating through regression methods in the context of an EIV model is that we obtain the estimator

$$\hat{\delta}_* = (Z^{*\prime}Z^*)^{-1}Z^{*\prime}y$$

and behave as if its distribution is given by

$$\hat{\delta}_* \sim N[\delta, \sigma_{00}(Z^{*\prime}Z^*)^{-1}],$$

the quantity σ_{00} being estimated by $\tilde{\sigma}_{00}^*$. Clearly, the procedure above is inappropriate; if error-free variables were available then we would obtain

$$\hat{\delta} = (Z'Z)^{-1}Z'y, \qquad \hat{\delta} \sim N[\delta, \sigma_{00}(Z'Z)^{-1}],$$

and we would estimate σ_{00} by $\tilde{\sigma}_{00}$. It is clear that, since

$$\plim_{T\to\infty} \frac{Z^{*\prime}Z^*}{T} = (\text{p})\lim_{T\to\infty} \frac{Z'Z}{T} + P = Q + P,$$

we would expect the diagonal elements of $(Z'Z)^{-1}$ to be not less (typically, strictly greater) than the diagonal elements of $(Z^{*\prime}Z^*)^{-1}$. Since we would expect (at least in the limit) $\tilde{\sigma}_{00}^* \geq \tilde{\sigma}_{00}$, and $(Z'Z)^{-1} - (Z^{*\prime}Z^*)^{-1}$ to be at least positive semidefinite, is there anything we can say about

$$\tilde{\sigma}^{*2}(Z^{*\prime}Z^*)^{-1} - \tilde{\sigma}^2(Z'Z)^{-1}?$$

Again resorting to the probability limit calculations, we find

$$\plim_{T\to\infty}\left[\tilde{\sigma}_{00}^*\left(\frac{Z^{*\prime}Z^*}{T}\right)^{-1} - \tilde{\sigma}_{00}\left(\frac{Z'Z}{T}\right)^{-1}\right]$$
$$= (\sigma_{00} + \phi - \omega)(Q + P)^{-1} - \sigma_{00}Q^{-1}.$$

Developing the matter further, we consider the simultaneous factorization of the matrices $Q + P$ and Q. We have,

$$Q + P = H'\Lambda H, \qquad Q = H'H, \tag{105}$$

where

$$\Lambda = \text{diag}(\lambda_1, \ _2, \ldots, \lambda_{r+s}),$$

the λ_i being the (positive) characteristic roots of $Q + P$ in the metric of Q arranged in decreasing order. In fact, it may be shown that the first r roots are greater than unity, while the last s are equal to unity. Consequently,

$$(\sigma_{00} + \phi - \omega)(Q + P)^{-1} - \sigma_{00} Q^{-1}$$
$$= H^{-1}[(\sigma_{00} + \phi - \omega)\Lambda^{-1} - \sigma_{00} I]H'^{-1},$$

and we see that the crucial comparison is that between $(\sigma_{00} + \phi - \omega)\Lambda_1^{-1}$ and $\sigma_{00} I_r$, where $\Lambda_1 = \text{diag}(\lambda_1, \lambda_2, \ldots, \lambda_r)$, ie, Λ_1 contains the roots that are greater than unity. But it is now quite obvious that this issue cannot be settled unambiguously; it is clear that $\sigma_{00}(\Lambda_1^{-1} - I_r)$ is negative definite and crucially depends on σ_{00}; $(\phi - \omega)\Lambda_1^{-1}$, on the other hand, is positive definite and depends on δ. Thus, for some parametric configurations one component will dominate, while for other parametric configurations another component may, and for still other parametric configurations the matrix difference will be indefinite. *Consequently we cannot say unambiguously whether, in the general case, the diagonal elements of the difference*

$$\tilde{\sigma}_{00}^*(Z^{*\prime}Z^*)^{-1} - \tilde{\sigma}_{00}(Z'Z)^{-1}$$

will be nonnegative or nonpositive.

We may summarize the preceding discussion in

Theorem 4. *Consider the model of Theorem 3 and the estimators*

$$\tilde{\delta}_* = (Z^{*\prime}Z^*)^{-1}Z^{*\prime}y, \qquad \tilde{\delta} = (Z'Z)^{-1}Z'y,$$

where in this case data have been centered about their respective sample means (and thus the constant term does not appear in $\tilde{\delta}_*$ or $\tilde{\delta}$). Consider, further

$$\tilde{u} = y - Z\tilde{\delta}, \qquad \tilde{u}^* = y - Z^*\tilde{\delta}^*, \qquad \tilde{\sigma}_{00} = \frac{\tilde{u}'\tilde{u}}{T}, \qquad \tilde{\sigma}_{00}^* = \frac{\tilde{u}^{*\prime}\tilde{u}^*}{T}.$$

Then the following statements are true.

(i)
$$\plim_{T \to \infty} \left(\frac{R^2}{1 - R^2} - \frac{R^{*2}}{1 - R^{*2}} \right) \geq 0,$$

where

$$\frac{R^2}{1 - R^2} = \frac{\tilde{\delta}'Z'Z\tilde{\delta}}{\tilde{u}'\tilde{u}}, \qquad \frac{R^{*2}}{1 - R^{*2}} = \frac{\tilde{\delta}'_* Z^{*\prime}Z^*\tilde{\delta}_*}{\tilde{u}^{*\prime}\tilde{u}^*}.$$

(ii)
$$\plim_{T \to \infty}(\tilde{\sigma}_{00}^* - \tilde{\sigma}_{00}) = \phi - \omega \geq 0,$$

where ϕ, ω have been defined immediately after (102).

(iii)
$$\plim_{T \to \infty} \left[\frac{Z^{*\prime}Z^*}{T} - \frac{Z'Z}{T} \right]$$

is a positive semidefinite matrix.

(iv) *No unambiguous statement can be made regarding the question whether*

$$\text{plim}_{T \to \infty} \left[\tilde{\sigma}_{00}^* \left(\frac{Z^{*\prime} Z^*}{T} \right)^{-1} - \tilde{\sigma}_{00} \left(\frac{Z^\prime Z}{T} \right)^{-1} \right]$$

is positive semidefinite, negative semidefinite, or indefinite.

(v) *No unambiguous statement may be made regarding the t-ratios of* OLS *estimated parameters in an* EIV *context relative to those that would prevail if error free observations were available.*

QUESTIONS AND PROBLEMS

1. In the Cobb–Douglas example $Q_t = A K_t^\alpha L_t^\beta e^{u_t}$, $t = 1, 2, \ldots, T$, suppose the standard assumptions hold but that K_t is not available. Suppose instead we have data on the capital stock, K_t^*, and that *it is known* that $K_t = \gamma K_t^*$. What are the properties of the estimators of the constant term, and the coefficients of $\ln K_t^*$, $\ln L_t$ in the regression of $\ln Q_t$ on $\ln K_t^*$, $\ln L_t$?

2. In the discussion of goodness of fit and misspecification analysis show that if, of the two models $y = X_1 \beta + u_1$, $y = X_2 \beta + u_2$, the first is known to be the "true" model, then

$$\text{plim}_{T \to \infty} \left(\frac{\hat{u}_2' \hat{u}_2}{T} \right) \geq \text{plim}_{T \to \infty} \left(\frac{\hat{u}_1' \hat{u}_1}{T} \right)$$

where

$$\hat{u}_2 = [I - X_2(X_2' X_2)^{-1} X_2'] y, \qquad \hat{u}_1 = [I - X_1(X_1' X_1)^{-1} X_1'] y.$$

[*Hint*: $\text{plim}_{T \to \infty} (u_1'[I - X_2(X_2' X_2)^{-1} X_2']u_1/T)$
$- \text{plim}_{T \to \infty} (u_1'[I - X_1(X_1' X_1)^{-1} X_1']u_1/T) = 0.$]

3. Verify the representation in Equation (12). [*Hint*: from Proposition 32 of *Mathematics for Econometrics* recall that

$$\begin{bmatrix} A_{11} & A_{12} \\ A_{21} & A_{22} \end{bmatrix}^{-1}$$

$$= \begin{bmatrix} (A_{11} - A_{12} A_{22}^{-1} A_{21})^{-1} & -A_{11}^{-1} A_{12}(A_{22} - A_{21} A_{11} A_{21})^{-1} \\ -A_{22}^{-1} A_{21}(A_{11} - A_{12} A_{22}^{-1} A_{21})^{-1} & (A_{22} - A_{21} A_{11}^{-1} A_{21})^{-1} \end{bmatrix},$$

and that if A is a symmetric matrix then so is its inverse. From Proposition 33 note also that if A is nonsingular and symmetric then

$$[A + caa']^{-1} = A^{-1} - \psi A^{-1} aa' A^{-1}, \qquad \psi = \frac{c}{1 - ca' A^{-1} a},$$

provided $1 - ca' A^{-1} a \neq 0$, a being a suitably dimensioned vector.]

4. Verify the representation in Equation (14).

5. Verify the representation in Equation (27).

6. In the discussion of principal components show that $\beta = \sum_{i=1}^{n} \gamma_i a_{.i}$, where $a_{.i}$, $i = 1, 2, \ldots, n$, are the characteristic vectors of $X'X$ and γ is the reparametrization of the unknown parameters of the GLM as exhibited in Equation (32).

7. Verify the representation in Equation (44) i.e.,

$$[X_k'(I - X_*(X_*'X_*)^{-1}X_*')X_k]^{-1}$$

$$= (X_k'X_k)^{-1} + (X_k'X_k)^{-1}X_k'X_*[X_*'(I - X_k(X_k'X_k)^{-1}X_k')X_*]^{-1}X_*'X_k(X_k'X_k)^{-1}.$$

[Hint: (i) Put $X_k'X_k = SS'$ for some nonsingular matrix S; (ii) show that the non-zero characteristic roots of $S^{-1}X_k'X_*(X_*'X_*)^{-1}X_*'X_kS'^{-1} = A$ are exactly those of $|rX_*'X_* - X_*'X_k(X_k'X_k)^{-1}X_k'X_*| = 0$; (iii) let V be the matrix of characteristic vectors corresponding to the roots of the equation in (ii) above

$$R = \text{diag}(r_1, r_2, \ldots, r_{n-k}), \qquad r_1 \geq r_2 \geq \cdots r_{n-k}$$

and impose the normalization $V'X_*'X_*V = R^{-1}$; (iv) let Q, R^* be the matrices of characteristic vectors and roots of A as defined in (ii) and note that $A = QR^*Q' = Q_* RQ_*'$, where we partition $Q = (Q_*, Q_{2k-n})$, $R^* = \text{diag}(R, 0)$; (v) show that $Q_* = S^{-1}X_k'X_*V$; (vi) finally, show that

$$(I - R)^{-1} - I = (I - R)^{-1}R = V^{-1}[X_*'(I - X_k(X_k'X_k)^{-1}X_k')X_*]^{-1}V'^{-1}.]$$

8. In the representation of Equation (52) verify that: (i) the OLS estimator corresponds to the choice $D = I$; (ii) the PC estimator corresponds to the choice

$$D = \begin{bmatrix} I_k & 0 \\ 0 & 0 \end{bmatrix};$$

(iii) the RR estimator corresponds to the choice $D = (R + kI)^{-1/2}R^{1/2}$; (iv) the GRR estimator corresponds to the choice $D = (R + K)^{-1/2}R^{1/2}$.

9. In connection with the Wald estimator of Section 3.2, what are its properties if the observations $(y_1, x_1), (y_2, x_2), \ldots, (y_T, x_T)$ are ranked in increasing order of magnitude for x_t?

10. In the EIV model, show that whether the dependent variable is observed with or without error does not in any way affect the nature of the problem. In particular, show that errors of observation in the dependent variable are equivalent to specifying that the structural equation contains an additive error term.

11. Verify that the vectors $\{s_t : t = 1, 2, \ldots, T\}$ of Equation (59) are independently nonidentically distributed with mean zero.

12. Verify that, provided

$$u_{t.}' \sim N(0, \Sigma), \qquad \Sigma = \begin{bmatrix} \sigma_{00} & 0 \\ 0 & \Sigma_{22} \end{bmatrix}, \qquad u_{t.} = (u_{t0}, u_{t.}^*),$$

$E[u_{t.}^* u_t^* \beta' u_{t.}^* u_{t.}^* \beta] = \beta' \Sigma_{22} \beta \Sigma_{22} + 2\Sigma_{22} \beta \beta' \Sigma_{22}$. [Hint: if $x = (x_1, x_2, \ldots, x_n)' \sim N(0, \Sigma)$ then $E(x_r x_s x_k x_j) = \sigma_{rs}\sigma_{kj} + \sigma_{rk}\sigma_{sj} + \sigma_{rj}\sigma_{sk}$.]

13. Consider the model $y_t = \alpha x_t + \beta w_t + u_t$, where w_t is observed with error, i.e., $w_t^* = w_t + u_{t1}$. In this case show that the direction of inconsistency for $\tilde{\alpha}$ is indeterminate. [*Hint*: choose, for arbitrary q, Q that is positive definite such that

$$Q = \frac{1}{2}\begin{bmatrix} 1 & q \\ q & \frac{1}{2} \end{bmatrix}, \qquad \Sigma_{22} = \frac{1}{4},$$

and consider the two cases $q > 0, q < 0$.]

14. What specific condition under which we operate ensures that

$$\plim_{T \to \infty} \frac{1}{T} u'Z^*(Z^{*\prime}Z^*)^{-1}Z^{*\prime}Z\delta = 0?$$

15. Show that $Q - Q(Q + R)^{-1}Q$ is positive definite. [*Hint*: $Q - Q(Q + R)^{-1}Q = Q[Q^{-1} - (Q + R)^{-1}]Q$, where Q is positive definite.]

16. In the factorization of (105) show that the first r diagonal elements of Λ are greater than unity while the remaining s are unity.

17. Suppose we operate with the model of Theorem 3 and we take Σ_0 to be some fixed nonnull matrix. Suppose further that the information that some variables are observed with errors turns out to be false, and indeed it is in fact true that $W = W^*$. Have we lost anything by processing the data (y, X, W^*) as if W^* contained errors of observation? [*Hint*: $\hat{\beta}$ is part of a characteristic vector of

$$A = \frac{1}{T}(y, W^*)'M(y, W^*), \qquad M = I - X(X'X)^{-1}X',$$

in the metric of Σ_0. What is the characteristic vector corresponding to the smallest characteristic root of $\plim_{T \to \infty} A$ in metric of Σ_0?]

Systems of Simultaneous Equations 6

1 Introduction

In previous chapters we examined extensively the GLM under a variety of circumstances. A common feature of these discussions was a certain aspect of unidirectionality. Generally, variations in the explanatory (right-hand) variables were transmitted to the dependent (left-hand) variable, but not vice versa. Another common feature was that the explanatory variables were assumed to be independent of or, minimally, uncorrelated with the error term of the model. Only in the EIV model was this condition violated.

What peculiarly characterizes simultaneous equations models is that these two features are typically absent, so that the error terms of a given equation are not necessarily uncorrelated with the explanatory variables, and variations in the dependent variable of a given equation may be (and typically are) transmitted to (some of) the explanatory variables occurring in that equation. The prototype of the simultaneous equations model in econometrics is the general equilibrium system of economic theory.

In what follows we shall examine the structure of the general linear structural econometric model (GLSEM), establish that OLS fails, in general, to give consistent estimators, and examine certain extensions of OLS that are particularly suitable for the GLSEM.

2 The Simultaneous Equations Model (SEM): Definitions, Conventions, and Notation

2.1 The Nature of the Problem

Consider the simplest of simple Keynesian models of national income determination,

$$c_t = \alpha_0 + \alpha_1 y_t + u_t,$$
$$y_t = c_t + A_t,$$

where c_t and y_t are (respectively) consumption and income at time t, A_t is "autonomous expenditure," u_t is a random error, and α_0 and α_1 are parameters to be estimated. In its own way this is a simple general equilibrium system in which consumption and income are simultaneously determined. Autonomous expenditures, A_t, such as investment and/or government expenditures, are given outside the model. The first equation is a behavioral equation purporting to describe the consumption function. The second is the familiar national income identity.

We may assume that the error term of the consumption function is i.i.d. with zero mean and finite variance. *Still the logic of the model* implies that u_t cannot be independent of (or uncorrelated with) y_t. This will become evident if we solve the system to obtain

$$c_t = \frac{\alpha_0}{1 - \alpha_1} + \frac{\alpha_1}{1 - \alpha_1} A_t + \frac{1}{1 - \alpha_1} u_t,$$

$$y_t = \frac{\alpha_0}{1 - \alpha_1} + \frac{1}{1 - \alpha_1} A_t + \frac{1}{1 - \alpha_1} u_t,$$

which shows that

$$\mathrm{Cov}(y_t, u_t) = \frac{1}{1 - \alpha_1} \sigma^2$$

The reader should note that in this simple model we have most of the essential features of SEM, viz.:

(i) some equations in the model may be written with an error component while others may be written without;
(ii) some explanatory variables may not be independent of or uncorrelated with the error terms of the equation in which they appear;
(iii) the transmission of variations may be bidirectional i.e., some explanatory variables may be affected by the variations of the dependent variable whose variation they may, in part, determine.

These conditions represent a situation generally not encountered in the context of the GLM and thus require special treatment.

2.2 Definition of GLSEM

Definition 1. The model

$$y_t. B^* = x_t. C + u_t., \qquad t = 1, 2, \ldots, T,$$

where $y_t.$ is $m \times 1$, $x_t.$ is $G \times 1$, B^* is $m \times m$, C is $G \times m$ and $u_t.$ is a random vector of dimension m, is said to be the *general linear structural econometric model* (GLSEM). The vector $y_t.$ contains the variables whose values at time t are determined by the system, of which the set of equations above is presumed to be a description. They are said to be the *jointly dependent* or *endogenous variables.* The vector $x_{t\cdot}$ contains the variables whose values at time t are determined outside the system at time t. They are said to be the *predetermined variables* of the model. The matrices B^*, C contain the unknown (structural) parameters of the model. Finally, the vector $u_{t\cdot}$ represents the random disturbances at time t.

Definition 2. The form of the model as exhibited in Definition 1 is said to be the *structural form* of the model.

Definition 3. The vector of predetermined variables may be further decomposed into *lagged dependent* or *lagged endogenous* and *exogenous variables.* Thus, generally, we may write

$$x_{t\cdot} = (y_{t-1\cdot}, y_{t-2\cdot}, \dots, y_{t-k\cdot}, p_{t\cdot})$$

on the assertion that the model contains at most lags of order k. The variables $y_{t-i\cdot}$, $i = 1, 2, \dots, k$, are the *lagged endogenous* variables, while $p_{t\cdot}$, which is an s-element vector, contains the *exogenous* variables.

A variable is said to be *exogenous* if its behavior is determined entirely outside the model. Specifically exogenous variables are assumed to be independent of (minimally uncorrelated with) the error terms of the GLSEM, no matter what the latter's probabilistic specification.

Remark 1. The classification of variables is motivated by two sets of considerations whose requirements are not always compatible. From the point of view of economic theory an endogenous variable is one whose behavior is determined by the model and an exogenous variable is one whose behavior is determined outside the model. This is clear enough. From an estimation point of view, however, what is relevant is whether, in a given equation, an explanatory variable is or is not independent of (minimally uncorrelated with) the equation's error term. Thus, in this context, we would want to classify variables accordingly. In the typical case, current endogenous variables are correlated with the error terms but lagged endogenous variables are not. Hence the classification of variables into current endogenous and predetermined makes a great deal of sense. When we go beyond the typical case in which we assume that the error vector

$$\{u'_{t\cdot}: t = 1, 2, \dots\}$$

is i.i.d. this classification loses its usefulness. Thus, we fall back to the econometrically relevant classification into those variables that are independent of structural errors and those that are not. The reader will do well to have a full and clear understanding of this aspect of the classification scheme.

The equations in the typical GLSEM are basically of three types.

(i) *Behavioral equations.* These purport to describe the behavior of economic agents. They are typically stochastic in nature, i.e., they contain an error term whose function is to delimit the departure of the agents' behavior from the hypothesized functional form that purports to describe it. A typical example would be, say, the marginal productivity conditions that purport to describe factor employment on the part of profit maximizing firms.

(ii) *Technical or institutional relations.* Typical examples are production functions or tax functions. Such relations originate with the technology and the legal institutional framework of the economy, and are, typically, written as stochastic equations although the stochastic justification here is somewhat more convoluted. Production and other technology-induced relationships are written as stochastic functions for the same reason given in the case of behavioral equations. There are, indeed, many individually infinitesimal factors that impinge on production beyond the typically specified inputs of capital and labor. Their collective effect is captured by the error term. In the case of relations that originate with the legal system, such as tax functions, the justification is shifted to an aggregation argument. After all, the tax tables are specific, and, given all individuals' taxable income, tax liability is a matter of definition. However, we do not know all individuals' taxable income from published sources. Hence by writing a tax function for aggregate income we are committing an aggregation error, and what is essentially a definitional relation between taxable income and tax liability now becomes an approximate relationship. Hence, the stochastic nature of such equations.

(iii) *Identities.* These are essentially definitional equations. All they do is to define new symbols. They can be eliminated without any difficulty at the cost of more ponderous notation. Examples are the usual national income identity or the definition of total consumption as the sum of durable, nondurable, and services consumption. Or consider the definition of total employment as the sum of employment in manufacturing, services, government and other sectors, and so on. *Identities are nonstochastic equations.*

Remark 2. We see that fundamentally the equations of a GLSEM are of two types, *stochastic and nonstochastic.* The nonstochastic equations are definitional equations such that if the components of the right-hand side are given we immediately have exactly the left-hand side. The existence of nonstochastic equations poses a minor problem in the specification of the probabilistic properties of the error vector

$$u_{t\cdot} = (u_{t1}, u_{t2}, \ldots, u_{tm}).$$

If, say, the last two equations of the model are identities then we must have

$$u_{t, m-1} = u_{tm} = 0$$

identically for all t. Hence, in specifying the covariance matrix of this vector we cannot assert that it is a positive definite matrix *because its last two columns and rows contain only zeros.* But, as we remarked earlier, identities can be suppressed (substituted out) at no cost to the substance of the model. For this reason we introduce

Convention 1. In dealing with the formal aspects of the specification and estimation of the GLSEM *we shall always assume that it contains no identities (or that, if does, they have been substituted out).*

2.3 Basic Conditions Under Which the GLSEM is Estimated

When dealing with the GLM the set of basic conditions under which we operate is a rather simple one and contains assumptions on the error terms, the explanatory variables, and the relation between the two. In the context of the GLSEM the situation is considerably more complex. Consider again the model

$$ y_{t\cdot}.B^* = x_{t\cdot}.C + u_{t\cdot}. \tag{1} $$

and suppose that by analogy to the GLM we assert:

(A.1) $\{u'_{t\cdot}: t = 1, 2, \ldots\}$ is a sequence of i.i.d. random variables with

$$ E(u'_{t\cdot}) = 0, \qquad \text{Cov}(u'_{t\cdot}) = \Sigma $$

and Σ is positive definite;

(A.2) the set of exogenous variables $P = (p_{tj}), j = 1, 2, \ldots, s, t = 1, 2, \ldots, T$, is such that

$$ \text{rank}(P) = s, \qquad (\text{p})\lim_{T \to \infty} \frac{P'P}{T} = F, \qquad |F| \neq 0; $$

(A.3) the elements of P and $u'_{t\cdot}$ are mutually independent.

In the context of the GLM this set was sufficient to ensure that the OLS estimators had a number of desirable properties. Is this so in the present context?

First we observe that even before we deal with the issue of the properties of OLS estimators, the model in (1) is ambiguous in the following sense. The inference problem entails the estimation of the matrices B^* and C from the observations

$$ Y = (y_{ti}), \qquad X = (x_{tj}), \qquad i = 1, 2, \ldots, m, j = 1, 2, \ldots, G, t = 1, 2, \ldots, T. \tag{2} $$

But if H is any nonsingular matrix and we postmultiply (1) by H we find

$$ y_{t\cdot}.B^{**} - x_{t\cdot}.C^* = u^*_{t\cdot}, \tag{3} $$

where

$$ B^{**} = B^*H, \qquad C^* = CH, \qquad u^*_{t\cdot} = u_{t\cdot}.H. \tag{4} $$

The errors of the model in (3) as well as its exogeneous variables obey the conditions in (A.1), (A.2), and (A.3). Moreover, if Y and X in (2) are compatible with the model in (1) then they are compatible with the model in (3) as well!

Hence if all we knew about the model is the specification in (1) subject to the conditions in (A.1), (A.2), and (A.3) we would not be able to differentiate, given the observations Y and X, between the models in (1) and (3) since both are *observationally equivalent. This is the identification problem in econometrics*, and we shall return to it in some detail below.

For the moment suffice it to say that for the situation in (3) and (4) to be obviated we must require that some restrictions be placed on the elements of B^* and C. Typically, we impose the restriction that some elements of B^* and C are known, a priori, to be zero. These are the so called *identifying (zero) restrictions*. Thus, we must add

(A.4) (Identifying restrictions). Some elements of B^* and C are known a priori to be zero.

Finally, we would expect that the phenomenon to which the model refers is such that, if the predetermined variables are specified, then the conditional mean of the current endogenous variables given the predetermined ones is uniquely determined. This implies the condition

(A.5) The matrix B^* is nonsingular.

Using the condition in (A.5) we find

$$y_{t\cdot} = x_{t\cdot} \Pi + v_{t\cdot} \tag{5}$$

where

$$\Pi = CD, \qquad v_{t\cdot} = u_{t\cdot} D, \qquad D = B^{*-1}. \tag{6}$$

Definition 4. The representation in (5) is said to be the *reduced form of the model*.

Remark 3. The essential difference between the structural and the reduced forms of the GLSEM is that in the reduced form all right-hand (explanatory) variables must be predetermined, while in the structural form some or all of the right-hand ("explanatory") variables in a given equation may be current endogenous. Of course it may well be that all explanatory variables for a given structural equation are predetermined. This is not excluded; *what is excluded is that current endogenous variables appear as explanatory variables in a reduced form equation*.

Remark 4. Given the restrictions in (A.4) we know that not all variables appear in every equation. It is thus always possible to write structural equations in the form, say

$$b_{11}^* y_{t1} = -\sum_{i=2}^{m_1+1} b_{i1}^* y_{ti} + \sum_{j=1}^{G_1} c_{ji} x_{tj} + u_{t1}. \tag{7}$$

If

$$b_{11}^* \neq 0$$

then we can certainly divide through by b_{11}^* and thus have the representation

$$y_{t1} = \sum_{i=2}^{m_1+1} b_{i1} y_{ti} + \sum_{j=1}^{G_1} c_{j1} x_{tj} + u_{t1}. \tag{8}$$

In (8) we have implicitly assumed that only m_1 current endogenous variables $(y_2, y_3, \ldots, y_{m_1+1})$ and G_1 predetermined variables $(x_1, x_2, \ldots, x_{G_1})$ appear as explanatory (right-hand) variables in the first equation. Indeed, (8) may well be a more natural way in which one may write the equations of an econometric model. We are thus led to

Convention 2 (Normalization). It is possible to write the GLSEM in such a way that in the ith structural equation y_{ti} appears with a coefficient of unity, $i = 1, 2, \ldots, m$.

Remark 5. It is a consequence of Convention 2 that we can write

$$B^* = I - B \tag{9}$$

in such a way that

$$b_{ii} = 0, \qquad i = 1, 2, \ldots, m, \tag{10}$$

where b_{ij} is the (i, j) element of B.

Convention 3. The ith structural equation contains as explanatory (right-hand) variables m_i current endogenous and G_i predetermined variables. In particular the variables are so numbered that the first structural equation contains, as explanatory variables, $y_2, y_3, \ldots, y_{m_1+1}$ and $x_1, x_2, \ldots, x_{G_1}$.

To implement these conventions it is useful to introduce the selection matrices L_{i1}, L_{i2} as follows:

$$YL_{i1} = Y_i, \qquad XL_{i2} = X_i,$$

where: L_{i1} selects from the matrix Y the columns containing the observations on *those current endogenous variables that appear as explanatory variables in the ith structural equation*; similarly, L_{i2} *selects from the matrix X the columns containing the observations on those predetermined variables that appear as explanatory variables in the ith structural equation.*

Remark 6. Note that L_{i1} is $m \times m_i$ and L_{i2} is $G \times G_i$. Moreover the elements of these matrices are either zero, or one in such a way that *each column contains exactly one element equal to unity, the others being null.* Note further that

$$\text{rank}(L_{i1}) = m_i, \quad \text{rank}(L_{i2}) = G_i, \quad i = 1, 2, \ldots, m.$$

Given the conventions and the notation introduced above we may write the observations on the structural model as

$$Y = YB + XC + U, \tag{11}$$

where, evidently,

$$U = (u_{ti}), \quad t = 1, 2, \ldots, T, \; i = 1, 2, \ldots, m,$$

and on the reduced form as

$$Y = X\Pi + V, \tag{12}$$

where

$$\Pi = CD, \quad V = UD, \quad D = (I - B)^{-1}. \tag{13}$$

The observations on the ith structural equation are

$$y_{\cdot i} = Yb_{\cdot i} + Xc_{\cdot i} + u_{\cdot i}, \quad i = 1, 2, \ldots, m, \tag{14}$$

where $y_{\cdot i}, b_{\cdot i}, c_{\cdot i}, u_{\cdot i}$ are the ith column of (respectively) the matrices Y, B, C, U.

Using Convention 3 and the selection matrix notation, we can write for the first structural equation

$$y_{\cdot 1} = Y_1 \beta_{\cdot 1} + X_1 \gamma_{\cdot 1} + u_{\cdot 1}$$

where, of course,

$$b_{\cdot 1} = \begin{pmatrix} 0 \\ \beta_{\cdot 1} \\ 0 \end{pmatrix}, \quad c_{\cdot 1} = \begin{pmatrix} \gamma_{\cdot i} \\ 0 \end{pmatrix}, \tag{15}$$

and Y_1 contains the columns $y_{\cdot 2}, y_{\cdot 3}, \ldots, y_{\cdot m_1 + 1}$ while X_1 contains the columns $x_{\cdot 1}, x_{\cdot 2}, \ldots, x_{\cdot G_1}$.

Given the numbering of variables in Convention 3 we have that the selection matrices for the first equation have the simple form

$$L_{11} = \begin{pmatrix} 0 \\ I \\ 0 \end{pmatrix}, \quad L_{12} = \begin{pmatrix} I \\ 0 \end{pmatrix}, \tag{16}$$

where the first zero in L_{11} represents an m_1-element row of zeros, the identity matrix is of order m_1, and the last zero represents a null matrix of dimension $(m - m_1 - 1) \times m_1$. Similarly, in L_{12} the identity matrix is of order G_1, and the zero represents a null matrix of dimension $(G - G_1) \times G_1$.

Moreover we see that given the identifying restrictions

$$Yb_{\cdot 1} = Y_1 \beta_{\cdot 1}$$

and since

$$YL_{11} = Y_1,$$

we must have

$$Yb_{\cdot 1} = YL_{11}\beta_{\cdot 1},$$

which implies

$$L_{11}\beta_{\cdot 1} = b_{\cdot 1}. \tag{17}$$

A similar argument will show that

$$L_{12}\gamma_{\cdot 1} = c_{\cdot 1}. \tag{18}$$

Given the numbering of variables in Convention 3, it is easy to write down the specific form of L_{11} and L_{12}. It is trivial to show, consequently, the validity of (17) and (18). But the specific form for L_{i1}, L_{i2}, $i \neq 1$, cannot be given unless more is said regarding the explanatory variables appearing in the ith structural equation. Nevertheless, if by $\beta_{\cdot i}$, $\gamma_{\cdot i}$ we denote the coefficients of (respectively) the explanatory current endogenous and predetermined variables appearing in the ith equation we have

$$y_{\cdot i} = Y_i \beta_{\cdot i} + X_i \gamma_{\cdot i} + u_{\cdot i}, \qquad i = 1, 2, \ldots, m. \tag{19}$$

Since

$$Y_i \beta_{\cdot i} = Yb_{\cdot i}, \qquad X_i \gamma_{\cdot i} = Xc_{\cdot i}, \qquad i = 1, 2, \ldots, m, \tag{20}$$

and since

$$Y_i = YL_{i1}, \qquad X_i = XL_{i2}, \qquad i = 1, 2, \ldots, m, \tag{21}$$

we easily conclude that

$$L_{i1}\beta_{\cdot i} = b_{\cdot i}, \qquad L_{i2}\gamma_{\cdot i} = c_{\cdot i}, \qquad i = 1, 2, \ldots, m, \tag{22}$$

which establishes the necessary and useful connection between the elements of the matrices B and C and the structural parameters actually to be estimated.

3 The Identification Problem

Perhaps the simplest and, intuitively most transparent way of grasping the nature of the identification problem is to consider the standard textbook model of supply and demand. Thus

$$
\begin{aligned}
q_t^D &= \alpha + \beta p_t + u_{t1}, \\
q_t^S &= a + b p_t + u_{t2}, \\
q_t^S &= q_t^D.
\end{aligned} \tag{23}
$$

The first equation is the demand function, the second is the supply function, and the third is the standard competitive condition for the clearing of markets. The u_{ti}, $i = 1, 2$ represent zero mean structural errors.

Typically, we have at our disposal information concerning the *transaction quantity and price*

$$\{(q_t, p_t): t = 1, 2, \ldots, T\}.$$

The question is: can we, on the basis of this information, identify the supply and demand functions? Evidently, the answer is no! This is perhaps best brought out pictorially.

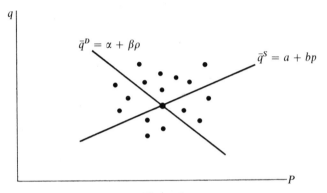

$$\bar{q}^D = \alpha + \beta p \qquad \bar{q}^S = a + bp$$

Figure 1

In Figure 1 the two lines represent the demand and supply function as they would be in the absence of the structural errors, u_{t1}, u_{t2}. The competitive equilibrium position of quantity and price would be at the point of their intersection. However, because of the presence of the zero mean structural errors the observed transaction price and quantity, in any given sample, would be represented by the cluster of points around the nonstochastic equilibrium. Clearly, given any such sample we cannot reliably estimate one line, let alone two!

In fact, the problem is worse than what is implied by the last statement. For it is not merely the efficiency of the estimator that is at stake. Rather, it is the basic logic of the inference procedure that is doubtful in the present context. For, given the market clearing condition in (23), *any pair of observations (q_t, p_t) that satisfies the equations there will also satisfy a model similar to that in (23) whose equations are linear combinations of the supply and demand functions in that model!* If that is the case, then our claim that by regressing q on p (or by whatever other means) we find "the" supply or "the" demand function characteristic of this particular market is without merit whatsoever.

So, intuitively, the identification problem arises because the equations of a given model are not sufficiently different from each other and thus may be confused with each other and/or linear combinations of such equations.

One might say that it is the task of identifying restrictions to "differentiate" the equations of the model sufficiently so that they may not be confused with each other and/or linear combinations of the equations of the system as a whole. In the case of Figure 1, if the demand function were known to depend on income, say y_t, and supply on rainfall, say r_t, then it would no longer be possible to mistake the supply for the demand function or a linear combination of the two.

With these brief preliminaries aside we now examine the formal aspects of the identification problem. In this connection it is useful, but not necessary, to introduce the additional assumption

(A.6) $u'_t. \sim N(0, \Sigma)$,

and thus write the (log) likelihood function of the T observations on the system in (1) as

$$L(B^*, C, \Sigma; Y, X) = -\frac{Tm}{2}\ln(2\pi) - \frac{T}{2}\ln|\Sigma| + \frac{T}{2}\ln|B^{*\prime}B^*|$$

$$-\frac{1}{2}\left\{\sum_{t=1}^{T}(y_t.B^* - x_t.C)\Sigma^{-1}(y_t.B^* - x_t.C)'\right\}. \quad (24)$$

The term

$$\frac{T}{2}\ln|B^{*\prime}B^*|$$

is simply the Jacobian when we consider (1) to be a transformation from $u'_t.$ to $y'_t.$.

Notice further that in (24) the sum can be simplified; thus,

$$(y_t.B^* - x_t.C)\Sigma^{-1}(y_t.B^* - x_t.C)' = (y_t. - x_t.\Pi)B^*\Sigma^{-1}B^{*\prime}(y_t. - x_t.\Pi)', \quad (25)$$

and hence the likelihood function can also be written

$$L(\Pi, \Sigma; Y, X) = -\frac{Tm}{2}\ln(2\pi) - \frac{T}{2}\ln|B^{*\prime-1}\Sigma B^{*-1}|$$

$$-\frac{1}{2}\operatorname{tr}(Y - X\Pi)B^*\Sigma^{-1}B^{*\prime-1}(Y - X\Pi)', \quad (26)$$

which is recognized as the likelihood function corresponding to the reduced form.

In view of the fact that we are not aiming at a complete discussion of the identification problem we shall be rather brief in developing the definitional framework and the derivation of the relevant results. The reader desiring greater detail is referred to Fisher [16]. Building on the intuitive notions examined earlier in this section we have

Definition 5. Suppose equation (1) describes an economic system. The triplet (B^*, C, Σ) is said to be a *structure* if all constituent elements of the three matrices above are known.

Definition 6. Suppose Equation (1) together with assumptions (A.1) through (A.5) describe an economic system. The structure (B^*, C, Σ) is said to be *admissible* if it satisfies all (known) *a priori* restrictions.

Definition 7. In the context of Definition 6 a *model* is the set of all admissible structures.

Definition 8. Two (admissible) structures are *observationally equivalent* if and only if they lead to the same likelihood function.

We have immediately

Proposition 1. *Consider the system of Equation* (1) *together with assumptions* (A.1) *through* (A.6). *Two* (*admissible*) *structures, say,*

$$(B^*_1, C_1, \Sigma_1), \qquad (B^*_2, C_2, \Sigma_2),$$

are observationally equivalent if and only if there exists a nonsingular matrix H such that

$$B^*_2 = B^*_1 H, \qquad C_2 = C_1 H, \qquad \Sigma_2 = H' \Sigma_1 H. \qquad (27)$$

PROOF. It is evident from the preceding discussion, and in particular from (26), that the likelihood function depends on the structure only through

$$\Pi = CD, \qquad \Omega = D' \Sigma D, \qquad D = B^{*-1}. \qquad (28)$$

Thus, for the two structures above suppose

$$\Pi_1 = \Pi_2, \qquad \Omega_1 = \Omega_2. \qquad (29)$$

It is evident that in such a case there must exist a nonsingular H such that

$$B^*_2 = B^*_1 H, \qquad C_2 = C_1 H, \qquad \Sigma_2 = H' \Sigma_1 H. \qquad (30)$$

Conversely, suppose that the conditions above hold; then clearly

$$\Pi_1 = \Pi_2, \qquad \Omega_1 = \Omega_2. \quad \text{q.e.d.} \qquad (31)$$

Finally we have

Definition 9. Let (B^*_1, C_1, Σ_1) be an admissible structure and consider the (nonsingular) transformation

$$B^*_2 = B^*_1 H, \qquad C_2 = C_1 H, \qquad \Sigma_2 = H' \Sigma_1 H.$$

The (matrix of the) transformation H is said to be *admissible* if the transformed structure is admissible, i.e., if (B^*_2, C_2, Σ_2) satisfies all (known) *a priori* restrictions.

Remark 7. The concepts of model, structure, and admissible transformation are connected as follows. Referring to Equation (1) and assumptions (A.1) through (A.6), let (B^*, C, Σ) be an admissible structure, and let $\mathcal{H}$ be the set of all admissible transformations. A model, then, is the set of structures

$$M = \{(B^*H, CH, H'\Sigma H): H \in \mathcal{H}\}.$$

What is meant by identification is quite easily stated in the context we have created above.

Definition 10. Let

$$S = (B^*, C, \Sigma)$$

be a structure and H an admissible transformation. The latter is said to be *trivial with respect to the ith equation* if and only if the ith column of H, say $h_{\cdot i}$, obeys

$$h_{\cdot i} = ke_{\cdot i}$$

where k is a nonnull scalar and $e_{\cdot i}$ is an m-element column vector all of whose elements are zero, save the ith, which is unity.

Remark 8. Notice that if an admissible transformation is trivial with respect to the ith equation, then

 (i) the transformed structure obeys all (known) *a priori* restrictions, and
(ii) in the transformed structure the parameters of the ith equation are given by

$$kb^*_{\cdot i}, \qquad kc_{\cdot i}, \qquad k \neq 0,$$

and the ith row and column of the transformed covariance matrix are the corresponding row and column of Σ multiplied by k.

Thus, being trivial with respect to the ith equation means that the only effect of the transformation (on the ith equation) is to multiply it by a nonnull scalar.

We thus have

Definition 11. Let Equation (1) together with assumptions (A.1) through (A.5) describe an economic system. The ith equation is said to be *identified* if and only if all admissible structures are connected by admissible transformations H, which are trivial with respect to the ith equation. The entire system is said to be *identified* if and only if every equation is identified, i.e., if all admissible transformations H are diagonal.

Remark 9. While the definition of identifiability above is intuitively quite transparent it is not very operational and hence not useful in practice. We need to derive more palpable criteria for identification. Because the derivation of such results may carry us well beyond the scope of this book some of the argumentation (proofs) will be omitted. We have

Theorem 1. *Consider the system*

$$y_t . B^* = x_t . C + u_t, \qquad t = 1, 2, \ldots, T,$$

together with assumptions (A.1) through (A.5). By Convention 3 (but not imposing normalization) the parameters relevant to the first structural equation are

$$b_{.1}^* = \begin{pmatrix} \beta_{.1}^0 \\ 0 \end{pmatrix}, \qquad c_{.1} = \begin{pmatrix} \gamma_{.1} \\ 0 \end{pmatrix},$$

where $\beta_{.1}^0$, $\gamma_{.1}$ are, respectively, $(m_1 + 1)$- and G_1-element (column) vectors.
   Partition

$$B^* = \begin{bmatrix} \beta_{.1}^0 & B_{12}^* \\ 0 & B_{22}^* \end{bmatrix}, \qquad C = \begin{bmatrix} \gamma_{.1} & C_{12} \\ 0 & C_{22} \end{bmatrix} \tag{32}$$

In such a way that B_{12}^* is $(m_1 + 1) \times (m - 1)$, B_{22}^* is $(m - m_1 - 1) \times (m - 1)$, C_{12} is $G_1 \times (m - 1)$, C_{22} is $G_1^* \times (m - 1)$, where

$$G_1^* = G - G_1.$$

Then, the first structural equation is identified if and only if

$$\operatorname{rank}\begin{pmatrix} B_{22}^* \\ C_{22} \end{pmatrix} = m - 1. \tag{33}$$

PROOF. See Dhrymes [10].

   While the preceding is perhaps the most appealing characterization of the identifiability criterion in that it puts the matter in the proper structural context, the characterization most often found in textbooks refers to the reduced form. We shall examine this aspect in somewhat greater detail because of its evident connection with certain estimation techniques to be examined in subsequent discussion.
   Identification and the Reduced Form. Consider the reduced form parameter matrix

$$\Pi = CD, \qquad D = B^{*-1}, \tag{34}$$

and note that we may also write the relation above as

$$\Pi B^* = C. \tag{35}$$

We may then pose the following question. Suppose the elements of Π are given; is it possible to recover from that the structural parameters appearing in, say, the first structural equation?
   The component of (35) relevant for answering this question is

$$\Pi b_{.1}^* = c_{.1}. \tag{36}$$

Imposing the *a priori* restrictions on $b_{\cdot 1}^*$ and $c_{\cdot 1}$ and partitioning Π conformably we have

$$\begin{bmatrix} \Pi_{11} & \Pi_{12} \\ \Pi_{21} & \Pi_{22} \end{bmatrix} \begin{pmatrix} \beta_{\cdot 1}^0 \\ 0 \end{pmatrix} = \begin{pmatrix} \gamma_{\cdot 1} \\ 0 \end{pmatrix}, \tag{37}$$

where, evidently, Π_{11} is $G_1 \times (m_1 + 1)$, Π_{12} is $G_1 \times (m - m_1 - 1)$ Π_{21} is $G_1^* \times (m_1 + 1)$, and Π_{22} is $G_1^* \times (m - m_1 - 1)$. Notice, incidentally, that Π_{21} is the matrix of the coefficients of those predetermined variables excluded from the first structural equation, in the reduced form representation of all $(m_1 + 1)$ jointly dependent variables contained therein.

We may rewrite (37) in more suggestive form as

$$\begin{aligned} \Pi_{11}\beta_{\cdot 1}^0 &= \gamma_1, \\ \Pi_{21}\beta_{\cdot 1}^0 &= 0, \end{aligned} \tag{38}$$

and we see that we are dealing with a decomposable system. If the second set of equations can be solved for $\beta_{\cdot 1}^0$ we can then trivially solve the first set in (38) for $\gamma_{\cdot 1}$! It can be shown that identification is crucially linked to the solubility of the second set of equations in (38). In particular, we have

Theorem 2. *Consider the conditions of Theorem 1. The first structural equation is identified if and only if*

$$\text{rank}(\Pi_{21}) = m_1,$$

where Π_{21} is the matrix defined in the partition of (37).

PROOF. Let (B^*, C, Σ), $(\bar{B}^*, \bar{C}, \bar{\Sigma})$ be two admissible structures. Since the reduced form is invariant with respect to admissible structures we must have, in the obvious notation,

$$\begin{aligned} \Pi_{11}\beta_{\cdot 1}^0 &= \gamma_{\cdot 1}, & \Pi_{11}\bar{\beta}_{\cdot 1}^0 &= \bar{\gamma}_{\cdot 1}, \\ \Pi_{21}\beta_{\cdot 1}^0 &= 0, & \Pi_{21}\bar{\beta}_{\cdot 1}^0 &= 0. \end{aligned}$$

Now suppose

$$\text{rank}(\Pi_{21}) = m_1.$$

Since Π_{21} is $G_1^* \times (m_1 + 1)$ and (evidently)

$$G - G_1 \geq m_1,$$

we have that the dimension of its (column) null space is unity (see Proposition 5 of *Mathematics for Econometrics*). But this means that all solutions of

$$\Pi_{21}\phi = 0 \tag{39}$$

are of the form

$$\phi = k\beta_{\cdot 1}^0,$$

where $k \neq 0$ and $\beta_{\cdot 1}^0$ is a solution of (39). Hence we conclude that all admissible transformations are trivial with respect to the first structural equation and thus the equation is identified. Conversely, suppose that the first equation is identified. This means that all admissible transformations are trivial with respect to the ith equation. Let $(B^*, C, \Sigma), (\bar{B}^*, \bar{C}, \bar{\Sigma})$ be any two admissible structures. Then we must have that

$$\bar{\beta}_{\cdot 1}^0 = k\beta_{\cdot 1}^0, \qquad \bar{\gamma}_{\cdot 1} = k\gamma_{\cdot 1}.$$

Since the structures are admissible they must satisfy all *a priori* restrictions. In particular, they must satisfy

$$\Pi_{21}\bar{\beta}_{\cdot 1}^0 = 0, \qquad \Pi_{21}\beta_{\cdot 1}^0 = 0.$$

But this shows that all solutions of (39) may be generated through scalar multiplication of a basic solution; in turn this means that the dimension of the null (column) space of Π_{21} is unity and thus

$$\text{rank}(\Pi_{21}) = m_1. \quad \text{q.e.d.}$$

Remark 10. What the theorem states is that given the reduced form we can "reconstruct" the structural parameters of, say, this first equation to within a scalar multiple if the identification condition

$$\text{rank}(\Pi_{21}) = m_1$$

holds.

Since, in view of Convention 2 (normalization), we agree that in the ith equation the ith variable has unit coefficient, it follows that the identification condition and the normalization convention enable us to recover uniquely the structural parameters of the first equation given knowledge of the reduced form.

To see how this works consider again the set of equations

$$\Pi_{21}\beta_{\cdot 1}^0 = 0.$$

By the normalization rule, we can write

$$\Pi_{*21}\beta_{\cdot 1} = \pi_{*G_1^*}, \tag{40}$$

where $\pi_{*G_1^*}$ is the first column of Π_{21}, Π_{*21} consists of the remaining columns, and we have put

$$\beta_{\cdot 1}^0 = (1, -\beta_{\cdot 1}')' \tag{41}$$

The identification condition implies

$$\text{rank}(\Pi_{*21}) = m_1, \tag{42}$$

and we note that Π_{*21} is $G_1^* \times m_1$ so that it is of full rank. Hence, Equation (40) has a unique solution and thus $\beta_{\cdot 1}^0$ is uniquely determined from (41).

Remark 11. Evidently, exactly the same argument applies to all structural equations and not merely the first. To see that let us recall that for the ith structural equation the analog of (36) is

$$\Pi b_{\cdot i}^* = c_{\cdot i}$$

and, moreover,

$$b_{\cdot i}^* = L_{i1}^* \beta_{\cdot i}^0 \quad c_{\cdot i} = L_{i2} \gamma_{\cdot i},$$

where L_{i1}^* is the selection matrix that corresponds to the $m_i + 1$ current endogenous variables appearing in the ith structural equation. Substituting we find

$$\Pi L_{i1}^* \beta_{\cdot i}^0 = L_{i2} \gamma_{\cdot i}, \qquad i = 1, 2, \ldots, m, \tag{43}$$

and we note that from the definition of L_{i1}^*,

$$\Pi L_{i1}^* = \Pi_i, \tag{44}$$

where the right-hand matrix Π_i corresponds to the endogenous variables appearing in the ith structural equation. Again, from the definition of L_{i2} we see that

$$L_{i2}' \Pi_i = \Pi_{i1}$$

yields the submatrix of Π_i corresponding to the predetermined variables appearing in the ith structural equation. Since L_{i2} is a $G \times G_i$ matrix of rank G_i, the columns of which contain all zeros except that in each column one element is unity, it follows that there exists a similar matrix, say L_{i3}, of dimension $G \times G_i^*$ ($G_i^* = G - G_i$) and rank G_i^* such that

$$L_i^* = (L_{i2}, L_{i3}) \tag{45}$$

is a $G \times G$ nonsingular matrix and

$$L_{i2}' L_{i3} = 0.$$

Notice also (Problem 1) that

$$L_{i2}' L_{i2} = I_G.$$

It is evident that

$$L_{i3}' \Pi_i = \Pi_{i2},$$

where Π_{i2} contains the coefficients of those predetermined variables *excluded* from the ith structural equation.

Premultiplying (43) by $L_i^{*\prime}$ and bearing in mind (44) we have

$$L_i^{*\prime} \Pi_i \beta_{\cdot i}^0 = L_i^{*\prime} L_{i2} \gamma_{\cdot i},$$

which, in view of the preceding, can also be written as the two sets of equations

$$\Pi_{i1} \beta_{\cdot i}^0 = \gamma_{\cdot i}, \tag{46}$$
$$\Pi_{i2} \beta_{\cdot i}^0 = 0.$$

We note that Π_{i2} is $G_i^* \times (m_i + 1)$, and, repeating the steps of Theorem 2, we can show that the ith structural equation is identified if and only if

$$\text{rank}(\Pi_{i2}) = m_i, \qquad i = 1, 2, \ldots, m. \tag{47}$$

By the normalization convention we can set one of the elements of $\beta_{\cdot i}^0$ equal to unity. By analogy with (42), if we suppress from (47) the column corresponding to the ith current endogenous variable (y_i) we shall further conclude that

$$\text{rank}(\Pi_{*i2}) = m_i, \qquad i = 1, 2, \ldots, m, \tag{48}$$

is the necessary and sufficient condition for identification, given Convention 2. In the preceding, of course, Π_{*i2} is the submatrix of Π_{i2} resulting when from the latter we suppress the column corresponding to y_i.

An immediate consequence of the preceding is

Proposition 2. *Subject to Convention 2 the relation between the reduced form and the structural parameters is*

$$\begin{bmatrix} \Pi_{*i1} & I \\ \Pi_{*i2} & 0 \end{bmatrix} \begin{pmatrix} \beta_{\cdot i} \\ \gamma_{\cdot i} \end{pmatrix} = \pi_{\cdot i}, \qquad i = 1, 2, \ldots, m,$$

where Π_{*i1}, Π_{*i2} are the submatrices of Π_{i1}, Π_{i2}, respectively, resulting when from the latter we suppress the column corresponding to y_i, and $\pi_{\cdot i}$ is the ith column of Π. Moreover, if the ith structural equation is identified,

$$\text{rank}\begin{bmatrix} \Pi_{*i1} & I \\ \Pi_{*i2} & 0 \end{bmatrix} = m_i + G_i. \tag{49}$$

PROOF. The first part of the proposition is evident by imposing the normalization on (46). For the second part we note that the matrix in (49) is $G \times (G_i + m_i)$ and hence that the proposition asserts that, if the ith structural equation is identified, the matrix in (49) has full column rank. Suppose not. Then there exist vectors a and b, of m_i and G_i elements, respectively, at least one of which is nonnull, such that

$$\Pi_{*i1}a + b = 0, \qquad \Pi_{*i2}a = 0.$$

By the identification condition, however, the second equation is satisfied only by

$$a = 0.$$

The first equation then implies

$$b = 0.$$

This is a contradiction.   q.e.d.

Remark 12. The identification condition in (48)—or in (47)— is said to be the *rank condition* and is both a necessary and a sufficient condition. On the other hand the matrix in (48) is $G_i^* \times m_i$ and hence the rank condition implies

$$G \geq G_i + m_i \qquad (G_i^* = G - G_i).$$

The condition above, which is only necessary, is referred to as the *order condition for identification*, and is the only one that can, in practice, be checked prior to estimation. The order condition provides the basis for the following.

Definition 12. The *i*th structural equation (subject to Convention 2—normalization) is said to be *identified* if (and only if)

$$\text{rank}(\Pi_{*i2}) = m_i.$$

Otherwise, it is said to be *not identified*. It is said to be *just identified* if it is identified and in addition

$$G = G_i + m_i.$$

It is said to be *overidentified* if it is identified and in addition

$$G > G_i + m_i.$$

Evidently, if

$$G < G_i + m_i$$

the rank condition cannot be satisfied and the equation is said to be *underidentified*.

4 Estimation of the GLSEM

4.1 Failure of OLS Methods

Consider the *i*th structural equation of the GLSEM after all *a priori* restrictions and the normalization convention have been imposed. This is Equation (19), which we write more compactly as

$$y_{\cdot i} = Z_i \delta_{\cdot i} + u_{\cdot i}, \qquad i = 1, 2, \ldots, m, \tag{50}$$

where

$$Z_i = (Y_i, X_i), \qquad \delta_{\cdot i} = (\beta'_{\cdot i}, \gamma'_{\cdot i})'.$$

The OLS estimator of $\delta_{\cdot i}$ is

$$\tilde{\delta}_{\cdot i} = (Z_i' Z_i)^{-1} Z_i' y_{\cdot i}. \tag{51}$$

We have

Theorem 3. *The* OLS *estimator of the parameters of the ith structural equation, as exhibited in* (51), *is inconsistent.*

PROOF. Substituting from (50) we find

$$\tilde{\delta}_{\cdot i} = \delta_{\cdot i} + (Z_i' Z_i)^{-1} Z_i' u_{\cdot i}$$

and we must determine its probability limit. But

$$Z_i' Z_i = \begin{bmatrix} Y_i' Y_i & Y_i' X_i \\ X_i' Y_i & X_i' X_i \end{bmatrix}$$

and from the reduced form representation we have

$$Y_i = X \Pi_{*i} + V_i, \qquad i = 1, 2, \ldots, m,$$
$$Y_i' Y_i = \Pi_{*i}' X' X \Pi_{*i} + V_i' V_i + \Pi_{*i}' X' V_i + V_i' X \Pi_{*i},$$
$$Y_i' X_i = \Pi_{*i}' X' X_i + V_i' X_i = \Pi_{*i}' X' X L_{i2} + V_i' X_i.$$

In view of the fact that

$$\underset{T \to \infty}{\text{plim}} \left(\frac{X' V_i}{T} \right) = 0, \qquad \underset{T \to \infty}{(\text{p})\text{lim}} \left(\frac{X' X}{T} \right) = M \tag{52}$$

with M positive definite[1] we immediately conclude

$$\underset{T \to \infty}{\text{plim}} \left(\frac{Z_i' Z_i}{T} \right) = \begin{bmatrix} \Pi_{*i}' M \Pi_{*i} + \Omega_i & \Pi_{*i}' M L_{i2} \\ L_{i2}' M \Pi_{*i} & L_{i2}' M L_{i2} \end{bmatrix}$$

$$= [\Pi_{*i}, L_{i2}]' M [\Pi_{*i}, L_{i2}] + \begin{bmatrix} \Omega_i & 0 \\ 0 & 0 \end{bmatrix}$$

where, evidently Ω_i is the submatrix of the reduced form covariance matrix corresponding to Y_i. Put

$$S_i = [\Pi_{*i}, L_{i2}] \tag{54}$$

and note that S_i is $G \times (m_i + G_i)$; moreover, in view of Proposition 2 it is of full column rank if the ith equation is identified. Hence

$$\underset{T \to \infty}{\text{plim}} \left(\frac{Z_i' Z_i}{T} \right) = S_i' M S_i + \begin{bmatrix} \Omega_i & 0 \\ 0 & 0 \end{bmatrix}. \tag{55}$$

But

$$S_i' M S_i$$

is a positive definite matrix if the ith equation is identified. Hence we conclude that

$$\underset{T \to \infty}{\text{plim}} \left(\frac{Z_i' Z_i}{T} \right)$$

is nonsingular.

---

[1] This is, strictly speaking, not directly implied by the condition (A.2), unless the model is static, i.e., it does not contain lagged endogenous variables. If the model is dynamic we would need a further assumption regarding the stability of the system. But this aspect lies outside the scope of the present volume and, thus, will not be examined. The reader may either take this on faith or simply regard the discussion as confined to static models.

Now,

$$Z_i' u_{\cdot i} = \begin{pmatrix} Y_i' u_{\cdot i} \\ X_i' u_{\cdot i} \end{pmatrix}$$

and

$$Y_i' u_{\cdot i} = \Pi_{*i}' X' u_{\cdot i} + V_i' u_{\cdot i}$$

Noting that

$$V_i = U D_i$$

where D_i is the submatrix of

$$D = (I - B)^{-1}$$

(or of B^{*-1} if normalization had not been imposed) corresponding to V_i we conclude

$$\operatorname*{plim}_{T \to \infty} \left(\frac{Z' u_{\cdot i}}{T} \right) = \begin{pmatrix} D_i' \sigma_{\cdot i} \\ 0 \end{pmatrix}, \tag{56}$$

where $\sigma_{\cdot i}$ is the ith column of Σ, the covariance matrix of the structural error. Hence, the probability limit of the OLS estimator exists and is given by

$$\operatorname*{plim}_{T \to \infty} \tilde{\delta}_{\cdot i} = \delta_{\cdot i} + \left\{ S_i' M S_i + \begin{bmatrix} \Omega_i & 0 \\ 0 & 0 \end{bmatrix} \right\}^{-1} \begin{pmatrix} D_i' \sigma_{\cdot i} \\ 0 \end{pmatrix} \tag{57}$$

Since, in general,

$$D_i' \sigma_{\cdot i} \neq 0,$$

we conclude

$$\operatorname*{plim}_{T \to \infty} \tilde{\delta}_{\cdot i} \neq \delta_{\cdot i}, \qquad i = 1, 2, \ldots, m. \quad \text{q.e.d.}$$

Remark 13. The proof of the theorem makes quite transparent why OLS "does not work," i.e., that OLS estimators are inconsistent in this context. The "reason" is the term

$$D_i' \sigma_{\cdot i},$$

which is the covariance between the structural error and the explanatory current endogenous variables. By so doing it also points to the possibility of alternative estimators that will be consistent.

4.2 Two Stage Least Squares (2SLS)

This is a method suggested by Theil [32] and Bassmann [3] and may be motivated totally by the preceding remark. Since the problem arises "because" of the correlation of Y_i and $u_{\cdot i}$, and since, by the reduced form representation

$$Y_i = X \Pi_{*i} + V_i,$$

the component of Y_i correlated with $u_{\cdot i}$ is V_i, the suggestion is self-evident that, somehow, we should "get rid" of V_i.

If we estimate the reduced form by OLS we know from previous discussion that such estimators would be consistent. Thus, we can write

$$Y = X\tilde{\Pi} + \tilde{V}, \tag{58}$$

where, evidently,

$$\tilde{\Pi} = (X'X)^{-1}X'Y, \qquad \tilde{V} = Y - \tilde{Y}, \qquad \tilde{Y} = X\tilde{\Pi}. \tag{59}$$

Again from OLS theory we know that

$$X'\tilde{V} = 0 \tag{60}$$

Thus, we can write the ith structural equation as

$$y_{\cdot i} = \tilde{Z}_i \delta_{\cdot i} + u_{\cdot i} + \tilde{V}_i \beta_{\cdot i}, \qquad \tilde{Z}_i = (\tilde{Y}_i, X_i), \tag{61}$$

where, obviously, $\tilde{Y}_i$ is the appropriate submatrix of $\tilde{Y}$ as defined in (59).

The 2SLS estimator consists of the OLS estimator of $\delta_{\cdot i}$ in the context of (60), i.e.,

$$\tilde{\delta}_{\cdot i} = (\tilde{Z}_i' \tilde{Z}_i)^{-1} \tilde{Z}_i' y_{\cdot i}. \tag{62}$$

We have

Theorem 4. The 2SLS estimator of the structural parameters of the ith equation as exhibited in (62) is consistent, provided the equation is identified.

PROOF. Substituting from (61) in (62) we have

$$\tilde{\delta}_{\cdot i} = \delta_{\cdot i} + (\tilde{Z}_i' \tilde{Z}_i)^{-1} \tilde{Z}_i' u_{\cdot i}.$$

But we note that

$$\tilde{Z}_i = (\tilde{Y}_i, X_i) = (X\tilde{\Pi}_{*i}, XL_{i2}) = X\tilde{S}_i, \qquad \tilde{S}_i = (\tilde{\Pi}_{*i}, L_{i2}) \tag{63}$$

and we can write

$$\tilde{\delta}_{\cdot i} = \delta_{\cdot i} + \left[\tilde{S}_i'\left(\frac{X'X}{T}\right)\tilde{S}_i\right]^{-1} \tilde{S}_i'\left(\frac{X'u_{\cdot i}}{T}\right). \tag{64}$$

Since by the standard assumptions

$$\plim_{T \to \infty} \left(\frac{X'u_{\cdot i}}{T}\right) = 0$$

and

$$\plim_{T \to \infty} \tilde{S}_i'\left(\frac{X'X}{T}\right)\tilde{S}_i = S_i' M S_i,$$

which is a nonsingular matrix, we conclude

$$\plim_{T \to \infty} \tilde{\delta}_{\cdot i} = \delta_{\cdot i}, \qquad i = 1, 2, \ldots, m. \quad \text{q.e.d.}$$

Remark 14. The term "two stage least squares" now becomes quite transparent. Conceptually, we may think of this procedure as proceeding in two steps. In the first, the stochastic component of Y_i that is correlated with $u_{\cdot i}$ is removed by application of OLS methods to the *reduced form*. In the second stage, we obtain consistent estimators of the structural parameters by applying OLS to the structural equation as exhibited in (61), treating

$$u_{\cdot i} + \tilde{V}_i \beta_i$$

as the "error term."

It should be stressed that while this way is excellent for conceptualizing the "reason" why 2SLS works, it is not very convenient for actually computing the estimator from a given sample nor is it the most useful viewpoint in the light of other procedures to be developed below.

Remark 15. The proof of the theorem provides a rather transparent link between the existence of the 2SLS estimator and the identification conditions. We see that

$$\frac{\tilde{Z}_i' \tilde{Z}_i}{T} = \tilde{S}_i' \tilde{M} \tilde{S}_i, \qquad \tilde{M} = \left(\frac{X'X}{T} \right),$$

and the inverse of this matrix is required for the existence of the 2SLS estimator. The identification condition assures us that the matrix

$$S_i' M S_i$$

is nonsingular. Since $\tilde{S}_i$ is a consistent estimator of S_i it follows that the probability that the matrix to be inverted is singular is negligible.

An Alternative Derivation of 2SLS. While the derivation of 2SLS in the preceding discussion is particularly useful in giving a good intuitive grasp of the method, it is not suitable when our interest lies in finding the limitations of 2SLS and/or in devising more efficient procedures. We shall now produce an alternative that fills these needs rather well.

Since $X'X$ is a positive definite matrix there exists a nonsingular matrix R such that

$$X'X = RR'. \tag{65}$$

Transform the ith structural equation through premultiplication by $R^{-1}X'$ to obtain

$$R^{-1}X'y_{\cdot i} = R^{-1}X'Z_i \delta_{\cdot i} + R^{-1}X'u_{\cdot i}, \qquad i = 1, 2, \ldots, m. \tag{66}$$

Put

$$w_{\cdot i} = R^{-1}X'y_{\cdot i}, \qquad Q_i = R^{-1}X'Z_i, \qquad r_{\cdot i} = R^{-1}X'u_{\cdot i}, \qquad i = 1, 2, \ldots, m, \tag{67}$$

and thus rewrite the system as

$$w_{\cdot i} = Q_i \delta_{\cdot i} + r_{\cdot i}, \qquad i = 1, 2, \ldots, m. \tag{68}$$

We observe that, in the context of the transformed system,

$$E(r_{\cdot i}| X_i) = 0, \qquad \text{Cov}(r_{\cdot i}|X) = \sigma_{ii}I, \qquad \plim_{T \to \infty} \frac{Q_i' r_{\cdot i}}{T} = 0, \qquad (69)$$

and thus each "structural" equation appears to obey the standard conditions for the GLM; consequently, one expects that OLS methods will be at least consistent. Pursuing this line of reasoning we obtain the OLS estimator of $\delta_{\cdot i}$ in the context of (68). Thus,

$$\tilde{\delta}_{\cdot i} = (Q_i' Q_i)^{-1} Q' w_{\cdot i}, \qquad i = 1, 2, \ldots, m. \qquad (70)$$

We have

Theorem 5. *The OLS estimator of $\delta_{\cdot i}$, in the context of the system in (68), is the 2SLS estimator of $\delta_{\cdot i}$ and as such is consistent.*

PROOF. We note that we can write

$$Q_i = R^{-1}X'[\tilde{Z}_i + (\tilde{V}_i, 0)] = R^{-1}X'X\tilde{S}_i = R'\tilde{S}_i. \qquad (71)$$

Hence, upon substitution in (70) we find

$$\tilde{\delta}_{\cdot i} = (\tilde{S}_i' RR'\tilde{S}_i)^{-1}\tilde{S}_i' X' y_{\cdot i}, \qquad i = 1, 2, \ldots, m.$$

In view of (65) and the representation of the 2SLS estimator in the proof of Theorem 4 the equivalence of the estimators in (70) and (62) is evident. Consistency then follows from Theorem 4.   q.e.d.

   Remark 16. Note that the transformation in (66) is dimension reducing. The vector $w_{\cdot i}$ is $G \times 1$; the matrix Q_i is $G \times (G_i + m_i)$ and $r_{\cdot i}$ is $G \times 1$. Moreover, we see that 2SLS has an interpretation as an OLS estimator in the context of a transformed system.
   In general we are not interested in only one structural equation out of the many that comprise the typical GLSEM, but rather in all of them. Thus, it would be convenient to have a simple representation of the systemwide 2SLS estimator. For this a more suitable notation is required.
   Put

$$w = \begin{pmatrix} w_{\cdot 1} \\ w_{\cdot 2} \\ \vdots \\ w_{\cdot m} \end{pmatrix}, \qquad Q = \text{diag}(Q_1, Q_2, \ldots, Q_m)$$

$$r = \begin{pmatrix} r_{\cdot 1} \\ r_{\cdot 2} \\ \vdots \\ r_{\cdot m} \end{pmatrix}, \qquad \delta = \begin{pmatrix} \delta_{\cdot 1} \\ \delta_{\cdot 2} \\ \vdots \\ \delta_{\cdot m} \end{pmatrix},$$

and write the entire system as

$$w = Q\delta + r. \tag{72}$$

We have immediately

Corollary 1. *The 2SLS estimator for the system as a whole is*

$$\tilde{\delta}_{2SLS} = (Q'Q)^{-1}Q'w.$$

PROOF. In view of the definitions in (71) the ith subvector of $\tilde{\delta}_{2SLS}$ is given by

$$(Q_i'Q_i)^{-1}Q_i'w_{\cdot i},$$

which is the 2SLS estimator of the parameters in the ith structural equation.

q.e.d.

4.3 Three Stage Least Squares (3SLS)

The system in (72) "looks" like a GLM, and the 2SLS estimator of the structural parameters is the OLS estimator of δ in the context of (72). But we know from the theory developed earlier that whether OLS is efficient or not depends, in the context of the GLM, on whether the covariance matrix of the errors is of the form

$$\phi I$$

where ϕ is a positive scalar. Is this so with respect to (72)?

To answer this question note that

$$\text{Cov}(r) = E(rr') = E\begin{bmatrix} r_{\cdot 1}r_{\cdot 1}' & \cdots & r_{\cdot 1}r_{\cdot m}' \\ r_{\cdot m}r_{\cdot 1}' & \cdots & r_{\cdot m}r_{\cdot m}' \end{bmatrix},$$

where $r_{\cdot i}r_{\cdot j}'$ is a square matrix of order G.

The typical block is

$$E(r_{\cdot i}r_{\cdot j}') = E[R^{-1}X'u_{\cdot i}u_{\cdot j}'XR'^{-1}]$$
$$= \sigma_{ij}R^{-1}X'XR'^{-1} = \sigma_{ij}I.$$

Consequently,

$$\text{Cov}(r) = \Sigma \otimes I = \Phi, \tag{73}$$

which is, evidently, *not* of the form ϕI. Thus, we would expect that Aitken methods applied to (72) will produce a more efficient estimator.

Accordingly, suppose that a consistent estimator is available for Σ, say $\tilde{\Sigma}$; then the feasible Aitken estimator is

$$\hat{\delta} = (Q'\tilde{\Phi}^{-1}Q)^{-1}Q'\tilde{\Phi}^{-1}w, \tag{74}$$

where, of course,

$$\tilde{\Phi} = \tilde{\Sigma} \otimes I. \tag{75}$$

Remark 17. We note that the identity matrix involved in the definition of Φ in (73) is of order G so that Φ is $mG \times mG$.

Definition 13. The estimator in (74) is said to be the *three stage least squares* (3SLS) *estimator* of the structural parameters of the GLSEM.

Remark 18. The term "three stage least squares" may be given the following explanation as to its origin. The procedure may be thought to consist of three "stages". In the first stage we "purge" the explanatory current endogenous variables of their component that is correlated with the structural error term. In the second stage we obtain consistent estimators of the elements of the covariance matrix Σ, through the 2SLS residuals. In the third stage we obtain the estimators of the system's structural parameters.

Needless to say, in computations we do not actually use a three stage procedure. It is all done in one "stage" in accordance with the relation in (74) where for the elements of $\tilde{\Sigma}$ we put

$$\tilde{\sigma}_{ij} = \frac{\tilde{u}'_{\cdot i}\tilde{u}_{\cdot j}}{T}, \qquad i, j = 1, 2, \ldots, m$$

$$\tilde{u}_{\cdot i} = [I - Z_i(\tilde{S}'_i X' X \tilde{S}_i)^{-1} \tilde{S}'_i X'] y_{\cdot i}$$

Remark 19. In comparing the 2SLS with the 3SLS estimator we note that 2SLS is only trivially a systemwide estimator, in the sense that it can be obtained by obtaining *seriatim* estimators of the structural coefficients in *each* equation. The 3SLS estimator, on the other hand, is intrinsically a systemwide estimator in the sense that we cannot obtain separately estimators of the structural parameters in only one structural equation. We obtain simultaneously estimators for all of the system's parameters. Notice that, as a consequence of this feature, misspecification errors tend to be localized with 2SLS but tend to be propagated with 3SLS. For example if a relevant predetermined variable is left out of the ith structural equation, this will affect *only* the properties of the 2SLS estimator for the ith equation, provided the variable appears elsewhere in the system. *However, the 3SLS estimator for all equations will be affected by this omission.*

To complete this section we ought to show that the 3SLS estimator is consistent. We have, first,

Proposition 3. *The estimator $\tilde{\sigma}_{ij}$ as given in (76) is consistent for the (i, j) element of the covariance matrix*

$$\Sigma = (\sigma_{ij}).$$

PROOF. By definition

$$\tilde{u}_{\cdot i} = [I - Z_i(\tilde{S}'_i X' X \tilde{S}_i)^{-1} \tilde{S}'_i X'] u_{\cdot i}.$$

Hence

$$T\tilde{\sigma}_{ij} = u'_{.i}u_{.j} + u'_{.i}X\tilde{S}_i(\tilde{S}'_iX'X\tilde{S}_i)^{-1}Z'_iZ_j(\tilde{S}'_jX'X\tilde{S}_j)^{-1}\tilde{S}'_jX'u_{.j}$$
$$- u'_{.i}X\tilde{S}_i(\tilde{S}'_iXX\tilde{S}_i)^{-1}Z'_iu_{.j} - u'_{.i}Z_j(\tilde{S}'_jX'X\tilde{S}_j)^{-1}\tilde{S}'_jX'u_{.j}. \quad (76)$$

Dividing through by T and taking probability limits we note that all terms, save the first, vanish. Hence

$$\plim_{T\to\infty} \tilde{\sigma}_{ij} = \plim_{T\to\infty} \left(\frac{u'_{.i}u_{.j}}{T}\right) = \sigma_{ij}. \quad \text{q.e.d.}$$

We may now prove

Theorem 6. *The* 3SLS *estimator as exhibited in* (74) *is consistent.*

PROOF. Substituting from (72) in (74) we find

$$\hat{\delta} = \delta + (Q'\tilde{\Phi}^{-1}Q)^{-1}Q'r.$$

In view of (71) we have

$$Q = (I \otimes R')\tilde{S}, \qquad \tilde{S} = \text{diag}(\tilde{S}_1, \tilde{S}_2, \ldots, \tilde{S}_m),$$
$$r = (I \otimes R^{-1})(I \otimes X')u, \qquad u = (u'_{.1}, u'_{.2}, \ldots, u'_{.m})'. \quad (77)$$

Consequently

$$\hat{\delta} - \delta = \left[\tilde{S}'\left(\frac{\tilde{\Sigma}^{-1} \otimes X'X}{T}\right)\tilde{S}\right]^{-1} \tilde{S}'(\tilde{\Sigma}^{-1} \otimes I)\left[\frac{(I \otimes X')u}{T}\right] \quad (78)$$

Thus, since

$$\plim_{T\to\infty} \tilde{S}'\left(\frac{\tilde{\Sigma}^{-1} \otimes X'X}{T}\right)\tilde{S} = S'(\Sigma^{-1} \otimes M)S$$

is a positive definite matrix if every equation in the system is identified, we conclude

$$\plim_{T\to\infty} \hat{\delta} = \delta. \quad \text{q.e.d.}$$

4.4 Asymptotic Properties of 2SLS and 3SLS

Although the consistency of these two estimators has been shown in earlier sections, this alone is not sufficient for empirical applications. In carrying out tests of hypotheses and other inference-related procedures we need a distribution theory.

Unfortunately, because of the highly nonlinear (in the structural errors) nature of such estimators their distribution for small samples is not known— except for a few highly special cases—*even if we were to specify the distribution of the structural errors.*

What is available is large sample (asymptotic) theory. We shall study this aspect jointly for the two estimators since in both cases we have exactly the same type of problem. We recall here some results from Chapter 8.

(i) If $f(\cdot)$ is continuous and $\{\alpha_n: n = 0, 1, 2, \ldots\}$ is a sequence converging in probability to α, then $\{f(\alpha_n): n = 0, 1, 2, \ldots\}$ converges in probability to $f(\alpha)$, provided the latter is defined.

(ii) If $\{(\alpha_n, \beta_n): n = 0, 1, 2, \ldots\}$ are two sequences such that the first converges in probability to α and the second converges in distribution to β, then $\{\alpha_n \beta_n: n = 0, 1, 2, \ldots\}$ converges in distribution to $\alpha\beta$.

Now, for the 2SLS estimator, following a sequence of substitutions similar to those leading to (78) we find

$$\sqrt{T}(\tilde{\delta} - \delta)_{2SLS} = \left[\tilde{S}'\left(\frac{I \otimes X'X}{T}\right)\tilde{S}\right]^{-1}\tilde{S}'\frac{(I \otimes X')u}{\sqrt{T}}. \tag{79}$$

Using (i) and (ii) above we easily conclude that, asymptotically,

$$\sqrt{T}(\tilde{\delta} - \delta)_{2SLS} \sim [S'(I \otimes M)S]^{-1}S'\frac{(I \otimes X')u}{\sqrt{T}},$$

$$\sqrt{T}(\tilde{\delta} - \delta)_{3SLS} \sim [S'(\Sigma^{-1} \otimes M)S]^{-1}S'(\Sigma^{-1} \otimes I)\frac{(I \otimes X')u}{\sqrt{T}}, \tag{80}$$

$$S = \operatorname{diag}(S_1, S_2, \ldots, S_m).$$

From (80) it is evident that, asymptotically, the 2SLS and 3SLS estimators are linear transformation of the variable to which

$$\frac{(I \otimes X')u}{\sqrt{T}}$$

converges, and hence the limiting distributions of the two estimators are easily derivable from the limiting distribution of the quantity above.

Deriving asymptotic distributions, however, is beyond the scope of this book. Thus, we shall solve the problem only in the highly special case, where

(a) all predetermined variables are exogenous, and
(b) the structural errors are jointly normal.

Now, if the structural errors obey

$$u'_{t\cdot} \sim N(0, \Sigma)$$

and are i.i.d. random variables, then the error vector u, defined in (77), obeys

$$u \sim N(0, \Sigma \otimes I_T).$$

Consequently, for each T we have that

$$\frac{(I \otimes X')u}{\sqrt{T}} \sim N\left[0, \Sigma \otimes \left(\frac{X'X}{T}\right)\right]. \tag{81}$$

Evidently, as T approaches infinity,

$$\frac{(I \otimes X')u}{\sqrt{T}} \sim N(0, \Sigma \otimes M),$$

where the relation in (81) is to be interpreted *as convergence in distribution.* It follows immediately then that

$$\sqrt{T}(\hat{\delta} - \delta)_{2\text{SLS}} \sim N(0, C_2), \tag{82}$$

$$\sqrt{T}(\hat{\delta} - \delta)_{3\text{SLS}} \sim N(0, C_3), \tag{83}$$

where

$$\begin{aligned} C_2 &= (S^{*\prime}S^*)^{-1}S^{*\prime}\Phi S^*(S^{*\prime}S^*)^{-1}, \quad S^* = (I \otimes \bar{R}')S, \\ C_3 &= (S^{*\prime}\Phi^{-1}S^*)^{-1}, \quad M = \bar{R}\bar{R}', \quad \Phi = \Sigma \otimes I_G. \end{aligned} \tag{84}$$

Remark 20. Although the results in (82) through (84) have been proved for a highly special case, it can be shown that they will hold under the standard assumptions (A.1) through (A.5), provided we add a condition on the stability of the model when the latter is dynamic.

It is a simple matter now to prove

Proposition 4. *The* 3SLS *estimator is asymptotically efficient relative to the* 2SLS *estimator. When*

(i) $\sigma_{ij} = 0, i \neq j$, *and*
(ii) *all equations are just identified,*

the two estimators are asymptotically equivalent—in fact, they coincide for every sample size.

PROOF. To prove the first part, for a suitable matrix H let

$$(S^{*\prime}S^*)^{-1}S^{*\prime} = (S^{*\prime}\Phi^{-1}S^*)^{-1}S^{*\prime}\Phi^{-1} + H \tag{85}$$

and note that

$$HS^* = 0. \tag{86}$$

Postmultiply (85) by Φ to obtain

$$(S^{*\prime}S^*)^{-1}S^{*\prime}\Phi = (S^{*\prime}\Phi^{-1}S^*)^{-1}S^{*\prime} + H\Phi. \tag{87}$$

Postmultiply (87) by the transpose of (85), giving

$$(S^{*\prime}S^*)^{-1}S^{*\prime}\Phi S^*(S^{*\prime}S^*)^{-1} = (S^{*\prime}\Phi^{-1}S^*)^{-1} + H\Phi H' \tag{88}$$

A comparison with (84) shows that the first part of the Proposition is proved since

$$H\Phi H' \qquad (89)$$

is a positive semidefinite matrix.

To prove the second part we note that if $\sigma_{ij} = 0$ then

$$(S^{*\prime}\Phi^{-1}S^*)^{-1}S^{*\prime}\Phi^{-1}$$

is a block diagonal matrix with typical block

$$(S_i^{*\prime}S_i^*)^{-1}S_i^{*\prime},$$

and thus coincides with

$$(S^{*\prime}S^*)^{-1}S^{*\prime}.$$

But this shows that in this case

$$H = 0,$$

which, in view of (88), shows the two estimators to be equivalent. If all equations of the system are just identified, then S^* is a nonsingular matrix, so that

$$(S^{*\prime}S^*)^{-1}S^{*\prime} = S^{*-1},$$

$$(S^{*\prime}\Phi^{-1}S^*)^{-1}S^{*\prime}\Phi^{-1} = S^{*-1}\Phi S^{*\prime-1}S^{*\prime}\Phi^{-1} = S^{*-1},$$

which again implies

$$H = 0.$$

Finally, carring out similar calculations with respect to the sample analogs of the matrices in (85) will lead to similar conclusions regarding the numerical equivalence of the two estimators when (i) and/or (ii) hold.   q.e.d.

We may summarize the preceding discussion in

Theorem 7. *Consider the* GLSEM *as exhibited in the statement of Theorem 1, together with assumptions* (A.1) *through* (A.5). *If the model is dynamic suppose that further suitable* (stability) *conditions are imposed. Subject to the normalization convention the following statements are true:*

(i) *the 2SLS estimator obeys, asymptotically,*

$$\sqrt{T}(\hat{\delta} - \delta)_{2SLS} \sim N(0, C_2), \qquad C_2 = (S^{*\prime}S^*)^{-1}S^{*\prime}\Phi S^*(S^{*\prime}S^*)^{-1},$$

   where S^* *is as defined in* (84)

(ii) *the 3SLS estimator obeys, asymptotically,*

$$\sqrt{T}(\hat{\delta} - \delta)_{3SLS} \sim N(0, C_3), \qquad C_3 = (S^{*\prime}\Phi^{-1}S^*)^{-1};$$

(iii) *the matrix difference* $C_2 - C_3$ *is positive semidefinite, i.e., 3SLS is, asymptotically, efficient relative to 2SLS;*

(iv) *if* $\Sigma = (\sigma_{ij})$, $\sigma_{ij} = 0$, $i \neq j$, *then* 2SLS *and* 3SLS *are asymptotically equivalent in the sense that* $C_2 = C_3$;
(v) *if every equation in the system is just identified,* $C_2 = C_3$;
(vi) *if the conditions in* (iv) *are taken into account during estimation,* 2SLS *and* 3SLS *estimators are numerically equivalent for every sample size*;
(vii) *if every equation of the system is just identified then* 2SLS *and* 3SLS *estimators are numerically equivalent for every sample size.*

To conclude this section we show how the parameters of the asymptotic distribution may be estimated consistently and how they may be employed in inference-related procedures.

Referring to both 2SLS and 3SLS we see that they involve two basic matrices, viz., S^* and Φ. It is clear that in either case we may estimate the latter consistently by

$$\tilde{\Phi} = \tilde{\Sigma} \otimes I, \qquad \tilde{\Sigma} = (\tilde{\sigma}_{ij}), \qquad \tilde{\sigma}_{ij} = \left(\frac{1}{T}\right)\tilde{u}'_{\cdot i}\tilde{u}_{\cdot j}, \tag{90}$$

where the $\tilde{u}_{\cdot i}$, $i = 1, 2, \ldots, m$, are, say, the 2SLS residuals, i.e.,

$$\tilde{u}_{\cdot i} = y_{\cdot i} - Z_i \tilde{\delta}_{\cdot i}, \qquad i = 1, 2, \ldots, m, \tag{91}$$

and $\tilde{\delta}_{\cdot i}$ is the 2SLS estimator of the structural parameters in the ith structural equation. We observe in particular that the covariance matrix corresponding to the latter's parameters is consistently estimated by

$$\tilde{\sigma}_{ii}(\tilde{S}_i^{*\prime}\tilde{S}_i^*)^{-1} = \tilde{\sigma}_{ii}\left[\tilde{S}_i'\left(\frac{X'X}{T}\right)\tilde{S}_i\right]^{-1} = T\tilde{\sigma}_{ii}(\tilde{S}_i'X'X\tilde{S}_i)^{-1} \qquad \tilde{S}_i = (\tilde{\Pi}_{*i}, L_{i2}), \tag{92}$$

where L_{i2} is the selection matrix corresponding to the predetermined variables appearing in the ith structural equation and $\tilde{\Pi}_{*i}$ is the OLS estimator of the reduced form representation of the *explanatory* current endogenous variables contained therein.

Similarly, the "cross" covariance matrix between the parameter estimators for the ith and jth equation is consistently estimated by

$$\tilde{\sigma}_{ij}(\tilde{S}_i^{*\prime}\tilde{S}_i^*)^{-1}\tilde{S}_i^{*\prime}\tilde{S}_j^*(\tilde{S}_j^{*\prime}\tilde{S}_j^*)^{-1} = \tilde{\sigma}_{ij}\left[\tilde{S}_i'\left(\frac{X'X}{T}\right)\tilde{S}_i\right]^{-1}\tilde{S}_i'\left(\frac{X'X}{T}\right)\tilde{S}_j$$

$$\times \left[\tilde{S}_j'\left(\frac{X'X}{T}\right)\tilde{S}_j\right]^{-1} = T\tilde{\sigma}_{ij}(\tilde{S}_i'X'X\tilde{S}_i)^{-1}\tilde{S}_i'X'X\tilde{S}_j(\tilde{S}_j'X'X\tilde{S}_j)^{-1}. \tag{93}$$

For 3SLS the covariance matrix of the estimator of the parameters of a single equation cannot be easily exhibited, but the consistent estimator of the covariance matrix for all of the system's parameters is simply

$$\tilde{C}_3 = (\tilde{S}^{*\prime}\tilde{\Phi}^{-1}\tilde{S}^*)^{-1} = \left[\tilde{S}'\left(\frac{\tilde{\Sigma}^{-1} \otimes X'X}{T}\right)\tilde{S}\right]^{-1}$$

$$= T[\tilde{S}'(\tilde{\Sigma}^{-1} \otimes X'X)\tilde{S}]^{-1} = T\tilde{C}_3^*. \tag{94}$$

It is clear that tests of significance or other inference-related procedures can be based on (consistent estimators of) the parameters of the asymptotic distribution. Thus, for example, suppose $\tilde{\delta}_{ki}$ is the 2SLS estimator of the kth parameter in the ith structural equation and δ_{ki}^0 is the value specified by the null hypothesis. Let

$$T\tilde{q}_{kk}^i$$

be the kth diagonal element of the covariance matrix estimator as exhibited in (92). We may form the ratio

$$\sqrt{T}(\tilde{\delta}_{ki} - \delta_{ki}^0)/\sqrt{T\tilde{q}_{kk}^i},$$

which may easily be shown to converge in distribution to the $N(0, 1)$ variable. Hence cancelling the factor $\sqrt{T}$ we conclude that, asymptotically,

$$\frac{\tilde{\delta}_{ki} - \delta_{ki}^0}{\sqrt{\tilde{q}_{kk}^i}} \sim N(0, 1). \tag{95}$$

Hence, tests on single parameters can be based on the standard normal distribution. *Such tests, however, are exact only asymptotically. For finite samples they are only approximately valid.* To carry out tests of significance on groups of parameters or linear functions involving many parameters we proceed as follows. Let A be a suitable matrix and let the null hypothesis to be tested be

$$H_0: \quad A\delta = A\delta^\circ.$$

Note that the consistent estimator of the 2SLS covariance matrix for the entire system may be written as

$$
\begin{aligned}
\tilde{C}_2 &= (\tilde{S}*'\tilde{S}*)^{-1}\tilde{S}*'\tilde{\Phi}\tilde{S}*(\tilde{S}*'\tilde{S}*)^{-1} \\
&= T[\tilde{S}'(I \otimes X'X)\tilde{S}]^{-1}\tilde{S}'(\tilde{\Sigma} \otimes X'X)\tilde{S}[\tilde{S}'(I \otimes X'X)\tilde{S}]^{-1} \\
&= T\tilde{C}_2^*.
\end{aligned} \tag{96}
$$

Since

$$\sqrt{T}A(\tilde{\delta} - \delta^0)_{\text{2SLS}} \sim N(0, AC_2A'). \tag{97}$$

we have

$$T[A(\tilde{\delta} - \delta)]'[A(T\tilde{C}_2^*)A']^{-1}[A(\tilde{\delta} - \delta^0)] \sim \chi^2_{\text{rank}(A)}.$$

But cancelling a factor of T yields

$$[A(\tilde{\delta} - \delta^0)_{\text{2SLS}}]'[A\tilde{C}_2^*A']^{-1}[A(\tilde{\delta} - \delta^0)_{\text{2SLS}}] \sim \chi^2_{\text{rank}(A)}. \tag{98}$$

Similarly, for 3SLS estimators we obtain

$$[A(\tilde{\delta} - \delta^0)_{\text{3SLS}}][(A\tilde{C}_3^*A')]^{-1}[A(\tilde{\delta} - \delta^0)_{\text{3SLS}}] \sim \chi^2_{\text{rank}(A)}, \tag{99}$$

so that tests involving groups of coefficients estimated by 2SLS or 3SLS methods can be carried out using the chi-square distribution.

Remark 21. Note that tests involving single parameters are special cases of the procedures entailed by the representations in (98) and (99). If, for example, we are interested in δ_{ki}, then A would be a single *row* vector having all zero elements except for that corresponding to δ_{ki}, which would be unity. Thus, in this case

$$A(\tilde{\delta} - \delta^0)_{2SLS} = (\tilde{\delta}_{ki} - \delta_{ki}^0).$$

Similarly,

$$A\tilde{C}_2^* A' = q_{kk}^i, \qquad \text{rank}(A) = 1,$$

and hence (98) reduces to

$$\frac{(\tilde{\delta}_{ki} - \delta_{ki}^0)^2}{\tilde{q}_{kk}^i} \sim \chi_1^2,$$

which simply gives the asymptotic distribution of the square of the left-hand side of (95).

Remark 22. The typical computer output for 2SLS and 3SLS estimation gives, along with parameter estimates, standard errors; on demand, it will also provide estimates for the entire covariance matrix. The user should bear in mind that such estimates are, in fact, $\tilde{C}_2^*$ for 2SLS and $\tilde{C}_3^*$ for 3SLS. Thus, if one forms the usual "t-ratios," i.e., the ratio of the estimated coefficient minus the value specified by the null hypothesis, to the standard error, such ratios have asymptotically the unit normal distribution. Similarly, if one forms the usual F-ratios such quantities will have asymptotically a chi-square distribution.

We may summarize this by stating that if one operates with the output of 2SLS and/or 3SLS computer programs in the "same way one operates with the GLM" then one would be carrying out tests which are exact asymptotically except that what in the GLM context would have been a t-test is now a unit normal test and what would have been an F-test is now a chi-square test.

5 Prediction from the GLSEM

Prediction problems arise as follows in the context of the standard GLSEM considered earlier:

$$y_{t.} = y_{t.}B + x_{t.}C + u_{t.}, \qquad t = 1, 2, \ldots, T.$$

Given the sample, we estimate the unknown elements of B and C, thus obtaining (consistent) estimates $\tilde{B}$ and $\tilde{C}$. If the predetermined variables of the model are solely exogenous, and if the values to be assumed by such variables at some future time $T + \tau$, $\tau > 0$, are specified, the prediction problem consists of predicting, in some suitably optimal sense, the values to be assumed by the endogenous variables of the model at time $T + \tau$.

The first thing to note is that the structural equations cannot, *per se*, be used in prediction or forecasting. For if we put, say,

$$\tilde{y}_{T+\tau\cdot} = \tilde{y}_{T+\tau\cdot}\tilde{B} + x_{T+\tau\cdot}\tilde{C} \tag{100}$$

where $x_{T+\tau\cdot}$ is the prespecified vector of the exogenous variables, and $\tilde{B}$ and $\tilde{C}$ are the consistent estimates of the parameter matrices B and C, we see that the variables to be forecast, $\tilde{y}_{T+\tau\cdot}$, appear on both sides of the equation; hence, it is not possible, without further manipulation, to determine the forecasts of the dependent variables, $\tilde{y}_{T+\tau\cdot}$, given the values to be assumed by the exogenous variables, i.e., $x_{T+\tau\cdot}$.

On the other hand we can operate with the reduced form of the model

$$y_{t\cdot} = x_{t\cdot}\Pi + v_{t\cdot}, \qquad \Pi = CD, \qquad D = (I - B)^{-1}$$

If an estimate for Π is available, say $\tilde{\Pi}$, we can then easily predict

$$\hat{y}_{T+\tau\cdot} = x_{T+\tau\cdot}\tilde{\Pi}. \tag{101}$$

Whether the prediction in (101) is equivalent to that implied by (100) and, if not, whether it is "better" or "worse" than that implied by (100), depends on how the reduced form matrix is estimated. To see the connection between "prediction efficiency" and "estimation efficiency" let us proceed a bit more systematically.

If we predict according to (101) we commit the forecast error (or prediction error)

$$e_{T+\tau\cdot} = y_{T+\tau\cdot} - \hat{y}_{T+\tau\cdot} = v_{T+\tau\cdot} - x_{T+\tau\cdot}(\tilde{\Pi} - \Pi) \tag{102}$$

and we see that the prediction error consists of two components, the reduced form error $v_{T+\tau\cdot}$ and a component $x_{T+\tau\cdot}(\tilde{\Pi} - \Pi)$ due to the fact that Π is not known but is estimated through $\tilde{\Pi}$. It should be noted that if the exogenous variables are not known but are "guessed at", say by $x^*_{T+\tau\cdot}$, then we should have another component due to the failure of the exogenous variable specification to coincide with what actually transpires.

In view of the standard assumption (A.1)—the i.i.d. assumption relating to the structural errors—we note that $v_{T+\tau\cdot}$, the reduced form error, is independent of the term containing $\tilde{\Pi}$ since the latter is estimated from a sample that runs only through "period" T, while $\tau > 0$. Hence the covariance of the forecast error will have only two components. To see this more precisely, rewrite (102) as

$$e'_{T+\tau\cdot} = v'_{T+\tau\cdot} - (I \otimes x_{T+\tau\cdot})(\tilde{\pi} - \pi) \tag{103}$$

where

$$\pi = (\pi'_{\cdot 1}, \pi'_{\cdot 1}, \ldots, \pi'_{\cdot m})' \tag{104}$$

and $\pi_{\cdot i}, i = 1, 2, \ldots, m$, is the ith column of Π.

For the special case where Π is estimated directly from the reduced form by OLS we easily establish that

$$E(\tilde{\pi}) = \pi, \qquad \text{Cov}(\tilde{\pi}) = \Omega \otimes (X'X)^{-1}, \qquad \Omega = D'\Sigma D, \qquad (105)$$

Ω being the covariance matrix of the reduced form errors. Thus, we may deduce unambiguously for this special case

$$E(e'_{T+\tau \cdot}) = 0, \qquad \text{Cov}(e'_{T+\tau \cdot}) = \Omega + (\Omega \otimes x_{T+\tau \cdot}(X'X)^{-1}x'_{T+\tau \cdot}). \qquad (106)$$

But even in the general case where Π is estimated indirectly through the estimators of the structural parameters, so that

$$\tilde{\Pi} = \tilde{C}\tilde{D}, \qquad \tilde{D} = (I - \tilde{B})^{-1},$$

we can write approximately

$$\text{``Cov}(e'_{T+\tau \cdot})\text{''} = \Omega + (I \otimes x_{T+\tau \cdot}) \text{``Cov}(\tilde{\pi})\text{''} (I \otimes x'_{T+\tau \cdot}), \qquad (107)$$

where now "$\text{Cov}(\tilde{\pi})$" indicates an approximation using the asymptotic distribution of the indirect (restricted) reduced form estimator—assuming that we can derive it. In (106) or (107) it is clear that the first component, Ω, is invariant with respect to the estimator one uses for Π. The second component, however, is a quadratic form in the covariance matrix of the elements of $\tilde{\Pi}$. Hence, how efficiently one estimates Π will determine, given $x_{T+\tau \cdot}$, how efficiently one forecasts the dependent variables.

As is implicit in the preceding discussion there are at least two obvious methods for estimating Π. One is to estimate it directly through the reduced form, by OLS methods, ignoring all *a priori* restrictions on the elements of the structural coefficient matrices B and C. The other basic method is to use the 2SLS or 3SLS estimators of B and C to form the estimator

$$\tilde{\Pi} = \tilde{C}\tilde{D}, \qquad \tilde{D} = (I - \tilde{B})^{-1}, \qquad (108)$$

which embodies all prior restrictions on the structural parameters. In order to compare the merits of the two procedures it is necessary to determine the asymptotic properties of the estimator in (108), a task to which we now turn. Thus, consider

$$\tilde{\Pi} - \Pi = \tilde{C}\tilde{D} - CD = \tilde{C}\tilde{D} - \tilde{C}D + \tilde{C}D - CD$$
$$= \tilde{C}\tilde{D}(D^{-1} - \tilde{D}^{-1})D + (\tilde{C} - C)D.$$

Bearing in mind that

$$D^{-1} - \tilde{D}^{-1} = \tilde{B} - B$$

we see that we can write

$$\tilde{\Pi} - \tilde{\Pi} = (\tilde{\Pi}, I)\binom{\tilde{B} - B}{\tilde{C} - C}D. \qquad (109)$$

In order to determine the asymptotic distribution of the elements of the left member of (109) it is convenient, as a matter of notation, to express them in column form. By definition the ith column of

$$(\tilde{\Pi}, I)\begin{pmatrix} \tilde{B} - B \\ \tilde{C} - C \end{pmatrix}$$

is given by

$$(\tilde{\Pi}, I)\begin{pmatrix} \tilde{b}_{\cdot i} - b_{\cdot i} \\ \tilde{c}_{\cdot i} - c_{\cdot i} \end{pmatrix}, \qquad i = 1, 2, \ldots, m.$$

Using the selection matrix notation of Convention 3 we have

$$\begin{pmatrix} \tilde{b}_{\cdot i} - b_{\cdot i} \\ \tilde{c}_{\cdot i} - c_{\cdot i} \end{pmatrix} = \begin{bmatrix} L_{i1} & 0 \\ 0 & L_{i2} \end{bmatrix}\begin{pmatrix} \tilde{\beta}_{\cdot i} - \beta_{\cdot i} \\ \tilde{\gamma}_{\cdot i} - \gamma_{\cdot i} \end{pmatrix} = \begin{bmatrix} L_{i1} & 0 \\ 0 & L_{i2} \end{bmatrix}(\tilde{\delta}_{\cdot i} - \delta_{\cdot i}),$$

$$i = 1, 2, \ldots, m. \tag{110}$$

Consequently, we can write

$$(\tilde{\Pi}, I)\begin{pmatrix} \tilde{b}_{\cdot i} - b_{\cdot i} \\ \tilde{c}_{\cdot i} - c_{\cdot i} \end{pmatrix} = (\tilde{\Pi}, I)\begin{bmatrix} L_{i1} & 0 \\ 0 & L_{i2} \end{bmatrix}(\tilde{\delta}_{\cdot i} - \delta_{\cdot i}) = \tilde{S}_i(\tilde{\delta}_{\cdot i} - \delta_{\cdot i}),$$

$$i = 1, 2, \ldots, m. \tag{111}$$

With this notation we have the representation

$$(\tilde{\pi} - \pi) = (D' \otimes I)\tilde{S}(\tilde{\delta} - \delta), \tag{112}$$

where, for example,

$$S = \mathrm{diag}(S_1, S_2, \ldots, S_m)\pi = (\pi'_{\cdot 1}, \pi'_{\cdot 2}, \ldots, \pi'_{\cdot m})' \tag{113}$$

and $\pi_{\cdot i}$, $i = 1, 2, \ldots, m$, is the ith column of Π.

Since $\tilde{S}$ converges in probability to S and $\sqrt{T}(\tilde{\delta} - \delta)$ has a well-defined limiting distribution we conclude that, asymptotically,

$$\sqrt{T}(\tilde{\pi} - \pi) \sim (D' \otimes I)S\sqrt{T}(\tilde{\delta} - \delta). \tag{114}$$

Thus, asymptotically, the estimator of the restricted reduced form induced by a given structural estimator behaves like a linear transformation of the latter. Hence, there is no novel problem to be solved in dealing with the asymptotic distribution of the restricted reduced form (RRF) as it is induced by a specified estimator of the structural form.

To facilitate future discussion we introduce a number of definitions.

Definition 14. Consider the GLSEM together with the standard assumptions (A.1) through (A.5) and Convention 3

$$Y = YB + XC + U,$$

and consider further its reduced form

$$Y = X\Pi + V, \qquad \Pi = CD, \qquad V = UD, \qquad D = (I - B)^{-1}.$$

An estimator of the reduced form matrix Π that does not take into account the restrictions placed on C and B is said to be an *unrestricted reduced form* (URF) estimator. An estimator of the reduced form matrix Π defined by

$$\tilde{\Pi} = \tilde{C}\tilde{D}, \qquad \tilde{D} = (I - \tilde{B})^{-1},$$

where $\tilde{C}$ and $\tilde{B}$ refer to structural estimators that take into account all *a priori* restrictions, is said to be a *restricted reduced form* (RRF) *estimator induced by the given structural estimator.*

We begin the formal discussion of the comparative merits of the various methods of forecasting from an econometric model by establishing the properties of the URF estimator briefly alluded to in Equation (105).

We observe that the reduced form of the GLSEM is a system of GLMs

$$y_{\cdot i} = X\pi_{\cdot i} + v_{\cdot i}, \qquad i = 1, 2, \ldots, m, \tag{115}$$

each model containing exactly the same variables. We recall from Chapter 3 that in such a case the OLS estimator is efficient irrespective of the properties of the error terms. Putting

$$y = (I \otimes X)\pi + v \tag{116}$$

where

$$y = (y'_{\cdot 1}, y'_{\cdot 2}, \ldots, y'_{\cdot m})', \qquad v = (v'_{\cdot 1}, v'_{\cdot 2}, \ldots, v'_{\cdot m})', \qquad v_{\cdot i} = Ud_{\cdot i}, \tag{117}$$

and $d_{\cdot i}$ is the ith column of D, the OLS estimator is given by

$$\tilde{\pi} = [(I \otimes X)'(I \otimes X)]^{-1}(I \otimes X')y = \pi + [I \otimes (X'X)^{-1}](I \otimes X')v. \tag{118}$$

Thus

$$\sqrt{T}(\tilde{\pi} - \pi) = \left[I \otimes \left(\frac{X'X}{T}\right)^{-1}\right]\frac{1}{\sqrt{T}}(I \otimes X')v. \tag{119}$$

Since

$$v = (D' \otimes I)u, \qquad u = (u'_{\cdot 1}, u'_{\cdot 2}, \ldots, u'_{\cdot m})' \tag{120}$$

we can rewrite (119) as

$$\sqrt{T}(\tilde{\pi} - \pi) = \left[I \otimes \left(\frac{X'X}{T}\right)^{-1}\right](D' \otimes I)\frac{I}{\sqrt{T}}(I \otimes X')u. \tag{121}$$

Since by assumption (or implication)

$$(\mathrm{p})\lim_{T \to \infty} \frac{X'X}{T} = M$$

exists as a positive definite matrix we see by comparison with Equations (80) that the problem involved in the determination of the asymptotic distribution of the URF estimator is exactly the same as that encountered in the determination of the limiting distribution of the 2SLS and 3SLS estimators. Consequently, using exactly the same arguments we conclude that

$$\sqrt{T}(\tilde{\pi} - \pi)_{\text{URF}} \sim N(0, F_0), \qquad F_0 = (D' \otimes \bar{R}'^{-1})\Phi(D \otimes \bar{R}^{-1}). \quad (122)$$

Returning now to Equation (114) and noting from Equation (84) the definition of S^* we can rewrite the former as

$$\sqrt{T}(\tilde{\pi} - \pi)_{\text{RRF}} \sim (D' \otimes \bar{R}'^{-1})S^*\sqrt{T}(\tilde{\delta} - \delta). \quad (123)$$

If the structural parameters have been estimated by 2SLS or 3SLS the asymptotic distribution of the left member of (123) is quite easily obtained from the discussion of Section 4.4 as follows.

$$\sqrt{T}(\tilde{\pi} - \pi)_{\text{RRF(2SLS)}} \sim N(0, F_2),$$
$$\sqrt{T}(\tilde{\pi} - \pi)_{\text{RRF(3SLS)}} \sim N(0, F_3), \qquad (124)$$

where

$$F_2 = (D' \otimes \bar{R}'^{-1})S^*(S^{*'}S^*)^{-1}S^{*'}\Phi S^*(S^{*'}S^*)^{-1}S^{*'}(D \otimes \bar{R}^{-1}),$$
$$F_3 = (D' \otimes \bar{R}'^{-1})S^*(S'\Phi^{-1}S^*)^{-1}S^{*'}(D \otimes \bar{R}^{-1}). \qquad (125)$$

We may summarize the preceding in

Theorem 8. *Consider the GLSEM and its associated reduced form under the conditions specified in Theorem 7. Then the following statements are true: asymptotically*

(i) $\sqrt{T}(\tilde{\pi} - \pi)_{\text{URF}} \sim N(0, F_0)$,

(ii) $\sqrt{T}(\tilde{\pi} - \pi)_{\text{RRF(SLS)}} \sim N(0, F_2)$,

(iii) $\sqrt{T}(\tilde{\pi} - \pi)_{\text{RRF(3SLS)}} \sim N(0, F_3)$,

where

$$F_0 = (D' \otimes \bar{R}'^{-1})\Phi(D \otimes \bar{R}^{-1}),$$
$$F_2 = (D' \otimes \bar{R}'^{-1})S^*(S^{*'}S^*)^{-1}S^{*'}\Phi S^*(S^{*'}S^*)^{-1}S^{*'}(D \otimes \bar{R}^{-1}).$$
$$F_3 = (D' \otimes \bar{R}'^{-1})S^*(S^{*'}\Phi^{-1}S^*)^{-1}S^{*'}(D \otimes \bar{R}^{-1}),$$

With the results of Theorem 8 at hand it is now rather simple to determine the relative (asymptotic) efficiency of the various reduced form estimators, since this exercise involves the determination of whether the matrix differences

$$F_0 - F_3, \qquad F_0 - F_2, \qquad F_2 - F_3$$

are positive (semi)definite or indefinite. If we make an unambiguous determination about the relative efficiencies of various reduced form estimators, then by the previous discussion, we would have answered the question regarding the merits of forecasting from the unrestricted and the restricted reduced form of the GLSEM. We have

Theorem 9. *Consider the matrices F_i, $i = 0, 2, 3$, defined in Theorem 8. The following statements are true:*

(i) $F_0 - F_3$ *is positive semidefinite;*
(ii) $F_2 - F_3$ *is positive semidefinite;*
(iii) $F_0 - F_2$ *is indefinite, except in highly special cases.*

PROOF. For (i) we note that

$$F_0 - F_3 = (D' \otimes \bar{R}'^{-1})[\Phi - S^*(S^{*\prime}\Phi^{-1}S^*)^{-1}S^{*\prime}](D \otimes \bar{R}^{-1}).$$

Since the matrix $(D \otimes \bar{R}^{-1})$ is clearly nonsingular we need only examine the matrix in square brackets. Now consider the characteristics roots of

$$S^*(S^{*\prime}\Phi^{-1}S^*)^{-1}S^{*\prime}$$

in the metric of Φ, i.e.,

$$|\lambda\Phi - S^*(S^{*\prime}\Phi^{-1}S^{*\prime})^{-1}S^{*\prime}| = 0. \tag{126}$$

The nonzero roots of (126), however, are exactly the roots of

$$|\Phi||\lambda I - (S^{*\prime}\Phi^{-1}S^*)^{-1}S^{*\prime}\Phi^{-1}S^*| = 0.$$

The latter, however, has $K = \sum_{i=1}^m (m_i + G_i)$ roots, all of which are unity. Thus, the matrix of the characteristic roots of (126) is

$$\Lambda = \begin{bmatrix} I_K & 0 \\ 0 & 0 \end{bmatrix}. \tag{127}$$

From Proposition 63 of *Mathematics for Econometrics* we know that there exists a nonsingular matrix, say P, such that

$$\Phi = P'P, \qquad S^*(S^{*\prime}\Phi^{-1}S^*)^{-1}S^{*\prime} = P'\Lambda P.$$

Hence

$$\Phi - S^*(S^{*\prime}\Phi^{-1}S^*)^{-1}S^{*\prime} = P'P - P'\Lambda P = P'\begin{bmatrix} 0 & 0 \\ 0 & I_{mG-K} \end{bmatrix}P,$$

which is clearly positive semidefinite.

For (ii) we note that

$$F_2 - F_3 = (D' \otimes R'^{-1})S^*[C_2 - C_3]S^{*\prime}(D \otimes \bar{R}^{-1}),$$

where C_2 and C_3 are, respectively, the covariance matrices of the asymptotic distribution of the 2SLS and 3SLS estimators. Its validity, therefore, follows immediately from (iii) of Theorem 7.

For (iii) we need to evaluate the difference

$$F_0 - F_2 = (D' \otimes \bar{R}'^{-1})[\Phi - A\Phi A](D \otimes \bar{R}^{-1}),$$

where

$$A = S^*(S^{*\prime}S^*)^{-1}S^{*\prime}.$$

As before we need only examine the matrix in square brackets. Since A is a symmetric idempotent matrix of dimension mG and rank K the matrix of its characteristic roots is also given by (127). Let T be the orthogonal matrix of the associated characteristic vectors. We may, thus, write

$$\Phi - A\Phi A = \Phi - T\Lambda T'\Phi T\Lambda T' = T[\Phi^* - \Lambda\Phi^*\Lambda]T' \qquad (128)$$

where

$$\Phi^* = T'\Phi T$$

and Λ is as in (127).

Whether the matrix on the furthest left side of (128) is semidefinite or indefinite depends only on whether the matrix in square brackets on the furthest right side of (128) has these properties. Partition

$$\Phi^* = \begin{bmatrix} \Phi_{11}^* & \Phi_{12}^* \\ \Phi_{21}^* & \Phi_{22}^* \end{bmatrix}$$

in such a way that Φ_{11}^* is $K \times K$, Φ_{22}^* $(mG - K) \times (mG - K)$, and so on. In view of the definition of Λ in (127) we have

$$\Phi^* - \Lambda\Phi^*\Lambda = \begin{bmatrix} 0 & \Phi_{12}^* \\ \Phi_{21}^* & \Phi_{22}^* \end{bmatrix}. \qquad (129)$$

The matrix in (129), however, is indefinite unless Φ_{12}^* is of rank zero, i.e., it is the zero matrix. First note that if

$$\Phi_{12}^* = 0$$

then

$$\Phi_{21}^* = \Phi_{12}^{*\prime} = 0,$$

and, since Φ^* is positive definite, so is Φ_{22}^*. Hence the matrix in (129) would be, in such a case, positive semidefinite. On the other hand, suppose Φ_{12}^* is not of rank zero and consider two appropriately dimensioned vectors, α and β. We find

$$(\alpha', \beta') \begin{bmatrix} 0 & \Phi_{12}^* \\ \Phi_{21}^* & \Phi_{22}^* \end{bmatrix} \begin{pmatrix} \alpha \\ \beta \end{pmatrix} = 2\alpha'\Phi_{12}^*\beta + \beta'\Phi_{22}^*\beta. \qquad (130)$$

We shall now show that for some nonnull vector $\begin{pmatrix} \alpha \\ \beta \end{pmatrix}$ the right member of (130) is positive, while for others it is negative. This will show that the matrix is indefinite. Thus let

$$\alpha = 0, \qquad \beta \neq 0.$$

For this choice clearly the right number is positive, in view of the fact that Φ_{22}^* is positive definite. Since Φ_{12}^* is not of rank zero evidently there exists a vector $\alpha \neq 0$, such that

$$\Phi_{21}^*\alpha \neq 0.$$

Choose

$$\beta = -\Phi_{22}^{*-1}\Phi_{21}^{*}\alpha$$

and note that for this choice the right side becomes

$$-\alpha'\Phi_{12}^{*}\Phi_{22}^{*-1}\Phi_{21}^{*}\alpha < 0,$$

thus completing the proof of the theorem.   q.e.d.

Corollary 2. *If all equations of the system are just identified then*

$$F_0 = F_2 = F_3.$$

PROOF. Obvious since for such a case

$$K = mG, \qquad A = I, \qquad C_2 = C_3.$$

Corollary 3. *If*

$$\Sigma = (\sigma_{ij}) \quad and \quad \sigma_{ij} = 0 \quad for \ i \neq j,$$

then

$$F_0 - F_2$$

is positive semidefinite.

PROOF. From (iv) of Theorem 7 we know that in such a case $C_2 = C_3$, and hence

$$F_2 = F_3.$$

The corollary then follows from (i) of Theorem 8. An alternative proof is given in Problem 21 at the end of this chapter.   q.e.d.

Remark 23. In (iii) of the statement of Theorem 9 it is claimed that $F_0 - F_2$ is indefinite "except in highly special cases," but this is not elaborated. On the other hand, in the proof of the theorem it becomes obvious that the matrix difference above is positive semidefinite if and only if $\Phi_{12}^{*} = 0$. Otherwise it is indefinite. From Problem 20 at the end of this chapter we have the representation

$$\Phi_{12}^{*} = [\sigma_{ij} T_1^{(i)'} T_2^{(j)}], \qquad i, j = 1, 2, \ldots, m, \tag{131}$$

where $T_1^{(i)}$ contains the characteristic vectors of

$$A_i = S_i^{*'}(S_i^{*'}S_i)^{-1}S_i^{*'}, \qquad i = 1, 2, \ldots, m,$$

corresponding to its *nonzero* (unit) roots, while $T_2^{(j)}$ contains the characteristic vectors of A_j corresponding to its *zero roots*. The representation in (131) makes the validity of the claim above quite transparent. In general

$$T_1^{(i)'} T_2^{(j)} \neq 0, \qquad i \neq j,$$

except in highly special cases. If, e.g., the ith and jth equations are *just identified*, $A_i = A_j = I$, and in such a case $T_2^{(j)} = T_2^{(i)}$. Moreover, if $\sigma_{ij} = 0$, $i \neq j$, then clearly $\Phi_{12}^* = 0$. Or if $\sigma_{ij} \neq 0$, $i \neq j$, only when the ith and jth equations are just identified (and otherwise $\sigma_{ij} = 0$, $i \neq j$), then again $\Phi_{12}^* = 0$. In the general case, however, one would expect that $\Phi_{12}^* \neq 0$, and hence that the URF estimator would not be inefficient relative to the RRF estimator induced by the 2SLS estimator of the structural parameters.

Remark 24. Since, implicit in the discussion of previous chapters is the premise that the more valid restrictions on parameters we take into account in the estimation phase the more efficient the resulting estimator, the reader may be puzzled at the content of Remark 23, which appears to be counter-intuitive. This seeming contradiction is easily dispelled, however, if we note that in estimating the GLSEM we have two kinds of information at our disposal, sample information given by the data matrices (Y, X) and prior information in the form of zero restrictions on the elements of the structural coefficient matrices B and C.

Now, the URF estimator of Π takes into account all relevant sample information but none of the prior information in estimating every one of its elements. The 2SLS induced RRF estimator of Π takes into account all prior information but not necessarily all relevant sample information since it ignores the fact that the structural errors of different equations may be correlated. In view of the fact that the two estimators ignore different parts of the total information available it is not surprising that we cannot rank them unambiguously.

There is no such ambiguity regarding the 3SLS induced RRF estimator since the latter takes into account all relevant information—sample as well as prior information.

Remark 25. The results of Theorem 9 and the discussion in the preceding remarks relate to the technical issue of whether the URF estimator of Π or 2SLS or 3SLS induced RRF estimators are to be preferred if the sole criterion is forecasting efficiency and the data at hand is generated by the model we have specified. The conclusion, on these premises, is that while the 3SLS induced RRF dominates the others, no unambiguous statement can be made in the comparison of the 2SLS induced RRF and URF estimators.

On the other hand, in practice the user of such methods may have reasons for preferring the 2SLS induced RRF estimator that go beyond the technical criterion of forecasting efficiency.

6 The GLSEM and Undersized Samples

In previous sections a universal assumption—whether it was made explicitly or not—was

$$T > G \tag{132}$$

viz., that the number of observations is greater than the number of pre-determined variables contained in the GLSEM. This condition is essential, e.g., for obtaining the URF estimator of Π for the "first" stage of 2SLS or 3SLS. It is frequently the case, in practice, that this condition is violated—hence the term "undersized samples." The question, then, is how to proceed in such situations. A number of procedures have been advanced, such as the use of some of the principal components in obtaining the "first" stage and thereafter proceeding in the "usual" way. But such procedures are not entirely satisfactory. By far, the most appealing procedure is the method of iterated instrumental variables (IIV). The key to this method lies in the observation that even though the condition in (132) is violated for the system as a whole, in fact each structural equation, typically, contains a very small number of parameters. In particular, in all of the large models extant, whether for the United States or other industrial economies, we have

$$T > m_i + G_i, \qquad i = 1, 2, \ldots, m. \tag{133}$$

But this means that we can, typically, estimate consistently, by instrumental variables (IV) methods, the parameters of each structural equation. Confining the set of possible instruments to the set of predetermined variables contained in the matrix X, we note that by assumption all such variables "qualify" as instruments, for any given structural equation, since in general

$$\plim_{T \to \infty} \frac{X'U}{T} = 0$$

and the matrix

$$\plim_{T \to \infty} \frac{X'Z_i}{T}$$

will contain a (square) submatrix of order $(m_i + G_i)$ that is nonsingular.

Let P_i be a (suitable) matrix of instruments corresponding[2] to Z_i and consider the instrumental variables (IV) estimators

$$\tilde{\delta}_{\cdot i} = (P_i' Z_i)^{-1} P_i' y_{\cdot i}, \qquad i = 1, 2, \ldots, m. \tag{134}$$

From (134) we can thus derive a RRF estimator of the matrix Π as

$$\tilde{\Pi} = \tilde{C}\tilde{D}, \qquad \tilde{D} = (I - \tilde{B})^{-1},$$

where the nonnull (unknown) elements of C and B have been estimated by (134). Consequently, we can define

$$\tilde{Y} = X\tilde{\Pi} \tag{135}$$

and thus derive the "instruments"

$$\tilde{Z}_i = (\tilde{Y}_i, X_i) = X\tilde{S}_i, \qquad \tilde{S}_i = (\tilde{\Pi}_{*i}, L_{i2}), \qquad i = 1, 2, \ldots, m, \tag{136}$$

---

[2] Note that this means that P_i is a $T \times (m_i + G_i)$ submatrix of X.

it being understood that $\tilde{Y}_i$ (and hence $\tilde{\Pi}_{*i}$) are appropriate submatrices of $\tilde{Y}$ and $\tilde{\Pi}$ appearing in (135). The IIV estimator is defined by

$$\hat{\delta}_{\cdot i} = (\tilde{Z}_i' Z_i)^{-1} \tilde{Z}_i' y_{\cdot i}, \qquad i = 1, 2, \ldots, m. \tag{137}$$

We may now prove

Theorem 10. *Consider the GLSEM as in Theorem 7, but suppose that*

$$T < G \quad and \quad T > m_i + G_i, \qquad i = 1, 2, \ldots, m.$$

Then the IIV estimator of the structural parameters as exhibited in Equation (137) has the same asymptotic distribution as 2SLS.

PROOF. We will actually prove a somewhat broader result, viz., that any estimator of the form (137) where $\tilde{S}_i$ is a *consistent* estimator of S_i has the same asymptotic distribution as 2SLS. From (137) we note that upon substitution for $y_{\cdot i}$ we have

$$\hat{\delta}_{\cdot i} = \delta_{\cdot i} + (\tilde{Z}_i' Z_i)^{-1} \tilde{Z}_i' u_{\cdot i}, \qquad i = 1, 2, \ldots, m.$$

Thus, for the system as a whole we have

$$\sqrt{T}(\hat{\delta} - \delta)_{\text{IIV}} = \left[\frac{\tilde{S}'(I \otimes X')Z^*}{T} \right]^{-1} \tilde{S}' \frac{(I \otimes X')u}{\sqrt{T}}, \tag{138}$$

where

$$Z^* = \text{diag}(Z_1, Z_2, \ldots, Z_m) \tag{139}$$

and

$$\tilde{S} = \text{diag}(\tilde{S}_1, \tilde{S}_2, \ldots, \tilde{S}_m),$$

with the $\tilde{S}_i$ as defined in Equation (136). It may be shown that the (inverse of the) first matrix on the right side of (138) converges in probability to $[S'(I \otimes M)S]$.[1] Consequently, we conclude that, asymptotically,

$$\sqrt{T}(\hat{\delta} - \delta)_{\text{IIV}} \sim [S'(I \otimes M)S]^{-1} S' \frac{(I \otimes X')u}{\sqrt{T}}.$$

A comparison with the first set of equations in (80) shows that the IIV and 2SLS estimators are asymptotically equivalent.   q.e.d.

Remark 26. As is evident from the proof, any estimator of the form (137) will have the same asymptotic distribution as the 2SLS estimator, provided

$$\underset{T \to \infty}{\text{plim}}\ \tilde{S}_i = S_i, \qquad i = 1, 2, \ldots, m.$$

Thus, it is a completely incidental feature that the estimator of $\tilde{\Pi}$ is obtained indirectly through IV estimators of the unknown structural parameters in C and B. The particular derivation employed earlier had the primary purpose of demonstrating the feasibility of such estimators.

Remark 27. Evidently, the choice of P_i in (34) is arbitrary. Since there are G predetermined variables there are at most

$$\binom{G}{m_i + G_i}$$

ways of choosing the columns of P_i. It is an obviously good practice to choose G_i of its columns to be X_i, i.e., the predetermined variables actually contained in the *i*th structural equation. This will reduce the number of ways in which the columns of P_i can be chosen to

$$\binom{G_i^*}{m_i},$$

where

$$G_i^* = G - G_i, \qquad i = 1, 2, \ldots, m.$$

The other m_i columns may be chosen with a view to obtaining maximal correlation—in some loose sense—with the variables to be "replaced." i.e., the columns of Y_i. However, it does not seem to be a particularly fruitful exercise to concern ourselves too much with this aspect. If one is concerned with the small sample consequences of an inept choice of instruments, by far the most sensible approach would be to repeat this procedure. This means that once the estimators in (137) are obtained we recompute (135) and (136) and obtain a new estimator of the structural parameters. As the theorem makes clear this does not result in asymptotic gain but serves to limit the small sample consequences of an "inept" choice of initial instruments.

Remark 28. The estimator in Theorem 10 is a limited information estimator, in that it fails to take into account possible correlation among the error terms attaching to the various structural equations of the system. To distinguish it from estimators that take this aspect into account it is convenient to term it the *limited information iterated instrumental variables* (LIIV) *estimator*. The estimator that takes into account the correlation structure of the system's error terms may be termed the *full information iterated instrumental variables* (FIIV) *estimator*. The FIIV estimator is, in fact, an instrumental variables version of 3SLS. As before, suppose that

$$T < G,$$

but that the condition in (133)—minimally—holds. Proceeding as before, we obtain consistent estimators of the S_i, $i = 1, 2, \ldots, m$, and also compute the residuals

$$\tilde{u}_{\cdot i} = y_{\cdot i} - Z_i \tilde{\delta}_{\cdot i}$$

and the estimator of variances and covariances

$$\tilde{\sigma}_{ij} = \left(\frac{1}{T}\right)\tilde{u}'_{\cdot i}\tilde{u}_{\cdot j}, \qquad i, j = 1, 2, \ldots, m.$$

It would appear that a minimal condition for the nonsingularity of

$$\tilde{\Sigma} = (\tilde{\sigma}_{ij}) \tag{140}$$

in the absence of specific prior restrictions on Σ is

$$T > m.$$

The FIIV estimator is consequently defined by

$$\tilde{\delta} = [\tilde{S}'(I \otimes X')\tilde{\Phi}^{-1}Z^*]^{-1}\tilde{S}'(I \otimes X')\tilde{\Phi}^{-1}y, \tag{141}$$

where

$$\tilde{\Phi} = \tilde{\Sigma} \otimes I$$

and $\tilde{\Sigma}$ is given by (140).

Provided there are sufficient observations so that the matrix to be inverted in (141) is, in fact, invertible, one can show that the estimator given therein is asymptotically equivalent to 3SLS, as is demonstrated in the following.

Theorem 11. *Consider the GLSEM as in Theorem* 10 *but suppose*

$$T > m.$$

Then, provided the estimator in (141) (FIIV *estimator*) *exists, it is asymptotically equivalent to* 3SLS.

PROOF. Substituting for y in (141) we find

$$\sqrt{T}(\hat{\delta} - \delta)_{\text{FIIV}} = \left[\tilde{S}'(\tilde{\Sigma}^{-1} \otimes I)\frac{(I \otimes X')Z^*}{T}\right]^{-1}\tilde{S}'(\tilde{\Sigma}^{-1} \otimes I)\frac{(I \otimes X')u}{\sqrt{T}}.$$

In view of the consistency of $\tilde{S}$ and $\tilde{\Sigma}$ as estimators of S and Σ respectively, we conclude that asymptotically

$$\sqrt{T}(\hat{\delta} - \delta)_{\text{FIIV}} \sim [S'(\Sigma^{-1} \otimes M)S]^{-1}S'(\Sigma^{-1} \otimes I)\frac{(I \otimes X')u}{\sqrt{T}}.$$

A comparison with the second set of equations in (80) shows the asymptotic equivalence of the FIIV and 3SLS estimators.   q.e.d.

Remark 29. Evidently the same comments regarding the initial choice of instruments, and the repetition of the procedure in order to limit the consequence of an inappropriate initial choice made in connection with the LIIV, apply in this instance as well.

7 Maximum Likelihood (ML) Estimators

Maximum Likelihood methods of estimating the parameters of the GLSEM differ from 2SLS and 3SLS methods (which are essentially extensions of OLS methods) in two important respects:

(i) a specific distribution needs to be specified for the structural errors;
(ii) a normalization convention need not be imposed at the outset of the estimation procedure.

Giving effect to (i)—as we did in Section 3 of the present chapter when we discussed the identification problem—we may assert (A.6) of that section, i.e., that

$$u'_{t\cdot} \sim N(0, \Sigma),$$

and thus obtain the log likelihood function as in Equation (24) that is repeated here for clarity:

$$L(B^*, C, \Sigma; Y, X) = -\frac{Tm}{2}\ln(2\pi) - \frac{T}{2}\ln|\Sigma| + \frac{T}{2}\ln|B^{*\prime}B^*|$$
$$-\frac{1}{2}\left\{\sum_{t=1}^{T}(y_{t\cdot}B^* - x_{t\cdot}C)\Sigma^{-1}(y_{t\cdot}B^* - x_{t\cdot}C)'\right\}.$$

Note that by (ii) we need not impose a normalization rule on the system, so in the context above we *do not* have

$$B^* = I - B,$$

and the diagonal elements of B^* are not necessarily equal to unity. Needless to say other identifying (zero) restrictions imposed by condition (A.4) of Section 2 remain in effect.

The full information maximum likelihood (FIML) estimator of the parameters of the GLSEM is obtained by maximizing the likelihood function above with respect to the unknown elements of B^*, C, and Σ. The resulting equations are highly nonlinear and can only be solved by iteration. Little, if anything, is to be gained by going through this exercise. Suffice to say that *although the FIML and 3SLS estimators do not, generally, coincide numerically they are equivalent, in that, subject to the same normalization, their asymptotic distributions are identical.* In practice FIML estimators are employed rather infrequently owing to their computational complexity; empirical econometricians reveal a definite preference for 3SLS methods whenever full information estimators are used.

If not all of the restrictions imposed by (A.4) are observed in estimation the resulting estimators are termed limited information maximum likelihood (LIML) estimators. Thus, there is a great variety of estimators that may be properly termed LIML estimators. In practice, however, the term is almost universally used to connote the single equation maximum likelihood

(SELIML) estimator, which we will sketch in the discussion to follow. The SELIML estimator, which we will "abbreviate" to LIML except when extreme clarity is required, produces estimators for the parameters of the ith structural equation by *observing only those restrictions, imposed by* (A.4), *that relate to the ith structural equation.*

Precisely, what one does is to seek a transformation of the GLSEM that isolates the parameters of interest, say those in the first equation ($b_{\cdot 1}^*$, $c_{\cdot 1}$, σ_{11}), and at the same time makes the likelihood function easily manipulable with respect to the parameters of other equations. To see how this may be accomplished put

$$A = \begin{pmatrix} B^* \\ -C \end{pmatrix}, \qquad M^* = \begin{pmatrix} Z'Z \\ T \end{pmatrix}, \qquad z_t = (y_t, x_t),$$

where Z is the $T \times (m + G)$ matrix whose tth row is z_t. and rewrite the sum in the log likelihood function as

$$-\frac{T}{2}\{\operatorname{tr} \Sigma^{-1} A' M^* A\}.$$

Consider now the transformation of the GLSEM by the matrix

$$H = \begin{bmatrix} 1 & -\Sigma_{11}^{-1}\Sigma_{12}H_{22} \\ 0 & H_{22} \end{bmatrix}, \tag{143}$$

where

$$\Sigma = \begin{bmatrix} \Sigma_{11} & \Sigma_{12} \\ \Sigma_{21} & \Sigma_{22} \end{bmatrix}, \qquad \Sigma_{11} = \sigma_{11}, \qquad H_{22}H_{22}' = (\Sigma_{22} - \Sigma_{21}\Sigma_{11}^{-1}\Sigma_{12})^{-1},$$

so that Σ_{11} is a trivial "matrix" consisting of σ_{11}, which is the variance of the error in the structural equation of interest, viz., the first.

The transformation yields

$$z_t.AH = u_t.H, \qquad t = 1, 2, \ldots, T,$$

and it can be shown quite easily that the log likelihood function of the transformed model is

$$L(A^*, \Sigma^*; Y, X) = -\frac{Tm}{2}\ln(2\pi) - \frac{T}{2}\ln|\Sigma^*| + \frac{T}{2}\ln|H'B^{*\prime}B^*H|$$

$$-\frac{T}{2}\operatorname{tr}\{\Sigma^{*-1} A^{*\prime} M^* A^*\}, \tag{144}$$

where

$$A^* = AH, \qquad \Sigma^* = H'\Sigma H.$$

If we now partition

$$A = (a_{\cdot 1}, A_1)$$

so that

$$a_{.1} = (b_{.1}^{*\prime}, -c_{.1}')'$$

and thus contains the structural parameters of interest, we note that

$$A^* = (a_{.1}, A_1 H_{22} - a_{.1} \Sigma_{11}^{-1} \Sigma_{12} H_{22}) = (a_{.1}, A_1^*),$$

so that, indeed, $a_{.1}$ has not been disturbed by the transformation. Similarly,

$$\Sigma^* = H'\Sigma H = \begin{bmatrix} \sigma_{11} & 0 \\ 0 & I \end{bmatrix}, \qquad B^*H = [b_{.1}^*, B_1^{**}],$$

and we see that in some sense the transformation has detached the equation of interest from the rest of the system, in that its error term is independent of the error terms in the other equations.

Giving effect to the specific consequences of the transformation enables us to write (144) in the more convenient form

$$L(a_{.1}, \sigma_{11}; A_1^*, Y, X)$$

$$= -\frac{Tm}{2} \ln(2\pi) - \frac{T}{2} \ln \sigma_{11} + \frac{T}{2} \ln \begin{vmatrix} b_{.1}^{*\prime} b_{.1}^* & b_{.1}^{*\prime} B_1^{**} \\ B_1^{**\prime} b_{.1}^* & B_1^{**\prime} B_1^{**} \end{vmatrix}$$

$$- \frac{T}{2} \operatorname{tr}(A_1^{*\prime} M^* A_1^*) - \frac{T}{2\sigma_{11}} a_{.1}' M^* a_{.1}. \tag{145}$$

Admittedly, this is a simple form separating as much as possible those parameters we are interested in, viz., $(a_{.1}, \sigma_{11})$, from those in which we are not.

The method now proceeds by maximizing (145) with respect to the parameters of no interest, viz., A_1^*, without regard to the restrictions that assumption (A.4) places on A and hence on A_1^*. *This is the sense in which we have a limited information estimator.* It may be shown[3] that if the maximizing values of A_1^* thus obtained are inserted in (145) we have the so-called *concentrated likelihood function*

$$L^*(a_{.1}, \sigma_{11}; Y, X)$$

$$= \frac{T}{2} - \frac{mT}{2} [\ln(2\pi) + 1] - \frac{T}{2} \ln|W| - \frac{T}{2} \ln \sigma_{11} + \frac{T}{2} \ln(b_{.1}^{*\prime} W b_{.1}^*)$$

$$- \frac{T}{2\sigma_{11}} (a_{.1}' M^* a_{.1}), \tag{146}$$

where

$$M^* = \begin{bmatrix} M_{yy}^* & M_{yx}^* \\ M_{xy}^* & M_{xx}^* \end{bmatrix}, \qquad W = M_{yy}^* - M_{yx}^* M_{xx}^{*-1} M_{xy}^*, \tag{147}$$

---

[3] The details of this operation are clearly beyond the scope of this volume. The interested reader may consult Dhrymes [9] or Koopmans and Hood [23].

and, W is evidently the second moment matrix of the residuals from the OLS estimated reduced form of the entire system. It is only at this stage that the (zero) restrictions imposed by (A.4) on the first structural equation are imposed—*but still we need not impose the normalization convention.*

Giving effect to the zero restrictions yields

$$b^*_{\cdot 1} = (\beta^{0\prime}_{\cdot 1}, 0)', \qquad a_{\cdot 1} = (\beta^{0\prime}_{\cdot 1}, 0, -\gamma'_{\cdot 1}, 0)' \tag{148}$$

in the standard notation employed in earlier sections. Now partitioning W conformably with $b^*_{\cdot 1}$ and M^* conformably with $a_{\cdot 1}$, one obtains

$$W = \begin{bmatrix} W_{11} & W_{12} \\ W_{21} & W_{22} \end{bmatrix}, \qquad M^* = \begin{bmatrix} M^*_{11} & M^*_{12} & M^*_{13} & M^*_{14} \\ M^*_{21} & M^*_{22} & M^*_{23} & M^*_{24} \\ M^*_{31} & M^*_{32} & M^*_{33} & M^*_{34} \\ M^*_{41} & M^*_{42} & M^*_{43} & M^*_{44} \end{bmatrix}, \tag{149}$$

where, evidently, W_{11} corresponds to the second moment matrix of the reduced form residuals for the current endogenous variables appearing in the first structural equation (i.e., the variables $y_1, y_2, \ldots, y_{m_1+1}$), M^*_{11} is the second moment matrix of these variables, M^*_{22} is the second moment matrix of the remaining current endogenous variables, M^*_{33} is the second moment matrix of the predetermined variables appearing in the first structural equation, M^*_{44} is the second moment matrix of the excluded predetermined variables, and the off-diagonal blocks are defined accordingly. Using the partitions of (148) and (149) the concentrated log likelihood function simplifies further to

$$L^*(\beta^0_{\cdot 1}, \gamma_{\cdot 1}, \sigma_{11}; Y, X) = \frac{T}{2} - \frac{Tm}{2}[\ln(2\pi) + 1] - \frac{T}{2}\ln|W| - \frac{T}{2}\ln\sigma_{11}$$

$$+ \frac{T}{2}\ln(\beta^{0\prime}_{\cdot 1}W_{11}\beta^0_{\cdot 1})$$

$$- \frac{T}{2\sigma_{11}}(\beta^{0\prime}_{\cdot 1}M^*_{11}\beta^0_{\cdot 1} - 2\gamma'_{\cdot 1}M^*_{31}\beta^0_{\cdot 1} + \gamma'_{\cdot 1}M^*_{33}\gamma_{\cdot 1}).$$

$$\tag{150}$$

If we maximize (150) partially with respect to $\gamma_{\cdot 1}$ we obtain for the maximizing value

$$\tilde{\gamma}_{\cdot 1} = M^{*-1}_{33}M^*_{31}\beta^0_{\cdot 1}. \tag{151}$$

Defining

$$W^*_{11} = M^*_{11} - M^*_{13}M^{*-1}_{33}M^*_{31} \tag{152}$$

and substituting (151) in (150) we obtain

$$L^*(\beta_{\cdot 1}^0, \sigma_{11}; Y, X)$$

$$= -\frac{Tm}{2}[\ln(2\pi) + 1] + \frac{T}{2} - \frac{T}{2}\ln|W| - \frac{T}{2}\ln\sigma_{11}$$

$$+ \frac{T}{2}\ln(\beta_{\cdot 1}^{0\prime}W_{11}\beta_{\cdot 1}^0) - \frac{T}{2\sigma_{11}}\beta_{\cdot 1}^{0\prime}W_{11}^*\beta_{\cdot 1}^0. \tag{153}$$

Further maximizing (153) with respect to σ_{11} we easily find that

$$\tilde{\sigma}_{11} = \beta_{\cdot 1}^{0\prime}W_{11}^*\beta_{\cdot 1}^0. \tag{154}$$

Inserting (154) in (153) we obtain the final form of the *concentrated likelihood function*,

$$L^*(\beta_{\cdot 1}^0; Y, X) = -\frac{Tm}{2}[\ln(2\pi) + 1] - \frac{T}{2}\ln|W| - \frac{T}{2}\ln\left(\frac{\beta_{\cdot 1}^{0\prime}W_{11}^*\beta_{\cdot 1}^0}{\beta_{\cdot 1}^{0\prime}W_{11}\beta_{\cdot 1}^0}\right), \tag{155}$$

which shows immediately that in order to maximize it we must *minimize* the quantity

$$\frac{\beta_{\cdot 1}^{0\prime}W_{11}^*\beta_{\cdot 1}^0}{\beta_{\cdot 1}^{0\prime}W_{11}\beta_{\cdot 1}^0}.$$

It is evident that this is the *ratio of two quadratic forms*. The matrices of the two quadratic forms W_{11}^* and W_{11} are, respectively, the second moment matrices of the residuals of the regression of the current endogenous variables appearing in the first structural equation on the predetermined variables appearing therein and on all the predetermined variables of the system. Since regressing on more variables can never increase the residuals of the regression we immediately conclude that

$$W_{11}^* - W_{11}$$

is a positive semidefinite matrix and furthermore that

$$(\beta_{\cdot 1}^{0\prime}W_{11}^*\beta_{\cdot 1}^0/\beta_{\cdot 1}^{0\prime}W_{11}\beta_{\cdot 1}^0) \geq 1. \tag{156}$$

Since the ratio in (156) has a lower bound it makes sense to minimize it. The reader will do well to recall the errors in variables (EIV) estimator of Chapter 5, where we obtained an estimator by minimizing the ratio of two quadratic forms. The procedure is to simultaneously decompose

$$W_{11} = P'P, \qquad W_{11}^* = P'\Lambda P,$$

where Λ is the diagonal matrix of the characteristic roots of W^*_{11} in the metric of W_{11} (see *Mathematics for Econometrics*). Putting

$$\xi = P\beta^0_{\cdot 1} \tag{157}$$

the ratio in (156) yields immediately

$$\lambda_{min} \leq \frac{\xi'\Lambda\xi}{\xi'\xi} \leq \lambda_{max}. \tag{158}$$

If $\tilde{\beta}^0_{\cdot 1}$ is chosen to be the characteristic vector corresponding to λ_{min}, i.e., the smallest characteristic root of W^*_{11} in the metric of W_{11}, we see that it must obey

$$W^*_{11}\tilde{\beta}^0_{\cdot 1} = \lambda_{min} W_{11}\tilde{\beta}^0_{\cdot 1}. \tag{159}$$

Premultiplying by $\tilde{\beta}^{0\prime}_{\cdot 1}$, we find

$$\frac{\tilde{\beta}^{0\prime}_{\cdot 1} W^*_{11}\tilde{\beta}_{\cdot 1}}{\tilde{\beta}^{0\prime}_{\cdot 1} W_{11}\tilde{\beta}^{0\prime}_{\cdot 1}} = \lambda_{min},$$

which, in view of (158), shows that $\tilde{\beta}^0_{\cdot 1}$ is the vector that gives the *maximum maximorum* of the concentrated likelihood function as exhibited in (155).

By a backward sequence, substituting $\tilde{\beta}^0_{\cdot 1}$ for $\beta^0_{\cdot 1}$ in (154) and (151), we obtain the solution to the problem of maximizing the likelihood function relevant to the LIML estimator of the parameters of the first equation as exhibited in (150). At this stage it may occur to the perceptive reader that we appear to have solved the estimation problem without recourse to a normalization convention. This, however, is evidently not so. A look at (159) will suffice; characteristic vectors are unique only up to a scalar multiple. Thus, if $\tilde{\beta}^0_{\cdot 1}$ satisfies (159), so could $c\tilde{\beta}^0_{\cdot 1}$ for any scalar $c \neq 0$. *In the context of* **SELIML** *estimation the normalization convention is needed only at this stage in order for us to make a proper choice from among the infinitude of admissible characteristic vectors in* (159). Thus, e.g., the standard normalization will lead to the choice

$$\tilde{\beta}^0_{\cdot 1} = \begin{pmatrix} 1 \\ -\tilde{\beta}_{\cdot 1} \end{pmatrix}, \tag{160}$$

and this will fix uniquely $\tilde{\beta}^0_{\cdot 1}$ and hence $\tilde{\sigma}_{11}$ and $\tilde{\gamma}_{\cdot 1}$. Operating with (160) as our standard of reference, if another normalization is desired the resulting estimator of $\beta^0_{\cdot 1}$ would be

$$\tilde{\tilde{\beta}}^0_{\cdot 1} = c\tilde{\beta}^0_{\cdot 1}, \tag{161}$$

where c is an appropriate constant. The implied estimators of the other parameters under the alternative normalization would be

$$\tilde{\tilde{\gamma}}_{\cdot 1} = c\tilde{\gamma}_{\cdot 1}, \qquad \tilde{\tilde{\sigma}}_{11} = c^2\tilde{\sigma}_{11}, \tag{162}$$

and we see that in the SELIML *context normalization is a mere detail. This is to be contrasted with the* 2SLS *estimator where normalization has to be imposed ab initio and the numerical estimates of parameters under alternative normalizations are not necessarily related to each other in the simple form exhibited in* (161) *and* (162).

Finally, it may be shown that under the same normalization 2SLS and SELIML estimators are equivalent in the sense that they have the same asymptotic distributions. They may, and generally do, differ numerically for given finite samples.

QUESTIONS AND PROBLEMS

1. Verify that for every i the selection matrices L_{i1}, L_{i2} of Remark 6 have the rank ascribed to them. [*Hint*: $L'_{i1} L_{i1} = I_{m_i}$, $L'_{i2} L_{i2} = I_{G_i}$.]

2. In connection with the model in (23) let $\{(q_t, p_t): t = 1, 2, \ldots, T\}$ be a sample that has been generated by that model. Consider now, for $\lambda \in [0, 1]$, the "model"

$$q_t^{*D} = \lambda q_t^D + (1 - \lambda)q_t^s, \qquad q_t^{*S} = q_t^S, \qquad q_t^{*D} = q_t^{*S}.$$

Show that the sample satisfies this model as well.

3. In connection with Equation (24) use the result of Proposition 4 in Chapter 8 to verify that the Jacobian of the transformation from u_t to y_t in $y_t . B^* = x_t . C + u_t$. is $|B^{*\prime} B^*|^{1/2}$.

4. Verify directly from the relevant definitions that (26) represents the (log) likelihood function of the observations (Y, X).

5. Verify in the proof of Proposition 1 that if $\Pi_1 = \Pi_2, \Omega_1 = \Omega_2$, then there exists a nonsingular matrix, H, such that $B_2^* = B_1^* H, C_2 = C_1 H, \Sigma_2 = H' \Sigma_1 H$. [*Hint*: take $H = D_1 D_2^{-1}$.]

6. In Equation (71) prove that $Q_i = R' \tilde{S}_i$.

7. Explain in detail why, upon division by T, the last three components of the right side of (76) have zero probability limits.

8. In the proof of Proposition 4 verify that:
   (i) if $\Sigma = \text{diag}(\sigma_{11}, \sigma_{22}, \ldots, \sigma_{mm})$ then $(S^{*\prime} \Phi^{-1} S^*)^{-1} S^{*\prime} \Phi^{-1} = (S^{*\prime} S^*)^{-1} S^{*\prime}$;
   (ii) if every equation satisfies the rank condition for identification, and each is just identified, then S^* is a nonsingular matrix;
   (iii) if (i) is taken into account in estimation, $\tilde{\delta}_{2SLS} = \hat{\delta}_{3SLS}$;
   (iv) if (ii) holds then the conclusion in (iii) holds as well.

9. (Decomposable systems). Consider the GLSEM subject to the conditions (A.1) through (A.5) and the normalization convention. Suppose further that B is an upper triangular matrix and that $\Sigma = \text{diag}(\sigma_{11}, \sigma_{22}, \ldots, \sigma_{mm})$. Show that the OLS estimator of *structural parameters* is consistent and is generally (asymptotically) efficient as compared with the 2SLS estimator. [*Hint*: for the ith structural

equation, the OLS estimator is given by $(Z_i'Z_i)^{-1}Z_i'y_{\cdot i}$, while the 2SLS estimator is given (in the notation of Theorem 4) by $(\tilde{Z}_i'\tilde{Z}_i)^{-1}\tilde{Z}_i'y_{\cdot i}$.]

10. In (102) verify, as a matter of notation, that the row vector $e_{T+\tau \cdot}$ can be written in column form as $v_{T+\tau \cdot}' - (I \otimes x_{T+\tau \cdot})(\tilde{\pi} - \pi)$. [Hint: the ith element of $e_{T+\tau \cdot}$ is given by $e_{T+\tau,i} = v_{T+\tau,i} - x_{T+\tau \cdot}(\tilde{\pi}_{\cdot i} - \pi_{\cdot i}), i = 1, 2, \dots, m.$]

11. In Equation (111) verify that

$$(\tilde{\Pi}, I)\begin{bmatrix} L_{i1} & 0 \\ 0 & L_{i2} \end{bmatrix} = \tilde{S}_i = (\tilde{\Pi}_{*i}, L_{i2}).$$

[Hint: $YL_{i1} = Y_i$.]

12. Verify Equation (112).

13. Complete the argument in the transition from Equation (121) to (122).

14. Verify that the nonzero roots of $|\lambda \Phi - S^*(S^{*\prime}\Phi^{-1}S^*)^{*\prime}S^*| = 0$ are exactly those of $|\lambda I - (S^{*\prime}\Phi^{-1}S^*)^{-1}S^{*\prime}\Phi^{-1}S^*| = 0$. [Hint: in the first equation factor out Φ and recall that if A is $m \times n$ and B is $n \times m$, $m \geq n$, then the nonzero roots of $|\lambda I - AB| = 0$ are exactly those of $|\lambda I - BA| = 0$.]

15. Give a direct proof for part i. of Theorem 9. [Hint: Put $\Phi = S^*(S^{*\prime}\Phi^{-1}S^*)^{-1}S^{*\prime} + H$ and postmultiply by Φ^{-1}.]

16. Verify that if A is $m \times n$ and $\text{rank}(A) = 0$ then A is the zero matrix.

17. Define $A = S^*(S^{*\prime}S^*)^{-1}S^{*\prime}$, $S^* = (I \otimes \bar{R}')S$, where S is as defined in (54), (80), and (84), and show that

(i) A is symmetric idempotent,

(ii) $\text{rank}(A) = \sum_{i=1}^{m}(m_i + G_i)$.

18. Show that the characteristic vectors of A as defined in Problem 17 are derivable from the characteristic vectors of $A_i = S_i^*(S_i^{*\prime}S_i^*)^{-1}S_i^{*\prime}$, $i = 1, 2, \dots, m$, and derive the appropriate expression for the matrix of characteristic vectors of A. [Hint: if $t_{\cdot i}^{(s)}$ is the ith characteristic vector of A_s then $e_{\cdot s} \otimes t_{\cdot i}^{(s)}$ is a characteristic vector of A, where $e_{\cdot s}$ is an m-element column vector all of whose elements are zero save the sth, which is unity.]

19. Let $T^{(i)}$ be the matrix of characteristic vectors of A_i and partition $T^{(i)} = (T_1^{(i)}, T_2^{(i)})$, $i = 1, 2, \dots, m$, in such a way that $T_1^{(i)}$ corresponds to the nonzero roots and $T_2^{(i)}$ to the zero roots of A_i, $i = 1, 2, \dots, m$. If the matrix T of Equation (128) is partitioned by $T = (T_1, T_2)$ conformably with the partition of Λ in (127) show that

$$T_1 = [e_{\cdot 1} \otimes T_1^{(1)}, e_{\cdot 2} \otimes T_1^{(2)}, \dots, e_{\cdot m} \otimes T_1^{(m)}],$$
$$T_2 = [e_{\cdot 1} \otimes T_2^{(1)}, e_{\cdot 2} \otimes T_2^{(2)}, \dots, e_{\cdot m} \otimes T_2^{(m)}].$$

20. Use the results of Problem 19 to show that Φ_{12}^* of Equation (129) may be expressed as $\Phi_{12}^* = [\sigma_{ij}T_1^{(i)\prime}T_2^{(j)}]$, $i, j = 1, 2, \dots, m$. [Hint: $\Phi_{12}^* = T_1'\Phi T_2$.]

21. From the representation of Φ_{12}^* in Problem 20, give an alternative (direct) proof of Corollary 3.

22. With Z^* as defined in Equation (139) show that

$$\plim_{T \to \infty} \frac{\tilde{S}'(I \otimes X')Z^*}{T} = S'(I \otimes M)S$$

where $M = (\text{p})\lim_{T \to \infty} (X'X/T)$.

23. Verify that $T > m$ is a necessary condition for the nonsingularity of $\tilde{\Sigma} = (\tilde{\sigma}_{ij})$, $\tilde{\sigma}_{ij} = (1/T)\tilde{u}'_{\cdot i}\tilde{u}_{\cdot j}$, $\tilde{u}_{\cdot i} = y_{\cdot i} - Z_i \tilde{\delta}_{\cdot i}$. [$Hint$: $\tilde{\Sigma} = (1/T)[Y - Y\tilde{B} - X\tilde{C}]'[Y - Y\tilde{B} - X\tilde{C}]$.]

7 Discrete Choice Models: Logit and Probit Analysis

1 Introduction

In the discussion of all preceding chapters we had dealt with the case where the dependent variables were continuous and, in principle, could assume any value in $[0, \infty)$ or even $(-\infty, \infty)$—upon centering. In fact, all economic variables are bounded, but the unbounded nature of the set defining the range of the dependent variables above does no violence to the nature of the problem under investigation. Thus, if we are examining the aggregate consumption function it does no violence to the intrinsic nature of the problem to put

$$c_t = \alpha + \beta y_t + u_t,$$

where c and y stand for (the logarithm of) consumption and income, respectively, and u is the unobservable error term. Thus, assuming, for example, that u is proportional to a unit normal variable does not create any conceptual problems, even though it does imply that, with some small probability, consumption can exceed any prespecified magnitude, however large.

Since the assumptions above may simplify the problem significantly and at the same time give difficulty only over a small range of the random variable, we are quite prepared to tolerate them. On the other hand, consider the following: Suppose we are interested in the problem of married women's participation in the labor force. For any individual i the "dependent variable" can assume only two values, according as to whether she participates in the labor force or not. If we were to model this in the usual general linear model (GLM) fashion we would have to write

$$y_i = x_i.\beta + u_i,$$

where

$$y_i \begin{cases} = 1 & \text{if } i\text{th individual participates in labor force} \\ = 0 & \text{otherwise.} \end{cases}$$

The variables in the k-element (row) vector x_i. record the considerations that impinge on the participation decision, such as, for example, age, education, age of children if any, income expected from employment, income of husband, and so on. It is clear that in this case considerable care has to be exercised in specifying the functional form as well as the probability structure of the error term.

In this type of model the standard formulations given in earlier chapters fail and new methods have to be devised.

2 The Nature of Discrete Choice Models

Consider an individual faced with the problem of getting from point A to point B. Let the setting be an urban one and the distance between the two points be not too great.

In many contexts the individual may have five options: to walk, to drive his own car, to go by taxi, to go by bus, or to go by rail rapid transit. Perhaps for completeness we ought to add a sixth option, namely, not to make the trip.

In studying the behavior of individuals faced with the choice of a mode of transport, we have a situation which is appreciably different from that encountered when we wish to study, for example, their expenditures on food consumed at home. The dependent variable there, "expenditure on food consumed at home," can assume many values and, in fact, we do no great violence to reality if we consider it to be a continuous one. Hence, if we have a sufficiently large sample of households or individuals of varying socioeconomic attributes, there is no presumption that the observations on the dependent variable will cluster about a small number of points. This is to be contrasted to the mode of transport example, above, where the choice set contained, at most, six alternatives.

Often the choice set is (maximally) restricted to two elements. There are many instances of problems involving binary choice. Thus, in studying high school youths we may be interested in whether they go to college or not. In studying the behavior of individuals over a certain age, we may be interested in whether they enter the labor force or not.

In general, there are many phenomena of interest in economics in which the dependent variable is defined by the choice of individuals over a set containing a finite, and generally rather small, number of alternatives. If the choice involves *only two alternatives* we have a *dichotomous choice* model. If it involves *more than two (but a finite and generally rather small number of)* *alternatives*, we have a model of *polytomous choice*.

3   Formulation of Dichotomous Choice Models

It is natural in the discrete choice context to be interested in the probability that the jth alternative is chosen. Knowledge of this will generally permit many useful applications. For example, suppose that in a carefully delimited region one is interested in planning public or private recreation facilities. Having information on the probability that individuals, of certain socioeconomic attributes, will utilize the jth recreational facility will permit estimation of expected use. Evidently, this will contribute to proper design of capacity and types of recreational facilities.

Let us see how one can proceed in the case of the dichotomous choice model. Thus, let p be the probability that an event E occurs. Evidently

$$q = 1 - p$$

is the probability that $\bar{E}$ (the complement of E) occurs, i.e., that E does not occur. In particular, E may be the event that a high school youth enters college, or that a person enters the labor force; in this case, $\bar{E}$ is the event that a high school youth does not enter college, or that an individual does not enter the labor force.

Since p is a number that lies between zero and one, care must be exercised in the specification of the model. It is natural to think of p as simply an ordinate of a cumulative distribution function (cdf) and thus write

$$p = F(t), \tag{1}$$

$F(\cdot)$ being a distribution function. If $f(\cdot)$ is the associated density function we have

$$p = \int_{-\infty}^{t} f(\zeta) \, d\zeta. \tag{2}$$

While (2) gives the generic law governing the probability of choosing alternatives, the expression will have to be particularized to the specific phenomenon and/or individual as the case may be. We may do so by expressing the upper limit t as a function of the attributes of the alternatives involved and the individual making the choice. Thus, we may put

$$t = x_{i\cdot}\beta, \tag{3}$$

where

$$x_{i\cdot} = (x_{i1}, x_{i2}, \ldots, x_{ik})$$

is a vector of attributes of the alternatives under consideration and/or the ith individual.

This may strike the reader as a somewhat inelegant approach in that it would appear that we should write

$$p_i = \int_{-\infty}^{x_{i\cdot}\beta} f(\zeta) \, d\zeta = F(x_{i\cdot}\beta),$$

so that we should have to determine as many probabilities as there are individuals. This, in fact, is what we must do, but we need only *determine one probability function*. This is made more explicit if we revert to the notation of the intermediary event E. Thus, define the variables

$$y_i \begin{cases} = 1 & \text{if } i\text{th individual chooses alternative corresponding to event } E \\ = 0 & \text{otherwise.} \end{cases}$$

If we have a sample of n individuals, then the probabilistic aspects of the sample are fully described by

$$\Pr\{y_i = 1\} = F(x_i.\beta), \qquad \Pr\{y_i = 0\} = 1 - F(x_i.\beta). \qquad (4)$$

Evidently, there are a great number of ways in which the *cdf*, $F(\cdot)$, may be chosen. Notice, also, that the left-hand member of (4) cannot be observed. It can, however, be estimated when the ith individual has been repeatedly confronted with the same choice problem and the alternative chosen is known. For such a case we should define

$$y_{it} \begin{cases} = 1 & \text{if at "time } t\text{" } i\text{th individual chooses alternative} \\ & \text{corresponding to event } E \\ = 0 & \text{otherwise,} \end{cases}$$

where $i = 1, 2, \ldots, n; t = 1, 2, \ldots, T$.

Evidently, since the vector $x_i.$ corresponds to the attributes of the alternatives and/or the ith individual, its elements do not depend on t, i.e., the place in the sequence of T choices at which the tth choice is made.

Consequently, we can compute

$$\hat{p}_i = \frac{1}{T} \sum_{t=1}^{T} y_{it}. \qquad (5)$$

The rationale for this is that we can treat the ith individual's T choices as a sequence of Bernoulli trials in which the probability of the event E occurring is

$$F(x_i.\beta)$$

and the probability of $\bar{E}$ occurring is

$$1 - F(x_i.\beta).$$

The substantive implication of this assertion is that the T choices exercised by the ith individual are mutually independent. Consequently, we can view this as a binomial process and, thus, estimate the "probability of success" by (5) since T is fixed and the random aspect is the value assumed by the variables $y_{it}, t = 1, 2, \ldots, T$.

The case just considered, $T > 1$, is often referred to as the case of "many observations per cell," and illuminates the terms "probit" and "logit" analysis as follows. It is intuitively clear that $\hat{p}_i$ is an estimate of the ordinate

$$F(x_i.\beta).$$

If we assume that $F(\cdot)$ *is the standard normal distribution*, then we can define the probit of $\hat{p}_i$ by finding the argument to which it corresponds, i.e.,

$$\text{Probit}(\hat{p}_i) = \hat{t}_i + 5, \tag{6}$$

where $\hat{t}_i$ is a number such that $F(\hat{t}_i) = \hat{p}_i$; 5 is added in order to prevent the right member of (6) from being negative—for all practical purposes. Thus, the probit of $\hat{p}_i$ is nothing but the value of the inverse function for $F(x_i.\beta)$, properly centered.

Similarly, if $F(\cdot)$ is the standard logistic distribution function, then we can define the logit of $\hat{p}_i$ by

$$\text{Logit}(\hat{p}_i) = \ln\left(\frac{\hat{p}_i}{1 - \hat{p}_i}\right) = \hat{t}_i. \tag{7}$$

This is so since the standard logistic distribution (*cdf*) is given by

$$F(t) = \frac{1}{1 + e^{-t}}. \tag{8}$$

Thus

$$\ln\left(\frac{F(t)}{1 - F(t)}\right) = \ln(e^t) = t.$$

Notice that whether we are dealing with logit or probit we are always involved in "inverting" the *cdf*, $F(\cdot)$. That is, given an ordinate, say $F(t_i)$, we find the argument t_i corresponding to it. For the normal *cdf* this can be done by simply looking up one of the many available tables of the normal distribution. For the logistic distribution, the function inverse to $F(\cdot)$ is explicitly available and is simply given by

$$t_i = \ln\left[\frac{F(t_i)}{1 - F(t_i)}\right].$$

In the case of probit one defines

$$z_i = \hat{t}_i + 5, \qquad i = 1, 2, \ldots, n, \tag{9}$$

and writes

$$z_i = x_i.\beta + \text{error}, \qquad i = 1, 2, \ldots, n, \tag{10}$$

while in the case of logit one defines

$$z_i = \hat{t}_i \tag{11}$$

and writes

$$z_i = x_i.\beta + \text{error}. \tag{12}$$

One refers to (10) as the probit analysis model and to (12) as the logit analysis model, i.e., probit analysis commonly refers to the use of the normal *cdf* and logit analysis commonly refers to the use of the logistic *cdf*.

What is the justification for the error term in (10) or (12)? Well, we note that the quantity $\hat{p}_i$ in (5) can be thought of as an estimate of

$$F(x_i.\beta) = \bar{p}_i. \tag{13}$$

Let $h(\cdot)$ be the function inverse to $F(\cdot)$. Then

$$h(\bar{p}_i) = x_i.\beta. \tag{14}$$

In general, the quantity in (5) can be written as

$$\hat{p}_i = \bar{p}_i + \text{error}, \tag{15}$$

the error having mean zero. By definition, in the probit case we have

$$z_i = h(\hat{p}_i) + 5 = h(\bar{p}_i + \text{error}) + 5.$$

If T is relatively large so that the error in (15) is, in some sense, small relative to $\bar{p}_i$, then by the mean value theorem we can write

$$h(\bar{p}_i + \text{error}) = h(\bar{p}_i) + h'(p_i^*)\text{error}, \tag{16}$$

where p_i^* is a point between $\bar{p}_i$ and $\hat{p}_i$. Hence, in some vague sense the error in (10) is $h'(p_i^*)$ times the sampling error in (15).

The situation is essentially similar in the logit case, except that now the function $h(\cdot)$ can be written out explicitly. Thus,

$$h(p) = \ln\left(\frac{p}{1-p}\right).$$

In this case,

$$h'(p) = \frac{1}{p(1-p)}, \tag{17}$$

so that (12) can be written more basically as

$$z_i = h(\hat{p}_i) = h(\bar{p}_i) + h'(p_i^*)\text{error} = x_i.\beta + h'(p_i^*)\text{error} \tag{18}$$

Now $h'(p_i^*)$ is given by (17), p_i^* is a point between $\bar{p}_i$ and $\hat{p}_i$, and the error in (18) is exactly the sampling error in (15).

Since (15) represents the estimate from a set of independent Bernoulli trials, it is easy to see that its expectation is zero and its variance is $\bar{p}_i(1 - \bar{p}_i)$. If the choices made *by the n individuals in the sample are mutually independent* then it might appear reasonable to consider (10) or (12) *as heteroskedastic general linear models* and to estimate their unknown parameter—the vector β—by suitable techniques. We shall not deal with this now, but we will return to estimation issues at a later stage.

4 A Behavioral Justification for the Dichotomous Choice Model

In the preceding section we had formulated the dichotomous choice model without specific reference to the standard choice apparatus of economic theory. Such formulation, however, is not difficult to achieve. We begin by postulating a utility function

$$U(w, r, \eta; \theta),$$

where w is a vector that corresponds to the attributes of the two alternatives; it will be convenient here to enter separately the vector corresponding to the attributes of the individual exercising choice, and this is indicated by the vector r. The utility function is *taken to be random over the alternatives* so that the random variable η is to be distinguished according as to whether we deal, in this case, with alternative one or alternative two. The vector θ is a vector of unknown parameters. It involves little loss of generality to write

$$U(w, r, \eta; \theta) = u(w, r; \theta) + \varepsilon, \tag{19}$$

where ε is a zero mean error term and is distinguished according as to whether we refer to alternative one or alternative two. Evidently, alternative one (the alternative corresponding to the event E of previous sections) is chosen if

$$U(w, r, \varepsilon_1; \theta_1) > U(w, r, \varepsilon_2; \theta_2), \tag{20}$$

where w is the vector of attributes and the θ_i, $i = 1, 2$, are the vectors of parameters corresponding to the two alternatives.

But (20) is equivalent to the statement that alternative one is chosen if

$$\varepsilon_2 - \varepsilon_1 < u(w, r; \theta_1) - u(w, r; \theta_2). \tag{21}$$

Then we ask: What is the probability that alternative one will be chosen? The answer is obvious from (21) once the distribution of the ε's is specified. For example, if the ε_i, $i = 1, 2$, are jointly normal, i.e.,

$$\varepsilon \sim N(0, \Sigma), \qquad \varepsilon = (\varepsilon_1, \varepsilon_2)', \qquad \Sigma = \begin{pmatrix} \sigma_{11} & \sigma_{22} \\ \sigma_{21} & \sigma_{22} \end{pmatrix}, \tag{22}$$

then

$$\varepsilon_2 - \varepsilon_1 \sim N(0, \sigma_{22} + \sigma_{11} - 2\sigma_{12}) \tag{23}$$

and the probability involved is simply

$$F[u(w, r; \theta_1) - u(w; r; \theta_2)],$$

where $F(\cdot)$ is the *cdf* of an $N(0, \sigma_{22} + \sigma_{11} - 2\sigma_{12})$ random variable.
  Suppose that

$$u(w, r; \theta_1) = w\alpha_1 + r\beta_1,$$

$$u(w, r; \theta_2) = w\alpha_2 + r\beta_2.$$

Then

$$u(w, r; \theta_1) - u(w, r; \theta_2) = (w, r)\begin{bmatrix} \alpha_1 - \alpha_2 \\ \beta_1 - \beta_2 \end{bmatrix}$$

If this refers to the ith individual, then put

$$x_{i\cdot} = (w, r_{i\cdot}), \qquad \beta = (\alpha'_1 - \alpha'_2, \beta'_1 - \beta'_2)' \tag{24}$$

so that $x_{i\cdot}$ refers to the attributes of the two alternatives and/or the attributes of the individual exercising choices and β is the corresponding parameter vector. In the notation of (24) we can say, in this context, that the alternative corresponding to the event E will be chosen by the ith individual with probability

$$F(x_{i\cdot} \beta).$$

But this is exactly what we have postulated *ab initio* in Section 4, except that there $F(\cdot)$ referred to the standard normal distribution and here it does not. But this is really a difference that is more apparent than real, since the scaling of the variables and coefficients is arbitrary. It merely suffices to define the parameter vector β as

$$\beta = \frac{1}{\sqrt{\sigma_{11} + \sigma_{22} - 2\sigma_{12}}} (\alpha'_1 - \alpha'_2, \beta'_1 - \beta'_2)'$$

in order to reduce $F(\cdot)$ to the standard normal distribution.

Thus, with the restriction that the *expected utility function is linear in the parameters* we have a plausible rationalization of the dichotomous choice model in its probit form in terms of the basic principles of economic theory.

A similar justification can be made for the logit form as well, but this is somewhat tedious and is relegated to the Appendix of this chapter.

5 Inapplicability of OLS Procedures

Not withstanding the discussion in the preceding section, it is often the case in empirical research that one postulates a model of the form

$$y_i = x_{i\cdot} \beta + u_i, \qquad i = 1, 2, \ldots, n, \tag{25}$$

where the symbols have the same meaning as before and u_i is here presumed to be the random variable of the general linear model (GLM). If we recall the discussion above, we immediately see that (25) is an inadmissible specification. The question then is why anyone would think of employing it in studying the problem of discrete choice.

To see how this may have eventuated recall the discussion surrounding Equations (1) through (3), and notice that we are interested in specifying the function that yields the ordinate of a *cdf*; moreover, note that from

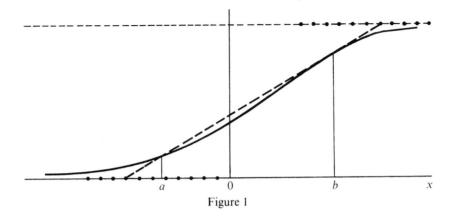

Figure 1

Equation (3) we have a link between certain variables of interest and the corresponding argument of the *cdf*. Suppose in (3) there is only one variable x and that t is an increasing function of x. Then, equation (1) could be represented as in Figure 1 above. For the typical *cdf* its middle range is well approximated by a linear function. Indeed, in Figure 1 a line is a good approximation over the range (a, b).

Since a line is determined by two points, it follows that if most of the observations in a sample are (equally) clustered about a and b, then a regression based on (25) would give reasonably good results, provided all the zeros have abscissas that are clustered about a and that all ones have abscissas that are clustered about b. Evidently, a regression based on (25) is a very cheap and convenient way of analysing the data and, under the circumstances noted above, would not entail serious inaccuracies. However, if most of the data on the dependent variable consist of ones and their abscissas are clustered about b, while the zeros have adscissas that range between a and 0, then a situation depicted in Figure 2 will be relevant and the OLS estimated (broken) line would tend to be steeper.

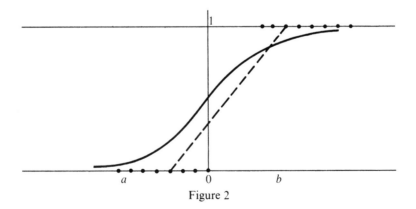

Figure 2

Evidently, in such a case the correspondence between the fitted function (indicated by the broken line) and the *cdf* is rather poor. The reader may also visualize data configurations that could lead to even poorer results.

On the formal side, if we are to consider (25) as a GLM, we should first observe that, since the dependent variable assumes the values zero or one,

$$u_i \begin{cases} = -x_i.\beta & \text{if } y_i = 0 \\ = 1 - x_i.\beta & \text{if } y_i = 1. \end{cases} \tag{26}$$

Thus, the expectation of the random term is

$$\begin{aligned} E(u_i) &= (1 - x_i.\beta)F(x_i.\beta) - [1 - F(x_i.\beta)]x_i.\beta \\ &= F(x_i.\beta) - x_i.\beta \end{aligned} \tag{27}$$

and it is seen to be nothing more than a measure of the departure of the *cdf*, $F(\cdot)$, from linearity! Similarly,

$$\begin{aligned} \text{Var}(u_i) &= (1 - x_i.\beta)^2 F(x_i.\beta) + (x_i.\beta)^2[1 - F(x_i.\beta)] - [E(u_i)]^2 \\ &= F(x_i.\beta)[1 - F(x_i.\beta)]. \end{aligned} \tag{28}$$

Thus, even if we assume that one individual's choice is independent of another's—so that the u_i are mutually independent—we would still have the problem of heteroskedasticity, since in general

$$F(x_i.\beta) \neq F(x_j.\beta), \qquad i \neq j.$$

It has been suggested that Aitken methods be used by estimating

$$\overline{F(x_i.\beta)[1 - F(x_i.\beta)]} = \hat{y}_i(1 - \hat{y}_i),$$

where

$$\hat{y}_i = x_i.\hat{\beta} \tag{29}$$

and the vector $\hat{\beta}$ has been estimated by OLS methods. It is apparent that this approach may fail since we have no guarantee that if we estimate $\hat{\beta}$ in (25) by OLS then the dependent variable in (29) will obey

$$0 < \hat{y}_i < 1. \tag{30}$$

Note that if (30) is violated, then Aitken-like methods will fail. Of course, we can impose the constraint in (30) on the estimation procedure, but then we no longer deal with simple techiques; if we are to engage in more complicated procedures, there are far better methods than constrained least squares. To recapitulate, while (25) coupled with OLS estimation of its parameters is a simple formulation of the discrete choice model, it has little beyond its simplicity to recommend it. It will yield "reasonable" results only in highly special circumstances, and with the data configurations one often en-counters in empirical work it is likely to lead to very poorly fitting probability functions.

An alternative often employed, viz., grouping data, is also unsatisfactory. Grouping will simply divide the sample in a (sufficiently large) number of groups. Let the groups be s in number and suppose the jth group has n_j members. Arrange the observations so that the index i has the range

$$(J_1, J_2, \ldots, J_s),$$

where

$$J_1 = (1, 2, \ldots, n_1),$$

$$J_r = \left(\sum_{k=1}^{r-1} n_k + 1, \sum_{k=1}^{r-1} n_k + 2, \ldots, \sum_{k=1}^{r} n_k\right), \qquad r = 2, 3, \ldots, s.$$

In this scheme J_j contains the number of observations corresponding to the jth group.

Now compute

$$p_j = \frac{1}{n_j} \sum_{i \in J_j} y_i, \qquad \bar{x}_{j\cdot} = \frac{1}{n_j} \sum_{i \in J_j} x_{i\cdot}, \qquad j = 1, 2, \ldots, s,$$

and regress the vector

$$p = (p_1, p_2, \ldots, p_s)'$$

on the variables in

$$\bar{x}_{j\cdot}, \qquad j = 1, 2, \ldots, s.$$

This is an attempt to create, by grouping, a situation analogous to the "many observations per cell" case examined earlier. Thus, dependent variables defined by

$$\text{Probit}(p_j), \qquad \text{Logit}(p_j), \qquad j = 1, 2, \ldots, s,$$

yield probit and logit analytic models applied to group probabilities and group means. Needless to say, we have purchased "respectability" of estimation at the cost of losing a great deal of information through grouping. Moreover, it is not clear just how this "group probability function" is to be interpreted or reconciled with the individual choice model we have examined in Section 4.

6 Maximum Likelihood Estimation

By far the most satisfactory means of estimating the parameters of the dichotomous choice model with the sort of data commonly available in economics is through maximum likelihood methods. The likelihood function for the type of model examined above (one observation per "cell") is

$$L^* = \prod_{i=1}^{n} F(x_{i\cdot} \beta)^{y_i} [1 - F(x_{i\cdot} \beta)]^{1 - y_i}. \tag{31}$$

The log likelihood function is, thus,

$$L = \sum_{i=1}^{n} y_i \ln F(x_i.\beta) + \sum_{i=1}^{n} (1 - y_i) \ln[1 - F(x_i.\beta)]. \tag{32}$$

The first- and second-order derivatives with respect to β are easily established as

$$\frac{\partial L}{\partial \beta} = \sum_{i=1}^{n} y_i \frac{f}{F} x_i. - \sum_{i=1}^{n} (1 - y_i) \left(\frac{f}{1 - F} \right) x_i., \tag{33}$$

$$\frac{\partial^2 L}{\partial \beta \, \partial \beta'} = \sum_{i=1}^{n} y_i \left[\frac{F x_i'.(\partial f/\partial \beta') - f^2 x_i'. x_i.}{F^2} \right]$$

$$- \sum_{i=1}^{n} (1 - y_i) \left[\frac{(1 - F) x_i'.(\partial f/\partial \beta') + f^2 x_i'. x_i.}{(1 - F)^2} \right] \tag{34}$$

If we specify the *cdf*, $F(\cdot)$, and set to zero the derivative of (33), we shall obtain an estimator of β. Of course, this will have to be done by numerical methods since the resulting equation will be highly nonlinear. Whether the estimator so obtained is the maximum likelihood (ML) or not will depend on whether the matrix of second derivatives in (34) is negative definite—or, more generally, whether the likelihood function is strictly convex—a sufficient condition for which is negative definiteness of the matrix in (34).

Under the condition above the solution to (33) set to zero would correspond to the *global maximum*, and hence, to the ML estimator. We shall outline these facts for the logistic distribution and relegate to the Appendix a more complete exposition as well as a similar discussion for the normal case. For the logistic, setting (33) to zero yields the normal equations

$$\frac{\partial L}{\partial \beta} = \sum_{i=1}^{n} y_i x_i. - \sum_{i=1}^{n} (1 + e^{-x_i.\beta})^{-1} x_i. = 0. \tag{35}$$

Evaluating the matrix of second-order derivatives yields

$$\frac{\partial^2 L}{\partial \beta \, \partial \beta'} = - \sum_{i=1}^{n} f(x_i.\beta) x_i'. x_i., \tag{36}$$

which, as shown in the Appendix, is a negative definite matrix. Thus, for the logistic case, the solution to (35), however obtained, is guaranteed to provide the ML estimator, provided the matrix of observations

$$P = \sum_{i=1}^{n} x_i'. x_i. \tag{37}$$

is nonsingular. Notice that P in (36) is the analog of the matrix

$$X'X$$

encountered in OLS estimation of the standard GLM.

7 Inference for Discrete Choice Models

In the discussion of the previous sections it became evident that estimation in the context of logit or probit models is extremely cumbersome. The same is true for inference. Dealing with various aspects of the GLM in earlier chapters we had followed the sequence of first discussing estimation and then various aspects of asymptotic distribution theory. We shall follow this custom here, except that our discussion will not be quite as extensive.

We begin with consistency. An outline of the proof that the ML estimator of β for the logit and probit models considered in the previous section, is consistent proceeds roughly as follows.

Let L be as in Equation (32) and define

$$L_n(\beta; y, x) = \frac{1}{n} \left\{ \sum_{i=1}^{n} [y_i \ln F(x_i.\beta) + (1 - y_i) \ln(1 - F(x_i.\beta)] \right\} \quad (38)$$

and observe that if $\hat{\beta}_{(n)}$ is the ML estimator, based on a sample of size n, then it obeys

$$L_n(\hat{\beta}_{(n)}; y, x) \ge L_n(\beta, y, x) \quad (39)$$

for all admissible parameter vectors β. Suppose the probability limit of the ML estimator exists and is given by

$$\operatorname*{plim}_{n \to \infty} \hat{\beta}_{(n)} = \bar{\beta}. \quad (40)$$

Define

$$\bar{L}(\beta) = \operatorname*{plim}_{n \to \infty} L_n(\beta; y, x) \quad (41)$$

Then, in view of (39) we have

$$\bar{L}(\bar{\beta}) \ge \bar{L}(\beta) \quad (42)$$

for all admissible β. On the other hand, for the true parameter vector, say β_*, we always have

$$\bar{L}(\beta_*) \ge \bar{L}(\beta) \quad (43)$$

for all admissible β. From (42) and (43) we conclude that

$$\bar{L}(\beta_*) = \bar{L}(\bar{\beta}), \quad (44)$$

and therefore that

$$\beta_* = \bar{\beta}, \quad (45)$$

which shows consistency.

As in other nonlinear contexts encountered earlier it is just not possible to obtain the distribution of the estimator $\hat{\beta}$ for any sample size. We may, however, generally obtain its limiting or asymptotic distribution as follows. Expand, by the mean value theorem,

$$\frac{\partial L}{\partial \beta}(\hat{\beta}) = \frac{\partial L}{\partial \beta}(\beta_*) + \frac{\partial^2 L}{\partial \beta \, \partial \beta'}(\beta_{**})(\hat{\beta} - \beta_*), \tag{46}$$

where, as before, β_* is the true parameter vector, $\hat{\beta}$ is the ML estimator, and β_{**} is a point intermediate between β_* and $\hat{\beta}$. Since the ML estimator obeys

$$\frac{\partial L}{\partial \beta}(\hat{\beta}) = 0,$$

we may rewrite (46) as

$$\sqrt{n}(\hat{\beta} - \beta_*) = -\left[\frac{1}{n} \frac{\partial^2 L}{\partial \beta \, \partial \beta'}(\beta_{**})\right]^{-1} \frac{1}{\sqrt{n}} \frac{\partial L}{\partial \beta}(\beta_*'). \tag{47}$$

It may be shown that the matrix in square brackets converges in probability to a well-defined matrix of constants, say

$$S^{-1} = \plim_{n \to \infty} \left[\frac{1}{n} \frac{\partial^2 L}{\partial \beta \, \partial \beta'}(\beta_*)\right]^{-1}. \tag{48}$$

Thus, we need only examine

$$\frac{1}{\sqrt{n}} \frac{\partial L}{\partial \beta}(\beta_*) = \frac{1}{\sqrt{n}} \sum_{i=1}^{n} [y_i - F(x_i.\beta_*)] \frac{f}{F(1 - F)} x_i.. \tag{49}$$

The only random variable in the right member of (49) is

$$y_i - F(x_i.\beta), \qquad i = 1, 2, \ldots, n,$$

which is a sequence of independent zero mean variables. The covariance matrix for each component term of the sum is therefore

$$\frac{f^2}{F(1 - F)} x_i'.x_i..$$

Thus, provided

$$S_* = \lim_{n \to \infty} \frac{1}{n} \sum_{i=1}^{n} \frac{f^2}{F(1 - F)} x_i'.x_i..$$

is a well-defined matrix, and certain other conditions hold, we may conclude by an appropriate central limit theorem that

$$\frac{1}{\sqrt{n}} \frac{\partial L}{\partial \beta}(\beta_*) \sim N(0, S_*). \tag{50}$$

It may further be verified that

$$S_* = S, \tag{51}$$

where S is as implicitly defined in (48). Thus we conclude that

$$\sqrt{n}(\hat{\beta} - \beta_*) \sim N(0, S^{-1}). \tag{52}$$

It is evident that S can be consistently estimated by

$$\hat{S} = \frac{1}{n} \frac{\partial^2 L}{\partial \beta \, \partial \beta'} (\hat{\beta}).$$

Thus, the inference problem here is completely resolved—in the sense that we have the means whereby we can test hypotheses on the parameters of the (dichotomous) logit and probit models.

8 Polytomous Choice Models

8.1 General Discussion

Since, in Sections 3 and 4, we gave an extensive discussion of the motivation for discrete choice models in the context of the dichotomous case, we need not repeat the background details here. Instead, it will suffice to note that the difference between this and the dichotomous case is that the choice set contains more than two alternatives. Code these alternatives as

$$c_j, \qquad j = 1, 2, \ldots, m.$$

For the ith individual, let y_i indicate his choice. The central problem is to model

$$\Pr\{y_i = c_j\}, \qquad j = 1, 2, \ldots, m,$$

and to estimate the underlying parameters. We may give a motivation to such a model quite parallel to that, in Section 4, for the dichotomous model. Thus, suppressing the subscript i, let U be the utility function of the individual and let it be written as

$$U(w, r, \eta; \theta) = u(w, r; \theta) + \varepsilon, \tag{53}$$

where, as before, w is a (row) vector corresponding to the attributes of the alternatives in the choice set, r is a (row) vector corresponding to the attributes of the individual exercising choice, θ is a set of unknown parameters, and ε (or η in the alternative form) is a random variable. Evaluating at the jth alternative we have

$$U(w, r, \eta_j; \theta_j) = u(w, r; \theta_j) + \varepsilon_j, \qquad j = 1, 2, \ldots, m. \tag{54}$$

Evidently, the individual will choose alternative j if

$$u(w, r; \theta_j) + \varepsilon_j - u(w, r; \theta_k) - \varepsilon_k \geq 0 \quad \text{for all } k \neq j.$$

Thus, the probability that the jth alternative will be chosen is given by

$$\Pr\{\varepsilon_k - \varepsilon_j \leq u(w, r; \theta_j) - u(w, r; \theta_k') \text{ for all } k \neq j\}.$$

It is clear that if

$$\varepsilon \sim N(0, I), \qquad \varepsilon = (\varepsilon_1, \varepsilon_2, \ldots, \varepsilon_m)',$$

we have the polytomous analog of the standard probit model. The formulation leading to the polytomous logistic model is examined in the Appendix to this chapter.

8.2 Estimation

We shall examine estimation only in the context of the logistic version of the polytomous choice model since the material becomes exceedingly cumbersome to exposit even when a representation of the *cdf* is available. In the case of the normal *cdf* for the polytomous choice model the situation is so cumbersome that little will be gained by a general exposition such as we presented in Section 6.

From the Appendix to this chapter we know that if we put y_i for the variable denoting the choice of the ith individual and if we code by $c_j, j = 1, 2, \ldots, m$, the m alternatives available, then for the logistic model

$$\Pr\{y_i = c_j\} = \frac{e^{x_i \cdot \beta_{\cdot j}}}{\sum_{k=1}^{m} e^{x_i \cdot \beta_{\cdot k}}}, \qquad j = 1, 2, \ldots, m. \tag{55}$$

Here, as before,

$$x_{i\cdot} = (w, r_{i\cdot}),$$

where the (row) vector w contains the attributes of the alternatives and the vector $r_{i\cdot}$ denotes the attributes of the ith individual. It is clear from (55) that there is a redundancy of parameters, since if

$$\beta_{\cdot j}, \qquad j = 1, 2, \ldots, m,$$

is a set of parameters satisfying (55), then

$$\beta_{\cdot j}^* = \beta_{\cdot j} + c,$$

where c is an arbitrary vector, will also satisfy (55). Thus, as in the analysis of variance model, only "contrasts", say

$$\beta_{\cdot j} - \beta_{\cdot m}, \qquad j = 1, 2, \ldots, m,$$

are identified.

But it is slightly preferable for purposes of exposition to impose the more neutral constraint

$$\sum_{j=1}^{m} \beta_{\cdot j} = 0. \tag{56}$$

Be that as it may, put

$$c_{ij} \begin{cases} = 1 & \text{if } i\text{th individual chooses } j\text{th alternative} \\ = 0 & \text{otherwise.} \end{cases}$$

$$F(x_i. \beta_{\cdot j}) = \frac{e^{x_i. \beta_{\cdot j}}}{\sum_{k=1}^{m} e^{x_i. \beta_{\cdot k}}} \tag{57}$$

and note that if we have a random sample of n observations, the likelihood function is simply

$$L^* = \prod_{i=1}^{n} \prod_{j=0}^{m} F(x_i. \beta_{\cdot j})^{c_{ij}} \tag{58}$$

and its logarithm is

$$L = \sum_{i=1}^{n} \sum_{j=1}^{m} c_{ij} \ln F(x_i. \beta_{\cdot j}). \tag{59}$$

Observe, further, that

$$\sum_{j=1}^{m} c_{ij} = 1. \tag{60}$$

Estimation here is carried out by maximizing (59) subject to (56).

As in the dichotomous case we can show that the likelihood function is convex, that the ML estimator is consistent, that the functions

$$\beta_{\cdot j} - \beta_{\cdot m}, \qquad j = 1, 2, \ldots, m - 1,$$

can be estimated consistently, and that their asymptotic distribution is given, *mutatis mutandis*, by (52).

Appendix

In this Appendix we discuss a number of issues too cumbersome to discuss in the text of the chapter.

A.1 A Random Choice Motivation for the Logistic Model

Let an individual have a utility function

$$U(w, r, \eta; \theta),$$

where, w is a row vector referring to the attributes of the m alternatives available to the individual for choice, r is a row vector containing the (relevant) attributes of the individual exercising choice, η is a random variable defined over alternatives, and θ is a vector of unknown parameters. Define

$$u(w, r; \theta) = E[U(w, r, \eta; \theta) | \text{given } w, r] \qquad (A.1)$$

and put

$$U(w, r, \eta; \theta) = u(w, r; \theta) + \varepsilon \qquad (A.2)$$

Thus, it involves little loss of generality to begin the discussion by postulating the utility function in (A.2) and asserting that the random term there, ε, is specific to each alternative. Thus, for example, evaluating the random utility function in (A.2) at the jth alternative yields

$$u(w, r; \theta_j) + \varepsilon_j$$

Evidently, the individual under consideration will choose the jth alternative if

$$u(w, r; \theta_j) + \varepsilon_j - u(w, r; \theta_k) - \varepsilon_k \geq 0 \quad \text{for all } k \neq j. \qquad (A.3)$$

Code the m alternatives

$$c_j, \quad j = 1, 2, \ldots, m,$$

and let y_i be the variable denoting choice for the ith individual. Thus, e.g.,

$$y_i = c_j$$

means that the ith individual chooses the jth alternative. In view of the preceeding we must have

$$\Pr\{y_i = c_j\} = \Pr\{\varepsilon_{ki} - \varepsilon_{ji} \leq u(w, r_{i\cdot}; \theta_j) \\ - u(w, r_{i\cdot}; \theta_k), \text{ all } k \neq j\}. \qquad (A.4)$$

Dropping the individual subscript i for simplicity of notation, suppose that the

$$\varepsilon_j, \quad j = 1, 2, \ldots, m,$$

are mutually independent (a restrictive assumption), each with density function

$$f(s) = e^{-s} e^{-e^{-s}}, \qquad (A.5)$$

where for notational simplicity we have substituted s for ε.

Remark A.1. Let q be an exponentially distributed random variable, i.e.,

$$q \sim e^{-q}, \quad q \in (0, \infty),$$

and put

$$v = \ln(q)^{-1} = -\ln q. \qquad (A.6)$$

Then

$$v \in (-\infty, \infty)$$

and the Jacobian of (A.6) is obtained from

$$q = e^{-v} \tag{A.7}$$

as

$$e^{-v}$$

We thus easily conclude that

$$v \sim e^{-v}e^{-e^{-v}}$$

and *we see that the distribution specified for the random terms in the utility function (A.2) is that of the logarithm of the inverse of an exponential variable!*
Now, given the assumptions in (A.5) and the discussion just preceding it, we see that we require the joint density of the random variables

$$\varepsilon_k - \varepsilon_j, \qquad k \neq j.$$

We proceed as follows: First we find the joint density of

$$\varepsilon_k, \qquad k = 1, 2, \ldots, m,$$

and then by a proper transformation we find the desired distribution. Evidently, the joint density of the ε_k is simply

$$f(s_1, s_2, \ldots, s_m) = \exp\left(-\sum_{i=1}^{m} s_i\right)\exp\left(-\sum_{i=1}^{m} e^{-s_i}\right). \tag{A.8}$$

Put

$$\begin{aligned}
s_j &= v_1, \\
s_1 &= v_1 + v_2, \\
&\vdots \\
s_{j-1} &= v_1 + v_j, \\
&\vdots \\
s_{j+1} &= v_1 + v_{j+1}, \\
&\vdots \\
s_m &= v_1 + v_m,
\end{aligned} \tag{A.9}$$

and note that the variables

$$v_i, \qquad i = 2, 3, \ldots, m,$$

correspond to the variables

$$\varepsilon_k - \varepsilon_j, \qquad k = 1, 2, \ldots, j-1, j+1, \ldots, m.$$

The Jacobian of the transformation in (A.9) is unity. Hence the joint density of the v_i, $i = 1, 2, \ldots, m$, is

$$f(v_1, v_2, \ldots, v_m) = \exp(-mv_1) \exp\left(-\sum_{i=2}^{m} v_i\right) \exp(-te^{-v_1}), \quad \text{(A.10a)}$$

where

$$t = 1 + \sum_{i=2}^{m} e^{-v_i}. \quad \text{(A.10b)}$$

In order to find the desired distribution we need only integrate out of (A.10a) the variable v_1. To this effect make the transformation

$$v = te^{-v_1}, \quad \text{(A.11)}$$

whose Jacobian is

$$v^{-1},$$

and note

$$\int_{-\infty}^{\infty} \exp(-mv_1) \exp\left(-\sum_{i=2}^{m} v_i\right) \exp(-te^{-v_1}) dv_1 \quad \text{(A.12)}$$

$$= t^{-m} \exp\left(-\sum_{i=2}^{m} v_i\right) \int_{0}^{\infty} v^{m-1} \exp(-v) \, dv$$

$$= \frac{(m-1)! \exp(-\sum_{i=2}^{m} v_i)}{[1 + \sum_{i=2}^{m} \exp(-v_i)]^m}.$$

Thus, the joint density of the variables of interest is given by (A.12). The *cdf* corresponding to (A.12) is easily shown to be

$$F(v_2, v_3, \ldots, v_m) = \frac{1}{1 + \sum_{i=2}^{m} e^{-v_i}}. \quad \text{(A.13)}$$

Consequently, and, again, for simplicity, writing

$$u(w, r; \theta_{\cdot j}) - u(w, r; \theta_{\cdot k}) = x(\theta_{\cdot j} - \theta_{\cdot k}), \qquad x = (w, r),$$

we can express (A.4) as

$$\Pr\{y_i = c_j\} = \frac{1}{1 + \sum_{k \neq j}^{m} e^{-x_i \cdot (\theta_{\cdot j} - \theta_{\cdot k})}} = \frac{e^{x_i \cdot \theta_{\cdot j}}}{\sum_{k=1}^{m} e^{x_i \cdot \theta_{\cdot k}}}, \qquad j = 1, 2, \ldots, m,$$

$$\text{(A.14)}$$

which is the polytomous generalization of the logistic, where the index i now refers to the ith individual. If

$$m = 2, \quad j = 1$$

then (A.14) becomes, say,

$$\frac{e^{x_i \cdot \theta_{\cdot 1}}}{e^{x_i \cdot \theta_{\cdot 1}} + e^{x_i \cdot \theta_{\cdot 2}}} = \frac{1}{1 + e^{-x_i \cdot \beta}}, \qquad \beta = \theta_{\cdot 1} - \theta_{\cdot 2}, \quad \text{(A.15)}$$

which is simply the logistic distribution for the dichotomous choice model first given in Equation (8) of the chapter with the understanding that

$$t = x_i. \beta.$$

The vectors $x_i.$ and β are related to the vectors of attributes w, r, and the basic parameters $\theta._j$, $j = 1, 2$, in the manner prescribed by Equation (24). It will always be understood in this Appendix that $x_i.$ contains k elements. Thus, we have given an economic theoretic foundation to both the dichotomous and polytomous logistic models of the discrete choice problem.

A.2 Convexity of the Likelihood Function

From the discussion of Section 3 we recall that the log likelihood function for the dichotomous choice model is

$$L = \sum_{i=1}^{n} \{y_i \ln F(x_i. \beta) + (1 - y_i) \ln[1 - F(x_i. \beta)]\} \qquad (A.16)$$

and that the first- and second-order partial derivatives are

$$\frac{\partial L}{\partial \beta} = \sum_{i=1}^{n} \left[y_i \frac{f}{F} - (1 - y_i) \frac{f}{1 - F} \right] x_i., \qquad (A.17)$$

$$\frac{\partial^2 L}{\partial \beta \, \partial \beta'} = \sum_{i=1}^{n} \left[y_i \frac{F(\partial f / \partial \beta') x_i. - f^2 x'_i. x_i.}{F^2} \right.$$

$$\left. - (1 - y_i) \frac{(1 - F)(\partial f / \partial \beta') x_i. + f^2 x'_i. x_i.}{(1 - F)^2} \right]. \qquad (A.18)$$

We shall now justify a number of representations given in the chapter and demonstrate that the log likelihood function in the case of both probit and logit is strictly convex. Convexity implies that if $\hat{\beta}$ is a solution of

$$\frac{\partial L}{\partial \beta} = 0, \qquad (A.19)$$

then it is the ML estimator. This, in particular, means that in finding a solution to (A.19) it is not necessary to begin the process with an initial consistent estimator and find the solution by iteration. Strict convexity means that the maximum is unique and, thus, there is no confusion between local and global maxima. Strict convexity is ensured by the condition that the matrix in (A 18) is negative definite for all (finite) $x_i.$ and $x_i. \beta$. We first note that for the logistic we have

$$f(t) = \frac{e^{-t}}{(1 + e^{-t})^2}, \qquad F(t) = \frac{1}{1 + e^{-t}}, \qquad 1 - F(t) = \frac{e^{-t}}{1 + e^{-t}}, \qquad (A.20)$$

$$\frac{f}{F} = 1 - F, \qquad f'(t) = -f F(1 - e^{-t})$$

Using (A.20) we easily establish

$$y_i \frac{f}{F} - (1 - y_i)\frac{f}{1 - F} = y_i(1 - F) - (1 - y_i)F = y_i - F, \quad \text{(A.21)}$$

which justifies Equation (35). Similarly, using (A.20) we find

$$y_i \frac{F(\partial f/\partial \beta')x_{i\cdot} - f^2 x'_{i\cdot} x_{i\cdot}}{F^2} - (1 - y_i)\frac{(1 - F)(\partial f/\partial \beta')x_{i\cdot} + f^2 x'_{i\cdot} x_{i\cdot}}{(1 - F)^2}$$

$$= -\left[y_i \frac{f^2 + fF^2(1 - e^{-x_{i\cdot}\beta})}{F^2} \right.$$

$$\left. + (1 - y_i)\frac{f^2 - fF(1 - F)(1 - e^{-x_{i\cdot}\beta})}{(1 - F)^2} \right] x'_{i\cdot} x_{i\cdot} \quad \text{(A.22)}$$

$$= -[y_i f + (1 - y_i)f]x'_{i\cdot} x_{i\cdot} = -f(x_{i\cdot}\beta)x'_{i\cdot} x_{i\cdot}.$$

Thus, for the logistic case we find

$$\frac{\partial^2 L}{\partial \beta\, \partial \beta'} = -\sum_{i=1}^{n} f(x_{i\cdot}\beta)x'_{i\cdot} x_{i\cdot}, \quad \text{(A.23)}$$

which justifies Equation (36). For the normal case, we note that

$$f(t) = (2\pi)^{-1/2}e^{-(1/2)t^2}, \qquad f'(t) = -tf(t). \quad \text{(A.24)}$$

Using (A.24) we can write (A.18), for the normal case, as

$$\frac{\partial^2 L}{\partial \beta\, \partial \beta'} = -\sum_{i=1}^{n} f\left[y_i \frac{f + (x_{i\cdot}\beta)F}{F^2} + (1 - y_i)\frac{f - (x_{i\cdot}\beta)(1 - F)}{(1 - F)^2} \right] x'_{i\cdot} x_{i\cdot}.$$

$$\text{(A.25)}$$

Since $f > 0$, $F^2 > 0$, $(1 - F)^2 > 0$ (except at $\pm\infty$), we shall proceed by first showing that the numerators of the two fractions in square brackets are everywhere positive. It is convenient to put

$$t = x_{i\cdot}\beta \quad \text{(A.26)}$$

and to write

$$S_1(t) = f + tF, \qquad S_2(t) = f - t(1 - F). \quad \text{(A.27)}$$

We first note

$$S_1(0) = f(0) > 0, \qquad S_2(t) = f(0) > 0,$$

$$\lim_{t \to -\infty} S_1(t) = 0, \qquad \lim_{t \to \infty} S_2(t) = 0, \quad \text{(A.28)}$$

$$S_1(-t) = S_2(t).$$

The last equation of (A.28) follows immediately if the reader notes that for the normal *cdf*

$$F(-t) = 1 - F(t). \quad \text{(A.29)}$$

Thus, to accomplish our first objective it is only necessary to show that

$$S_1(t) > 0 \qquad \text{for } t \in (-\infty, 0] \tag{A.30}$$

since it is apparent from (A.27) that

$$S_1(t) > 0 \quad \text{for } t \in [0, \infty). \tag{A.31}$$

But we note that

$$S_1'(t) = -tf + F + tf = F > 0 \quad \text{for } t \in (-\infty, 0]. \tag{A.32}$$

Consequently, $S_1(t)$ is monotone increasing within the range $(-\infty, 0]$ and, thus, (A.30) holds. Since

$$S_1(-t) = S_2(t), \tag{A.33}$$

it follows immediately that

$$S_2(t) > 0 \quad \text{for } t \in [0, \infty). \tag{A.34}$$

From the definition of $S_1(t)$ and $S_2(t)$ we conclude in view of the preceding discussion that

$$S_1(t) > 0, \qquad S_2(t) > 0 \quad \text{for } t \in (-\infty, \infty). \tag{A.35}$$

Since the quantity multiplying $x_i'.x_i.$ in (A.25) is simply

$$\alpha_i = \left[y_i \frac{f}{F^2} S_1(t) + (1 - y_i) \frac{f}{(1 - F)^2} S_2(t) \right] \tag{A.36}$$

we have that

$$\alpha_i > 0, \qquad i = 1, 2, \dots, n. \tag{A.37}$$

The condition above holds for all i, so long as

$$x_i.\beta = t, \qquad t \in (-\infty, \infty).$$

But that is all that is required.

To complete the proof we need to show in both the logistic and normal cases that if

$$P = X'X, \qquad X = \begin{pmatrix} x_1. \\ x_2. \\ \vdots \\ x_n. \end{pmatrix}, \tag{A.38}$$

is positive definite, as is always assumed, then

$$V = \sum_{i=1}^{n} \alpha_i x_i'.x_i. \tag{A.39}$$

is also positive definite. In the case of the normal *cdf* (probit) the α_i are given by (A.36), while in the case of the logistic *cdf* (logit) the α_i are given by

$$\alpha_i = f(x_i.\beta) > 0 \quad \text{for } x_i.\beta = t, \qquad t \in (-\infty, \infty) \tag{A.40}$$

We proceed somewhat formally. Let γ be $m \times 1$ and note that since P is positive definite, then for any nonnull vector γ

$$0 < \gamma'P\gamma = \sum_{i=1}^{n} \gamma'x_i'.x_i.\gamma = \sum_{i=1}^{n} \delta_i, \tag{A.41}$$

where

$$\delta_i = \gamma'x_i'.x_i.\gamma. \tag{A.42}$$

Since $x_i'.x_i.$ is positive semidefinite we have that

$$\delta_i \geq 0, \qquad i = 1, 2, \dots, n, \tag{A.43}$$

and, in view of (A.41), at least one of the δ_i is strictly positive.
    Now for nonnull γ we find

$$\gamma'V\gamma = \sum_{i=1}^{n} \alpha_i\delta_i > 0 \tag{A.44}$$

since at least one of the δ_i must be positive and *all* the α_i, $i = 1, 2, \dots, n$, are positive. We have therefore proved

Proposition A.1. *Consider the dichotomous choice model*

$$\Pr\{y_i = 1\} = F(x_i.\beta), \qquad i = 1, 2, \dots, n,$$

and the corresponding likelihood function as derived in Section 6. This likelihood function is strictly convex, whether $F(\cdot)$ is the logistic or the standard normal cdf, in the sense that the matrix of second-order partial derivatives is everywhere negative definite.

A.3 Convexity of the Likelihood Function in the Polytomous Logistic Case

In the discussion of the chapter the polytomous generalization of the choice problem was dealt with in terms of the multivariate logistic distribution. We recall that if y_i denotes the choice of individual i over a set of, say, m alternatives coded c_j, $j = 1, 2, \dots, m$, the probability that individual i will choose c_j was written as

$$\Pr\{y_i = c_j\} = \frac{e^{x_i.\beta^*_j}}{\sum_{k=1}^{m} e^{x_i.\beta^*_k}}, \qquad j = 1, 2, \dots, m. \tag{A.45}$$

It is apparent that the parameter vectors $\beta^*_{.j}$ cannot all be identified and, consequently, we need impose the constraint

$$\sum_{j=1}^{m} \beta^*_{.j} = 0. \tag{A.46}$$

Defining

$$\beta_{.j} = \beta^*_{.j} - \beta^*_{.m}, \qquad j = 1, 2, \ldots, m - 1, \tag{A.47}$$

we can rewrite (A.45) in the asymmetric form

$$\Pr\{y_i = c_j\} = \frac{e^{x_i . \beta_{.j}}}{1 + \sum_{k=1}^{m-1} e^{x_i . \beta_{.k}}}, \qquad j = 1, 2, \ldots, m - 1$$

$$= \frac{1}{1 + \sum_{k=1}^{m-1} e^{x_i . \beta_{.k}}}, \qquad j = m.$$

The log likelihood function may be written as

$$L = \sum_{i=1}^{n} \left[\sum_{j=1}^{m-1} c_{ij} x_i . \beta_{.j} - \sum_{j=1}^{m} c_{ij} \ln \left(1 + \sum_{k=1}^{m-1} e^{x_i . \beta_{.k}} \right) \right], \tag{A.48}$$

where

$$c_{ij} \begin{cases} = 1 & \text{if } y_i = c_j \\ = 0 & \text{otherwise.} \end{cases} \tag{A.49}$$

Notice that since every individual does make a choice,

$$\sum_{j=1}^{m} c_{ij} = 1, \qquad i = 1, 2, \ldots, n. \tag{A.50}$$

Hence the log likelihood function is simply

$$L = \sum_{i=1}^{n} \sum_{j=1}^{m-1} c_{ij} x_i . \beta_{.j} - \sum_{i=1}^{m} \ln \left(1 + \sum_{k=1}^{m-1} e^{x_i . \beta_{.k}} \right). \tag{A.51}$$

We easily establish

$$\frac{\partial L}{\partial \beta_{.r}} = \sum_{i=1}^{n} \left[c_{ir} - \frac{e^{x_i . \beta_{.r}}}{1 + \sum_{k=1}^{m-1} e^{x_i . \beta_{.k}}} \right] x_i ., \qquad r = 1, 2, \ldots, m - 1. \tag{A.52}$$

Moreover,

$$\frac{\partial^2 L}{\partial \beta_{.r} \partial \beta'_{.s}} \begin{cases} = \sum_{i=1}^{n} \frac{e^{x_i . \beta_{.r}} e^{x_i . \beta_{.s}}}{(1 + \sum_{k=1}^{m-1} e^{x_i . \beta_{.k}})^2} x'_i . x_i . & \text{if } r \neq s \\ = -\sum_{i=1}^{m} \left[\frac{e^{x_i . \beta_{.r}}}{1 + \sum_{k=1}^{m-1} e^{x_i . \beta_{.k}}} - \left(\frac{e^{x_i . \beta_{.r}}}{1 + \sum_{k=1}^{m-1} e^{x_i . \beta_{.k}}} \right)^2 \right] x'_i . x_i . & \text{if } r = s, \end{cases} \tag{A.53}$$

provided there are no further restrictions on the $\beta_{\cdot j}$, $j = 1, 2, \ldots, m - 1$. To prove the desired result, it is convenient to define

$$Q = (q_{ir}), \qquad q_{ir} = \frac{e^{x_i \cdot \beta_{\cdot r}}}{1 + \sum_{k=1}^{m-1} e^{x_i \cdot \beta_{\cdot k}}}, \tag{A.54}$$

and to notice that

$$\frac{\partial^2 L}{\partial \beta_{\cdot r} \partial \beta'_{\cdot s}} = \sum_{i=1}^{n} q_{ir} q_{is} x'_i \cdot x_i \cdot \qquad \text{if } r \neq s$$

$$- \sum_{i=1}^{n} (q_{ir} - q_{ir}^2) x'_i \cdot x_i \cdot \quad \text{if } r = s. \tag{A.55}$$

Put

$$\beta = (\beta'_{\cdot 1}, \beta'_{\cdot 2}, \ldots, \beta'_{\cdot m-1})' \tag{A.56}$$

and notice that our problem is to show that

$$\frac{\partial^2 L}{\partial \beta \, \partial \beta'}$$

is negative definite. Notice, further, that this matrix occurs naturally in block form and that its (r, s) block is given by (A.55). Define

$$\hat{q}_{i\cdot} = \operatorname{diag}(q_{i1}, q_{i2}, \ldots, q_{im-1}),$$
$$q_{i\cdot} = (q_{i1}, q_{i2}, \ldots, q_{im-1}), \tag{A.57}$$

and observe that the matrix of interest may be written in the convenient form

$$\frac{\partial^2 L}{\partial \beta \, \partial \beta'} = - \sum_{i=1}^{n} [(\hat{q}_{i\cdot} - q'_{i\cdot} q_{i\cdot}) \otimes x'_i \cdot x_i \cdot]. \tag{A.58}$$

Before we proceed we need the following preliminary

Proposition A.2. *For every* i, $i = 1, 2, \ldots, n$, *the matrix*

$$A_i = \hat{q}_{i\cdot} - q'_{i\cdot} q_{i\cdot}$$

is positive definite.

PROOF. The diagonal elements of $\hat{q}_{i\cdot}$ are strictly positive, and hence it is a positive definite matrix. Consider now the characteristic roots of $q'_{i\cdot} q_{i\cdot}$ in the metric of $\hat{q}_{i\cdot}$. It can be shown that all such roots are zero except one, which obeys

$$0 < \lambda_1 = \sum_{r=1}^{m-1} q_{ir} < 1. \tag{A.59}$$

By Proposition 61 of *Mathematics for Econometrics* there exists a non-singular matrix R such that

$$\hat{q}_i = RR', \qquad q'_{i\cdot}q_{i\cdot} = R\Lambda R',$$

where Λ is the matrix of characteristic roots of $q'_{i\cdot}q_{i\cdot}$ in the metric of $\hat{q}_{i\cdot}$. By the preceding,

$$\Lambda = \begin{bmatrix} \lambda_1 & 0 \\ 0 & 0 \end{bmatrix}.$$

Hence

$$A_i = \hat{q}_i - q_{i\cdot}q_{i\cdot} = R \begin{bmatrix} 1 - \lambda_1 & 0 \\ 0 & I \end{bmatrix} R', \qquad i = 1, 2, \ldots, n,$$

which is evidently positive definite for every i.   q.e.d.

We are now in a position to prove the strict convexity of the log likelihood function. We remind the reader that we deal with the situation in which the matrix A_i above is $(m - 1) \times (m - 1)$, the vector $x_{i\cdot}$ contains k elements, and the matrix

$$X = (x_{ig}), \qquad i = 1, 2, \ldots, n, \qquad g = 1, 2, \ldots, k,$$

is of rank k, i.e., $X'X$ is positive definite.
    We have

Proposition A.3. *The matrix in* (A.58) *is negative definite.*

PROOF. We shall prove the equivalent statement, that

$$-\frac{\partial^2 L}{\partial \beta \, \partial \beta'} = \sum_{i=1}^{n} (A_i \otimes B_i)$$

is a positive definite matrix, where

$$A_i = \hat{q}_{i\cdot} - q'_{i\cdot}q_{i\cdot}, \qquad B_i = x'_{i\cdot}x_{i\cdot}.$$

The strategy of the proof is as follows: For an appropriate positive (constant) scalar s define

$$A_i^* = A_i - sI$$

such that A_i^* is positive definite for all i, and rewrite

$$A_i \otimes B_i = (A_i^* \otimes B_i) + (sI \otimes B_i).$$

Then

$$\sum_{i=1}^{n} (A_i \otimes B_i) = \sum_{i=1}^{n} (A_i^* \otimes B_i) + (sI \otimes X'X).$$

Since

$$\sum_{i=1}^{n} x'_{i\cdot} x_{i\cdot} = X'X$$

is positive definite by assumption, it follows that

$$sI \otimes X'X$$

is positive definite. Moreover,

$$\sum_{i=1}^{n} (A_i^* \otimes B_i)$$

is at least positive semidefinite. But the sum of a positive semidefinite and a positive definite matrix is *positive definite*, thus demonstrating that

$$\frac{\partial^2 L}{\partial \beta \, \partial \beta'}$$

is negative definite and, consequently, completing the proof.

The crucial step in the proof is the existence of such a positive scalar s. To construct such a scalar let

$$v = \max_{i} \sum_{r=1}^{m-1} q_{ir} < 1$$

and set

$$s = \min_{i, r} q_{ir}(1 - v) - \delta, \tag{A.60}$$

where δ is any small preassigned positive scalar such that $s > 0$. This is possible since

$$1 - v > 0, \qquad q_{ir} > 0 \quad \text{for all } i, r.$$

It remains to show that for the choice of s as in (A.60) the matrices

$$A_i^* = (\hat{q}_{i\cdot} - sI) - q'_{i\cdot} q_{i\cdot}, \qquad i = 1, 2, \ldots, n,$$

are positive definite. Following exactly the same procedure as in the proof of Proposition A.2 we can show that the characteristic roots of $q'_{i\cdot} q_{i\cdot}$ in the metric of $\hat{q}_{i\cdot} - sI$ are all zero save one, which is

$$\lambda_{i1} = \sum_{r=1}^{m-1} \left(\frac{q_{ir}^2}{q_{ir} - s} \right), \qquad i = 1, 2, \ldots, n.$$

But, by the construction of s,

$$q_{ir} - s > q_{ir} - q_{ir}(1 - v) = q_{ir} v > 0.$$

Hence,

$$\lambda_{i1} < \sum_{r=1}^{m-1} \left(\frac{q_{ir}^2}{q_{ir} v} \right) = \frac{1}{v} \sum_{r=1}^{m-1} q_{ir} \le 1$$

and we can write, for some nonsingular matrix P_i^*,

$$\hat{q}_{i\cdot} - sI = P_i^* P_i^{*\prime}, \qquad q_{i\cdot}' q_{i\cdot} = P_i^* \begin{bmatrix} \lambda_{i1} & 0 \\ 0 & 0 \end{bmatrix} P_i^{*\prime}.$$

Finally,

$$A_i^* = P_i^* \begin{bmatrix} 1 - \lambda_{i1} & 0 \\ 0 & I \end{bmatrix} P_i^{*\prime}, \qquad i = 1, 2, \ldots, n,$$

which shows the A_i^* to be unambiguously positive definite matrices.   q.e.d.

Statistical and Probabilistic Background

8

1 Multivariate Density and Distribution Functions

1.1 Introduction

Most of the economic relationships that are studied empirically involve more than two variables. For example, the demand for food on the part of a given consumer would depend on the consumer's income, the price of food, and the prices of other commodities. Similarly, the demand for labor on the part of a firm would depend on anticipated output and relative factor prices. One can give many more such examples. What is common among them is that often the dependence is formulated to be linear in the parameters, leading to the so-called general linear model. Once the parameters of the model are estimated we are interested in the probability characteristics of the estimators. Since we are, typically, dealing with the estimators of more than one parameter simultaneously, it becomes important to develop the apparatus for studying multivariate relationships. This we shall do in the discussion to follow.

1.2 Multivariate Distributions

It is assumed that the reader is familiar with univariate density functions. Thus the exposition of certain aspects of the discussion below will be rather rudimentary, and certain basic concepts will be assumed to be known.

Definition 1. A *random variable* is a function from the sample space, say S, to the real line.

EXAMPLE 1. The sample space is the space of all conceivable outcomes of an experiment. Thus, in the game of rolling two dice and recording the faces showing, the sample space is the collection of pairs $S = \{(1, 1), (1, 2) \ldots,$ $(1, 6), (2, 1), (2, 2) \ldots, (2, 6) \ldots, (6, 1), (6, 2), \ldots, (6, 6)\}$. On the sample space we can define the random variable X that gives the sum of the faces showing. If we denote by $S_{ij} = (i, j)$ the typical element of S then the random variable X is defined on S and maps elements of S in the real line (more precisely, the integers 1 through 12) by the operation

$$X(S_{ij}) = i + j.$$

Thus, we must learn to distinguish between a random variable, which is a function, and the values assumed by the random variable, which are real numbers. It is often convenient to distinguish between the two by denoting the random variable by capital letters and the values assumed by the random variable by lower case letters. No doubt the reader has dealt with the density and distribution function of univariate random variables. (In the case of a discrete random variable one speaks of the mass function rather than the density function).

Similarly, for multivariate random variables we introduce the density function as follows. Let $\{X_1, X_2, \ldots, X_n\}$ be a set of (continuous) random variables. A nonnegative function $f(\cdot, \cdot, \ldots, \cdot)$ such that

$$F(x_1, x_2, \ldots, x_n) = \Pr\{X_1 \leq x_1, X_2 \leq x_2, \ldots, X_n \leq x_n\}$$

$$= \int_{-\infty}^{x_n} \cdots \int_{-\infty}^{x_1} f(\zeta_1, \zeta_2, \ldots, \zeta_n) d\zeta_1 \cdots d\zeta_n \qquad (1)$$

is said to be the *joint density function of* $X_1, X_2, \ldots, X_n$, and $F(\cdot, \cdot, \ldots, \cdot)$ is said to be their *joint distribution function*.

It is, of course, implicit in (1) that $f(\cdot, \cdot, \ldots, \cdot)$ is an integrable function. In addition, $f(\cdot, \cdot, \ldots, \cdot)$ must satisfy the normalization condition

$$\int_{-\infty}^{\infty} \cdots \int_{-\infty}^{\infty} f(\zeta_1, \zeta_2, \ldots, \zeta_n) d\zeta_1 \, d\zeta_2 \cdots d\zeta_n = 1. \qquad (2)$$

It is an immediate consequence of (1) and (2) that F has the following properties:

(i) $F(-\infty, -\infty, \ldots, -\infty) = 0$;
(ii) $F(+\infty, +\infty, \ldots, +\infty) = 1$;
(iii) F is nondecreasing in each of its arguments.

Remark 1. In (1) and (2) we are dealing with special cases in which f can be defined as the derivative of F; it is not always true that such a simple relation will hold but we will not have the occasion, in this book, to deal with nondifferentiable distribution functions.

Definition 2. Let the set $\{X_1, X_2, \ldots, X_n\}$ have the joint density function $f(\cdot, \cdot, \ldots, \cdot)$; the *marginal density of X_1* is given by

$$g_1(x_1) = \int_{-\infty}^{\infty} \cdots \int_{-\infty}^{\infty} f(x_1, \zeta_2, \zeta_3, \ldots, \zeta_n) d\zeta_2, d\zeta_3 \cdots d\zeta_n. \qquad (3)$$

Let

$$X^{(1)} = (x_1, x_2, \ldots, x_k), \quad k < n.$$

The marginal density function of this subset is defined by

$$g_k(x_1, x_2, \ldots, x_k)$$
$$= \int_{-\infty}^{\infty} \cdots \int_{-\infty}^{\infty} f(x_1, x_2, \ldots, x_k, \zeta_{k+1}, \zeta_{k+2}, \ldots, \zeta_n) d\zeta_{k+1}, d\zeta_{k+2}, \ldots, d\zeta_n$$

Remark 2. The marginal density of a subset of the set $\{X_1, X_2, \ldots, X_n\}$ enables us to make probability statements about elements in that subset without reference to the other elements, i.e., we average over the behavior of the other elements.

In contrast to marginal densities we can introduce various types of conditional densities. Recall from elementary probability theory that if A and B are two events then *the conditional probability of A given B is defined by*

$$\Pr(A|B) = \frac{\Pr(A \cap B)}{\Pr(B)},$$

provided $\Pr(B) \neq 0$. Thus, the fact that B has occurred may convey some information regarding the probability of occurrence of A. The situation is entirely similar when dealing with joint densities, i.e., the fact that something is given regarding one subset of random variables may convey some information regarding probability statements we can make about another subset. We are, thus, led to

Definition 3. Let $\{X_1, X_2, \ldots, X_n\}$ have the joint density $f(\cdot, \cdot, \ldots, \cdot)$, and partition the set by $X^{(1)}$, $X^{(2)}$ so that $X^{(1)} = (X_1, X_2, \ldots, X_k)$, $X^{(2)} = (X_{k+1}, X_{k+2}, \ldots, X_n)$. Let $g_{n-k}(\cdot, \cdot, \ldots, \cdot)$ be the marginal density of $X^{(2)}$. Then the *conditional density of $X^{(1)}$ given $X^{(2)}$* is defined as

$$h(x_1, x_2, \ldots, x_k | x_{k+1}, x_{k+2}, \ldots, x_n) = \frac{f(x_1, x_2, \ldots, x_n)}{g_{n-k}(x_{k+1}, \ldots, x_n)},$$

provided the denominator does not vanish. The denominator is, of course, the marginal density of the elements of $X^{(2)}$.

The introduction of conditional densities affords us a rather simple way of defining independence between two sets of random variables. Recall again from elementary probability theory that two events A and B are said to be independent if and only if $\Pr(A|B) = \Pr(A)$. This immediately suggests that the joint probability of A and B, $\Pr(A \cap B)$, obeys

$$\Pr(A \cap B) = \Pr(A)\Pr(B).$$

The situation is entirely similar in the current context. We have

Definition 4. Let $(X_1, X_2, \ldots, X_n)$ and $X^{(1)}$, $X^{(2)}$ be as in Definition 3; then the two sets are said to be *mutually independent* if and only if the conditional density of $X^{(1)}$ given $X^{(2)}$ is equal to the marginal density of $X^{(1)}$. More generally, the elements of $(X_1, X_2, \ldots, X_n)$ are said to be mutually independent if and only if their joint density is equal to the product of their marginal densities.

1.3 Expectation and Covariance Operators

In a univariate context the reader was exposed to the expectation and variance operators. In particular, if X is a scalar (univariate) random variable with density function $f(\cdot)$, then the expectation operator E is defined by

$$E(X) = \int_{-\infty}^{\infty} x f(x) dx = \mu \tag{4}$$

The variance operator Var is defined by

$$\mathrm{Var}(X) \equiv E(X - \mu)^2 = \int_{-\infty}^{\infty} (x - \mu)^2 f(x) dx = \sigma^2. \tag{5}$$

It is a property of the expectation operator that if a_i, $i = 1, 2, \ldots, n$, are fixed constants and if X_i are random variables with expectations (means) μ_i, $i = 1, 2, \ldots, n$, respectively, then

$$E\left[\sum_{i=1}^{n} a_i X_i\right] = \sum_{i=1}^{n} a_i E(X_i) = \sum_{i=1}^{n} a_i \mu_i. \tag{6}$$

If, in addition, the X_i are mutually independent,

$$\mathrm{Var}\left(\sum_{i=1}^{n} a_i X_i\right) = \sum_{i=1}^{n} a_i^2 \, \mathrm{Var}(X_i) = \sum_{i=1}^{n} a_i^2 \sigma_i^2, \tag{7}$$

where, of course, $\mathrm{Var}(X_i) = \sigma_i^2$.

We shall now extend the definition of such operators to the multivariate case. Actually, the extension involves little more than repeated applications of the expectation operator as just explained above.

Definition 5. Let $X = (X_{ij})$ be a matrix whose elements, X_{ij}, are random variables. The *expectation of X* is defined as

$$E(X) = [E(X_{ij})] \tag{8}$$

Remark 3. The definition above makes abundantly clear that no new concept or operation is involved in defining the expectation of a random matrix; it is simply the matrix whose elements are the expectation of the elements of the matrix X.

An immediate consequence of Definition 5 is

Proposition 1. *Let A, B, C be matrices whose elements are nonstochastic, and let X be a random matrix such that the quantities AXB and $AX + C$ are defined. Then*

$$E(AXB) = AE(X)B, \qquad E(AX + C) = AE(X) + C.$$

PROOF. The typical elements of AXB and $AX + C$ are, respectively,

$$\sum_s \sum_k a_{ik} X_{ks} b_{sj} \quad \text{and} \quad \sum_k a_{ik} X_{kj} + c_{ij}.$$

By Equation (6) we have

$$E \sum_s \sum_k a_{ik} X_{ks} b_{sj} = \sum_s \sum_k a_{ik} E(X_{ks}) b_{sj},$$

$$E \sum_k a_{ik} X_{kj} + c_{ij} = \sum_k a_{ik} E(X_{kj}) + c_{ij}.$$

The right members of the equations above are, respectively, the (i, j) elements of $AE(X)B$ and $AE(X) + C$. q.e.d.

When we operate in a univariate context, the mean and variance are commonly employed summary characterizations of the random variable's density function. In the case of the normal distribution these two parameters completely determine the nature of the distribution; this, of course, is not true of all distributions. Still, even if we confine ourselves to the normal distribution, means and variances are by no means sufficient to determine the shape of a multivariate distribution, nor are they sufficient descriptions of the second moment properties of the distribution. Minimally, we also require *some measure of the covariation between pairs of (scalar) random variables.* This leads us to

Definition 6. Let $\{X_1, X_2, \ldots, X_n\}$ be a set of random variables and let $f(\cdot, \cdot, \ldots, \cdot)$ be the joint density function. The *covariance between X_i and X_j,* $i, j = 1, 2, \ldots, n$, is defined by

$$\text{Cov}(X_i, X_j) = E[(X_i - \mu_i)(X_j - \mu_j)],$$

where $\mu_i = E(X_i)$, $i = 1, 2, \ldots, n$.

Remark 4. The meaning of the expectation operator in a multivariate context is as follows: If $f(\cdot, \cdot, \ldots, \cdot)$ is the joint density as in Definition 6, then

$$\mu_i = \int_{-\infty}^{\infty} \cdots \int_{-\infty}^{\infty} \zeta_i f(\zeta_1, \zeta_2, \ldots, \zeta_n) d\zeta_1 \cdots d\zeta_n = E(X_i),$$

i.e., *it is determined with respect to the marginal distribution of X_i.* Similarly,

$$\text{Cov}(X_i, X_j) = \int_{-\infty}^{\infty} \cdots \int_{-\infty}^{\infty} (\zeta_i - \mu_i)(\zeta_j - \mu_j) f(\zeta_1, \zeta_2, \ldots, \zeta_n) d\zeta_1 \cdots d\zeta_n.$$

A convenient representation of the second moment properties of a multi-variate distribution is in terms of the covariance matrix.

Definition 7. Let $\{X_1, X_2, \ldots, X_n\}$ be a set of random variables as in Definition 6. Then the *covariance matrix (of its joint distribution)* is

$$\Sigma = (\sigma_{ij}), \qquad i, j = 1, 2, \ldots, n,$$

where

$$\sigma_{ij} = \text{Cov}(X_i, X_j).$$

A number of properties of a covariance matrix follow immediately from the definition. We have

Proposition 2. *Let* $\{X_1, X_2, \ldots, X_n\}$ *be a set of random variables as in Definition 6 and let* Σ *be the covariance matrix of their joint distribution. Then:*

(a) Σ *is symmetric;*
(b) Σ *is positive definite, unless the random variables are linearly dependent.*[1]

PROOF. The first part of the proposition is obvious since

$$\sigma_{ji} = \int_{-\infty}^{\infty} \cdots \int_{-\infty}^{\infty} (\zeta_j - \mu_j)(\zeta_i - \mu_i) f(\zeta_1, \zeta_2, \ldots, \zeta_n) d\zeta_1 \cdots d\zeta_n = \sigma_{ij}.$$

To prove the second part we proceed as follows. Let α be any n-element vector of constants and $\mu = (\mu_1, \mu_2, \ldots, \mu_n)'$ be the vector of means. Consider the scalar random variable

$$Z = \alpha'(X - \mu), \qquad X = (X_1, X_2, \ldots, X_n)'. \tag{9}$$

We have

$$E(Z) = 0, \qquad \text{Var}(Z) = E(Z^2) = E[\alpha'(X - \mu)(X - \mu)'\alpha] = \alpha'\Sigma\alpha. \tag{10}$$

We now have to show that if α is not the null vector $\alpha'\Sigma\alpha > 0$ unless the elements of the vector X are linearly dependent. Thus, consider any α such that $\alpha \neq 0$. Since $\alpha'\Sigma\alpha$ is the variance of the random variable Z it is *nonnegative*. If $\alpha'\Sigma\alpha = 0$, then we conclude that Z is nonstochastic, i.e., it is a constant, say $Z = c$. Since $E(Z) = 0$ we thus have that $Z = 0$. Since α is nonnull, it has at

---

[1] A set of random variables $\{X_1, X_2, \ldots, X_n\}$ is said to be *linearly dependent* if there exists a set of constants $c_1, \ldots, c_n$ not all of which are zero such that $\sum_{i=1}^{n} c_i X_i = c_0$ and c_0 is a suitable constant. When this is so we can express one of these random variables (exactly) as a linear combination of the remaining ones. If the variables are not linearly dependent, they are said to be *linearly independent*. We should stress that this is *not the same* as mutual independence in the sense of Definition 4. If a set of variables exhibits mutual independence in the sense of Definition 4, then the variables are also linearly independent. The converse, however, is not necessarily true.

least one nonzero element. Without loss of generality suppose $\alpha_n \neq 0$. Then we can write

$$X_n - \mu_n = -\frac{1}{\alpha_n} \sum_{i=1}^{n-1} \alpha_i (X_i - \mu_i), \tag{11}$$

which shows that $\alpha'\Sigma\alpha = 0$ implies the linear dependence in (11). Consequently, if no linear dependence exists, $\alpha'\Sigma\alpha > 0$ for $\alpha \neq 0$, which shows that Σ is positive definite.   q.e.d.

Frequently, it is convenient to have a measure of the covariation of two random variables that does not depend on the units of measurement—as is the case for the covariance as given in Definition 7.
We have

Definition 8. Let $X = \{X_1, X_2, \ldots, X_n\}$ be a set of random variables, and let $\Sigma = (\sigma_{ij})$ be their covariance matrix, as in Definition 7. The (*simple*) *correlation coefficient between* X_i *and* X_j, $i, j = 1, 2, \ldots, n$, is defined by

$$\rho_{ij} = \frac{\sigma_{ij}}{\sqrt{\sigma_{ii}\sigma_{jj}}}.$$

The correlation matrix is, thus,

$$R = (\rho_{ij}) = S\Sigma S,$$

where

$$S = \mathrm{diag}(\sigma_{11}^{-1/2}, \sigma_{22}^{-1/2}, \ldots, \sigma_{nn}^{-1/2}).$$

It is an obvious consequence of Definition 8 and Proposition 2 that R is symmetric and positive definite, unless the elements of X are linearly dependent.

1.4 A Mathematical Digression

Suppose it is given that $x = (x_1, x_2, \ldots, x_n)'$ is a random vector[2] with mean μ and covariance matrix Σ. Frequently, we shall be interested in various aspects of the distribution of, say, a linear transformation of x. In particular, if A and b are, respectively, $k \times n$ and $k \times 1$, we may consider

$$y = Ax + b.$$

---

[2] Henceforth, we shall abandon the convention of denoting random variables by capital letters and the values they assume by lower case letters. We shall, instead, indiscriminately denote random variables by lower case letters.

We can easily prove

Proposition 3. *Let* $x = (x_1, x_2, \ldots, x_n)'$ *be a random vector with mean μ and covariance matrix Σ. Then*

$$y = Ax + b$$

has mean $A\mu + b$ and covariance matrix $A\Sigma A'$, where A is $k \times n$ and b is $k \times 1$.

PROOF. From Proposition 1 we easily conclude that

$$E(y) = AE(x) + b = A\mu + b.$$

From Problem 4 we have that

$$Cov(y) = E(y - E(y))(y - E(y))' = E[A(x - \mu)(x - \mu)'A'] = A\Sigma A'.$$

The last equality follows, of course, by Proposition 1.   q.e.d.

Typically, however, we shall be interested in a more general problem; thus, given that x has the density function $f(\cdot)$ and $y = h(x)$, what is the density function of y? In order to solve this problem we need the following auxiliary concepts.

Definition 9. Let x be $n \times 1$ and consider the transformation

$$y = h(x) = (h_1(x), h_2(x), \ldots, h_n(x))',$$

where y is $n \times 1$. Suppose the inverse transformation

$$x = g(y) = (g_1(y), g_2(y), \ldots, g_n(y))'$$

exists, where $h_i(\cdot)$ and $g_i(\cdot)$ are (scalar) functions of n variables. The matrix

$$\left[\frac{\partial x_i}{\partial y_j}\right], \quad i = 1, 2, \ldots, n, \ j = 1, 2, \ldots, n,$$

is said to be the *Jacobian matrix of the transformation* (from x to y); the absolute value of the determinant of this matrix is said to be the *Jacobian of the transformation* and is denoted by $J(y)$.

We have, then, the following useful result, which is given without formal proof.

Proposition 4. Let x be an n-element random vector the joint density function of whose elements is $f(\cdot)$. Let $y = h(x)$, where y is an n-element vector. Suppose the inverse function $x = g(y)$ exists, and let $J(y)$ be the Jacobian of the transformation from x to y. Then the joint density of the elements of y is given by

$$\phi(y) = f[g(y)]J(y).$$

PROOF. Although a formal proof is not appropriate in a book of this type, perhaps the following informal discussion will clarify the situation.

Let ω be a subset of the n-dimensional Euclidean space $\mathbb{E}_n$. Then

$$\Pr\{x \in \omega\} = \int_{\omega} f(x)dx$$

i.e., the probability that the random vector x will assume a value in the set ω is found by integrating its joint density function over ω. Let

$$\omega' = \{y \,|\, y = h(x), x \in \omega\}$$

i.e., ω' is the image of ω under the transformation $h(\cdot)$. The notation above is to be read as: ω' is the set of all points y such that $y = h(x)$ and x belongs to the set ω.

It is clear that if x_0 is a point in ω, i.e., if $x_0 \in \omega$, and $y_0 = h(x_0)$, then $y_0 \in \omega'$. Conversely, if a point y_0 belongs to the set ω' then the point $x_0 = g(y_0)$ has the property $x_0 \in \omega$. But this means that

$$\Pr\{y \in \omega'\} = \Pr\{x \in \omega\}$$

Now by definition

$$\Pr\{x \in \omega\} = \int_{\omega} f(x)dx,$$

$$\Pr\{y \in \omega'\} = \int_{\omega'} \phi(y)dy,$$

where $\phi(\cdot)$ is the density function of y. Suppose now in the integral $\int_{\omega} f(x)dx$ we make the change in variable $x = g(y)$. By the usual rules of changing variables in multiple integrals,

$$\int_{\omega} f(x)dx = \int_{\omega'} f[g(y)]J(y)dy.$$

But this simply says that

$$\Pr\{y \in \omega'\} = \int_{\omega'} f[g(y)]J(y)dy.$$

If we can show that this holds for all suitable sets ω' then the relation above defines

$$\phi(y) = f[g(y)]J(y)$$

as the density function of y, which is what is stated by Proposition 4.   q.e.d.

2 The Multivariate Normal Distribution

2.1 Joint, Marginal, and Conditional Density Functions

It is assumed that the reader is thoroughly familiar with the standard univariate normal distribution. Thus, we recall that if x is a normally distributed (scalar) random variable with mean zero and variance one, we write

$$x \sim N(0, 1),$$

and its density function is given by

$$f(x) = (2\pi)^{-1/2} \exp(-\tfrac{1}{2}x^2).$$

Now, let x_i, $i = 1, 2, \ldots, n$, be mutually independent identically distributed random variables such that

$$x_i \sim N(0, 1), \qquad i = 1, 2, \ldots, n.$$

It is clear from Definition 4 that their joint density is

$$f(x_1, x_2, \ldots, x_n) = (2\pi)^{-n/2} \exp(-\tfrac{1}{2}x'x), \qquad x = (x_1, x_2, \ldots, x_n)'. \quad (12)$$

Consider now the transformation

$$y = Ax + b,$$

where A is an $n \times n$ nonsingular matrix. By assumption,

$$\mathrm{Cov}(x) = I,$$

I being the identity matrix of order n. From Proposition 3 we have

$$E(y) = b, \qquad \mathrm{Cov}(y) = AA'.$$

The inverse of the transformation above is

$$x = A^{-1}(y - b).$$

The Jacobian matrix is

$$\left[\frac{\partial x_i}{\partial y_j}\right] = A^{-1}.$$

But, then, (12) in conjunction with Proposition 4 implies

$$\phi(y) = (2\pi)^{-n/2}|A'A|^{-1/2} \exp\{-\tfrac{1}{2}(y - b)'A'^{-1}A^{-1}(y - b)\}. \quad (13)$$

To put this in more standard notation, let

$$b = \mu, \qquad AA' = \Sigma, \quad (14)$$

and note that μ is the mean vector of y and Σ its covariance matrix. We have therefore proved the following.

Proposition 5. *Let* $\{x_i: i = 1, 2, \ldots, n\}$ *be a set of mutually independent identically distributed random variables such that*

$$x_i \sim N(0, 1), \qquad i = 1, 2, \ldots, n,$$

and consider

$$y = Ax + \mu,$$

where A is an $n \times n$ *nonstochastic, nonsingular matrix and* μ *a nonstochastic n-element vector. Then the joint distribution of the elements of y is given by*

$$\phi(y) = (2\pi)^{-n/2} |\Sigma|^{-1/2} \exp\{-\tfrac{1}{2}(y - \mu)'\Sigma^{-1}(y - \mu)\}. \tag{15}$$

We now define the multivariate normal distribution by

Definition 10. Let y be an n-element random vector the joint distribution of whose elements is given by (15). Then y is said to have the *multivariate normal distribution with mean vector* μ *and covariance matrix* Σ. This fact is denoted by

$$y \sim N(\mu, \Sigma).$$

Remark 5. Readers may wonder why we had employed the somewhat elaborate procedure of Proposition 5 in defining the multivariate normal distribution. They may ask: why did we not define the multivariate normal with mean μ and covariance matrix Σ by (15) without any preliminaries? The reason is that, if we did so define it, it would have been incumbent on us to show, by integration, that μ is indeed the mean vector, and Σ is indeed the covariance matrix. In the development we chose to follow we have built upon elementary facts, and the mean and covariance parameters of y were derived by the relatively trivial operations entailed by Proposition 3. In this fashion all pertinent facts about the multivariate normal were deduced from the properties of mutually independent univariate standard normal random variables.

Having now established the nature of the multivariate normal let us consider some of the properties of jointly normal random variables. In particular we should be interested in the joint density of linear combinations of random variables; we should also examine marginal and conditional densities. We have

Proposition 6. *Let* $y \sim N(\mu, \Sigma)$, *let y be* $n \times 1$, *and let B be a nonsingular nonstochastic* $n \times n$ *matrix. Then*

$$z = By$$

has the distribution

$$z \sim N(B\mu, B\Sigma B').$$

PROOF. Using Proposition 3 we have that

$$y = B^{-1}z,$$

and the Jacobian of the transformation is

$$J = |BB'|^{-1/2}.$$

Inserting in the joint density of the elements of y, as given, say, in (15) we find

$$\phi(z) = (2\pi)^{-n/2}|\Sigma|^{-1/2}|BB'|^{-1/2} \exp\{-\tfrac{1}{2}(B^{-1}z - \mu)'\Sigma^{-1}(B^{-1}z - \mu)\}$$
$$= (2\pi)^{-n/2}|B\Sigma B'|^{-1/2} \exp\{-\tfrac{1}{2}(z - B\mu)'(B\Sigma B')^{-1}(z - B\mu)\} \qquad (16)$$

which is recognized as a multivariate normal density with mean $B\mu$ and covariance matrix $B\Sigma B'$.   q.e.d.

Remark 6. It ought to be pointed out that there is a difference between Propositions 5 and 6: in the first we proved that a linear transformation of *mutually independent* $N(0, 1)$ *variables* is normally distributed; in the second we proved that a linear transformation of jointly normal variables with arbitrary (positive definite) covariance matrix is normally distributed. Proposition 6, of course, implies Proposition 5.

As pointed out in previous sections, associated with the joint density of a set of random variables are various marginal and conditional density functions. It is interesting that both types of densities associated with the normal distribution are also normal.

Proposition 7. *Let* $y \sim N(\mu, \Sigma)$, *let* Σ *be positive definite, and let* y *be* $n \times 1$ *and partition*

$$y = \begin{pmatrix} y^1 \\ y^2 \end{pmatrix}$$

so that y^1 *has* k *elements and* y^2 *has* $n - k$ *elements. Partition*

$$\mu = \begin{pmatrix} \mu^1 \\ \mu^2 \end{pmatrix},$$

conformably with y *and put*

$$\Sigma = \begin{bmatrix} \Sigma_{11} & \Sigma_{12} \\ \Sigma_{21} & \Sigma_{22} \end{bmatrix},$$

so that Σ_{11} *is* $k \times k$, Σ_{22} *is* $(n - k) \times (n - k)$, *and* Σ_{12} *is* $k \times (n - k)$. *Then*

$$y^1 \sim N(\mu^1, \Sigma_{11}).$$

PROOF. Since Σ is positive definite, so is Σ^{-1}; thus there exists a lower triangular matrix T such that

$$\Sigma^{-1} = T'T.$$

Partition T conformably with Σ^{-1}, i.e.,

$$T = \begin{pmatrix} T_1 & 0 \\ T_2 & T_3 \end{pmatrix},$$

and observe that

$$T^{-1} = \begin{bmatrix} T_1^{-1} & 0 \\ -T_3^{-1}T_2T_1^{-1} & T_3^{-1} \end{bmatrix}, \qquad \Sigma_{11} = T_1^{-1}T_1'^{-1}.$$

Now, consider

$$z = T(y - \mu)$$

and conclude by Proposition 6 that

$$z \sim N(0, I).$$

Consequently, the elements of z are mutually independent, as is apparent from the discussion surrounding equation (12). But partitioning z conformably with y we have

$$z = \begin{pmatrix} z^1 \\ z^2 \end{pmatrix} = \begin{pmatrix} T_1(y^1 - \mu^1) \\ T_2(y^1 - \mu^1) + T_3(y^2 - \mu^2) \end{pmatrix}$$

and we see that y^1 can be expressed solely in terms of z^1. Since $z \sim N(0, I)$, it is clear that the marginal distribution of z^1 is $N(0, I_k)$. Since y^1 can be expressed solely in terms of z^1, its marginal distribution can be derived from that of z^1. But we note

$$y^1 = T_1^{-1}z^1 + \mu^1.$$

By Proposition 6 we have

$$y^1 \sim N(\mu^1, \Sigma_{11}). \quad \text{q.e.d.}$$

Remark 7. The preceding shows, in particular, that if $y \sim N(\mu, \Sigma)$, then any element of y, say the ith, has the marginal distribution $y_i \sim N(\mu_i, \sigma_{ii})$, $i = 1, 2, \ldots, n$.

The conditional distributions associated with the multivariate normal are obtained as follows.

Proposition 8. *Let $y \sim N(\mu, \Sigma)$ and partition as in Proposition 7. Then the conditional density of y^1 given y^2 is*

$$N[\mu^1 + \Sigma_{12}\Sigma_{22}^{-1}(y^2 - \mu^2), \Sigma_{11} - \Sigma_{12}\Sigma_{22}^{-1}\Sigma_{21}].$$

PROOF. By Proposition 7 the marginal density of y^2 is $N(\mu^2, \Sigma_{22})$. By definition, the conditional density of y^1 given y^2 is

$$\frac{N(\mu, \Sigma)}{N(\mu^2, \Sigma_{22})} = \frac{(2\pi)^{-n/2} \exp\{-\frac{1}{2}(y - \mu)'\Sigma^{-1}(y - \mu)\}|\Sigma|^{-1/2}}{(2\pi)^{-(n-k)/2} \exp\{-\frac{1}{2}(y^2 - \mu^2)'\Sigma_{22}^{-1}(y^2 - \mu^2)\}|\Sigma_{22}|^{-1/2}}$$

$$= (2\pi)^{-k/2} \exp\{-\frac{1}{2}[(y - \mu)'\Sigma^{-1}(y - \mu) - (y^2 - \mu^2)'\Sigma_{22}^{-1}$$

$$\times (y^2 - \mu^2)]\}|\Sigma|^{-1/2}|\Sigma_{22}|^{1/2}.$$

To evaluate the exponential above we note

$$
\Sigma = \begin{bmatrix} I & \Sigma_{12}\Sigma_{22}^{-1} \\ 0 & I \end{bmatrix} \begin{bmatrix} I & -\Sigma_{12}\Sigma_{22}^{-1} \\ 0 & I \end{bmatrix} \begin{bmatrix} \Sigma_{11} & \Sigma_{12} \\ \Sigma_{21} & \Sigma_{22} \end{bmatrix}
$$

$$
\times \begin{bmatrix} I & 0 \\ -\Sigma_{22}^{-1}\Sigma_{21} & I \end{bmatrix} \begin{bmatrix} I & 0 \\ \Sigma_{22}^{-1}\Sigma_{21} & I \end{bmatrix}.
$$

This is so since

$$
\begin{bmatrix} I & \Sigma_{12}\Sigma_{22}^{-1} \\ 0 & I \end{bmatrix} \begin{bmatrix} I & -\Sigma_{12}\Sigma_{22}^{-1} \\ 0 & I \end{bmatrix} = \begin{bmatrix} I & 0 \\ 0 & I \end{bmatrix}.
$$

From the middle three matrices we find

$$
\begin{bmatrix} I & -\Sigma_{12}\Sigma_{22}^{-1} \\ 0 & I \end{bmatrix} \begin{bmatrix} \Sigma_{11} & \Sigma_{12} \\ \Sigma_{21} & \Sigma_{22} \end{bmatrix} \begin{bmatrix} I & 0 \\ -\Sigma_{22}^{-1}\Sigma_{21} & I \end{bmatrix} = \begin{bmatrix} \Sigma_{11} - \Sigma_{12}\Sigma_{22}^{-1}\Sigma_{21} & 0 \\ 0 & \Sigma_{22} \end{bmatrix}.
$$

Putting $\Sigma_{11 \cdot 2} = \Sigma_{11} - \Sigma_{12}\Sigma_{22}^{-1}\Sigma_{21}$, we thus determine $|\Sigma| = |\Sigma_{22}||\Sigma_{11 \cdot 2}|$ and

$$
\Sigma^{-1} = \begin{bmatrix} I & 0 \\ -\Sigma_{22}^{-1}\Sigma_{21} & I \end{bmatrix} \begin{bmatrix} \Sigma_{11 \cdot 2}^{-1} & 0 \\ 0 & \Sigma_{22}^{-1} \end{bmatrix} \begin{bmatrix} I & -\Sigma_{12}\Sigma_{22}^{-1} \\ 0 & I \end{bmatrix}.
$$

Consequently,

$$
\begin{aligned}
&(y - \mu)'\Sigma^{-1}(y - \mu) \\
&= [y^1 - \mu^1 - \Sigma_{12}\Sigma_{22}^{-1}(y^2 - \mu^2)]'\Sigma_{11 \cdot 2}^{-1} [y^1 - \mu^1 - \Sigma_{12}\Sigma_{22}^{-1}(y^2 - \mu^2)] \\
&\quad + (y^2 - \mu^2)'\Sigma_{22}^{-1}(y^2 - \mu^2).
\end{aligned}
$$

Using these relations we see immediately that

$$
\frac{N(\mu, \Sigma)}{N(\mu^2, \Sigma_{22})} = (2\pi)^{-k/2}|\Sigma_{11 \cdot 2}|^{-1/2} \exp\{-\tfrac{1}{2}(y^1 - v^1)'\Sigma_{11 \cdot 2}^{-1}(y^1 - v^1)\},
$$

where

$$
v^1 = \mu^1 + \Sigma_{12}\Sigma_{22}^{-1}(y^2 - \mu^2).
$$

But this is recognized as the $N(v^1, \Sigma_{11 \cdot 2})$ distribution.   q.e.d.

Remark 8. Often, $\Sigma_{12}\Sigma_{22}^{-1}$ is termed the *matrix of regression coefficients of* y^1 *on* y^2, and v^1, the mean of the conditional distribution, is said to be the *regression of* y^1 *on* y^2.

Given the results of Proposition 8, it is now a simple matter to give necessary and sufficient conditions for two sets of jointly normal variables to be mutually independent. We have

Proposition 9. *Let* $y \sim N(\mu, \Sigma)$ *and partition* y, μ, *and* Σ *by*

$$y = \begin{pmatrix} y^1 \\ y^2 \end{pmatrix}, \quad \mu = \begin{pmatrix} \mu^1 \\ \mu^2 \end{pmatrix}, \quad \Sigma = \begin{bmatrix} \Sigma_{11} & \Sigma_{12} \\ \Sigma_{21} & \Sigma_{22} \end{bmatrix}$$

so that y^1 *is* $k \times 1$, y^2 *is* $(n-k) \times 1$, Σ_{11} *is* $k \times k$, Σ_{22} *is* $(n-k) \times (n-k)$, *etc. Then* y^1 *and* y^2 *are mutually independent if and only if* $\Sigma_{12} = 0$.

PROOF. Since Σ is symmetric, $\Sigma'_{21} = \Sigma_{12}$ so that $\Sigma_{12} = 0$ implies $\Sigma_{21} = 0$. If $\Sigma_{12} = 0$, Proposition 8 implies that the conditional distribution of y^1 given y^2 is $N(\mu^1, \Sigma_{11})$; on the other hand, Proposition 7 implies—whether $\Sigma_{12} = 0$ or not—that the marginal distribution of y^1 is $N(\mu^1, \Sigma_{11})$. Thus, by Definition 4 the two sets of random variables are mutually independent. Conversely, if y^1 and y^2 are mutually independent, and if $y_i \in y^1$ and $y_j \in y^2$, it follows from a general property of independent random variables that $\mathrm{Cov}(y_i, y_j) = 0$. But this shows that the typical element of Σ_{12} is zero; thus $\Sigma'_{21} = \Sigma_{12} = 0$.

q.e.d.

Remark 9. It is, of course, true quite generally that if two random variables are mutually independent their covariance vanishes; thus, independence implies lack of correlation. The *converse, however, is not generally true; lack of correlation does not imply independence for all types of (joint) distributions.* As Proposition 9 makes clear *this is the case for the normal distribution.* We give an example that illustrates this fact.

EXAMPLE 2. Let x be a random variable that assumes the values $\pm\frac{1}{4}, \pm\frac{1}{2}, \pm 1$ each with probability $\frac{1}{6}$. Let $y = x^2$; then y assumes the values $\frac{1}{16}, \frac{1}{4}, 1$ each with probability $\frac{1}{3}$. The joint mass function of (x, y) is given by

$$\mathrm{Pr}\{x = -\tfrac{1}{4}, y = \tfrac{1}{16}\} = \mathrm{Pr}\{x = \tfrac{1}{4}, y = \tfrac{1}{16}\} = \tfrac{1}{6},$$

$$\mathrm{Pr}\{x = -\tfrac{1}{2}, y = \tfrac{1}{4}\} = \mathrm{Pr}\{x = \tfrac{1}{2}, y = \tfrac{1}{4}\} = \tfrac{1}{6},$$

$$\mathrm{Pr}\{x = -1, y = 1\} = \mathrm{Pr}\{x = 1, y = 1\} = \tfrac{1}{6}.$$

We have, by construction,

$$E(x) = 0,$$

Since

$$\mathrm{Cov}(x, y) = E(xy) - E(x)E(y),$$

we see that

$$\mathrm{Cov}(x, y) = E(xy) = 0.$$

Thus, even though the two random variables are functionally dependent (i.e., one is an exactly specified function of the other) their covariance, and hence their correlation, is zero.

Earlier on we had occasion to define a type of correlation coefficient, the simple correlation coefficient between x_i and x_j as $\rho_{ij} = \sigma_{ij}/\sqrt{\sigma_{ii}\sigma_{jj}}$, σ_{ij} being the covariance between x_i and x_j. There are, however, other definitions of correlation coefficients. One may think of ρ_{ij} as being defined with respect to *the marginal distribution* of the pair (x_i, x_j). If we define correlation coefficients with respect to conditional distributions we are led to

Definition 11. Let $y \sim N(\mu, \Sigma)$, and partition

$$y = \begin{pmatrix} y^1 \\ y^2 \end{pmatrix}, \quad \mu = \begin{pmatrix} \mu^1 \\ \mu^2 \end{pmatrix}, \quad \Sigma = \begin{pmatrix} \Sigma_{11} & \Sigma_{12} \\ \Sigma_{21} & \Sigma_{22} \end{pmatrix}$$

as in Proposition 9. Let $y_i, y_j \in y^1$; the partial correlation coefficient between y_i and y_j given $y_{k+1}, \ldots, y_n$ is defined as the simple correlation coefficient between y_i and y_j *in the context of the conditional distribution of y^1 given y^2,* and is denoted by $\rho_{ij \cdot k+1, k+2, \ldots, n}$.

Remark 10. As the preceding definition makes abundantly clear, there are as many partial correlation coefficients between y_i and y_j as there are conditional distributions of (y_i, y_j), i.e., by varying the set of conditioning variables we obtain a number of conditional distributions of the pair (y_i, y_j). For each such conditional distribution we can compute a distinct partial correlation coefficient.

Before we conclude this section we ought to observe that the preceding discussion does not address itself to the following question. Suppose $x \sim N(\mu, \Sigma)$ and Σ is positive definite. Let A be a nonstochastic matrix of order $s \times n$ ($s < n$) and rank$(A) = s$. What is the distribution of Ax? The reader ought to note that *Proposition 6 requires that the matrix A, above, be nonsingular. Since A is not square, however, it cannot possibly be nonsingular. Thus, Proposition 6 does not apply.* We have, however,

Proposition 10. *Let A be a nonstochastic $s \times n$ matrix $s < n$ with* rank$(A) = s$ *and suppose $x \sim N(\mu, \Sigma)$ with Σ nonsingular. Then*

$$y \sim N(A\mu, A\Sigma A').$$

where

$$y = Ax.$$

PROOF. Since the rows of A are linearly independent, there exists an $(n - s) \times n$ matrix, say A_*, such that

$$B = \begin{pmatrix} A \\ A_* \end{pmatrix}$$

is nonsingular. By Proposition 6 we have $Bx \sim N(B\mu, B\Sigma B')$. In particular,

$$Bx = \begin{pmatrix} y \\ z \end{pmatrix}$$

where, $z = A_* x$. Thus,

$$B\mu = \begin{pmatrix} A\mu \\ A_*\mu \end{pmatrix}, \qquad B\Sigma B' = \begin{bmatrix} A\Sigma A' & A\Sigma A'_* \\ A_*\Sigma A' & A_*\Sigma A'_* \end{bmatrix}.$$

By Proposition 7 it follows that $y \sim N(A\mu, A\Sigma A')$.　q.e.d.

The reader should carefully note the limitations of the result above. Thus, e.g., suppose that A is $s \times n$ $(s < n)$ but rank$(A) < s$. Putting $y = Ax$ we note, that, from elementary considerations, Cov$(y) = A\Sigma A'$. But $A\Sigma A'$ is an $s \times s$ matrix of rank less than s, i.e., it is a singular matrix. Or, alternatively, suppose that A is as above but $s > n$. In this case, proceeding as before, we find Cov$(y) = A\Sigma A'$. The matrix $A\Sigma A'$ is $s \times s$ but its rank cannot exceed n; although singular covariance matrices do not occur routinely in econometrics, nonetheless they occur frequently enough to warrant some discussion. We examine the problems posed by this incidence at the end of this section, so that those not interested in pursuing the matter may skip the discussion without loss of continuity.

2.2 The Moment Generating Function

There is another interesting fact about normal variables that should be kept in mind, viz., if all (nontrivial) linear combinations of a random vector y are normally distributed then the joint distribution of the elements of this vector is normal. In order to prove this, we first introduce the moment generating function associated with the multivariate normal distribution. We recall

Definition 12. Let x be a random variable, s a real number, and $f(\cdot)$ be the density function of x; then

$$M_x(s) = E(e^{sx}) = \int_{-\infty}^{\infty} e^{sx} f(x)dx \qquad (17)$$

is said to be the *moment generating function associated with the random variable* x, provided the integral exists and is continuous at $s = 0$.

We also recall that if $x \sim N(0, 1)$, then

$$M_x(t) = \int_{-\infty}^{\infty} \frac{1}{\sqrt{2\pi}} e^{-(1/2)x^2 + tx} \, dx = e^{(1/2)t^2} \qquad (18)$$

Building on the relation in (18) we shall now derive the moment generating function of the multivariate normal distribution with mean μ and covariance matrix Σ. We have

Proposition 11. *Let* $y \sim N(\mu, \Sigma)$. *Then the moment generating function associated with it is*

$$M_y(t) = e^{t'\mu + (1/2)t'\Sigma t}.$$

PROOF. Since Σ is positive definite there exists a matrix T such that

$$T'T = \Sigma^{-1}.$$

Consider now

$$x = T(y - \mu) \tag{19}$$

and observe that $x \sim N(0, I)$. Thus the (n) elements of x are mutually independent and each is $N(0, 1)$. Let $s = (s_1, s_2, \ldots, s_n)'$ be a vector of real numbers. The moment generating function of the vector x is, by definition,

$$M_x(s) = E(e^{s'x}) = \prod_{i=1}^{n} E(e^{s_i x_i}) = e^{(1/2)s's}.$$

The second equality above follows by the independence of the x_i and the last follows from (18).

Now, the moment generating function of y is, by definition,

$$M_y(t) = E(e^{t'y}) = E[e^{t'T^{-1}x} e^{t'\mu}] = e^{t'\mu} M_x(T'^{-1}t) = e^{t'\mu + (1/2)t'\Sigma t}. \tag{20}$$

The second equality follows by reversing the transformation in (19), i.e., by substituting x for y using (19); the third equality follows since $e^{t'\mu}$ is not random; the last equality follows from the definition of the moment generating function of x.   q.e.d.

Remark 11. The proposition shows that the moment generating function of $y \sim N(\mu, \Sigma)$ is an exponential. The exponent is quadratic in the auxiliary variable t. The coefficients in the linear terms are the means, while the quadratic terms are a quadratic form in the covariance matrix.

The usefulness of the moment generating function is twofold. First, it affords a somewhat routine way of computing various moments of the distribution. For example, suppose we wish to obtain

$$E(x_1 - \mu_1)(x_2 - \mu_2)(x_3 - \mu_3)(x_4 - \mu_4),$$

where $x = (x_1, x_2, \ldots, x_n)'$ is $N(\mu, \Sigma)$. It is not simple to compute this directly. On the other hand note that $y = x - \mu$ is $N(0, \Sigma)$. Thus, the moment generating function of y is

$$M_y(t) = e^{(1/2)t'\Sigma t}$$

and

$$E \prod_{i=1}^{4} (x_i - \mu_i) = E(y_1 y_2 y_3 y_4).$$

If we differentiate the expression defining $M_y(t)$ with respect to t_1, t_2, t_3, t_4 and then evaluate the resulting expression at $t_i = 0, i = 1, 2, \ldots, n$, we find

$$\frac{\partial^4 M_y(t)}{\partial t_1 \partial t_2 \partial t_3 \partial t_4} = (\sigma_{12}\sigma_{34} + \sigma_{13}\sigma_{24} + \sigma_{23}\sigma_{14})e^{(1/2)t'\Sigma t}$$

$$+ [\sigma_{12}\sigma_3.t\sigma_4.t + (\sigma_{23}\sigma_1.t + \sigma_{13}\sigma_2.t)\sigma_4.t + \sigma_{24}\sigma_1.t\sigma_3.t$$

$$+ \sigma_{14}\sigma_2.t\sigma_1.t + \sigma_2.t\sigma_3.t\sigma_4.t]e^{(1/2)t'\Sigma t}$$

where $\sigma_i.$ is the ith row of Σ.

Evaluating the above at $t_i = 0, i = 1, 2, \ldots, n$, we find

$$\sigma_{12}\sigma_{34} + \sigma_{13}\sigma_{24} + \sigma_{23}\sigma_{14}.$$

On the other hand, from the definition we have

$$\frac{\partial^4 M_y(t)}{\partial t_1 \partial t_2 \partial t_3 \partial t_4} = \int_{-\infty}^{\infty} \cdots \int_{-\infty}^{\infty} \frac{\partial^4}{\partial t_1 \partial t_2 \partial t_3 \partial t_4} e^{t'y} f(y)dy$$

$$= \int_{-\infty}^{\infty} \cdots \int_{-\infty}^{\infty} y_1 y_2 y_3 y_4 f(y)dy.$$

after evaluation of the derivative at $t_i = 0, i = 1, 2, \ldots, n$. Comparing the two expressions we conclude

$$E(y_1 y_2 y_3 y_4) = \sigma_{12}\sigma_{34} + \sigma_{13}\sigma_{24} + \sigma_{23}\sigma_{14}.$$

Second, it may be shown that there is a one–one correspondence between density and moment generating functions, provided the latter exist. Thus, if we are given a moment generating function we can determine the joint density function associated with it. It is this second aspect that we shall find useful below.

Proposition 12. *Let* x *be an n-element random variable with mean μ and covariance matrix Σ; if every (nontrivial) linear combination $\alpha'x$ is normally distributed, then*

$$x \sim N(\mu, \Sigma).$$

PROOF. For any arbitrary (nonnull) vector α, let

$$z_\alpha = \alpha'x$$

and observe

$$E(z_\alpha) = \alpha'\mu, \qquad \text{Var}(z_\alpha) = \alpha'\Sigma\alpha.$$

By hypothesis, for every such α, z_α is normal; consequently, its moment generating function is

$$M_{z_\alpha}(s) = E(e^{sz_\alpha}) = e^{s\alpha'\mu + (1/2)s^2\alpha'\Sigma\alpha}. \tag{21}$$

But

$$E(e^{sz_\alpha}) = E(e^{s\alpha'x}) = M_x(s\alpha).$$

Since s, α are arbitrary, we conclude that for $t = s\alpha$

$$M_x(t) = E(e^{t'x}) = e^{t'\mu + (1/2)t'\Sigma t}. \quad \text{q.e.d.}$$

In the preceding discussion we have examined extensively the standard (nondegenerate) multivariate normal distribution. Occasionally in empirical research we encounter situations in which the random variables of interest will have a singular distribution. For this reason an exposition of what this entails is in order. Thus, we introduce

Definition 13. An m-element random vector ξ with mean v and covariance matrix Φ, of rank n ($n \leq m$), is said to be *normally distributed* (with mean v and covariance matrix Φ) if and only if there exists a representation

$$\xi = Ax + b$$

where A and b are nonstochastic, A is $m \times n$ of rank n, and

$$x \sim N(\mu, \Sigma),$$

Σ being positive definite, and

$$v = A\mu + b, \qquad \Phi = A\Sigma A'.$$

Remark 12. The definition states, in effect, the convention that any full rank linear transformation of a normal vector with nondegenerate distribution is normally distributed. Notice also that Definition 13 represents a natural generalization of Definition 10, a fact that becomes obvious if (above) we take $b = 0$, $A = I$.

Remark 13. Although Definition 13 assigns to the vector ξ the normal distribution, the reader should not think of ξ as having *the density function of Equation* (15) *if* $n < m$; *such a density function does not exist, owing to the fact that Φ is singular, and the representation in* (15) *would require that Φ be inverted and that we divide by its determinant, which is zero.* Thus, the assignment of the normal distribution to ξ is best understood as a convention. We shall conclude this section by obtaining two results stated in Propositions 13 and 14 below. The first is, in fact, an amplification and extension of Proposition 2; the second is a generalization of Proposition 6.

Proposition 13. *Let ξ be an m-element random vector such that*

$$E(\xi) = v, \qquad \text{Cov}(\xi) = \Phi,$$

the matrix Φ being of rank $n \leq m$. Then there exists a matrix A and a vector a, A being $m \times n$ of rank n and a being $m \times 1$ such that

$$\xi = Ax + a,$$

where x is an n-element vector with mean μ, positive definite covariance matrix Σ, and

$$v = A\mu + a, \qquad \Phi = A\Sigma A'.$$

PROOF. Since Φ is positive semidefinite of rank n, it has n positive characteristic roots. Put

$$\Lambda_n = \text{diag}(\lambda_1, \lambda_2, \ldots, \lambda_n),$$

where λ_i is the ith characteristic root. Let Q_n be the matrix of the characteristic vectors corresponding to the nonzero roots and Q_* the matrix of the $(m - n)$ characteristic vectors corresponding to the zero roots. Then we have

$$Q'\Phi Q = \begin{bmatrix} \Lambda_n & 0 \\ 0 & 0 \end{bmatrix}, \qquad Q = (Q_n, Q_*).$$

Define

$$z = \begin{pmatrix} x \\ y \end{pmatrix} = \begin{pmatrix} Q'_n \xi \\ Q'_* \xi \end{pmatrix}$$

and observe that

$$E(x) = Q'_n v, \qquad \text{Cov}(x) = Q'_n \,\text{Cov}(\xi)Q_n = \Lambda_n,$$
$$E(y) = Q'_* v, \qquad \text{Cov}(y) = Q'_* \,\text{Cov}(\xi)Q_* = 0.$$

Thus y is, in effect, a nonstochastic vector. Reversing the transformation we obtain

$$\xi = Qz = Q_n x + Q_* y = Q_n x + a, \qquad (22)$$

where $a = Q_* Q'_* v$. Noting that Q_n is $m \times n$ of rank n, that x has a nonsingular covariance matrix, and taking $A = Q_n$, we have

$$\xi = Ax + a \quad \text{q.e.d.}$$

Remark 14. The result just proved amplifies substantially the considerations advanced in the proof of Proposition 2. What we have shown is that if an m-element random vector, ξ, has a singular covariance matrix (of rank $n < m$), it can be represented as a *linear* (more precisely *affine*) *transformation of a random vector, say x, which has a nondegenerate distribution and whose dimension (i.e., the number of its elements) is equal to the rank of the singular covariance matrix of ξ.* In particular, what this means is that the density of ξ, if it exists, must assign zero probability to values of ξ *not obeying*

$$\xi = Ax + a.$$

But since ξ lies in a space of m dimensions (Euclidean m-space) and $m > n$ it follows the density must assign zero probability to "nearly all" values of ξ. Hence the density must be zero almost everywhere. This, perhaps, is best understood if the reader considers the bivariate case. Let (x_1, x_2) have a non-degenerate distribution and let $f(\cdot, \cdot)$ be its density function. What is the probability assigned to the set $\{(x_1, x_2): a_1 x_1 + a_2 x_2 = 0, a_2 \neq 0\}$? This is a line through the origin and it is clear that the integral of a bivariate density over a line is zero, just as the "integral" of a univariate density over a point is zero.

We are now in a position to prove a generalization of Propositions 6 and 10.

Proposition 14. *Let* x *be an* n-*element random vector*

$$x \sim N(\mu, \Sigma),$$

Σ *being positive definite. Let* A *be* $m \times n$, *nonstochastic, and of rank* r. *Let* a *be a nonstochastic* m-*element vector. Then,*

$$y = Ax + a$$

has the distribution

$$y \sim N(A\mu + a, A\Sigma A').$$

PROOF. There are three possibilities:

(i) $m = n = r$;
(ii) $m \neq n, r = \min(m, n)$;
(iii) $m \neq n, r < \min(m, n)$.

The conclusion for (i) was proved in Proposition 6. Thus, we need consider (ii) and (iii). To this effect we refer the reader to Proposition 15 of *Mathematics for Econometrics*, which gives the *rank factorization theorem*. The latter states, for the case under consideration, that if A is $m \times n$ of rank r, then there exist matrices C_1 and C_2, which are, respectively, $m \times r$ and $r \times n$ *(both of rank* r*)*, such that $A = C_1 C_2$. Putting $z = C_2 x$ we immediately see that if we can prove that z has a nondegenerate (normal) distribution then, in effect, we have proved the substance of the proposition. Since C_2 is $r \times n, r \leq n$ there clearly exists a matrix C_3 such that

$$C = \begin{pmatrix} C_2 \\ C_3 \end{pmatrix}$$

is a nonsingular $n \times n$ matrix.

Define $z^* = Cx$ and by Proposition 6 conclude that $z^* \sim N(C\mu, C\Sigma C')$. Now partition

$$z^* = \begin{pmatrix} z \\ z_3 \end{pmatrix} = \begin{pmatrix} C_2 x \\ C_3 x \end{pmatrix}$$

and conclude, by Proposition 7, that the (marginal) distribution of z is given by $z \sim N(C_2 \mu, C_2 \Sigma C_2')$. Since $C_2 \Sigma C_2'$ is an $r \times r$ matrix of rank r, z therefore has a nondegenerate normal distribution. Consequently,

$$y = Ax + a = C_1 z + a, \tag{23}$$

and we must now determine the distribution of y. We observe that

$$E(y) = A\mu + a, \qquad \text{Cov}(y) = C_1(C_2 \Sigma C_2')C_1',$$

the covariance matrix being $m \times m$ of rank $r < m$. From Definition 13 and Equation (23) we conclude that

$$y \sim N(A\mu + a, A\Sigma A'). \qquad \text{q.e.d.}$$

Remark 15. Proposition 14 represents a significant generalization of Proposition 6. Thus, e.g., if A is a single (row) vector Proposition 6 does not allow us to deduce the distribution of the scalar

$$Ax.$$

Formally, A is $1 \times n$ of rank 1. The argument by which we deduced the distribution of z enables us to assert that the scalar Ax is distributed as $N(A\mu, A\Sigma A')$. Similarly, if A is $n \times n$ but *singular* Proposition 6 does not allow us to infer the distribution of Ax, whereas Proposition 14 does. Moreover, the latter offers a particularly simple way in which we can determine marginal distributions in the degenerate case, which we state below.

Proposition 15. *Let* ξ *be an m-element random vector having the degenerate normal distribution with mean* v *and covariance matrix* Φ, *of rank* $r < m$. *Partition*

$$\xi = \begin{pmatrix} \xi^{(1)} \\ \xi^{(2)} \end{pmatrix}, \qquad v = \begin{pmatrix} v^{(1)} \\ v^{(2)} \end{pmatrix}, \qquad \Phi = \begin{bmatrix} \Phi_{11} & \Phi_{12} \\ \Phi_{21} & \Phi_{22} \end{bmatrix}$$

such that $\xi^{(1)}, v^{(1)}$ *are both k-element vectors,* Φ_{11} *is* $k \times k$, Φ_{12} *is* $k \times (m-k)$, *and* Φ_{22} *is* $(m-k) \times (m-k)$. *Then the marginal distribution of* $\xi^{(1)}$ *is given by*

$$\xi^{(1)} \sim N(v^{(1)}, \Phi_{11}).$$

PROOF. By the preceding discussion ξ has the representation $\xi = Ax + a$, where A is a nonstochastic $m \times r$ matrix of rank r, a is a nonstochastic m-element vector, and x is an r element random vector having the nondegenerate distribution $x \sim N(\mu, \Sigma)$.

Define the $k \times m$ matrix $B = (I, 0)$, where I is an identity matrix of order k. In effect, we seek the distribution of

$$\xi^{(1)} = B\xi = A_1 x + a^{(1)},$$

where A_1 is the submatrix of A consisting of its first k rows and $a^{(1)}$ is the subvector of a consisting of its first k elements. By Proposition 14

$$\xi^{(1)} \sim N(A_1 \mu + a^{(1)}, A_1 \Sigma A_1').$$

But since

$$v = A\mu + b, \qquad \Phi = A\Sigma A',$$

we see that

$$v^{(1)} = A_1\mu + a^{(1)}, \qquad \Phi_{11} = A_1\Sigma A'_1. \quad \text{q.e.d.}$$

Corollary. *If $k > r$ the (marginal) distribution of $\xi^{(1)}$ is degenerate.*

PROOF. Obvious.

We conclude this section by raising and answering the following question. Let ξ, ζ be two random vectors having m_1, m_2 elements respectively. Suppose they are jointly normally distributed but their distribution is degenerate. What do we want to mean by the statement that they are mutually independent? This is answered by

Convention. Let ξ, ζ be two random variables such that

$$\xi = Ax + a, \qquad \zeta = Bx + b, \quad \text{and} \quad x \sim N(\mu, \Sigma),$$

where Σ is nonsingular. Then, ξ, ζ are said to be mutually independent if and only if

$$\text{Cov}(\xi, \zeta) = 0.$$

3 Point Estimation

It is assumed that the reader has been exposed to the basic elements of point estimation theory. Thus, it is assumed that the concepts of random sample, point estimation, and so on are known. What we shall do below will consist chiefly of a review, and an elucidation of the properties of unbiasedness, consistency, and efficiency.

In previous sections we have studied certain aspects of the theory of the multivariate normal distribution. In doing so we considered the mean and covariance parameters of the distribution and expressed the marginal and conditional densities in terms of such parameters, proceeding as if the parameters were known.

By contrast, in inference and in empirical work we are given a body of data and we hypothesize that they have been generated by a well-defined process whose parameters are unknown to us. It is apparent that the nature of the data conveys some information on the parameters. For example, if we are throwing a coin and in 100 tosses 80 show heads, we might easily conclude that the coin is "biased" since intuitively we would feel that 80 heads is "too many" if the coin were "perfectly balanced." Similarly, in more complicated situations the nature of the data will give us some indication about the parameters governing the data generating process. Thus, some data configurations will be more likely to be observed under one parametric configuration than another. Consequently, if we have such data we would be more likely to believe that a certain set of parameters governs the data generating process.

This is, essentially, the intuitive basis of the elementary estimation theory to be outlined below. We first introduce some terminology.

Definition 14. Let $\{x_i: i = 1, 2, \ldots, n\}$ be a set of mutually independent identically distributed (i.i.d.) random variables. The set $\{x_i: i = 1, 2, \ldots, n\}$ is then referred to as a *random sample*. If the elements of the set are not i.i.d., the set is referred to as a *sample*.

Definition 15. Let $\{x_i: i = 1, 2, \ldots, n\}$ be a sample and suppose the joint density of the sample is given by $f(x; \theta)$, where θ is a set of unknown parameters. A function of the data not containing θ, say $h(x)$, is said to be an *estimator*, or a *statistic*. The value assumed by the function for a *given sample* is said to be an *estimate*.

Remark 16. The definition of an estimator as simply a function of the sample (whether random or not) is of course very broad. Presumably the estimator, say $h(x)$ above, conveys some information about θ, in which case it is said to be an estimator of θ; or it may convey information on some function(s) of the elements of θ. But, basically, estimators are just functions of the sample (data) alone and as such they are random variables. Their density function may frequently be inferred from the density function that characterizes the data generating process. The properties of an estimator may be derived by examining the properties of its density function; if that is not known, it may be possible to examine the moments of the estimator, e.g., its mean and variance (or covariance matrix); if that is not possible, then perhaps we can examine the aspects above as the sample size becomes larger and larger. The latter is known as *asymptotic theory*. We shall not have occasion to deal extensively with the asymptotic theory of estimators and this aspect will not be developed here.

We have

Definition 16. Let $\{x_i: i = 1, 2, \ldots, n\}$ be a sample and let $f(x; \theta)$ be its joint density function; let $\hat{\theta} = h(x)$ be an estimator of θ. The estimator is said to be *unbiased* if

$$E(\hat{\theta}) = \theta.$$

The estimator is said to be *biased* if $E(\hat{\theta}) \neq \theta$, and its bias is defined by

$$b(\theta) = E(\hat{\theta}) - \theta.$$

EXAMPLE 3. Let $\{x_i: i = 1, 2, \ldots, n\}$ be a *random* sample with density function $f(x_i; \theta)$; let each x_i be a scalar and suppose

$$\theta = \begin{pmatrix} \mu \\ \sigma^2 \end{pmatrix},$$

i.e., the density of the x_i is characterized completely by two parameters, say the mean and variance. The vector

$$(\bar{x}, s^2)', \qquad \bar{x} = \frac{1}{n} \sum_{i=1}^{n} x_i, \qquad s^2 = \frac{1}{n} \sum_{i=1}^{n} (x_i - \bar{x})^2$$

is an estimator. It is an estimator of θ since its elements convey information on μ and σ^2 respectively. In particular, we have

$$E(\bar{x}) = \frac{1}{n} \sum_{i=1}^{n} E(x_i) = \mu,$$

$$E(s^2) = \frac{1}{n} \sum_{i=1}^{n} E[(x_i - \mu) - (\bar{x} - \mu)]^2 = \frac{n-1}{n} \sigma^2.$$

This shows that $\bar{x}$ is *an unbiased estimator* of μ, while s^2 is *a biased estimator of σ^2*, whose bias is $-(1/n)\sigma^2$. On the other hand, we see that as the sample size increases the bias diminishes to zero. In particular,

$$\lim_{n \to \infty} b(s^2) = 0$$

so that s^2 is an *asymptotically unbiased estimator of σ^2*.

EXAMPLE 4. Let $\{x_i: i = 1, 2, \ldots, n\}$ be a random sample with density function $f(x_i; \theta)$; suppose, however, that each x_i is an m-element random vector and $n > m$. The unknown parameters now are μ and Σ, the mean vector and covariance matrix respectively. Define

$$\bar{x} = \frac{1}{n} \sum_{i=1}^{n} x_i = \frac{1}{n} X'e, \qquad X = (x_{ij}), \qquad i = 1, 2, \ldots, n, j = 1, 2, \ldots, m,$$

$$A = \frac{1}{n} X'X - \bar{x}\bar{x}',$$

where $e = (1, 1, 1, \ldots, 1)'$, i.e., it is an n-element vector all of whose elements are unity, and $\bar{x}$ and A are estimators of μ and Σ respectively. In particular, we see

$$E(\bar{x}) = \frac{1}{n} \sum_{i=1}^{n} E(x_i) = \frac{1}{n} \sum_{i=1}^{n} \mu = \mu.$$

Adding and subtracting $\mu\bar{x}' + \mu'\bar{x} + \mu\mu'$ in the expression defining A we find

$$A = \frac{1}{n} (X - e\mu')'(X - e\mu') - (\bar{x} - \mu)(\bar{x} - \mu)'.$$

The (r, s) element of A is, thus,

$$a_{rs} = \frac{1}{n} \sum_{i=1}^{n} (x_{ri} - \mu_r)(x_{si} - \mu_s) - (\bar{x}_r - \mu_r)(\bar{x}_s - \mu_s). \qquad (24)$$

Since we are dealing with a random sample

$$E(x_{ri} - \mu_r)(x_{si} - \mu_s) = \sigma_{rs}, \qquad i = 1, 2, \ldots, n,$$

where, in the above, μ_r is the rth element of the mean vector μ and σ_{rs} is the (r, s) element of the covariance matrix Σ. We also note that

$$E(\bar{x}_r - \mu_r)(\bar{x}_s - \mu_s) = \frac{1}{n} \sigma_{rs}.$$

Thus,

$$E(a_{rs}) = \sigma_{rs} - \frac{1}{n} \sigma_{rs}$$

and

$$E(A) = \Sigma - \frac{1}{n} \Sigma. \qquad (25)$$

As before we note that *A is a biased estimator of Σ and its bias is*

$$b(A) = -\frac{1}{n} \Sigma$$

We also note that

$$\lim_{n \to \infty} b(A) = 0$$

so that, asymptotically, A is an unbiased estimator of Σ.

It bears repeating that an estimator is a function (of the sample); hence properties of estimators essentially refer to the procedure by which the sample data are processed in order to produce inferences regarding the constant but unknown parameters. An estimate, on the other hand, is the value assumed by the function for a given sample. As such it is a value assumed by a random variable. Nothing more can be fruitfully said about an estimate. All properties, such as unbiasedness as well as consistency and efficiency (to be discussed below), *pertain to the estimators, and not the estimates, i.e., they pertain to the manner in which data are being processed and not to the numerical values assumed by the estimator once a given sample is obtained.*

In this light unbiasedness is seen to have the following meaning. Suppose we wish to estimate a certain parameter θ. If we have repeated samples (for simplicity let us suppose they are all of the same size) then for each sample, say the ith, we shall obtain an estimate, say $\hat{\theta}_i$. What unbiasedness means intuitively is that, on the average, these estimates will be close to θ. Thus, if the number of samples (say k) is large then, $(1/k) \sum_{i=1}^{k} \hat{\theta}_{(i)} \approx \theta$. This is the intuitive meaning of unbiasedness. The formal meaning is, of course, quite evident. Since a sample is a set of random variables and an estimator is a function of the sample we see that an estimator is also a random variable. As such it will possess density and distribution functions. Being unbiased indicates that the mean of the distribution is the parameter being estimated.

This property is to be distinguished from that of *consistency*. We have

Definition 17. Let $\hat{\theta}_n$ be an estimator of a (scalar) parameter θ, based on a sample of size n. It is said to be a *consistent estimator of θ* if given any $\varepsilon, \delta > 0$ there exists a sample size n^* such that for all $n > n^*$

$$\Pr\{|\hat{\theta}_n - \theta| > \delta\} < \varepsilon.$$

Consistency is also refered to as *convergence in probability* and the operation above is also indicated by

$$\plim_{n \to \infty} \hat{\theta}_n = \theta,$$

which is read: "the probability limit of $\hat{\theta}_n$ is θ."

Remark 17. Consistency is a "large sample" property, i.e., it is a property that refers to a limiting behavior. While an estimator may be unbiased even when applied to a fixed sample size, consistency is a property that can be established only if the sample size is allowed to increase indefinitely. Intuitively, an estimator is consistent if by taking larger and larger samples we can make the probability that an estimate lies in some prespecified neighborhood of the parameter arbitrarily close to one. Notice that consistency does not rule out large deviations of the estimate from the true parameter, by increasing the size of the sample; it merely ensures that large deviations can be made less "likely" by taking a larger sample. As such it is useful in econometric work when dealing with cross-sectional samples, say of individual firms or households. Here we have the option of increasing the probability that estimates are close to the true parameters at the cost of increasing the size of the sample—assuming, of course, that the firms or households are structurally homogeneous. Consistency is, perhaps, less cogent a property with time series data, firstly because we cannot at will increase the sample size, and secondly because the longer the time span over which an economic phenomenon is observed the less likely that it will remain structurally unaltered. For certain types of markets, however, weekly or monthly observations may form quite appropriate samples for studying their characteristics. In such cases large numbers of observations can be accumulated over relatively short calendar time periods.

We can illustrate these considerations by means of some hypothetical examples. Suppose, for example, we wished to study the supply of agricultural commodities in the United States over the period 1800–1970. A common way of approaching this problem is to specify a "production function" connecting "inputs" and "outputs." Inputs are typically the land under cultivation, labor, and capital in various forms—such as farm machinery, irrigation facilities, etc. The production function, say the usual Cobb–Douglas or constant elasticity of substitution, will contain a number of unknown parameters. In order for us to make inferences about these parameters, using the data for the period (1800–1970), it appears intuitively plausible that we should expect such parameters not to have changed over the sample period. For if they had, then some observations would give information on one parametric configuration while other observations would give information about another. It is not clear what it is that we gain by combining the two sets unless some parameters are asserted to be common to the two subperiods. Thus, although we appear to have at our disposal 171 observations, this is illusory since the sample may not be structurally homogeneous. Parametric variation is, presumably, induced by the considerable "technical change" that has occurred in the process of producing agricultural commodities. Unless the manner in which change takes place is incorporated in the specification of the "production function" it is not possible to use the entire sample in making inferences (by the methods examined in this book) regarding the structure of agricultural production. In essence, what is being said is that if observations cannot be extended beyond a certain number (without violating the structural homogeneity of the phenomenon being studied), then in such a context consistency may well be an irrelevant property. On the other hand consider various submarkets of the money market, and suppose that markets clear quickly so that weekly observations are meaningful. In 4 years we can accumulate 200 observations and, in this context, consistency will exhibit greater cogency, since significant structural change is less likely to have transpired in such a relatively short period of (calendar) time.

It is sometimes thought that if an estimator is consistent, then it is also asymptotically unbiased. This is based on the intuitive description of a consistent estimator as one for which its probability density function collapses to a single point as the sample size approaches infinity. While this is a reasonable intuitive description of what consistency is, it does not imply asymptotic unbiasedness. This is so since the expectation of a random variable may be thought of as a weighted sum of the values assumed by the variable, the weights being derived from the density function. Thus, for asymptotic unbiasedness it is not enough to know that the "tails" of the density function shrink as the sample size tends to infinity. An example will clarify this issue.

EXAMPLE 5. Let $\{x_T: T = 1, 2, \ldots\}$ be a sequence of random variables having the following probability structure:

$$x_T = \alpha \quad \text{with probability } 1 - \frac{1}{T},$$

$$= T \quad \text{with probability } \frac{1}{T}.$$

It is clear that for given ε, $\delta > 0$ we can choose T large enough so that

$$\Pr\{|x_T - \alpha| > \delta\} < \varepsilon$$

In particular, choose T such that $1/T < \varepsilon$, or $T > 1/\varepsilon$. Thus x_T is a "consistent estimator" of α. On the other hand,

$$E(x_T) = \alpha\left(1 - \frac{1}{T}\right) + T\left(\frac{1}{T}\right) = \alpha + 1 - \frac{\alpha}{T}.$$

This shows that x_T is a biased estimator of α; moreover, the bias is

$$1 - \frac{\alpha}{T},$$

which does not vanish as $T \to \infty$. Consequently, x_T is not an asymptotically unbiased estimator of α, even though it is a consistent one.

Now that the concept of consistency has been defined, let us see how we can operate with it. The first question is: if we are given an estimator, how can we determine whether it is consistent, especially if, as is typically the case, its density function is not completely known? We recall

Proposition 16 (Chebyshev's inequality). *Let x be a (scalar) random variable with mean μ and variance σ^2 and let $k > 0$ be a real number. Then*

$$\Pr\{|x - \mu| > k\} \le \frac{\sigma^2}{k^2}$$

PROOF. Let $f(\cdot)$ be the density function of x; then

$$\sigma^2 = \int_{-\infty}^{\infty} (x - \mu)^2 f(x)dx$$

$$= \int_{-\infty}^{\mu-k} (x - \mu)^2 f(x)dx + \int_{\mu+k}^{\infty} (x - \mu)^2 f(x)dx + \int_{\mu-k}^{\mu+k} (x - \mu)^2 f(x)dx$$

$$\ge \int_{-\infty}^{\mu-k} (x - \mu)^2 f(x)dx + \int_{\mu+k}^{\infty} (x - \mu)^2 f(x)dx$$

$$\ge k^2 \left[\int_{-\infty}^{\mu-k} f(x)dx + \int_{\mu+k}^{\infty} f(x)dx\right] = k^2 \Pr\{|x - \mu| > k\},$$

which may also be written as

$$\Pr\{|x - \mu| > k\} \le \frac{\sigma^2}{k^2}. \quad \text{q.e.d.}$$

We now introduce another notion of convergence.

Definition 18. Let $\{x_i: i = 1, 2, \ldots\}$ be sequence of (scalar) random variables, let ζ be a constant, and suppose

$$\lim_{i \to \infty} E(x_i - \zeta)^2 = 0.$$

Then the sequence is said to *converge in quadratic mean* to ζ.

Remark 18. Since an estimator may be thought of as a sequence of random variables indexed by the sample size we have, through Definition 18, another way of defining convergence of an estimator to the parameter it seeks to estimate.

Proposition 17. *Let $\hat{\theta}_T$ be an estimator of the (scalar) parameter θ, and suppose $\hat{\theta}_T$ possesses a well-defined density function with mean $\bar{\theta}_T$ and variance σ_T^2. Then $\hat{\theta}_T$ converges in quadratic mean to θ if and only if its bias and variance vanish asymptotically. Moreover, if $\hat{\theta}_T$ converges to θ in quadratic mean then it also converges to θ in probability. The converse, however, is not generally true.*

PROOF. By definition

$$E(\hat{\theta}_T - \theta)^2 = [b(\hat{\theta}_T)]^2 + \text{Var}(\hat{\theta}_T), \qquad b(\hat{\theta}_T) = \bar{\theta}_T - \theta.$$

Moreover,

$$\lim_{T \to \infty} E(\hat{\theta}_T - \theta)^2 = \lim_{T \to \infty} [b(\hat{\theta}_T)]^2 + \lim_{T \to \infty} \text{Var}(\hat{\theta}_T) = 0$$

if and only if

$$\lim_{T \to \infty} b^2(\hat{\theta}_T) = \lim_{T \to \infty} \text{Var}(\hat{\theta}_T) = 0.$$

For the second half of the proposition we observe that a slight variation of the proof of Proposition 16 (Chebyshev's inequality) implies that for any $k > 0$,

$$\Pr\{|\hat{\theta}_T - \theta| > k\} \le \frac{E(\hat{\theta}_T - \theta)^2}{k^2}.$$

Since for any specified k and δ we can choose a T^* such that for $T > T^*$ $E(\hat{\theta}_T - \theta)^2 < \delta k^2$, consistency is proved.

On the other hand, consistency does not imply convergence in quadratic mean. An example will suffice. Thus, consider again the sequence in Example 5 and recall that $\text{plim}_{T \to \infty} x_T = \alpha$ in the example's notation. But

$$E(x_T - \alpha)^2 = 0 \cdot \left(1 - \frac{1}{T}\right) + (T - \alpha)^2 \cdot \frac{1}{T},$$

$$\lim_{T \to \infty} E(x_T - \alpha)^2 = \lim_{T \to \infty} T - 2\alpha,$$

which is, certainly, a nonzero quantity.   q.e.d.

Generalization of the results to estimators of vectors is straightforward.

Definition 19. Let $\hat{\theta}_T$ be an estimator of the (vector) parameter θ. Then $\hat{\theta}_T$ is said to be a *consistent estimator of* θ if given any δ, $\varepsilon > 0$ there exists a T^* such that, for $T > T^*$,

$$\Pr\{|\hat{\theta}_T - \theta| > \delta\} < \varepsilon,$$

where

$$|\hat{\theta}_T - \theta| = \left(\sum_{i=1}^{n}(\hat{\theta}_{Ti} - \theta_i)^2\right)^{1/2},$$

$\hat{\theta}_{Ti}$ and θ_i being (respectively) the ith component of $\hat{\theta}_T$ and θ. We then write

$$\operatorname*{plim}_{T \to \infty} \hat{\theta}_T = \theta.$$

We have the following generalization of Proposition 16.

Proposition 18. *Let x be a (vector) random variable with mean vector μ and (positive definite) covariance matrix Σ. Then, for any $k > 0$*

$$\Pr\{(x - \mu)'\Sigma^{-1}(x - \mu) > k\} \leq \frac{n}{k}.$$

PROOF. Since Σ^{-1} is positive definite there exists a nonsingular matrix R such that $\Sigma^{-1} = R'R$. Put $\xi = R(x - \mu)$ and notice that ξ is (an n-element) random vector with mean zero and covariance matrix I. Thus,

$$n = \mathrm{E}(x - \mu)'\Sigma^{-1}(x - \mu) = \mathrm{E}(\xi'\xi)$$

$$= \int_0^\infty \zeta f(\zeta)d\zeta = \int_0^k \zeta f(\zeta)d\zeta + \int_k^\infty \zeta f(\zeta)d\zeta$$

$$\geq \int_k^\infty \zeta f(\zeta)d\zeta \geq k \int_k^\infty f(\zeta)d\zeta = k \Pr\{\xi'\xi > k\},$$

where $f(\cdot)$ is the density function of $\xi'\xi$. Thus we conclude

$$\Pr\{(x - \mu)'\Sigma^{-1}(x - \mu) > k\} \leq \frac{n}{k}. \quad \text{q.e.d.}$$

As before, we can also define convergence in quadratic mean for vector estimators by

Definition 20. Let $\hat{\theta}_T$ be an estimator of the n-element parameter vector θ. Then $\hat{\theta}_T$ is said to *converge in quadratic mean to* θ if

$$\lim_{T \to \infty} \mathrm{E}|\hat{\theta}_T - \theta|^2 = 0.$$

An immediate generalization of Proposition 17 is

Proposition 19. *Let $\hat{\theta}_T$ be an estimator of the (vector) parameter θ and suppose $\hat{\theta}_T$ possesses a well defined density with mean (vector) $\bar{\theta}_T$ and covariance matrix Σ_T. Then $\hat{\theta}_T$ converges in quadratic mean to θ if and only if its bias and covariance matrix vanish asymptotically. Moreover, if $\hat{\theta}_T$ converges in quadratic mean to θ, then it also converges to θ in probability. The converse, however, is not generally true.*

PROOF. We have, adding and subtracting $\bar{\theta}_T$,

$$E|\hat{\theta}_T - \theta|^2 = E|\hat{\theta}_T - \bar{\theta}_T|^2 + |\bar{\theta}_T - \theta|^2 = \text{tr } \Sigma_T + |b(\hat{\theta}_T)|^2.$$

Since Σ_T is a covariance matrix, it vanishes if and only if tr Σ_T vanishes. The first part of the proposition is obvious from the last member of the equation above; note that the second term is the sum of squares of the biases, i.e., of the individual elements of the bias vector $b(\hat{\theta}_T)$.

For the second part of the proposition we observe from Problems 11 and 12 at the end of this chapter that, for any $\varepsilon > 0$,

$$\Pr\{|\hat{\theta}_T - \theta|^2 > \varepsilon\} \le \frac{E|\hat{\theta}_T - \theta|^2}{\varepsilon},$$

which shows consistency immediately. That consistency does not imply convergence in quadratic mean has been shown already in the last part of the proof of Proposition 17.   q.e.d.

We conclude the discussion of consistency by giving, without formal proof, a number of results from probability theory that pertain to operations with probability limits.

Proposition 20. *Let $\{\zeta_t : t = 1, 2, \ldots\}$ be a sequence of (vector) random variables converging in probability to the constant ζ. Let $\phi(\cdot)$ be a continuous function such that $\phi(\zeta)$ is well defined. Then $\{\phi(\zeta_t) : t = 1, 2, \ldots\}$ converges in probability to $\phi(\zeta)$, provided $\phi(\zeta)$ is defined, i.e.,*

$$\underset{t \to \infty}{\text{plim}}\ \phi(\zeta_t) = \phi(\zeta).$$

PROOF. Although a formal proof is well beyond the scope of this book the reader may find the following heuristic argument useful.

For clarity, let ζ be a k-element vector. We have to prove that given ε, $\delta > 0$ there exists a t^* such that for all $t > t^*$

$$\Pr\{|\phi(\zeta_t) - \phi(\zeta)| < \delta\} \ge 1 - \varepsilon.$$

Since $\phi(\cdot)$ is a continuous function, choose a $\delta_1 > 0$ such that $|\zeta_t - \zeta| < \delta_1$ implies $|\phi(\zeta_t) - \phi(\zeta)| < \delta$. Now, because ζ_t converges in probability to ζ, given $\delta_1, \varepsilon > 0$ there exists a t^* such that for all $t > t^*$

$$\Pr\{|\zeta_t - \zeta| < \delta_1\} \ge 1 - \varepsilon.$$

For all ζ_t obeying the inequality above we have

$$|\phi(\zeta_t) - \phi(\zeta)| < \delta$$

by the continuity of $\phi(\cdot)$, and, moreover, for all $t > t^*$

$$\Pr\{|\phi(\zeta_t) - \phi(\zeta)| < \delta\} > \Pr\{|\zeta_t - \zeta| < \delta_1\} \geq 1 - \varepsilon. \quad \text{q.e.d.}$$

EXAMPLE 6. Let $\{x_i : i = 1, 2, \ldots\}$, $\{y_i : i = 1, 2, \ldots\}$ be two sequences of (scalar) random variables such that the first converges in probability to ζ_1 and the second to ζ_2. Then the proposition above implies, for example, that

$$\plim_{i \to \infty} \frac{x_i}{y_i} = \frac{\zeta_1}{\zeta_2},$$

provided $\zeta_2 \neq 0$. Also,

$$\plim_{i \to \infty} x_i^2 = \zeta_1^2, \qquad \plim_{i \to \infty} x_i y_i = \zeta_1 \zeta_2,$$

and so on.

Proposition 21 (Khinchine's theorem). *Let* $\{x_t : t = 1, 2, \ldots\}$ *be a sequence of (scalar) independent identically distributed random variables such that* $E(x_t)$ $= \mu$, *exists. Then*

$$\plim_{T \to \infty} \frac{1}{T} \sum_{t=1}^{T} x_t = \mu.$$

PROOF. See Chung [6].

Remark 19. The result above is considerably more powerful than it might appear at first sight. Suppose in the sequence of the proposition above the rth moment exists. Then the proposition asserts that

$$\plim_{T \to \infty} \frac{1}{T} \sum_{t=1}^{T} x_t^r = \mu_r,$$

where

$$\mu_r = E(x_t^r).$$

This is so since $\{x_t^r : t = 1, 2, \ldots\}$ is a sequence of independent identically distributed random variables possessing a finite first moment. Similarly, if $\{x_t : t = 1, 2, \ldots\}$ is an n-element vector random variable with finite covariance matrix Σ and, for simplicity, mean zero, then the proposition asserts that

$$\plim_{T \to \infty} \frac{1}{T} \sum_{t=1}^{T} x_t x_t' = \Sigma.$$

The final property of estimators we ought to consider is that of efficiency. We have

Definition 21. Let C be a class of estimators of a parameter (vector) θ possessing certain properties; let $\hat{\theta}_1$, $\hat{\theta}_2 \in C$, both estimators possessing covariance matrices. Then $\hat{\theta}_1$ is said to be *efficient relative to* $\hat{\theta}_2$ if the difference

$$\text{Cov}(\hat{\theta}_2) - \text{Cov}(\hat{\theta}_1)$$

is positive semidefinite. If for *any* $\hat{\theta}_i \in C$

$$\text{Cov}(\hat{\theta}_i) - \text{Cov}(\hat{\theta}_1)$$

is positive semidefinite then $\hat{\theta}_1$ is said to be *best*, or *efficient, with respect to the class C*.

Remark 20. In most of our work C will be the class of linear (in the data) unbiased estimators; an efficient estimator with respect to this class is said to be a *best linear unbiased estimator* (BLUE).

4 Elements of Bayesian Inference

4.1 Prior and Posterior Distributions

In all of the preceding discussion the parameters characterizing the density (or mass) functions of random variables considered were treated as *fixed but unknown scalars or vectors*. Estimation was then defined as a process by which information was extracted from the data regarding these fixed, but unknown, constants (parameters).

In the Bayesian context the distinction between random variables and parameters is blurred. Without going into the philosophical issues entailed, it will suffice for the reader to place the matter in a decision theoretic context in order to reconcile the Bayesian and the "classical" view exposited earlier. Since parameters are not known the empirical investigator is acting under uncertainty when he attempts to estimate them. For each action he takes, i.e., *with each estimate he produces*, there corresponds an error since he is unlikely with a finite set of data to estimate the parameter precisely. With each error there is associated a cost (for having committed the error). In addition, the investigator may have some prior information regarding the parameter(s) under consideration and this information may be formalized as a density function over the parameter(s) in question.

If one is prepared to act in this fashion then a suitable framework may be to blur the sharp distinction between parameters and random variables. Unknown parameters, which are the object of decision making under uncertainty, may be treated as random variables and the process of inference (estimation in the earlier discussion) would be defined as a way in which we allow the data (observations) to alter our prior views regarding the parameter(s) in question.

An example will clarify these issues. Thus let $\{x_i: i = 1, 2, \ldots, n\}$ be a set of i.i.d. random variables each obeying $x_i \sim N(\mu, 1)$, the mean μ being unknown. In the framework discussed earlier in the chapter we use the sample information to estimate μ. Using, e.g., the standard maximum likelihood procedure we write the log likelihood function

$$L(\mu; x) = -\frac{n}{2} \ln(2\pi) - \frac{1}{2} \sum_{i=1}^{n} (x_i - \mu)^2, \qquad x = (x_1, x_2, \ldots, x_n)', \quad (26)$$

and maximizing with respect to μ we find

$$\hat{\mu} = \frac{1}{n} \sum_{i=1}^{n} x_i.$$

Here μ is treated as an unknown, but fixed, constant and the sample is used to obtain some information about it.

A Bayesian approach will be slightly different. In the first instance, μ will be treated as a random variable. Thus (26) will be interpreted as the *conditional (log) likelihood function, given* μ. Consequently, in order to characterize the problem fully we have to specify an initial distribution function for μ, which embodies our prior beliefs regarding this parameter. This is termed the *prior distribution* or *prior density function* for the parameter in question. Let this be $\psi(\mu)$. Then the *joint density of the data and the parameters* may be written as

$$p(x, \mu) = \{(2\pi)^{-(n/2)} \exp[-(\tfrac{1}{2})(x - \mu e)'(x - \mu e)]\}\psi(\mu), \qquad (27)$$

where e is an $n \times 1$ vector all of whose elements are unity.

The inferential process in Bayesian terms involves the transition from the prior density of μ to the density of μ given the sample information. *Thus from the joint density of x and μ in* (27) *we wish to determine* the conditional density of μ given x. If by $p(x)$ we denote the *marginal density of the data* then clearly the conditional density of μ given x is simply

$$p(\mu|x) = \frac{p(x, \mu)}{p(x)}, \qquad (28)$$

provided $p(x) \neq 0$, *where $p(\cdot)$ is a generic notation denoting a probability density function. The conditional density $p(\mu|x)$ is called the* posterior density.

Remark 21. The term Bayesian, used above, derives—as is evident from the discussion preceding—from Bayes' rule, viz., if A and B are two events then

$$\Pr(A|B) = \frac{\Pr(A \cap B)}{\Pr(B)}, \qquad \Pr(B) \neq 0,$$

and conversely

$$\Pr(B|A) = \frac{\Pr(A \cap B)}{\Pr(A)}, \qquad \Pr(A) \neq 0.$$

In the preceding we argue that the conditional density of the observations given the parameter is, say, $p(x|\mu)$. If the prior density on μ is specified as $\psi(\mu)$ then the joint density of the observations *and* the parameters is

$$p(x|\mu)\psi(\mu).$$

The object of the Bayesian inference procedure is the conditional density of μ, *given the observations.* From Bayes' rule we easily find

$$p(\mu|x) = \frac{p(x|\mu)\psi(\mu)}{p(x)},$$

where

$$p(x) = \int p(x|\mu)\psi(\mu)d\mu.$$

EXAMPLE 7. If $\{x_t: t = 1, 2, \ldots, n\}$ is a sequence of i.i.d. random $N(\mu, 1)$ variables and if the prior density of μ is $N(m, q^{-2})$ with *known m and q* then the joint density of the observations *and* the parameter is

$$p(x|\mu)\psi(\mu) = (2\pi)^{-(n+1)/2}q \exp\left\{-\frac{1}{2}\sum_{i=1}^{n}(x_i - \mu)^2 - \frac{q^2}{2}(\mu - m)^2\right\}.$$

The exponential may be written as

$$-\frac{1}{2}\left[\sum_{i=1}^{n}(x_i - \mu)^2 + q^2(\mu - m)^2\right]$$

$$= -\frac{1}{2}\left[\sum_{i=1}^{n}(x_i - \bar{x})^2 + n(\bar{x} - \mu)^2 + q^2(\mu - m)^2\right].$$

Moreover

$$n(\bar{x} - \mu)^2 + q^2(\mu - m)^2 = r(\mu - s)^2 - rs^2 + n\bar{x}^2 + q^2m^2,$$

where

$$r = n + q^2, \qquad s = \frac{n\bar{x} + q^2m}{r}.$$

In addition,

$$-rs^2 + n\bar{x}^2 + q^2m^2 = \frac{nq^2}{r}(\bar{x} - m)^2.$$

Consequently, the joint density can be written

$$p(x|\mu)\psi(\mu) = (2\pi)^{-n/2}qr^{-1/2} \exp\left\{-\frac{1}{2}\sum_{i=1}^{n}(x_i - \bar{x})^2 - \frac{nq^2}{2r}(\bar{x} - m)^2\right\}$$

$$\times (2\pi)^{-1/2}r^{1/2} \exp\left\{-\frac{r}{2}(\mu - s)^2\right\}.$$

Integrating out μ we shall find the marginal distribution of the observations. But

$$\int (2\pi)^{-1/2} r^{1/2} \exp\{-\tfrac{1}{2}r(\mu - s)^2\} \, d\mu = 1,$$

and we see that

$$p(x) = (2\pi)^{-n/2} q r^{-1/2} \exp\left\{-\frac{1}{2}\left[\sum_{i=1}^{n} (x_i - \bar{x})^2 + \frac{nq^2}{r}(\bar{x} - m)^2\right]\right\}.$$

Thus, the posterior density of μ is given by

$$p(\mu|x) = \frac{p(x|\mu)\psi(\mu)}{p(x)} = (2\pi)^{-1/2} r^{1/2} \exp\left\{-\frac{r}{2}(\mu - s)^2\right\},$$

which is recognized as a $N(s, 1/r)$ density.

Remark 22. Notice that the prior density has mean m while the posterior density has mean

$$s = \frac{n\bar{x} + q^2 m}{n + q^2},$$

which is a *weighted average (convex combination) of the prior and sample means*. Note also that the same situation prevails if we consider the inverse of the variances. Thus, the inverse of the variance of the prior distribution is q^2 while that of the posterior distribution is $n + q^2$. But this is simply the sum of the inverse of the variance of the prior distribution (q^2) and the inverse of the variance of the sample mean (n). *This is not always the case, nor is it always the case that prior and posterior densities are of the same form as in the preceding example.*

Remark 23. *When the prior density for a parameter is of the same form as the density function of the sample based estimator of this parameter we term it the* conjugate prior density.

The framework above *can be easily extended to the case of vector random variables and parameter vectors.*

Thus, if x is a random vector with density $f(\cdot; \theta)$ where θ is a vector of unknown parameters, and if $\{x_t : t = 1, 2, \ldots, T\}$ is a random sample and $\psi(\theta)$ is the prior density of θ, then in Bayesian terms the joint density of the parameters and the observations is

$$p(x, \theta) = \left[\prod_{t=1}^{T} f(x_t; \theta)\right]\psi(\theta),$$

while the posterior density of θ is

$$p(\theta|x) = \frac{p(x, \theta)}{p(x)}$$

with

$$p(x) = \int p(x, \theta)d\theta.$$

An example will serve well in illustrating the discussion above.

EXAMPLE 8. Let $\{x'_t. : t = 1, 2, \ldots, T\}$ be a random sample from a population characterized by a $N(\mu, \Sigma)$ *density with Σ known*. Given μ, the likelihood function of the observations is

$$(2\pi)^{-(nT/2)}|\Sigma|^{-T/2} \exp\{-\tfrac{1}{2} \operatorname{tr} \Sigma^{-1}(X - e\mu')'(X - e\mu')\},$$

where X is a $T \times n$ matrix whose tth row is $x_t.$, and e is a T-element column vector all of whose elements are unity.
If the prior density on μ belongs to the conjugate family, then

$$\mu \sim N(m_0, Q^{-1}),$$

where m_0 and Q are a known vector and matrix respectively. We may further simplify the exponential above by noting that if we put

$$\hat{\mu} = \frac{X'e}{T} = \frac{X'e}{e'e}$$

then

$$X - e\mu' = X - e\hat{\mu}' + e\hat{\mu}' - e\mu' = (X - e\hat{\mu}') + e(\hat{\mu} - \mu)'.$$

Consequently,

$$\operatorname{tr} \Sigma^{-1}(X - e\mu')'(X - e\mu')$$

$$= \operatorname{tr} \Sigma^{-1}X'\left(I - \frac{ee'}{e'e}\right)X + \operatorname{tr} T\Sigma^{-1}(\hat{\mu} - \mu)(\hat{\mu} - \mu)'.$$

The joint distribution of the observations and μ can now be written as

$$p(\mu, X) = (2\pi)^{-nT/2}(2\pi)^{-n/2}|\Sigma|^{-T/2}|Q|^{1/2}$$

$$\times \exp\left\{-\frac{1}{2}\left[\operatorname{tr} \Sigma^{-1}X'\left(I - \frac{ee'}{e'e}\right)X\right.\right.$$

$$\left.\left. + \operatorname{tr} T\Sigma^{-1}(\mu - \hat{\mu})(\mu - \hat{\mu})' + \operatorname{tr} Q(\mu - m_0)(\mu - m_0)'\right]\right\}. \quad (29)$$

As in the univariate case, combining the two last expressions involving μ yields

$$\operatorname{tr}(T\Sigma^{-1} + Q)(\mu - m_1)(\mu - m_1)' + \operatorname{tr}(T\Sigma^{-1}\hat{\mu}\hat{\mu}' + Qm_0m_0')$$

$$- \operatorname{tr}(T\Sigma^{-1} + Q)^{-1}(T\Sigma^{-1}\hat{\mu} + Qm_0)(T\Sigma^{-1}\hat{\mu} + Qm_0)'$$

where

$$m_1 = (T\Sigma^{-1} + Q)^{-1}(T\Sigma^{-1}\hat{\mu} + Qm_0).$$

Since, however,

$$(2\pi)^{-n/2}|T\Sigma^{-1} + Q|^{1/2} \exp\{-\tfrac{1}{2}\operatorname{tr}(T\Sigma^{-1} + Q)(\mu - m_1)(\mu - m_1)'$$

is recognized as the density of a $N(m_1, Q_1^{-1})$ variable with

$$Q_1 = (T\Sigma^{-1} + Q),$$

we deduce that its integral is unity.

Thus, if in (29) we integrate out μ we find

$$p(X) = (2\pi)^{-(nT/2)}|\Sigma|^{-T/2}|Q|^{1/2}|T\Sigma^{-1} + Q|^{-1/2}$$

$$\times \exp\left\{-\frac{1}{2}\left[\operatorname{tr}\Sigma^{-1}X'\left(I - \frac{ee'}{e'e}\right)X + \operatorname{tr}(T\Sigma^{-1}\hat{\mu}\hat{\mu}' + Qm_0 m_0')\right.\right.$$

$$\left.\left. - \operatorname{tr}(T\Sigma^{-1} + Q)^{-1}(T\Sigma^{-1}\hat{\mu} + Qm_0)(T\Sigma^{-1}\hat{\mu} + Qm_0)'\right]\right\}.$$

Consequently

$$p(\mu|X) = \frac{p(\mu, X)}{p(X)} = (2\pi)^{-n/2}|T\Sigma^{-1} + Q|^{1/2}$$

$$\times \exp\{-\tfrac{1}{2}\operatorname{tr}(T\Sigma^{-1} + Q)(\mu - m_1)(\mu - m_1)'\}$$

is the posterior distribution of μ, given the observations, and is, as we re-marked above, a $N(m_1, Q_1^{-1})$ density.

Remark 24. As in the univariate case, we see that the posterior distribution of the mean of a normal process with known covariance matrix and a con-jugate prior is *also multivariate normal* with a mean that is a weighted sum of the prior mean and the ML estimator of the mean, and with a covariance matrix whose inverse is the sum of inverses of the covariance matrices of the prior distribution and that of the distribution of the ML estimator of the mean.

4.2 Inference in a Bayesian Context

In the classical context, where the distinction between parameters and random variables is very sharp, by inference we mean the process by which observations on random variables are converted into information (estimates) regarding the unknown parameters. The information conveyed by the sample observations through the "estimator"—which is a certain function of the observations—is converted into a numerical "guess" regarding the value of the unknown parameter. Since estimates are values assumed by the esti-mator, which is a random variable, we can, in many instances, infer the distribution of the estimator. Based on the numerical value of the estimate and the distributional properties of the estimator we can often carry out tests of hypotheses regarding the unknown (true) value of the parameter in question.

In the Bayesian context the inferential problem is somewhat different. Emphasis on estimation and testing hypotheses disappears since parameters (in the classical context) are treated as random variables. *Inference in a Bayesian context is crucially related to the way in which data, observations, are allowed to modify our prior conceptions regarding the "unknown parameters" of the process under consideration. This, essentially, involves the transition from the prior to the posterior distribution.* It is not an integral part of Bayesian inference, given a body of data, to produce a number that "best" summarizes our information regarding an unknown parameter. Nonetheless, if the context is such that a numerical estimate is desired the problem is approached in the same way as one approaches any problem of decision making under uncertainty. One formulates *a loss function and one obtains a numerical estimate by minimizing expected loss.* The expectation is carried out with respect to the appropriate distribution. This may be either the prior distribution, if sample information is lacking, or the posterior distribution if sample information is available.

An example will clarify this matter.

EXAMPLE 9. Operating in a Bayesian context, suppose we are interested in "estimating" the unknown mean μ of a scalar random variable

$$x \sim N(\mu, \sigma^2),$$

where σ^2 is known.

As indicated earlier, *we proceed by treating μ as a random variable* with distribution, say

$$\mu \sim N(m_0, q_0^{-2}).$$

If we designate our estimator by m the error we commit is given by $\mu - m$. Let us denote the "cost" of this error by $c(\mu - m)$. To be concrete let this cost function be quadratic, i.e., let

$$c(\mu - m) = c_0 + c_1(\mu - m) + c_2(\mu - m)^2,$$

such that $c_2 > 0$. If it is desired that the function be symmetric in $\mu - m$, then we must have $c_1 = 0$. For the moment, let us not impose this condition.

If no sample information is available, then one way of proceeding might be to determine m by *minimizing expected cost, where the expectation is taken with respect to the prior density of μ.* Put

$$V(m, m_0, q_0) = E[c(\mu - m)] = c_0 + c_1(m_0 - m) + c_2(m_0 - m)^2 + c_2 q_0^{-2}.$$

Minimizing with respect to m yields

$$m_0^* = m_0 + \frac{c_1}{2c_2}.$$

The value of the minimand is thus

$$V(m_0^*, m_0, q_0) = c_0 + c_2 q_0^{-2} - \frac{c_1^2}{4c_2}.$$

Note that if the cost function is symmetric the minimand attains the value

$$c_0 + c_2 q_0^{-2}.$$

Suppose now we have a random sample $\{x_t : t = 1, 2, \ldots, T\}$. Proceeding as before we obtain the posterior density of μ, which is $N(m_1, q_1^{-2})$, where

$$m_1 = (T\sigma^{-2} + q_0^2)^{-1}(T\sigma^{-2}\bar{x} + q_0^2 m_0),$$

$$q_1^2 = (T\sigma^{-2} + q_0^2).$$

Our decision will, again, be based on the minimization of the *expected cost function, where now the expectation is taken with respect to the posterior distribution.* We thus determine the expected cost function

$$V(m, m_1, q_1) = c_0 + c_1(m_1 - m) + c_2(m_1 - m)^2 + c_2 q_1^{-2}.$$

Minimizing yields

$$m_1^* = m_1 + \frac{c_1}{2c_2}.$$

The value attained by the minimand is, thus,

$$V(m_1^*, m_1, q_1) = c_0 + c_2 q_1^{-2} - \frac{c_1^2}{4c_2}.$$

We may take the difference

$$V(m_0^*, m_0, q_0) - V(m_1^*, m_1, q_1) = c_2[q_0^{-2} - q_1^{-2}] = s(T)$$

to be a measure of the value of the information conveyed by the sample. We note that

$$s(T) = c_2 \left[\frac{1}{q_0^2} - \frac{1}{T\sigma^{-2} + q_0^2} \right] = \frac{c_2}{q_0^2 \left(1 + \frac{\sigma^2 q_0^2}{T} \right)} \tag{30}$$

and we verify that

$$s'(T) > 0, \qquad \lim_{T \to \infty} s(T) = \frac{c_2}{q_0^2},$$

so that sample information has a positive value. In fact, this may be used to determine the "optimal" sample size, when sampling is costly. We may reason as follows. In the absence of sampling we attain the minimand value

$$V(m_0^*, m_0, q_0).$$

With sampling we further reduce the minimand by

$$s(T) = \frac{c_2}{q_0^2 \left(1 + \dfrac{\sigma^2 q_0^2}{T}\right)}.$$

On the other hand, sampling may be costly, the cost of a sample of size T being, say,

$$b(T). \tag{31}$$

Hence the net gain from sampling is

$$\psi = s(T) - b(T).$$

Proceeding somewhat heuristically, we can maximize ψ with respect to T to obtain

$$\frac{\partial \psi}{\partial T} = \frac{\sigma^2 c_2}{(T + \sigma^2 q_0^2)^2} - b'(T) = 0. \tag{32}$$

For the special case where the sampling cost is linear,

$$b(T) = b_0 + b_1 T.$$

Solving (32) yields

$$T^* = \left[\left(\frac{\sigma^2 c_2}{b_1}\right)^{1/2} - q_0^2 \sigma^2\right], \tag{33}$$

i.e., it is the integer part of the quantity

$$\left(\frac{c_2 \sigma^2}{b_1}\right)^{1/2} - q_0^2 \sigma^2.$$

Thus, the "optimal" sample size will be larger the cheaper it is to sample— i.e., the lower is b_1— and also the higher the "penalty" for error—i.e., the larger is c_2. Conversely the larger is q_0^2 the smaller the "optimal" sample size. But a "large" q_0^2 means a small variance for the prior distribution. *Another way of putting this is to say that the more "certain" we are, a priori, about μ the less we would be inclined to sample.*

The effect of the variance of the process error, σ^2, on the "optimal" sample size depends directly on the sign of

$$\frac{1}{2} \left(\frac{c_2}{b_1 \sigma^2}\right)^{1/2} - q_0^2.$$

A similar exposition can be made for the vector case. Thus, consider

EXAMPLE 10. Let

$$x \sim N(\mu, \Sigma),$$

with known covariance matrix Σ but unknown mean vector μ. Suppose we wish to estimate μ but we have no sample observations. Instead we have the prior density

$$\mu \sim N(m_0, Q_0^{-1}).$$

With the quadratic cost function

$$C(m) = \alpha'(\mu - m) + (\mu - m)'A(\mu - m)$$

where α and A are, respectively, a known vector and a positive definite matrix, we have, taking expectations with respect to the prior density,

$$V(m, m_0, Q_0) = \alpha'(m_0 - m) + \text{tr } AQ_0^{-1} + (m_0 - m)'A(m_0 - m).$$

Minimizing we find

$$m_0^* = m_0 + \tfrac{1}{2}A^{-1}\alpha.$$

The value attained by the minimand is

$$V(m_0^*, m_0, Q_0) = \text{tr } AQ_0^{-1} - \tfrac{1}{4}\alpha'A^{-1}\alpha.$$

If $\{x_t' : t = 1, 2, \ldots, T\}$, a random sample of size T, is available then as before we can obtain the posterior distribution $\mu \sim N(m_1, Q_1^{-1})$, where

$$m_1 = (T\Sigma^{-1} + Q_0)^{-1}(T\Sigma^{-1}\bar{x} + Q_0 m_0), \qquad \bar{x} = \frac{1}{T}\sum_{t=1}^{T} x_t'.$$

$$Q_1 = T\Sigma^{-1} + Q_0.$$

Operating with the posterior distribution yields the "estimator"

$$m_1^* = m_1 + \tfrac{1}{2}A^{-1}\alpha$$

and the value attained by the minimand is

$$V(m_1^*, m_1, Q_1) = \text{tr } AQ_1^{-1} - \tfrac{1}{4}\alpha'A^{-1}\alpha'.$$

As before

$$s(T) = V(m_0^*, m_0, Q_0) - V(m_1^*, m_1, Q_1) = \text{tr } A[Q_0^{-1} - Q_1^{-1}]$$

may serve as an index of the value of sample information. Since $Q_0^{-1} - Q_1^{-1}$ is a positive definite matrix it easily follows that $s(T) > 0$ and, in addition, $\lim_{T \to \infty} s(T) = \text{tr } AQ_0^{-1}$, results which are quite analogous to the scalar case.

Remark 25. The discussion in this section is a very brief introduction to the elements of Bayesian inference theory. The reader who desires a fuller development of the topic may consult Raifa and Schleifer [27] or Zellner [36].

QUESTIONS AND PROBLEMS

1. Referring to Definition 2, what is the marginal density of X_2? What is the marginal density of X_2 and X_3 (i.e., the joint density of these two variables). What is the marginal density of $(X_1, X_2, \ldots, X_k)$?

2. Again referring to Definition 2, show that $g_1(\cdot)$ is a density function, i.e., that $g_1(\cdot)$ is nonnegative, $\int_{-\infty}^{\infty} g_1(\zeta_1)d\zeta_1 = 1$, and that $F_1(x_1) = \Pr\{X_1 \le x_1\} = \int_{-\infty}^{x_1} g_1(\zeta_1)d\zeta_1$.

3. Referring to Definition 4 show that if the conditional density of $X^{(1)}$ given $X^{(2)}$ is equal to the marginal density of $X^{(1)}$ then the conditional density of $X^{(2)}$ given $X^{(1)}$ is equal to the marginal density of $X^{(2)}$.

4. Referring to Proposition 2, show that, in fact, the covariance matrix is defined by

$$\Sigma = E(X - \mu)(X - \mu)',$$

where $X = (X_1, X_2, \ldots, X_n)'$ is the random vector whose covariance matrix is Σ. Also verify that Σ is always at least positive semidefinite.

5. Show that both the mean vector and covariance matrix of a random vector $X = (X_1, X_2, \ldots, X_n)'$ depend on the units in which the random variables are "measured." [*Hint*: consider $Y = DX$, where $D = \text{diag}(d_1, d_2, \ldots, d_n)$.]

6. Show that the correlation matrix of Definition 8 is independent of the units in which the elements of X are measured; show also that the elements of R lie in the interval $[-1, 1]$. [*Hint*: $\sigma_{ij}^2 \le \sigma_{ii}\sigma_{jj}$, using Schwarz' inequality for integrals.]

7. Referring to Definition 11, suppose $y_1, y_2 \in y^1$. What meaning do you give to

$$\rho_{12.5}, \rho_{12.57}, \rho_{12.579}, \rho_{12.345\ldots n}?$$

8. Show that if $(x_1, x_2)' \sim N(0, I)$, where I is a 2×2 identity matrix, then $x_1 \sim N(0, 1)$. Extend this to show that if

$$x = \begin{pmatrix} x^1 \\ x^2 \end{pmatrix},$$

x^1 being $k \times 1$, x being $n \times 1$, and $x \sim N(0, I)$, then the marginal distribution of x^1 is $N(0, I_k)$. [*Hint*: $\int_{-\infty}^{\infty} e^{-(1/2)\zeta^2} d\zeta = \sqrt{2\pi}$.]

9. Show that if $x \sim N(0, 1)$, its moment generating function is $M_x(s) = e^{(1/2)s^2}$. [*Hint*: $M_x(s) = E(e^{sx}) = (2\pi)^{-1/2} \int_{-\infty}^{\infty} e^{-(1/2)\zeta^2 + s\zeta} d\zeta = (2\pi)^{-1/2} \int_{-\infty}^{\infty} e^{-(1/2)(\zeta-s)^2} e^{(1/2)s^2} d\zeta$.]

10. Show that in Example 5 $E(\bar{x}_r - \mu_r)(\bar{x}_s - \mu_s) = (1/n)\sigma_{rs}$. [*Hint*: $\bar{x}_r = (1/n)\sum_{i=1}^{n} x_{ir}$, $\bar{x}_s = (1/n)\sum_{i=1}^{n} x_{is}$, $E(x_{ir}) = \mu_r$, $\text{Cov}(x_{ir}, x_{is}) = \sigma_{rs}$ for all i.]

11. Show that if $\hat{\theta}_T$ is an estimator of θ (not necessarily unbiased) and if $E(\hat{\theta}_T - \theta)^2$ exists then for any $k > 0$

$$\Pr\{|\hat{\theta}_T - \theta| > k\} \le \frac{E(\hat{\theta}_T - \theta)^2}{k^2}.$$

[*Hint*: $E(\hat{\theta}_T - \theta)^2 = \int_{-\infty}^{\infty} (\zeta - \theta)^2 f_T(\zeta)d\zeta$, where $f_T(\cdot)$ is the density of $\hat{\theta}_T$, and break up the interval into three parts—$(-\infty, \theta - k)$, $(\theta - k, \theta + k)$, $(\theta + k, \infty)$.]

12. Let x be a vector random variable ($n \times 1$) with mean μ and positive definite co-variance matrix Σ. Show that for $k > 0$

$$\Pr\{(x - \mu)'(x - \mu) > k\} \le \frac{\text{tr } \Sigma}{k}.$$

[*Hint*: $\mathrm{E}(x - \mu)'(x - \mu) = \text{tr } \Sigma$; define $\zeta = (x - \mu)'(x - \mu)$, which is a scalar random variable.]

13. Show that

$$X - e\hat{\mu}' = \left(I - \frac{ee'}{e'e}\right)X \quad \text{and} \quad \left(I - \frac{ee'}{e'e}\right)\left(I - \frac{ee'}{e'e}\right) = \left(I - \frac{ee'}{e'e}\right)$$

i.e., that $I - (ee'/e'e)$ is an idempotent matrix.

Tables for Testing Hypotheses on the Autoregressive Structure of the Errors in a GLM[1]

These tables are meant to facilitate the test of hypotheses regarding the autoregressive properties of the error term in a general linear model containing a constant term but no *lagged* dependent variables. The order of the autoregression can be at most four.

Tables I, Ia, and Ib contain upper and lower significance points at the 1%, 2.5%, and 5% level respectively, for testing that the first-order autocorrelation is zero. Table II contains upper and lower significance points at the 5% level for testing that the second-order autocorrelation is zero.

Similarly, Tables III and IV contain upper and lower significance points at the 5% level for tests on the third- and fourth-order autocorrelation.

Perhaps a word of explanation is in order regarding their use. The tables have been constructed on the basis of the properties of the statistics

$$d_j = \sum_{t=j+1}^{T} (\hat{u}_t - \hat{u}_{t-j}) / \sum_{t=1}^{T} \hat{u}_t^2, \quad j = 1, 2, 3, 4,$$

where the $\hat{u}_t$ are the residuals of a GLM containing a constant term but not containing lagged dependent variables. If X is the data matrix of the GLM, then

$$\hat{u} = [I - X(X'X)^{-1}X']u,$$

where

$$\hat{u} = (\hat{u}_1, \hat{u}_2, \hat{u}_3, \ldots, \hat{u}_T)'$$

---

[1] These tables are reproduced with the kind permission of the publisher Marcel Dekker Inc., and the author H. D. Vinod. Tables II, III, and IV first appeared in H. D. Vinod, "Generalization of the Durbin–Watson Statistic for Higher Order Autoregressive Processes," *Communications in Statistics*, vol. 2, 1973, pp. 115–144.

and it is assumed that

$$u \sim N(0, \sigma^2 I).$$

Hence, if we wish to test a first-order hypothesis i.e., that in

$$u_t = \rho_1 u_{t-1} + \varepsilon_t$$

we have

$$H_0: \quad \rho_1 = 0,$$

as against

$$H_1: \quad \rho_1 > 0,$$

we can use Tables I, Ia, or Ib exactly as we use the standard Durbin–Watson tables—indeed, they are the same.

If we wish to test for a *second-order autoregression of the special form*

$$u_t = \rho_2 u_{t-2} + \varepsilon_t$$

we can do so using the statistic d_2 and Table II in exactly the same fashion as one uses the standard Durbin–Watson tables.

Similarly, if we wish to test for a third-order autoregression of the special type

$$u_t = \rho_3 u_{t-3} + \varepsilon_t$$

or for a fourth-order autoregression of the special type

$$u_t = \rho_4 u_{t-4} + \varepsilon_t$$

we may do so using the statistics d_3 and d_4 and Tables III and IV respectively.

Again, the tables are used in the same fashion as the standard Durbin–Watson tables, i.e., we accept the hypothesis that the (relevant) autocorrelation coefficient is zero if the statistic d_3 or d_4 exceeds the appropriate upper significance point, and we accept the hypothesis that the (relevant) autocorrelation coefficient is positive if the statistic d_3 or d_4 is less than the lower significance point.

Now, it may be shown that if we have two autoregressions of order m and $m + 1$ respectively and if it is known that these two autoregressions have the same autocorrelations of order $1, 2, \ldots, m$, then a certain relationship must exist between the coefficients describing these autoregressions. In particular, it may be shown that if for a fourth-order autoregression, say

$$u_t = a_{41} u_{t-1} + a_{42} u_{t-2} + a_{43} u_{t-3} + a_{44} u_{t-4} + \varepsilon_t,$$

the autocorrelations of order 1, 2, 3 are zero, then

$$a_{41} = a_{42} = a_{43} = 0$$

and thus the process is of the special form

$$u_t = a_{44} u_{t-4} + \varepsilon_t \quad \text{and} \quad a_{44} = \rho_4,$$

i.e., a_{44} is the autocorrelation of order 4. Similarly, if for the third-order autoregression

$$u_t = a_{31}u_{t-1} + a_{32}u_{t-2} + a_{33}u_{t-3} + \varepsilon_t$$

it is known that the first two autocorrelations are zero, then

$$a_{31} = a_{32} = 0$$

so that the process is of the special form

$$u_t = a_{33}u_{t-3} + \varepsilon_t \quad \text{and} \quad a_{33} = \rho_3,$$

i.e., a_{33} is the autocorrelation of order 3. Finally, if for the second-order autoregression

$$u_t = a_{21}u_{t-1} + a_{22}u_{t-2} + \varepsilon_t$$

it is known that the first-order autocorrelation is zero then

$$a_{21} = 0$$

so that the process is of the special form

$$u_t = a_{22}u_{t-2} + \varepsilon_t \quad \text{and} \quad a_{22} = \rho_2,$$

i.e., a_{22} is the autocorrelation of order 2.

Vinod [34] uses these relations to suggest a somewhat controversial test for the case where we wish to test for autoregression in the error term of the GLM and are willing to limit the alternatives to, at most, the fourth-order autoregression

$$u_t = \sum_{i=1}^{4} a_{4i}u_{t-1} + \varepsilon_t.$$

The proposed test is as follows. First test that the first-order autocorrelation is zero, i.e.,

$$H_{01}: \quad \rho_1 = 0,$$

as against

$$H_{11}: \quad \rho_1 > 0,$$

using Tables I, Ia, or Ib. If H_{01} is accepted then test

$$H_{02}: \quad \rho_2 = 0,$$

as against

$$H_{12}: \quad \rho_2 > 0.$$

If H_{02} is also accepted then test

$$H_{03}: \quad \rho_3 = 0$$

as against

$$H_{13}: \quad \rho_3 > 0.$$

If H_{03} is accepted then test

$$H_{04}: \quad \rho_4 = 0$$

as against

$$H_{14}: \quad \rho_4 > 0.$$

There are a number of problems with this: first, the level of significance of the second, third, and fourth tests cannot be the stated ones, since we proceed to the ith test only conditionally upon accepting the null hypothesis in the $(i-1)$th test; second, if at any point we accept the alternative, it is not clear what we should conclude.

Presumably, if we accept H_{12} (at the second test) we should conclude that the process is at least second order, make allowance for this, in terms of search or Cochrane–Orcutt procedures, and then proceed to test using the residuals of the transformed equation.

An alternative to the tests suggested by Vinod [34] would be simply to regress the residuals $\hat{u}_t$ on $\hat{u}_{t-1}, \hat{u}_{t-2}, \hat{u}_{t-3}, \hat{u}_{t-4}$, thus obtaining the estimates

$$\hat{a}_{4i}, \quad i = 1, 2, \ldots, 4.$$

Since we desire to test

$$H_0: \quad a = 0,$$

as against

$$H_1: \quad a \neq 0,$$

where $a = (a_{41}, a_{42}, a_{43}, a_{44})'$, we may use the (asymptotic) distribution of $\hat{a}$ under the null hypothesis as well as the multiple comparison test, as given in the appendix to Chapter 2. Thus, testing the null hypothesis of no autocorrelation in the errors, i.e.,

$$H_0: \quad a = 0$$

as against

$$H_1: \quad a \neq 0,$$

is best approached through the asymptotic distribution, given by

$$\sqrt{T}\,\hat{a} \sim N(0, I).$$

This implies the chi-square and associated multiple comparison tests: accept H_0 if $T\hat{a}'\hat{a} \leq \chi^2_{\alpha;4}$, where $\chi^2_{\alpha;4}$ is the α significance point of a chi-square variable with four degrees of freedom; otherwise reject H_0 and accept any of the hypotheses whose acceptance is implied by the multiple compassion intervals

$$-(\chi^2_{\alpha;4}h'h)^{1/2} \leq \sqrt{T}\,h'\hat{a} \leq (\chi^2_{\alpha;4}h'h)^{1/2}.$$

Finally, we illustrate the use of these tables by an example. Suppose in a GLM with five bona fide explanatory variables and thirty observations we have the Durbin–Watson statistic

$$d = 1.610.$$

From Table I we see that the upper significance point for the 1% level is 1.606. Hence the hypothesis of no autocorrelation will be accepted. For the 2.5% level the upper significant point is 1.727; hence we will not accept it at this level. On the other hand the lower significance point is .999 so that the test is indeterminate. For the 5% level the upper significance point is 1.833 while the lower is 1.070; hence at the 5% level the test is indeterminate as well.

Table I First-Order Autoregression: Level of Significance 1%

n	$k' = 1$		$k' = 2$		$k' = 3$		$k' = 4$		$k' = 5$	
	d_L	d_U	d_L	d_U	d_L	d_U	d_L	d_U	d_L	d_U
15	.813	1.072	.701	1.254	.592	1.467	.488	1.707	.391	1.970
16	.845	1.088	.738	1.255	.633	1.449	.532	1.666	.436	1.903
17	.876	1.103	.773	1.257	.672	1.433	.575	1.633	.481	1.849
18	.903	1.118	.805	1.260	.708	1.424	.614	1.606	.523	1.805
19	.929	1.133	.836	1.265	.743	1.417	.651	1.585	.562	1.769
20	.953	1.148	.863	1.271	.774	1.412	.686	1.568	.599	1.738
21	.976	1.161	.890	1.277	.804	1.409	.718	1.554	.634	1.713
22	.998	1.174	.915	1.284	.832	1.407	.749	1.544	.667	1.691
23	1.018	1.187	.938	1.291	.858	1.407	.778	1.535	.699	1.674
24	1.037	1.199	.960	1.298	.883	1.407	.805	1.528	.728	1.659
25	1.056	1.211	.981	1.305	.906	1.409	.831	1.523	.756	1.646
26	1.072	1.222	1.001	1.311	.929	1.411	.855	1.519	.783	1.635
27	1.089	1.233	1.020	1.318	.949	1.413	.879	1.516	.808	1.625
28	1.104	1.244	1.037	1.325	.069	1.415	.900	1.513	.832	1.618
29	1.119	1.256	1.054	1.333	.988	1.419	.921	1.512	.855	1.612
30	1.133	1.263	1.070	1.339	1.006	1.422	.941	1.510	.877	1.606
31	1.147	1.273	1.086	1.346	1.023	1.414	.960	1.510	.897	1.601
32	1.160	1.282	1.100	1.352	1.040	1.428	.978	1.510	.917	1.597

n										
33	1.594	.936	1.510	.996	1.432	1.055	1.353	1.114	1.291	1.172
34	1.591	.954	1.511	1.012	1.435	1.071	1.364	1.127	1.299	1.185
35	1.589	.972	1.513	1.028	1.438	1.085	1.371	1.141	1.307	1.195
36	1.587	.988	1.513	1.043	1.442	1.098	1.376	1.153	1.315	1.206
37	1.586	1.004	1.514	1.058	1.446	1.111	1.382	1.164	1.322	1.216
38	1.585	1.019	1.515	1.072	1.450	1.124	1.388	1.176	1.330	1.227
39	1.585	1.034	1.517	1.086	1.453	1.137	1.393	1.187	1.337	1.237
40	1.583	1.048	1.518	1.099	1.456	1.149	1.399	1.197	1.344	1.246
45	1.584	1.111	1.528	1.156	1.474	1.201	1.423	1.245	1.375	1.288
50	1.587	1.164	1.538	1.205	1.491	1.245	1.445	1.285	1.403	1.324
55	1.592	1.210	1.548	1.248	1.506	1.284	1.466	1.320	1.427	1.356
60	1.598	1.249	1.558	1.283	1.520	1.316	1.484	1.350	1.448	1.383
65	1.605	1.283	1.568	1.314	1.534	1.346	1.500	1.376	1.468	1.407
70	1.611	1.313	1.578	1.342	1.546	1.372	1.515	1.400	1.485	1.429
75	1.617	1.340	1.587	1.368	1.557	1.395	1.529	1.422	1.501	1.448
80	1.623	1.364	1.595	1.390	1.568	1.416	1.541	1.441	1.515	1.466
85	1.630	1.387	1.603	1.411	1.577	1.435	1.553	1.458	1.528	1.482
90	1.636	1.407	1.611	1.429	1.587	1.472	1.563	1.474	1.540	1.497
95	1.641	1.425	1.619	1.446	1.596	1.468	1.573	1.489	1.551	1.510
100	1.647	1.441	1.625	1.462	1.604	1.482	1.582	1.502	1.562	1.523

n = number of observations, k' = number of explanatory variables (excluding the constant term)

Table Ia First-Order Autoregression: Level of Significance 2.5%

n	$k'=1$ d_L	$k'=1$ d_U	$k'=2$ d_L	$k'=2$ d_U	$k'=3$ d_L	$k'=3$ d_U	$k'=4$ d_L	$k'=4$ d_U	$k'=5$ d_L	$k'=5$ d_U
15	.949	1.222	.827	1.405	.705	1.615	.588	1.848	.476	2.099
16	.980	1.235	.864	1.403	.748	1.594	.636	1.807	.528	2.036
17	1.009	1.249	.899	1.403	.788	1.579	.679	1.773	.574	1.983
18	1.036	1.261	.981	1.405	.825	1.567	.720	1.746	.619	1.940
19	1.061	1.274	.960	1.408	.859	1.558	.759	1.724	.660	1.903
20	1.083	1.286	.987	1.411	.890	1.552	.793	1.705	.699	1.872
21	1.104	1.298	1.013	1.415	.920	1.546	.827	1.690	.734	1.845
22	1.124	1.300	1.037	1.419	.947	1.544	.858	1.678	.769	1.823
23	1.143	1.319	1.059	1.425	.973	1.541	.887	1.668	.801	1.804
24	1.161	1.329	1.079	1.429	.997	1.540	.914	1.659	.830	1.788
25	1.178	1.340	1.100	1.434	1.020	1.539	.939	1.652	.859	1.774
26	1.193	1.349	1.118	1.440	1.041	1.539	.963	1.647	.885	1.762
27	1.209	1.358	1.136	1.444	1.061	1.539	.986	1.642	.911	1.751
28	1.223	1.367	1.152	1.450	1.080	1.540	1.008	1.638	.935	1.742
29	1.230	1.375	1.168	1.455	1.098	1.541	1.028	1.634	.957	1.734
30	1.249	1.383	1.183	1.460	1.116	1.543	1.048	1.632	.978	1.727
31	1.262	1.391	1.197	1.465	1.132	1.544	1.065	1.630	.999	1.721
32	1.272	1.398	1.211	1.470	1.147	1.546	1.083	1.628	1.018	1.715

n	d_L	d_U	d_L	d_U	d_L	d_U	d_L	d_U	d_L	d_U
33	1.284	1.406	1.224	1.474	1.162	1.548	1.100	1.627	1.037	1.710
34	1.294	1.413	1.236	1.479	1.176	1.550	1.115	1.626	1.054	1.706
35	1.305	1.420	1.248	1.484	1.190	1.552	1.131	1.626	1.071	1.703
36	1.314	1.426	1.259	1.488	1.203	1.555	1.145	1.625	1.087	1.700
37	1.324	1.433	1.270	1.493	1.214	1.557	1.159	1.625	1.102	1.697
38	1.333	1.439	1.280	1.497	1.227	1.559	1.172	1.626	1.117	1.695
39	1.342	1.445	1.291	1.502	1.238	1.562	1.185	1.636	1.131	1.693
40	1.350	1.450	1.300	1.506	1.249	1.564	1.197	1.626	1.145	1.692
45	1.388	1.477	1.343	1.525	1.298	1.577	1.252	1.631	1.205	1.686
50	1.420	1.500	1.379	1.543	1.338	1.588	1.297	1.636	1.255	1.685
55	1.447	1.520	1.410	1.559	1.373	1.600	1.335	1.642	1.297	1.686
60	1.471	1.538	1.438	1.574	1.403	1.611	1.369	1.649	1.334	1.688
65	1.492	1.554	1.461	1.587	1.430	1.620	1.398	1.655	1.366	1.692
70	1.511	1.569	1.483	1.598	1.453	1.630	1.423	1.662	1.394	1.695
75	1.529	1.582	1.502	1.610	1.474	1.638	1.446	1.668	1.418	1.699
80	1.544	1.594	1.518	1.620	1.493	1.646	1.467	1.674	1.441	1.703
85	1.557	1.604	1.534	1.629	1.510	1.655	1.485	1.680	1.461	1.707
90	1.570	1.615	1.548	1.638	1.525	1.662	1.502	1.686	1.479	1.711
95	1.582	1.624	1.561	1.646	1.539	1.668	1.518	1.691	1.496	1.715
100	1.593	1.633	1.573	1.654	1.552	1.675	1.532	1.696	1.511	1.719

n = number of observations, k' = number of explanatory variables (excluding the constant term)

Table Ib First-Order Autoregression: Level of Significance 5%

n	$k' = 1$ d_L	$k' = 1$ d_U	$k' = 2$ d_L	$k' = 2$ d_U	$k' = 3$ d_L	$k' = 3$ d_U	$k' = 4$ d_L	$k' = 4$ d_U	$k' = 5$ d_L	$k' = 5$ d_U
15	1.077	1.361	.9453	1.543	.813	1.750	.684	1.977	.560	2.219
16	1.106	1.371	.9819	1.538	.856	1.728	.733	1.935	.614	2.157
17	1.133	1.381	1.015	1.536	.896	1.710	.778	1.900	.663	2.104
18	1.157	1.391	1.046	1.535	.933	1.696	.820	1.872	.709	2.060
19	1.181	1.401	1.074	1.536	.966	1.685	.858	1.848	.752	2.023
20	.917	1.104	.811	1.202	.705	1.322	.600	1.457	.504	1.611
21	1.221	1.420	1.125	1.538	1.026	1.669	.927	1.812	.828	1.964
22	1.240	1.429	1.147	1.541	1.053	1.664	.957	1.798	.862	1.940
23	1.257	1.437	1.168	1.544	1.078	1.660	.986	1.785	.894	1.920
24	1.273	1.446	1.118	1.546	1.101	1.656	1.013	1.775	.924	1.902
25	1.288	1.454	1.206	1.550	1.123	1.654	1.038	1.767	.9529	1.886
26	1.302	1.461	1.224	1.553	1.143	1.652	1.061	1.759	.9792	1.873
27	1.316	1.469	1.240	1.556	1.162	1.651	1.084	1.753	1.004	1.861
28	1.328	1.476	1.255	1.560	1.181	1.650	1.104	1.747	1.028	1.850
29	1.341	1.483	1.270	1.563	1.198	1.650	1.124	1.742	1.050	1.841
30	1.352	1.489	1.284	1.567	1.214	1.650	1.142	1.739	1.070	1.833
31	1.363	1.496	1.297	1.570	1.229	1.650	1.160	1.735	1.090	1.825
32	1.373	1.502	1.309	1.574	1.244	1.651	1.177	1.732	1.109	1.819

n										
33	1.384	1.509	1.321	1.577	1.257	1.651	1.193	1.730	1.127	1.813
34	1.393	1.514	1.333	1.580	1.271	1.652	1.208	1.728	1.144	1.807
35	1.402	1.519	1.343	1.584	1.283	1.653	1.222	1.726	1.160	1.803
36	1.410	1.524	1.354	1.587	1.296	1.654	1.236	1.725	1.175	1.799
37	1.419	1.530	1.364	1.590	1.307	1.655	1.249	1.723	1.190	1.795
38	1.427	1.535	1.373	1.594	1.318	1.656	1.261	1.723	1.204	1.792
39	1.435	1.539	1.382	1.597	1.328	1.658	1.274	1.721	1.218	1.789
40	1.442	1.544	1.391	1.600	1.339	1.659	1.285	1.721	1.231	1.786
45	1.475	1.566	1.430	1.615	1.383	1.666	1.336	1.720	1.287	1.776
50	1.503	1.585	1.463	1.629	1.421	1.674	1.378	1.721	1.335	1.771
55	1.528	1.602	1.491	1.640	1.452	1.682	1.414	1.724	1.374	1.768
60	1.549	1.616	1.515	1.652	1.480	1.689	1.444	1.727	1.408	1.767
65	1.567	1.630	1.536	1.662	1.503	1.696	1.471	1.731	1.438	1.767
70	1.583	1.641	1.554	1.672	1.524	1.703	1.494	1.735	1.464	1.768
75	1.598	1.652	1.571	1.680	1.543	1.709	1.515	1.739	1.487	1.770
80	1.611	1.662	1.586	1.688	1.560	1.715	1.533	1.743	1.507	1.771
85	1.624	1.671	1.600	1.696	1.575	1.721	1.551	1.747	1.525	1.774
90	1.634	1.680	1.612	1.703	1.589	1.726	1.566	1.751	1.542	1.776
95	1.645	1.687	1.623	1.709	1.602	1.732	1.580	1.755	1.557	1.778
100	1.654	1.695	1.634	1.715	1.613	1.736	1.592	1.758	1.571	1.781

n = number of observations, k' = number of explanatory variables (excluding the constant term)

Table II Second-Order Autoregression: Level of Significance 5%

n	$k' = 1$		$k' = 2$		$k' = 3$		$k' = 4$		$k' = 5$	
	d_2^L	d_2^U	d_2^L	d_2^U	d_2^L	d_2^U	d_2^L	d_2^U	d_2^L	d_2^U
15	.948	1.217	.818	1.374	.692	1.583	.567	1.784	.4528	2.040
16	.985	1.236	.861	1.380	.742	1.574	.621	1.744	.5124	1.995
17	1.019	1.255	.901	1.392	.786	1.566	.670	1.736	.5624	1.949
18	1.050	1.273	.937	1.400	.828	1.562	.716	1.712	.6134	1.916
19	1.077	1.289	.971	1.410	.865	1.559	.759	1.706	.6574	1.886
20	1.104	1.305	1.002	1.418	.901	1.557	.798	1.691	.7015	1.863
21	1.128	1.319	1.030	1.428	.934	1.557	.835	1.687	.7402	1.842
22	1.151	1.333	1.057	1.435	.964	1.557	.869	1.677	.7783	1.826
23	1.171	1.346	1.082	1.444	.992	1.558	.901	1.674	.8121	1.811
24	1.191	1.358	1.105	1.451	1.019	1.560	.930	1.668	.8460	1.798
25	1.209	1.370	1.126	1.459	1.044	1.561	.958	1.666	.8760	1.787
26	1.226	1.381	1.147	1.466	1.067	1.564	.985	1.662	.9051	1.778
27	1.242	1.392	1.666	1.473	1.089	1.566	1.010	1.662	.9319	1.770
28	1.258	1.401	1.184	1.480	1.110	1.569	1.033	1.659	.9582	1.763
29	1.272	1.411	1.200	1.487	1.129	1.572	1.055	1.659	.9819	1.757
30	1.286	1.420	1.217	1.493	1.147	1.575	1.075	1.658	1.005	1.753
31	1.299	1.429	1.232	1.499	1.164	1.578	1.095	1.658	1.026	1.748
32	1.311	1.437	1.247	1.505	1.181	1.580	1.114	1.658	1.047	1.745

n	$k'=1$		$k'=2$		$k'=3$		$k'=4$		$k'=5$	
	d_L	d_U	d_L	d_U	d_L	d_U	d_L	d_U	d_L	d_U
33	1.323	1.445	1.260	1.511	1.197	1.583	1.131	1.658	1.066	1.741
34	1.334	1.453	1.273	1.516	1.212	1.587	1.148	1.658	1.085	1.738
35	1.345	1.460	1.286	1.522	1.226	1.589	1.164	1.659	1.103	1.736
36	1.355	1.467	1.298	1.527	1.240	1.593	1.180	1.660	1.120	1.734
37	1.465	3.474	1.309	1.532	1.252	1.595	1.194	1.661	1.135	1.732
38	1.374	1.480	1.320	1.537	1.265	1.598	1.208	1.661	1.151	1.731
39	1.384	1.487	1.330	1.542	1.277	1.601	1.221	1.662	1.166	1.729
40	1.392	1.493	1.340	1.546	1.288	1.604	1.234	1.663	1.180	1.728
45	1.431	1.521	1.385	1.568	1.338	1.618	1.290	1.670	1.242	1.726
50	1.464	1.544	1.422	1.586	1.380	1.631	1.337	1.677	1.294	1.726
55	1.491	1.565	1.453	1.602	1.415	1.643	1.377	1.684	1.337	1.728
60	1.515	1.582	1.481	1.617	1.446	1.654	1.410	1.691	1.374	1.731
65	1.536	1.598	1.504	1.630	1.472	1.663	1.439	1.698	1.406	1.734
70	1.555	1.612	1.525	1.647	1.495	1.673	1.465	1.704	1.465	1.737
75	1.572	1.625	1.544	1.653	1.516	1.681	1.488	1.711	1.459	1.741
80	1.586	1.637	1.561	1.662	1.535	1.689	1.508	1.717	1.481	1.745
85	1.600	1.647	1.576	1.672	1.551	1.696	1.526	1.722	1.501	1.748
90	1.612	1.657	1.589	1.680	1.566	1.703	1.543	1.727	1.520	1.752
95	1.624	1.666	1.602	1.688	1.580	1.710	1.558	1.732	1.536	1.756
100	1.634	1.674	1.614	1.695	1.593	1.716	1.572	1.737	1.551	1.759

n = number of observations, k' = number of explanatory variables (excluding the constant term)

411

Table III Third-Order Autoregression: Level of Significance 5%

n	$k' = 1$		$k' = 2$		$k' = 3$		$k' = 4$		$k' = 5$	
	d_3^L	d_3^U	d_3^L	d_3^U	d_3^L	d_3^U	d_3^L	d_3^U	d_3^L	d_3^U
15	.832	1.075	.706	1.236	.579	1.376	.4743	1.577	.370	1.873
16	.873	1.105	.754	1.253	.632	1.399	.5232	1.575	.422	1.829
17	.913	1.132	.797	1.269	.681	1.401	.5774	1.578	.471	1.798
18	.949	1.157	.840	1.284	.727	1.405	.6274	1.562	.525	1.776
19	.980	1.180	.876	1.299	.769	1.420	.6698	1.564	.572	1.753
20	1.011	1.201	.911	1.313	.808	1.426	.7133	1.566	.615	1.736
21*	1.039	1.221	.943	1.327	.844	1.432	.7541	1.561	.660	1.724
22	1.065	1.239	.973	1.340	.878	1.443	.7885	1.563	.698	1.712
23	1.089	1.256	1.001	1.352	.909	1.449	.8234	1.556	.734	1.703
24	1.112	1.272	1.027	1.364	.938	1.456	.8562	1.565	.771	1.696
25	1.133	1.287	1.051	1.374	.966	1.465	.8852	1.567	.803	1.690
26	1.153	1.302	1.074	1.385	.992	1.471	.9142	1.571	.833	1.684
27	1.171	1.315	1.095	1.395	1.017	1.477	.9416	1.571	.863	1.681
28	1.189	1.328	1.116	1.405	1.039	1.485	.9657	1.574	.890	1.677
29	1.206	1.340	1.134	1.414	1.061	1.491	.9899	1.577	.916	1.675
30	1.221	1.352	1.153	1.423	1.081	1.496	1.012	1.579	.941	1.673
31	1.236	1.363	1.170	1.431	1.101	1.503	1.033	1.582	.944	1.672
32	1.250	1.373	1.186	1.439	1.119	1.508	1.054	1.585	.986	1.670

n	d_L	d_U	d_L	d_U	d_L	d_U	d_L	d_U	d_L	d_U
33	1.264	1.383	1.202	1.448	1.137	1.514	1.073	1.587	1.008	1.670
34	1.277	1.393	1.216	1.455	1.153	1.520	1.091	1.590	1.028	1.669
35	1.289	1.402	1.230	1.462	1.169	1.524	1.109	1.593	1.047	1.669
36	1.301	1.410	1.243	1.469	1.184	1.529	1.125	1.596	1.065	1.669
37	1.312	1.419	1.256	1.475	1.198	1.535	1.141	1.598	1.083	1.669
38	1.323	1.427	1.268	1.482	1.212	1.539	1.156	1.602	1.099	1.670
39	1.333	1.435	1.280	1.488	1.225	1.544	1.171	1.604	1.115	1.670
40	1.343	1.442	1.291	1.494	1.238	1.549	1.184	1.607	1.130	1.671
45	1.387	1.475	1.341	1.522	1.293	1.569	1.246	1.620	1.198	1.675
50	1.424	1.503	1.382	1.545	1.340	1.587	1.297	1.633	1.253	1.681
55	1.455	1.528	1.417	1.565	1.379	1.604	1.340	1.644	1.300	1.688
60	1.482	1.549	1.448	1.583	1.412	1.618	1.377	1.655	1.341	1.694
65	1.506	1.567	1.474	1.598	1.441	1.631	1.408	1.665	1.375	1.700
70	1.526	1.583	1.497	1.612	1.466	1.643	1.436	1.674	1.405	1.706
75	1.545	1.598	1.517	1.625	1.489	1.653	1.461	1.682	1.432	1.712
80	1.561	1.611	1.536	1.637	1.509	1.663	1.483	1.690	1.456	1.718
85	1.576	1.624	1.552	1.647	1.528	1.672	1.502	1.697	1.477	1.723
90	1.590	1.634	1.567	1.657	1.544	1.680	1.521	1.704	1.496	1.728
95	1.603	1.645	1.581	1.666	1.559	1.688	1.537	1.710	1.514	1.733
100	1.614	1.654	1.594	1.674	1.573	1.695	1.552	1.716	1.530	1.738

n = number of observations, k' = number of explanatory variables (excluding the constant term)

Table IV Fourth-Order Autoregression: Level of Significance 5%

n	$k'=1$		$k'=2$		$k'=3$		$k'=4$		$k'=5$	
	d_4^L	d_4^U	d_4^L	d_4^U	d_4^L	d_4^U	d_4^L	d_4^U	d_4^L	d_4^U
15	.727	.946	.609	1.082	.489	1.266	.3771	1.387	.300	1.568
16	.774	.981	.661	1.108	.548	1.275	.4345	1.381	.349	1.532
17	.812	1.016	.705	1.135	.597	1.289	.4877	1.421	.394	1.560
18	.853	1.047	.746	1.159	.642	1.302	.5371	1.422	.447	1.582
19	.890	1.107	.788	1.182	.684	1.314	.5833	1.424	.496	1.567
20	.924	1.102	.827	1.203	.727	1.327	.6263	1.428	.543	1.557
21	.953	1.126	.861	1.222	.765	1.339	.6682	1.446	.581	1.565
22	.983	1.148	.892	1.240	.801	1.350	.7079	1.450	.624	1.574
23	1.011	1.170	.923	1.257	.834	1.361	.7445	1.455	.663	1.568
24	1.037	1.189	.952	1.274	.866	1.371	.7788	1.459	.701	1.565
25	1.059	1.208	.979	1.288	.896	1.381	.8111	1.472	.732	1.571
26	1.082	1.225	1.003	1.303	.923	1.391	.8417	1.477	.766	1.576
27	1.103	1.241	1.028	1.316	.949	1.401	.8707	1.481	.798	1.575
28	1.123	1.257	1.050	1.328	.974	1.410	.8975	1.487	.827	1.575
29	1.141	1.271	1.070	1.341	.998	1.419	.9233	1.496	.853	1.580
30	1.159	1.285	1.090	1.352	1.020	1.427	.9480	1.501	.880	1.584
31	1.176	1.298	1.109	1.363	1.041	1.435	.9711	1.506	.905	1.586
32	1.192	1.311	1.127	1.373	1.061	1.443	.9926	1.511	.929	1.587

n										
33	1.206	1.322	1.144	1.383	1.080	1.450	1.014	1.517	.950	1.590
34	1.221	1.334	1.160	1.393	1.098	1.458	1.033	1.522	.972	1.594
35	1.235	1.344	1.176	1.402	1.115	1.465	1.052	1.527	.9931	1.596
36	1.248	1.355	1.191	1.410	1.131	1.471	1.070	1.532	1.013	1.598
37	1.260	1.365	1.204	1.419	1.147	1.478	1.088	1.537	1.030	1.602
38	1.272	1.374	1.218	1.427	1.162	1.484	1.104	1.542	1.048	1.650
39	1.284	1.383	1.231	1.435	1.176	1.490	1.120	1.546	1.066	1.607
40	1.295	1.392	1.243	1.442	1.190	1.496	1.135	1.550	1.082	1.609
45	1.344	1.431	1.298	1.475	1.250	1.523	1.202	1.571	1.154	1.623
50	1.385	1.463	1.343	1.503	1.300	1.546	1.257	1.589	1.214	1.635
55	1.420	1.491	1.381	1.528	1.343	1.566	1.303	1.605	1.264	1.646
60	1.449	1.515	1.415	1.549	1.379	1.583	1.343	1.619	1.307	1.656
65	1.475	1.536	1.443	1.567	1.410	1.599	1.377	1.632	1.344	1.666
70	1.498	1.555	1.468	1.583	1.438	1.613	1.407	1.643	1.377	1.675
75	1.518	1.571	1.491	1.598	1.463	1.626	1.434	1.654	1.405	1.583
80	1.537	1.586	1.511	1.611	1.484	1.637	1.457	1.663	1.431	1.691
85	1.553	1.600	1.529	1.623	1.504	1.648	1.479	1.673	1.453	1.698
90	1.568	1.612	1.545	1.634	1.522	1.658	1.498	1.681	1.474	1.705
95	1.582	1.624	1.560	1.645	1.538	1.666	1.515	1.688	1.493	1.711
100	1.594	1.634	1.573	1.654	1.552	1.675	1.531	1.695	1.510	1.717

n = number of observations, k' = number of explanatory variables (excluding the constant term)

References

[1] Anderson, T. W., "On the theory of testing serial correlation," *Skandinavisk Aktuarietidskrift* **31** (1948), 88–116.
[2] Anderson, T. W., *The Statistical Analysis of Time Series* (1971; Wiley, New York).
[3] Bassmann, R., "A generalized classical method of linear coefficients in a structural equation," *Econometrica* **25** (1957), 77–83.
[4] Bellman, Richard, *Introduction to Matrix Analysis* (1960; McGraw-Hill, New York).
[5] Beyer, W. H.(ed.), *Handbook of Tables for Probability and Statistics* (1960; The Chemical Rubber Co., Cleveland).
[6] Chung, K. L., *A Course in Probability Theory* (1968; Harcourt, Brace and World, New York).
[7] Cochrane, D., and G. H. Orcutt, "Applications of least squares to relations containing autocorrelated error terms," *Journal of the American Statistical Association* **44** (1949), 32–61.
[8] Dhrymes, Phoebus J., *Distributed Lags: Problems of Estimation and Formulation* (1971; Holden–Day, San Francisco).
[9] Dhrymes, Phoebus J., *Econometrics: Statistical Foundations and Applications* (1974; Springer–Verlag, New York).
[10] Dhrymes, Phoebus J., "Econometric models," in *Encyclopedia of Computer Science and Technology*, (J. Beizer, A. G. Holzman, and A. Kent, eds.), Vol. 8 (1977; Marcel Dekker, Inc., New York).
[11] Durbin, J., and G. S. Watson, "Testing for serial correlation in least squares regression, I," *Biometrika* **37** (1950), 408–428.
[12] Durbin, J., and G. S. Watson, "Testing for serial correlation in least squares regression, II," *Biometrika* **38** (1951), 159–178.
[13] Durbin, J., and G. S. Watson, "Testing for serial correlation in least squares regression, III," *Biometrika* **58** (1970), 1–19.
[14] Durbin, J., "Testing for serial correlation in least squares regressions when some of the regressors are lagged dependent variables," *Econometrica* **38** (1970), 410–421.
[15] Durbin, J., "An alternative to the bounds test for testing for serial correlation in least squares regression," *Econometrica* **38** (1970), 422–429.
[16] Fisher, F. M., *The Identification Problem in Econometrics* (1966; McGraw–Hill, New York).

[17] Geary, R. C., and C. E. V. Leser, "Significance Tests in multiple regression," *The American Statistician* **22** (1968), 20–21.

[18] Hadley, George, *Linear Algebra* (1961; Addison–Wesley, Reading, Mass.).

[19] Hoerl, A. E., "Optimum solution of many variables equations," *Chemical Engineering Progress* **55** (1959), 69–78.

[20] Hoerl, A. E., "Application of ridge analysis to regression problems," *Chemical Engineering Progress*, **58** (1962), 54–59.

[21] Hoerl, A. E., and R. W. Kennard, "Ridge regression: biased estimation of nonorthogonal problems," *Technometrics* **12** (1970), 55–69.

[22] Hoerl, A. E., and R. W. Kennard, "Ridge regression: applications to non-orthogonal problems," *Technometrics* **12** (1970), 69–82.

[23] Koopmans, T. C., and W. C. Hood: "The estimation of simultaneous linear economic relationships," in *Studies in Econometric Method* (W. C. Hood and T. C. Koopmans, eds.), Chapter 6. (1953; Wiley, New York).

[24] Marquardt, D. W., "Generalized inverses, ridge regression, biased linear and nonlinear estimation," *Technometrics* **12** (1970), 591–612.

[25] McDonald, G. C., and D. I. Galarneau: "A Monte Carlo evaluation of some ridge-type estimators," 1973; General Motors Research Laboratory Report, GMR-1322-B, Marven, Michigan).

[26] Newhouse, J. P., and S. D. Oman, "An evaluation of ridge estimators" Report No. R-716-PR, (1971; Rand Corporation, Santa Monica, California).

[27] Raifa, H. and R. Schiefer, *Applied Statistical Decision Theory* (1961; Harvard Business School, Boston).

[28] Sargan, J. D., "Wages and prices in the United Kingdom: a study in econometric methodology," in *Econometric Analysis for National Economic Planning* P. E. Hart *et al.*, eds) (1964; Butterworths, London).

[29] Scheffè, H., "A method for judging all contrasts in the analysis of variance," *Biometrika* **40** (1953), 87–104.

[30] Scheffè, H., *The Analysis of Variance* (1959; Wiley, New York).

[31] Scheffè, H., A note on a formulation of the S-method of multiple comparison," *Journal of the American Statistical Association* **72** (1977), 143–146. [See also in this issue R. A. Olshen's comments, together with Scheffè's rejoinders.]

[32] Theil, H., *Estimation and Simultaneous Correlation in Complete Equation Systems* (1953; Central Plan Bureau, The Hague).

[33] Vinod, H. D., "Ridge regression of signs and magnitudes of individual regression coefficients," (undated mimeograph).

[34] Vinod, H. D., "Generalization of the Durbin–Watson statistic for higher order autoregressive processes," *Communications in Statistics* **2** (1973), 115–144.

[35] Wald, A., "The fitting of straight lines if both variables are subject to error," *Annals of Mathematical Statistics* **11** (1940), 284–300.

[36] Zellner, Arnold, *Introduction to Bayesian Inference in Econometrics* (1971; Wiley, New York).

APPENDIX:

Mathematics for Econometrics

Vectors and Vector Spaces 1

In nearly all of the discussion to follow we shall deal with the set of *real* numbers. Occasionally, however, we shall deal with *complex* numbers as well. In order to avoid cumbersome repetition we shall denote the set we are dealing with by F and let the context elucidate whether we are speaking of real or complex numbers or both.

1.1 Complex Numbers

A *complex* number, say z, is denoted by

$$z = x + iy,$$

where, x and y are *real* numbers and the symbol i is defined by

$$i^2 = -1. \tag{1}$$

All other properties of the entity denoted by i are derivable from the basic definition in (1). For example,

$$i^4 = (i^2)(i^2) = (-1)(-1) = 1.$$

Similarly,

$$i^3 = (i^2)(i) = (-1)i = -i,$$

and so on.

It is important for the reader to grasp, and bear in mind, that a complex number is describable in terms of (an ordered) pair of real numbers.

Let

$$z_j = x_j + iy_j, \qquad j = 1, 2$$

be two complex numbers. We say

$$z_1 = z_2$$

if and only if

$$x_1 = x_2 \quad \text{and} \quad y_1 = y_2.$$

Operations with complex numbers are as follows.

Addition:

$$z_1 + z_2 = (x_1 + x_2) + i(y_1 + y_2).$$

Multiplication by a *real* scalar:

$$cz_1 = (cx_1) + i(cy_1).$$

Multiplication of two complex numbers:

$$z_1 z_2 = (x_1 x_2 - y_1 y_2) + i(x_1 y_2 + x_2 y_1).$$

Addition and multiplication are, evidently, associative and commutative, i.e., for complex $z_j, j = 1, 2, 3,$

$$z_1 + z_2 + z_3 = (z_1 + z_2) + z_3 \quad \text{and} \quad z_1 z_2 z_3 = (z_1 z_2) z_3,$$

$$z_1 + z_2 = z_2 + z_1 \quad \text{and} \quad z_1 z_2 = z_2 z_1,$$

and so on.

The *conjugate* of a complex number z is denoted by $\bar{z}$ and is defined by

$$\bar{z} = x - iy.$$

Associated with each complex number is its *modulus* or *length* or *absolute value*, which is a *real* number denoted by $|z|$ and defined by

$$|z| = (z\bar{z})^{1/2} = (x^2 + y^2)^{1/2}.$$

For the purpose of carrying out multiplication and division (an operation which we have not, as yet, defined) of complex numbers it is convenient to express them in polar form.

Polar Form Of Complex Numbers

Let z, a complex number, be represented in Figure 1 by the point (x_1, y_1)–its coordinates.

It is easily verified that the length of the line from the origin to the point (x_1, y_1) represents the modulus of z_1, which for convenience let us denote by r_1. Let the angle described by this line and the abscissa be denoted by θ_1.

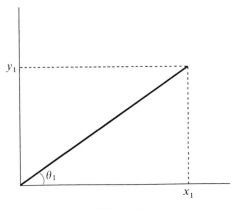

Figure 1

As is well known from elementary trigonometry we have

$$\cos \theta_1 = \frac{x_1}{r_1}, \qquad \sin \theta_1 = \frac{y_1}{r_1}. \tag{2}$$

Thus we may write the complex number as

$$z_1 = x_1 + iy_1 = r_1 \cos \theta_1 + ir_1 \sin \theta_1 = r_1(\cos \theta_1 + i \sin \theta_1).$$

Further, we may define the quantity

$$e^{i\theta_1} = \cos \theta_1 + i \sin \theta_1 \tag{3}$$

and thus write the complex number in the standard *polar form*

$$z_1 = r_1 e^{i\theta_1}. \tag{4}$$

In the representation above r_1 is the *modulus* and θ_1 the *argument* of the complex number z_1. It may be shown that the quantity $e^{i\theta_1}$ as defined in (3) has all the properties of real exponentials in so far as the operations of multiplication and division are concerned. If we confine the *argument* of a complex number to the range $[0, 2\pi)$ we have a unique correspondence between the (x, y) coordinates of a complex number and the modulus and argument needed to specify its polar form. Thus, for any complex number z the representations

$$z = x + iy, \qquad z = re^{i\theta},$$

where

$$r = (x^2 + y^2)^{1/2}, \qquad \cos \theta = \frac{x}{r}, \qquad \sin \theta = \frac{y}{r},$$

are completely equivalent.

Addition and division of complex numbers, in polar form, are extremely simple. Thus,

$$z_1 z_2 = (r_1 r_2)e^{i(\theta_1 + \theta_2)}$$

$$\frac{z_1}{z_2} = \left(\frac{r_1}{r_2}\right)e^{i(\theta_1 - \theta_2)},$$

provided $z_2 \neq 0$.

We may extend our discussion to *complex vectors*, i.e., ordered n-tuples of complex numbers. Thus

$$z = x + iy$$

where x and y are n-element (real) vectors (a concept to be defined immediately below) is a complex vector. As in the scalar case, two complex vectors z_1, z_2 are equal if and only if

$$x_1 = x_2, \qquad y_1 = y_2,$$

where now x_i, y_i, $i = 1, 2$, are n-element (column) vectors. The complex conjugate of the vector z is given by

$$\bar{z} = x - iy$$

and the modulus of the complex vector is defined by

$$(z'\bar{z})^{1/2} = [(x + iy)'(x - iy)]^{1/2} = (x'x + y'y)^{1/2},$$

the quantities $x'x, y'y$ being ordinary scalar products of two vectors. Addition and multiplication of complex vectors are defined by

$$z_1 + z_2 = (x_1 + x_2) + i(y_1 + y_2),$$
$$z_1' z_2 = (x_1' x_2 - y_1' y_2) + i(y_1' x_2 + x_1' y_2),$$
$$z_1 z_2' = (x_1 x_2' - y_1 y_2') + i(y_1 x_2' + x_1 y_2'),$$

where x_i, y_i, $i = 1, 2$, are real n-element column vectors. The notation, for example, x_1' or y_2' means that the vectors are written in *row form*, rather than the customary column form. Thus $x_1 x_2'$ is a matrix, while $x_1' x_2$ is a scalar. These concepts (vector, matrix) will be elucidated below. It is somewhat awkward to introduce them now; still, it is best to set forth what we need regarding complex numbers at the beginning.

1.2 Vectors

Let $a_i \in F, i = 1, 2, \ldots, n$; then the ordered n-tuple

$$a = \begin{pmatrix} a_1 \\ a_2 \\ \vdots \\ a_n \end{pmatrix}$$

is said to be an *n-dimensional vector*.

Notice that a scalar is a trivial case of a vector whose dimension is $n = 1$.

Customarily we write vectors as *columns*, so strictly speaking we should use the term *column vectors*. But this is cumbersome and will not be used unless required for clarity.

If the elements of a vector, a_i, $i = 1, 2, \ldots, n$, belong to F, we abbreviate this by writing

$$a \in F.$$

If a, b are two n-dimensional vectors and a, $b \in F$, we define their sum by

$$a + b = \begin{pmatrix} a_1 + b_1 \\ \vdots \\ a_n + b_n \end{pmatrix}.$$

If c is a scalar and belongs to F, then we define

$$ca = \begin{pmatrix} ca_1 \\ ca_2 \\ \vdots \\ ca_n \end{pmatrix}.$$

If a, b are two n-dimensional vectors with elements in F, their *inner product* (which is a scalar) is defined by

$$a'b = a_1 b_1 + a_2 b_2 + \cdots + a_n b_n.$$

The inner product of two vector is also called their *scalar product*.

Let $a_{(i)}$, $i = 1, 2, \ldots, k$, be n-dimensional vectors whose elements belong to F. Let c_i, $i = 1, 2, \ldots, k$, be scalars belonging to F. If

$$\sum_{i=1}^{k} c_i a_{(i)} = 0$$

implies that

$$c_i = 0, \qquad i = 1, 2, \ldots, k,$$

then the vectors $\{a_{(i)} : i = 1, 2, \ldots, k\}$ are said to be *linearly independent*, or to constitute a *linearly independent set*. If there exist scalars c_i, $i = 1, 2, \ldots, k$, not all of which are zero such that $\sum_{i=1}^{k} c_i a_{(i)} = 0$, then the vectors $\{a_{(i)} : i = 1, 2, \ldots, k\}$ are said to be *linearly dependent*, or to constitute a *linearly dependent set*.

Notice that *if a set of vectors is linearly dependent this means that one or more such vectors can be expressed as a linear combination of the remaining vectors. On the other hand if the set is linearly independent this is not possible.*

1.3 Vector Spaces

Let V_n be a set of n-dimensional vectors whose elements belong to F. Then V_n is said to be a vector space if and only if the following two conditions hold:

(i) for any scalar $c \in F$ and any vector $a \in V_n$,

$$ca \in V_n;$$

(ii) for any two vectors $a, b \in V_n$,

$$a + b \in V_n;$$

Evidently, the set of all n-dimensional vectors with elements in F is such a vector space, say S_n.
  Let

$$a_{(i)} \in S_n, \qquad i = 1, 2, \ldots, m, \; m \geq n.$$

If any vector in S_n, say b, can be written as

$$b = \sum_{i=1}^{m} c_i a_{(i)}$$

we say that the set $\{a_{(i)}: i = 1, 2, \ldots, m\}$ *spans* the vector space S_n.
  A *basis* for a vector space V_n is *a minimal set of linearly independent vectors that span V_n.*
  For the vector space S_n above it is evident that the set

$$\{e_{\cdot i}: i = 1, 2, \ldots, n\}$$

forms a *basis*, where $e_{\cdot i}$ is an n-dimensional vector all of whose elements are zero save the ith, which is unity. It is clear that if V_n is a vector space and

$$A = \{a_{(i)}: i = 1, 2, \ldots, m, \; a_{(i)} \in V_n\}$$

is a subset that spans V_n then there exists a subset of A that forms a basis for V_n. Moreover, if $\{a_{(i)}: i = 1, 2, \ldots, k, \, k < m\}$ is a linearly independent subset of A we can choose a basis that contains it. This is done by noting that since A spans V_n then, if it is linearly independent, it is a basis and we have the result. If it is not, then we simply eliminate some of its vectors that can be expressed as linear combinations of the remaining vectors. Since the remaining subset is linearly independent it can be made part of the basis.
  *A basis is not unique. But all bases for a given vector space contain the same number of vectors. This number is called the **dimension** of the vector space V_n and denoted by*

$$\dim(V_n).$$

Suppose $\dim(V_n) = n$. Then it may be shown that *any $n + i$ vectors in V_n are linearly dependent* for $i \geq 1$, and that no set containing less than n vectors can span V_n.

Let V_n be a vector space and P_n a subset of V_n, in the sense that if $b \in P_n$ then it is also true that $b \in V_n$. If P_n is also a vector space, then it is said to be a *subspace of* V_n, and all discussion regarding spanning or basis sets or dimension apply to P_n as well.

Finally, notice that if $\{a_{(i)} : i = 1, 2, \ldots, m\}$ is a basis for a vector space V_n, then every vector in V_n, say b, is uniquely expressible in terms of this basis. Thus, suppose we have two representations, say

$$b = \sum_{i=1}^{m} b_i^{(1)} a_{(i)} = \sum_{i=1}^{m} b_i^{(2)} a_{(i)},$$

where $b_i^{(1)}$, $b_i^{(2)}$ are appropriate sets of scalars. This implies

$$0 = \sum_{i=1}^{m} (b_i^{(1)} - b_i^{(2)}) a_{(i)}.$$

But a basis is a linearly independent set; hence we conclude

$$b_i^{(1)} = b_i^{(2)}, \qquad i = 1, 2, \ldots, m,$$

which shows uniqueness of representation.

2 Matrix Algebra

2.1 Basic Definitions

Definition 1. Let $a_{ij} \in F$, $i = 1, 2, \ldots, m$, $j = 1, 2, \ldots, n$. Then the ordered rectangular array

$$A = \begin{bmatrix} a_{11} & a_{12} & \cdots & a_{1n} \\ a_{21} & a_{22} & \cdots & a_{2n} \\ \vdots & \vdots & & \vdots \\ a_{m1} & a_{m2} & \cdots & a_{mn} \end{bmatrix} = [a_{ij}]$$

is said to be *a matrix of dimension $m \times n$.*

Remark 1. Note that the first subscript locates the *row* in which the typical element lies while the second subscript locates the *column*. For example a_{ks} denotes the element lying in the kth row and sth column of the matrix A. When writing a matrix we usually write down its typical element as well as its dimension. Thus

$$A = (a_{ij}), \qquad i = 1, 2, \ldots, m, j = 1, 2, \ldots, n,$$

denotes a matrix whose typical element is a_{ij} and which has m rows and n columns.

Convention 1. Occasionally we shall have reason to refer to the columns or rows of the matrix, individually. If A is a matrix we shall denote its jth column by $a_{.j}$, i.e.,

$$a_{.j} = \begin{pmatrix} a_{1j} \\ a_{2j} \\ \vdots \\ a_{mj} \end{pmatrix}$$

and its ith row by

$$a_{i\cdot} = [a_{i1}, a_{i2}, \ldots, a_{in}].$$

Definition 2. Let A be a matrix as in Definition 1. Its transpose, denoted by A', is defined to be the $n \times m$ matrix

$$A' = [a_{ji}], \quad j = 1, 2, \ldots, n, i = 1, 2, \ldots, m,$$

i.e., it is obtained by interchanging rows and columns.

Definition 3. Let A be as in Definition 1. If $m = n$, A is said to be a *square matrix*.

Definition 4. If A is a square matrix, it is said to be *symmetric* if and only if

$$A' = A$$

If A is a square matrix with, say, n rows and n columns, it is said to be a *diagonal matrix* if

$$a_{ij} = 0, \quad i \neq j;$$

In this case it is denoted by

$$A = \operatorname{diag}(a_{11}, a_{22}, \ldots, a_{nn}).$$

Remark 2. If A is square matrix then, evidently, it is not necessary to refer to the number of its rows and columns separately. If it has, say, n rows and n columns, we say that A is of *dimension* (or *order*) n.

Definition 5. Let A be a square matrix of order n. It is said to be an *upper triangular matrix* if

$$a_{ij} = 0, \quad i > j.$$

It is said to be a *lower triangular matrix* if

$$a_{ij} = 0, \quad i < j.$$

Remark 3. As the terms imply, for a *lower triangular matrix* all elements *above the main diagonal must be zero*, while for an *upper triangular* matrix all elements *below the main diagonal must be zero*.

Definition 6. The *identity* matrix of order n, denoted by I_n (the subscript is typically omitted), is a diagonal matrix all of whose nonnull elements are unity.

Definition 7. The *null* matrix of dimension $m \times n$ is a matrix all of whose elements are null (zeros).

Definition 8. Let A be a square matrix of order n. It is said to be an *idempotent matrix* if and only if

$$AA = A.$$

2.2 Basic Operations

Let A, B be two $m \times n$ matrices with elements in F, and let c be a scalar in F. Then we have:

(i) Scalar multiplication:

$$cA = [ca_{ij}].$$

(ii) Matrix addition:

$$A + B = [a_{ij} + b_{ij}].$$

Remark 4. Note that while scalar multiplication is defined for every matrix, matrix addition for A and B is *not* defined unless both have the *same dimensions*.

Let A be $m \times n$ and B be $q \times r$, both with elements in F; then we have:

(iii) Matrix multiplication:

$$AB = \left[\sum_{s=1}^{n} a_{is} b_{sj} \right] \quad provided \ n = q;$$

$$BA = \left[\sum_{k=1}^{r} b_{ik} a_{kj} \right] \quad provided \ r = m.$$

Remark 5. Notice that matrix multiplication is not defined for any two matrices A, B. They must satisfy certain conditions of dimensional conformability. Notice further that if the product

$$AB$$

is defined, the product

$$BA$$

need not be defined, and if it is, it is *not true that*

$$AB = BA$$

except in highly special cases.

Remark 6. If two matrices are such that a given operation between them is defined we shall say that they are *conformable with respect to that operation.* Thus, for example, if A is $m \times n$ and B is $n \times r$ we shall say that A and B are *conformable with respect to the operation of multiplying A on the right* by B. If A is $m \times n$ and B is $q \times m$ we shall say that A and B are *conformable with respect to the operation of multiplying A on the left by B.* Or if A and B are both $m \times n$ we shall say that A and B are *conformable with respect to matrix addition.* Since being precise is rather cumbersome, we often merely say that two matrices are conformable and we let the context define precisely the sense in which conformability is to be understood.

An immediate consequence of the definitions above is

Proposition 1. *Let A be $m \times n$, B be $n \times r$. The jth column of*

$$C = AB$$

is given by

$$c_{\cdot j} = \sum_{s=1}^{n} a_{\cdot s} b_{sj}, \qquad j = 1, 2, \ldots, r.$$

PROOF. Obvious from the definition of matrix multiplication.   q.e.d.

Proposition 2. *Let A be $m \times n$, B be $n \times r$. The ith row of*

$$C = AB$$

is given by

$$c_{i\cdot} = \sum_{q=1}^{n} a_{iq} b_{q\cdot}, \qquad i = 1, 2, \ldots, m.$$

PROOF. Obvious from the definition of matrix multiplication.   q.e.d.

Proposition 3. *Let A, B be $m \times n$, $n \times r$ respectively. Then*

$$C' = B'A',$$

where

$$C = AB.$$

PROOF. The typical element of C is given by

$$c_{ij} = \sum_{s=1}^{n} a_{is} b_{sj}.$$

By definition, the typical element of C', say c'_{ij}, is given by

$$c'_{ij} = c_{ji} = \sum_{s=1}^{n} a_{js} b_{si}.$$

But

$$a_{js} = a'_{sj}, \qquad b_{si} = b'_{is},$$

i.e., a_{js} is the (s, j) element of A', say a'_{sj}, and b_{si} is the (i, s) element of B', say b'_{is}. Consequently,

$$c'_{ij} = c_{ji} = \sum_{s=1}^{n} a_{js} b_{si} = \sum_{s=1}^{n} b'_{is} a'_{sj},$$

which shows that the (i, j) element of C' is the (i, j) element of $B'A'$.   q.e.d.

2.3 Rank and Inverse of a Matrix

Definition 9. Let A be $m \times n$. The *column rank* of A is the maximum number of linearly independent columns it contains. The *row rank* of A is the maximal number of linearly independent rows it contains.

Remark 7. It may be shown—but not here—that *the row rank of A is equal to its column rank.* Hence the concept of rank is unambiguous and we denote by

$$r(A)$$

the rank of A. Thus, if we are told that A is $m \times n$ we can immediately conclude that

$$r(A) \leq \min(m, n).$$

Definition 10. Let A be $m \times n, m \leq n$. We say that A is of *full rank* if and only if

$$r(A) = m.$$

Definition 11. Let A be a square matrix of order m. We say that A is non-singular if

$$r(A) = m.$$

Remark 8. An example of a nonsingular matrix is the diagonal matrix

$$A = \text{diag}(a_{11}, a_{22}, \ldots, a_{mm})$$

for which

$$a_{ii} \neq 0, \qquad i = 1, 2, \ldots, m.$$

We are now in a position to define a matrix operation that corresponds to that of division for scalars. Thus, for example, if $c \in F$ and $c \neq 0$, we know that for any $a \in F$

$$\frac{a}{c}$$

means the operation of defining

$$\frac{1}{c}$$

(the "inverse" of c) and multiplying that by a. The "inverse" of a scalar, say c, is another scalar, say b, such that

$$bc = cb = 1.$$

We have a similar operation for square matrices.

Matrix Inversion

Let A be a square matrix of order m. Its inverse, B, if it exists, is defined by the property

$$AB = BA = I,$$

where I is the identity matrix of order m.

Definition 12. Let A be a square matrix of order m. If its inverse exists, it is denoted by A^{-1}, and the matrix A is said to be *invertible*.

Remark 9. The terms *invertible, nonsingular,* and of full rank are synonymous for square matrices. This is made clear below.

Proposition 4. *Let A be a square matrix of order m. Then A is invertible if and only if*

$$r(A) = m.$$

PROOF. Suppose A is invertible; then there exists a square matrix B (of order m) such that

$$AB = I. \tag{5}$$

Let $c \neq 0$ be any m-element vector and note that (5) implies

$$ABc = c.$$

Since $c \neq 0$ we must have that

$$Ad = c, \qquad d = Bc \neq 0.$$

But this means that if c is any m-dimensional vector it can be expressed as a linear combination of the columns of A, which in turn means that the columns of A span the vector space S_m consisting of all m-dimensional vectors with elements in F. Since the dimension of this space is m it follows that the (m) columns of A are linearly independent; hence its rank is m.

Conversely, suppose that

$$\text{rank}(A) = m.$$

Then its columns form a basis for S_m. The vectors $\{e_{\cdot i} : i = 1, 2, \ldots, m\}$ such that $e_{\cdot i}$ is m-dimensional with all elements zero save the ith, which is unity, all belong to S_m. Thus we can write

$$e_{\cdot i} = Ab_{\cdot i} = \sum_{s=1}^{m} a_{\cdot s} b_{si}, \qquad i = 1, 2, \ldots, m.$$

The matrix

$$B = [b_{si}]$$

has the property[1]

$$AB = I. \quad \text{q.e.d.}$$

Corollary 1. *Let A be a square matrix of order m. If A is invertible then the following hold for its inverse B: B is of rank m; thus B also is invertible; the inverse of B is A.*

PROOF. Obvious from the definition of the inverse and the proposition. q.e.d.

It is useful here to introduce

Definition 13. Let A be $m \times n$. The *column space* of A, denoted by $C(A)$, is the set of vectors

$$C(A) = \{\xi : \xi = Ax\},$$

where x is n-dimensional with elements in F. Similarly, the *row space* of A, $R(A)$, is the set of (row) vectors

$$R(A) = \{\xi : \xi = yA\},$$

where y is a row vector of dimension m with elements in F.

Remark 10. It is clear that the column space of A is a vector space and that it is *spanned* by the columns of A. Moreover the dimension of this vector space is simply the rank of A. Similarly, the row space of A is a vector space spanned by its rows and the dimension of this space is also equal to the rank of A since the row rank of A is equal to its column rank.

Definition 14. Let A be $m \times n$. The (column) *null space* of A, denoted by $N(A)$, is the set

$$N(A) = \{x : Ax = 0\}.$$

---

[1] Strictly speaking we should also provide an argument based on the rows of A and on $BA = I$, but this is repetitious and is omitted for the sake of simplicity.

Remark 11. A similar definition can be made for the (row) null space of A.

Definition 15. Let A be $m \times n$ and consider its null space $N(A)$. This is a vector space; its dimension is termed the *nullity* of A and is denoted by

$$n(A).$$

We now have the important

Proposition 5. *Let A be $p \times q$. Then*

$$r(A) + n(A) = q.$$

PROOF. Suppose the nullity of A is $n(A) = n \leq q$, and let $\{\xi_i : i = 1, 2, \ldots, n\}$ be a basis for $N(A)$. Note that each ξ_i is a q-dimensional (column) vector with elements in F. We may extend this to a basis for S_q, the vector space of all q-dimensional vectors with elements in F; thus, let

$$\{\xi_1, \xi_2, \ldots, \xi_n, \zeta_1, \zeta_2, \ldots, \zeta_{q-n}\}$$

be such a basis. If x is any q-dimensional vector we can write, uniquely,

$$x = \sum_{i=1}^{n} c_i \xi_i + \sum_{j=1}^{q-n} f_j \zeta_j.$$

Now,

$$y = Ax \in C(A)$$

and, moreover,

$$y = \sum_{i=1}^{n} c_i A\xi_i + \sum_{j=1}^{q-n} f_j A\zeta_j = \sum_{j=1}^{q-n} f_j(A\zeta_j). \tag{6}$$

This is so since

$$A\xi_i = 0, \qquad i = 1, 2, \ldots, n,$$

owing to the fact that the ξ's are a basis for the null space of A.

But (6) means that the vectors

$$\{A\zeta_j : j = 1, 2, \ldots, q - n\}$$

span $C(A)$, since x, and hence y, is arbitrary. We claim that these vectors are linearly independent, and hence a basis for $C(A)$. For suppose not. Then there exist scalars, $g_j, j = 1, 2, \ldots, q - n$, not all of which are zero, such that

$$0 = \sum_{j=1}^{q-n} (A\zeta_j)g_j = A\left(\sum_{j=1}^{q-n} \zeta_j g_j\right). \tag{7}$$

From (7) we conclude that

$$\zeta = \sum_{j=1}^{q-n} \zeta_j g_j \tag{8}$$

lies in the null space of A since (7) means $A\zeta = 0$. As such $\zeta \in S_q$ and has a unique representation in terms of the basis of that vector space, say

$$\zeta = \sum_{i=1}^{n} d_i \xi_i + \sum_{j=1}^{q-n} k_j \zeta_j. \tag{9}$$

Moreover, since $\zeta \in N(A)$, we know that in (9)

$$k_j = 0, \qquad j = 1, 2, \ldots, q - n.$$

But (9) and (8) are incompatible in terms of the unique representation of ζ in the basis for S_q unless

$$g_j = 0, \qquad j = 1, 2, \ldots, q - n$$
$$d_i = 0, \qquad i = 1, 2, \ldots, n$$

This shows that (7) can be satisfied only by null $g_j, j = 1, 2, \ldots, q - n$; hence, the set $\{A\zeta_j : j = 1, 2, \ldots, q - n\}$ is linearly independent and, consequently, a basis for $C(A)$. Therefore, the dimension of $C(A)$, and thus the rank of A, obey

$$\dim[C(A)] = r(A) = q - n. \quad \text{q.e.d.}$$

A further useful result is the following.

Proposition 6. *Let A be $p \times q$, let B be a nonsingular matrix of order q, and put*

$$D = AB.$$

Then

$$r(D) = r(A).$$

PROOF. We shall show that

$$C(A) = C(D),$$

which is equivalent to the claim of the proposition. Suppose $y \in C(A)$. Then there exists a vector $x \in S_q$ such that $y = Ax$. Since B is nonsingular define the vector $\xi = B^{-1}x$. We note $D\xi = ABB^{-1}x = y$ *which shows that*

$$C(A) \subset C(D). \tag{10}$$

Conversely, suppose $z \in C(D)$. This means there exists a vector $\xi \in S_q$ such that $z = D\xi$. Define the vector $x = B\xi$ and note that

$$Ax = AB\xi = D\xi = z$$

so that $z \in C(A)$, which shows

$$C(D) \subset C(A). \tag{11}$$

But (10) and (11) together imply

$$C(A) = C(D). \quad \text{q.e.d.}$$

Finally we have

Proposition 7. *Let A be p × q and B q × r, and put*

$$D = AB.$$

Then

$$r(D) \leq \min[r(A), r(B)].$$

PROOF. Since $D = AB$ we note that if $x \in N(B)$ then $x \in N(D)$, whence we conclude

$$N(B) \subset N(D),$$

and thus that

$$n(B) \leq n(D). \tag{12}$$

But from

$$r(D) + n(D) = r,$$

$$r(B) + n(B) = r,$$

we find in view of (12)

$$r(D) \leq r(B). \tag{13}$$

Next suppose that $y \in C(D)$. This means that there exists a vector, say, $x \in S_r$, such that $y = Dx$ or $y = ABx = A(Bx)$ so that $y \in C(A)$. But this means that

$$C(D) \subset C(A),$$

or that

$$r(D) \leq r(A) \tag{14}$$

Together (13) and (14) imply

$$r(D) \leq \min[r(A), r(B)]. \quad \text{q.e.d.}$$

Remark 12. The preceding results can be stated in the following useful form: multiplying two (and therefore any finite number of) matrices results in a matrix whose rank cannot exceed the rank of the lowest ranked factor. The product of nonsingular matrices is nonsingular. Multiplying a matrix by a nonsingular matrix does not change its rank.

2.4 Hermite Forms and Rank Factorization

We begin with a few elementary aspects of matrix operations.

Definition 16. Let A be $m \times n$; any one of the following operations is said to be an *elementary transformation* of A:

  (i) interchanging two rows (or columns);
 (ii) multiplying the elements of a row (or column) by a (nonzero) scalar c;
(iii) multiplying the elements of a row (or column) by a (nonzero) scalar c
     and adding the result to another row (or column).

The operations above are said to be *elementary row (or column) operations*.

    Remark 13. The matrix performing operation (i) is the matrix obtained from the identity matrix by interchanging the two rows (or columns) in question; the matrix performing operation (ii) is obtained from the identity matrix by multiplying the corresponding row (or column) by the scalar c; finally, the matrix performing operation (iii) is obtained from the identity matrix by inserting in the row (column) corresponding to the row (column) we wish to add something to the scalar c in the position corresponding to the row (column) we wish to add to the given row (column). The matrices above are termed *elementary matrices*. An elementary *row* operation is performed on A by multiplying the corresponding elementary matrix, E, on the right by A, i.e.,

$$EA.$$

An elementary *column* operation is performed by

$$AE.$$

EXAMPLE 1. Let

$$A = \begin{bmatrix} a_{11} & a_{12} & a_{13} \\ a_{21} & a_{22} & a_{23} \\ a_{31} & a_{32} & a_{33} \end{bmatrix}$$

and suppose we want to interchange the position of the first and third rows (columns). Define

$$E_1 = \begin{bmatrix} 0 & 0 & 1 \\ 0 & 1 & 0 \\ 1 & 0 & 0 \end{bmatrix}.$$

Then

$$E_1 A = \begin{bmatrix} a_{31} & a_{32} & a_{33} \\ a_{21} & a_{22} & a_{23} \\ a_{11} & a_{12} & a_{13} \end{bmatrix}, \qquad AE_1 = \begin{bmatrix} a_{13} & a_{12} & a_{11} \\ a_{23} & a_{22} & a_{21} \\ a_{33} & a_{32} & a_{31} \end{bmatrix}.$$

Suppose we wish to multiply the second row (column) of A by the scalar c. Define

$$E_2 = \begin{bmatrix} 1 & 0 & 0 \\ 0 & c & 0 \\ 0 & 0 & 1 \end{bmatrix}.$$

Then

$$E_2 A = \begin{bmatrix} a_{11} & a_{12} & a_{13} \\ ca_{21} & ca_{22} & ca_{23} \\ a_{31} & a_{32} & a_{33} \end{bmatrix}, \qquad AE_2 = \begin{bmatrix} a_{11} & ca_{12} & a_{13} \\ a_{21} & ca_{22} & a_{23} \\ a_{31} & ca_{32} & a_{33} \end{bmatrix}$$

Finally, suppose we wish to add c times the first row (column) to the third row (column). Define

$$E_3 = \begin{bmatrix} 1 & 0 & 0 \\ 0 & 1 & 0 \\ c & 0 & 1 \end{bmatrix}$$

and note

$$E_3 A = \begin{bmatrix} a_{11} & a_{12} & a_{13} \\ a_{21} & a_{22} & a_{23} \\ ca_{11} + a_{31} & ca_{12} + a_{32} & ca_{13} + a_{33} \end{bmatrix}$$

$$AE_3 = \begin{bmatrix} a_{11} & a_{12} & ca_{11} + a_{13} \\ a_{21} & a_{22} & ca_{21} + a_{23} \\ a_{31} & a_{32} & ca_{31} + a_{33} \end{bmatrix}.$$

We have an immediate result.

Proposition 8. *Every matrix of elementary transformations is nonsingular and its inverse is a matrix of the same type.*

PROOF. For matrices of type E_1 it is clear that $E_1 E_1 = I$. The inverse of a matrix of type E_2 is of the same form but with c replaced by $1/c$. Similarly, the inverse of a matrix of type E_3 is of the same form but with c replaced by $-c$. q.e.d.

Definition 17. An $m \times n$ matrix C is said to be an (*upper*) *echelon matrix*, if

(i) it can be partitioned:

$$C = \begin{pmatrix} C_1 \\ 0 \end{pmatrix},$$

where C_1 is $r \times n$ $(r \leq n)$ and there is no row in C_1 consisting entirely of zeros,

(ii) the first nonzero element appearing in each row of C_1 is unity and, if the first nonzero element in row i is c_{ij}, then all other elements in column j are zero,

(iii) when the first nonzero element in the kth row of C_1 is c_{kj_k}, then $j_1 < j_2 < j_3 < \cdots < j_r$.

An immediate consequence of the definition is

Proposition 9. *Let A be $m \times n$; there exists a nonsingular $(m \times m)$ matrix B such that*

$$BA = C$$

and C is an (upper) echelon matrix.

PROOF. Consider the first column of A and suppose it contains a nonzero element (if not, consider the second column, etc.). Without loss of generality we suppose this to be a_{11} (if not, simply interchange rows so that it does become the first element). Multiply the first row by $1/a_{11}$. This is accomplished through multiplication on the left by a matrix of type E_2. Next multiply the first row of the resulting matrix by $-a_{s1}$ and add to the sth row. This is accomplished through multiplication on the left by a matrix of type E_3. Continuing in this fashion we make all elements in the first column zero save the first, which is unity. Repeat this operation for all other columns of A. In the end some rows may consist entirely of zeros. If they do not occur at the end interchange rows so that all zero rows occur at the end. This is done through multiplication on the left by a matrix of type E_1. The resulting matrix is, thus, in upper echelon form and has been obtained through multiplication on the left by a number of elementary matrices. Since the latter are nonsingular we have

$$BA = C,$$

where B is nonsingular and C is in upper echelon form.   q.e.d.

Proposition 10. *Let A be $m \times n$ and suppose it can be reduced to an (upper) echelon matrix*

$$BA = \begin{pmatrix} C_1 \\ 0 \end{pmatrix} = C$$

such that C_1 is $r \times n$. Then

$$\operatorname{rank}(A) = r.$$

PROOF. By construction, the rows of C_1 are linearly independent; thus,

$$\operatorname{rank}(C_1) = \operatorname{rank}(C) = r$$

and

$$\operatorname{rank}(C) \le \operatorname{rank}(A).$$

We also have

$$A = B^{-1}C.$$

Hence

$$\text{rank}(A) \leq \text{rank}(C)$$

which shows

$$\text{rank}(A) = \text{rank}(C) = r. \quad \text{q.e.d.}$$

Definition 18. An $n \times n$ matrix H^* is said to be in (*upper*) *Hermite form* if and only if:

(i) H^* is (upper) triangular;
(ii) the elements along the main diagonal of H^* are either zero or one;
(iii) if a main diagonal element of H^* is *zero*, then all elements in the *row* in which the element occurs are zero;
(iv) if a main diagonal element of H^* is unity then all other elements *in the column* in which the element occurs are zero.

Definition 19. An $n \times m$ matrix H is said to be in (*upper*) *Hermite canonical form* if

$$H = \begin{bmatrix} I & H_1 \\ 0 & 0 \end{bmatrix}.$$

Proposition 11. *Every matrix in Hermite form can be put in Hermite canonical form by elementary row and column operations.*

PROOF. Let H^* be a matrix in Hermite form; by interchanging rows we can put all zero rows at the end so that for some nonsingular matrix B_1 we have

$$B_1 H^* = \begin{pmatrix} H_1^* \\ 0 \end{pmatrix},$$

where the first nonzero element in each row of H_1^* is unity and H_1^* contains no zero rows. By interchanging columns we can place the (unit) first nonzero elements of the rows of H_1^* along the main diagonal, so that there exists a nonsingular matrix B_2 for which

$$B_1 H^* B_2 = \begin{bmatrix} I & H_1 \\ 0 & 0 \end{bmatrix}. \quad \text{q.e.d.}$$

Proposition 12. *The rank of a matrix H^*, in Hermite form, is equal to the dimension of the identity block in its Hermite canonical form.*

PROOF. Obvious.   q.e.d.

Proposition 13. *Every matrix H^* in Hermite form is idempotent (although it is obviously not necessarily symmetric).*

PROOF. We have to show that

$$H^* H^* = H^*.$$

Since H^* is *upper triangular* we know that H^*H^* is *also upper triangular*, and thus we need determine its (i, j) element for $i \leq j$. Now, the (i, j) element of H^*H^* is

$$(H^*H^*)_{ij} = \sum_{k=1}^{n} h_{ik}^* h_{kj}^* = \sum_{k=i}^{j} h_{ik}^* h_{kj}^*.$$

If $h_{ii}^* = 0$ then $h_{ij}^* = 0$ for all j; hence $(H^*H^*)_{ij} = 0$ for all j. If $h_{ii}^* = 1$ then

$$(H^*H^*)_{ij} = h_{ij}^* + h_{i,i+1}^* h_{i+1,j}^* + \cdots + h_{ij}^* h_{jj}^*.$$

Now $h_{i+1,i+1}^*$ is either zero or one; if zero then $h_{i+1,j}^* = 0$ for all j and hence the second term on the right side in the equation above is zero. If $h_{i+1,i+1}^* = 1$ then $h_{i,i+1}^* = 0$ so that again the second term is null. Similarly, if $h_{i+2,i+2}^* = 0$, then the third term is null; if $h_{i+2,i+2}^* = 1$ then $h_{i,i+2}^* = 0$ so that again the third term on the right side of the equation defining $(H^*H^*)_{ij}$ is zero. Finally, if $h_{jj}^* = 1$ then $h_{ij}^* = 0$, and if $h_{jj}^* = 0$ then again $h_{ij}^* = 0$. Consequently, it is always the case that

$$(H^*H^*)_{ij} = h_{ij}^*$$

and thus

$$H^*H^* = H^*. \quad \text{q.e.d.}$$

Rank Factorization

Proposition 14. *Let A be $n \times n$ of rank $r \leq n$. There exist nonsingular matrices Q_1, Q_2 such that*

$$Q_1^{-1}AQ_2^{-1} = \begin{bmatrix} I_r & 0 \\ 0 & 0 \end{bmatrix}.$$

PROOF. By Proposition 9 there exists a nonsingular matrix Q_1 such that

$$Q_1^{-1}A = \begin{pmatrix} A_1^* \\ 0 \end{pmatrix},$$

i.e., $Q_1^{-1}A$ is an (upper) echelon matrix. By transposition of columns we obtain

$$\begin{pmatrix} A_1^* \\ 0 \end{pmatrix} B_1 = \begin{bmatrix} I_r & A_1 \\ 0 & 0 \end{bmatrix}.$$

By elementary column operations we can eliminate A_1, i.e.,

$$\begin{bmatrix} I_r & A_1 \\ 0 & 0 \end{bmatrix} B_2 = \begin{bmatrix} I_r & 0 \\ 0 & 0 \end{bmatrix}.$$

Take $Q_2^{-1} = B_1 B_2$ and note that we have

$$Q_1^{-1}AQ_2^{-1} = \begin{bmatrix} I_r & 0 \\ 0 & 0 \end{bmatrix}. \quad \text{q.e.d.}$$

Proposition 15. *Let A be $m \times n$ ($m \leq n$) of rank $r \leq m$. There exists an $m \times r$ matrix C_1 of rank r and an $r \times n$ matrix C_2 of rank r such that*

$$A = C_1 C_2.$$

PROOF. Let

$$A_0 = \binom{A}{0},$$

A_0 is $n \times n$ of rank r. By Proposition 14 there exist nonsingular matrices Q_1, Q_2 such that

$$A_0 = Q_1 \begin{bmatrix} I_r & 0 \\ 0 & 0 \end{bmatrix} Q_2.$$

Partition

$$Q_1 = \begin{bmatrix} C_1 & C_{11} \\ C_{21} & C_{22} \end{bmatrix}, \qquad Q_2 = \binom{C_2}{C*}$$

so that C_1 is $m \times r$ and C_2 is $r \times n$ (of rank r). Thus

$$A_0 = \binom{C_1 C_2}{C_{21} C*},$$

whence

$$A = C_1 C_2.$$

Since

$$r = r(A) \leq \min[r(C_1), r(C_2)] = \min[r(C_1), r]$$

we must have that

$$r(C_1) = r. \quad \text{q.e.d.}$$

Remark 14. Proposition 15 is the so called *rank factorization theorem*.

2.5 Trace and Determinants

Associated with square matrices are two important scalar functions, the *trace* and the *determinant*.

Definition 20. Let A be a square matrix of order m. Its trace is denoted by $\text{tr}(A)$ and is defined by

$$\text{tr}(A) = \sum_{i=1}^{m} a_{ii}.$$

An immediate consequence of the definition is

Proposition 16. *Let A, B be two square matrices of order m. Then*

$$\mathrm{tr}(A + B) = \mathrm{tr}(A) + \mathrm{tr}(B),$$

$$\mathrm{tr}(AB) = \mathrm{tr}(BA).$$

PROOF. By definition the typical element of $A + B$ is

$$a_{ij} + b_{ij}.$$

Hence

$$\mathrm{tr}(A + B) = \sum_{i=1}^{m} (a_{ii} + b_{ii}) = \sum_{i=1}^{m} a_{ii} + \sum_{i=1}^{m} b_{ii} = \mathrm{tr}(A) + \mathrm{tr}(B).$$

Similarly, the typical element of AB is

$$\sum_{k=1}^{m} a_{ik} b_{kj}.$$

Hence

$$\mathrm{tr}(AB) = \sum_{i=1}^{m} \sum_{k=1}^{m} a_{ik} b_{ki}.$$

The typical element of BA is

$$\sum_{i=1}^{m} b_{ki} a_{ij}.$$

Hence

$$\mathrm{tr}(BA) = \sum_{k=1}^{m} \sum_{i=1}^{m} b_{ki} a_{ik} = \sum_{i=1}^{m} \sum_{k=1}^{m} a_{ik} b_{ki},$$

which shows

$$\mathrm{tr}(AB) = \mathrm{tr}(BA). \quad \text{q.e.d.}$$

Definition 21. Let A be a square matrix of order m; its *determinant*, denoted by $|A|$ or by det A, is given by

$$|A| = \sum (-1)^{s} a_{1j_{1}} a_{2j_{2}} \cdots a_{mj_{m}},$$

where: $j_1, j_2, \ldots, j_m$ is an arrangement (permutation) of the numbers $1, 2, \ldots, m$, and s is zero or one depending on whether the number of transpositions required to restore $j_1, j_2, \ldots, j_m$ to the natural sequence $1, 2, 3, \ldots, m$ is even or odd; the sum is taken over all possible such arrangements.

EXAMPLE 2. Consider the matrix

$$A = \begin{bmatrix} a_{11} & a_{12} & a_{13} \\ a_{21} & a_{22} & a_{23} \\ a_{31} & a_{32} & a_{33} \end{bmatrix}.$$

According to the definition its determinant is given by

$$\begin{aligned} |A| = &(-1)^{s_1}a_{11}a_{22}a_{33} + (-1)^{s_2}a_{11}a_{23}a_{32} \\ &+ (-1)^{s_3}a_{12}a_{21}a_{33} + (-1)^{s_4}a_{12}a_{23}a_{31} \\ &+ (-1)^{s_5}a_{13}a_{21}a_{32} + (-1)^{s_6}a_{13}a_{22}a_{31}. \end{aligned}$$

To determine s_1 we note that the second subscripts in the corresponding term are in natural order; hence $s_1 = 0$. For the second term we note that one transposition restores the second subscripts to the natural order; hence $s_2 = 1$. For the third term $s_3 = 1$. For the fourth term two transpositions are required; hence $s_4 = 0$. For the fifth term two transpositions are required; hence $s_5 = 0$. For the sixth term one transposition is required; hence $s_6 = 1$.

Remark 15. It should be noted that although Definition 21 is stated with the rows in natural order, a completely equivalent definition is one in which the columns are in natural order. Thus, for example, we could just as well have defined

$$|A| = \sum (-1)^d a_{i_1 1} a_{i_2 2} \cdots a_{i_m m}$$

where d is zero or one according as the number of transpositions required to restore $i_1, i_2, \ldots, i_m$ to the natural order $1, 2, 3, \ldots, m$ is even or odd.

EXAMPLE 3. Consider the matrix A of Example 2 and obtain the determinant in accordance with Remark 15. Thus

$$\begin{aligned} |A| = &(-1)^{d_1}a_{11}a_{22}a_{33} + (-1)^{d_2}a_{11}a_{32}a_{23} \\ &+ (-1)^{d_3}a_{21}a_{12}a_{33} + (-1)^{d_4}a_{21}a_{32}a_{13} \\ &+ (-1)^{d_5}a_{31}a_{12}a_{23} + (-1)^{d_6}a_{31}a_{22}a_{13}. \end{aligned}$$

It is easily determined that $d_1 = 0, d_2 = 1, d_3 = 1, d_4 = 0, d_5 = 0, d_6 = 1$. Noting, in comparison with Example 2, that $s_1 = d_1, s_2 = d_2, s_3 = d_3, s_4 = d_5, s_5 = d_4, s_6 = d_6$, we see that we have exactly the same terms.

An immediate consequence of the definition is

Proposition 17. *Let A be a square matrix of order m. Then*

$$|A'| = |A|.$$

PROOF. Obvious from Definition 21 and Remark 15   q.e.d.

Proposition 18. *Let A be a square matrix of order m and consider the matrix B that is obtained by interchanging the kth and rth rows of A ($k \leq r$). Then*

$$|B| = -|A|.$$

PROOF. By definition,

$$\begin{aligned} |B| &= \sum (-1)^s b_{1j_1} b_{2j_2} \cdots b_{mj_m}, \\ |A| &= \sum (-1)^s a_{1j_1} a_{2j_2} \cdots a_{mj_m}. \end{aligned} \tag{15}$$

However, each term in $|B|$ (except possibly for sign) is exactly the same as in $|A|$ but for the interchange of the kth and rth rows. Thus, for example, we can write

$$|B| = \sum (-1)^s a_{1j_1} \cdots a_{k-1\,j_{k-1}} a_{rj_k} a_{k+1\,j_{k+1}}$$
$$\cdots a_{r-1\,j_{r-1}} a_{kj_r} a_{r+1\,j_{r+1}} \cdots a_{mj_m}. \tag{16}$$

Now if we restore the first subscripts in (16) to their natural order we will have an expression like the one for $|A|$ in (15), except that in (16) we would require an odd number of additional transpositions to restore the second subscripts to the natural order $1, 2, \ldots, m$. Hence, the sign of each term in (16) is exactly the opposite of the corresponding term in (15). Hence

$$|B| = -|A|. \quad \text{q.e.d.}$$

Proposition 19. *Let A be a square matrix of order m and suppose it has two identical rows. Then*

$$|A| = 0.$$

PROOF. Let B be the matrix obtained by interchanging the two identical rows. Then by Proposition 18,

$$|B| = -|A|. \tag{17}$$

Since these two rows are identical $B = A$, and thus

$$|B| = |A|. \tag{18}$$

But (17) and (18) imply

$$|A| = 0. \quad \text{q.e.d.}$$

Proposition 20. *Let A be a square matrix of order m, and suppose all elements in its ith row are zero. Then*

$$|A| = 0.$$

PROOF. By the definition of a determinant we have

$$|A| = \sum (-1)^s a_{1j_1} a_{2j_2} a_{3j_3} \cdots a_{mj_m},$$

and it is clear that every term above contains an element from the ith row, viz., a_{ij_i}. Hence all terms vanish and thus

$$|A| = 0. \quad \text{q.e.d.}$$

Remark 16. It is clear that in any of the propositions regarding determinants we may substitute "column" for "row" without disturbing the conclusion. This is clearly demonstrated by Remark 15 and the example following. Thus, while most of the propositions are framed in terms of rows an equivalent result would hold in terms of columns.

Proposition 21. *Let A be a square matrix of order m. Let B be the matrix obtained when we multiply the ith row by a scalar k. Then*

$$|B| = k|A|.$$

PROOF. By Definition 21,

$$|B| = \sum (-1)^s b_{1j} b_{2j_1} \cdots b_{mj_m}$$
$$= k \sum (-1)^s a_{1j_m} a_{2j_m} \cdots a_{mj_m} = k|A|.$$

This is so since

$$b_{sj_s} = a_{sj_s} \quad \text{for } s \neq i,$$
$$= ka_{sj_s} \quad \text{for } s = i. \quad \text{q.e.d.}$$

Proposition 22. *Let A be a square matrix of order m. Let B be the matrix obtained when to the rth row of A we add k times its sth row. Then*

$$|B| = |A|.$$

PROOF. By Definition 21,

$$|B| = \sum (-1)^s b_{1j_1} b_{2j_2} \cdots b_{mj_m}$$
$$= \sum (-1)^s a_{1j_1} \cdots a_{r-1j_{r-1}} (a_{rj_r} + ka_{sj_r}) \cdots a_{mj_m}$$
$$= \sum (-1)^s a_{1j_1} \cdots a_{r-1j_{r-1}} a_{rj_r} \cdots a_{mj_n} + k \sum (-1)^s a_{1j_1}$$
$$\cdots a_{r-1j_{r-1}} a_{sj_r} \cdots a_{mj_m}.$$

The first term on the right side of the equation above gives $|A|$ and the second term represents k times the determinant of a matrix having two identical rows. By Proposition 19, that determinant is zero. Hence

$$|B| = |A|. \quad \text{q.e.d.}$$

Remark 17. It is evident, by a simple extension of the argument above, that if to the rth row (or column) of A we add a linear combination of the remaining rows (or columns) we do not affect the determinant of A.

Remark 18. While Definition 21 is intuitively illuminating and, indeed, leads rather easily to the derivation of certain important properties of the determinant, it is not particularly convenient for computational purposes. We shall give below a number of useful alternatives for evaluating determinants.

Definition 22. Let A be a square matrix of order m and let B_{ij} be the matrix obtained by deleting from A its ith row and jth column. The quantity

$$A_{ij} = (-1)^{i+j}|B_{ij}|$$

is said to be the *cofactor* of the element a_{ij} of A. The matrix B_{ij} is said to be an $(m-1)$-*order minor* of A.

Proposition 23 (Expansion by cofactors). *Let A be a square matrix of order m. Then*

$$|A| = \sum_{j=1}^{m} a_{ij}A_{ij}, \qquad |A| = \sum_{i=1}^{m} a_{ij}A_{ij}.$$

PROOF. For definiteness we shall prove this for a specific value of i (for the expansion by cofactors in a given row). By the definition of a determinant,

$$|A| = \sum (-1)^s a_{1j_1}a_{2j_2}\cdots a_{mj_m}.$$

This can also be written more suggestively as follows:

$$
\begin{aligned}
|A| &= a_{11}\sum(-1)^s a_{2j_2}\cdots a_{mj_m} + a_{12}\sum(-1)^s a_{2j_2}\cdots a_{mj_m}\\
&\quad + \cdots + a_{1m}\sum(-1)^s a_{2j_2}\cdots a_{mj_m}\\
&= a_{11}f_{11} + a_{12}f_{12} + a_{1m}f_{1m}, \qquad (19)
\end{aligned}
$$

where, for example,

$$f_{1r} = \sum(-1)^s a_{2j_2}\cdots a_{mj_m},$$

and the numbers $j_2, j_3, \ldots, j_m$ represent some arrangement (permulation) of the integers $1, 2, \ldots, m$, *excluding the integer $r \le m$*. But it is clear that, except (possibly) for sign, f_{1r} is simply the determinant of an $(m-1)$ order minor of A obtained by deleting its first row and rth column. In that determinant s would be zero or one depending on whether the number of transpositions required to restore $j_2, j_3, \ldots, j_m$ to the natural order $1, 2, \ldots, r-1, r+1, \ldots, m$ is even or odd. In (19), however, the corresponding s would be zero or one depending on whether the number of transpositions required to restore $r, j_2, j_3, \ldots, j_m$ to the natural order $1, 2, \ldots, r-1, r, r+1, \ldots, m$ is even or odd. But for $r > 1$, this would be exactly $r-1$ more than before, and would be exactly the same if $r = 1$. Thus,

$$f_{11} = |B_{11}|, \qquad f_{12} = (-1)|B_{12}|,$$

$$f_{13} = (-1)^2|B_{13}|, \qquad \ldots, \qquad f_{1m} = (-1)^{m-1}|B_{1m}|.$$

Noting that

$$A_{1j} = (-1)^{j+1}|B_{1j}|$$

and that

$$(-1)^{j+1} = (-1)^{j-1}$$

we conclude that

$$f_{1j} = A_{1j}.$$

A similar argument can be made for the expansion along any row—and not merely the first—as well as expansion along any column.   q.e.d.

Remark 19. Expansion by cofactors is a very useful way of evaluating a determinant, and the one most commonly used. For certain instances, however, another method—the Laplace expansion—is preferable. Its proof, however, is cumbersome and will, thus, be omitted.

Definition 23. Let A be a square matrix of order m. Let P be an n-order $(n < m)$ minor formed by the rows $i_1, i_2, \ldots, i_n$ and the columns $j_1, j_2, \ldots, j_n$ of A, and let Q be the $(m - n)$-order minor formed by taking the remaining rows and columns. Then Q is said to be the *complementary minor of P* (and conversely P is said to be the complementary minor of Q). Moreover,

$$M = (-1)^{\sum_{i=1}^{n}(i_r+j_r)}|Q|$$

is said to be the *complementary cofactor of P*.

Proposition 24 (Laplace expansion). *Let A be a square matrix of order m. Let $P(i_1, i_2, \ldots, i_n | j_1, j_2, \ldots, j_n)$ be an n-order minor of A formed by rows $i_1, i_2, \ldots, i_n$ and columns $j_1, j_2, \ldots, j_n$, $n < m$. Let M be its associated complementary cofactor. Then*

$$A = \sum_{j_1 < j_2 < \cdots < j_n} |P(i_1, i_2, \ldots, i_n | j_1, j_2, \ldots, j_n)| M,$$

where the sum is taken over all possible choices, the number of which is in fact

$$\binom{m}{n},$$

for the columns of P. Similarly,

$$|A| = \sum_{i_1 < i_2 < \cdots < i_n} P(i_1, i_2, \ldots, i_n | j_1, j_2, \ldots, j_n) M,$$

the sum now chosen over all $\binom{m}{n}$ ways in which the rows of P may be chosen.

PROOF. The proof, while conceptually simple, is rather cumbersome, and not particularly instructive. The interested reader is referred to Hadley [18].   q.e.d.

Remark 20. The first representation in Proposition 23 refers to an expansion by n rows, the second to an expansion by n columns. It is simple to see that this method is a generalization of the method of expansion by cofactors.

The usefulness of the Laplace expansion lies chiefly in the evaluation of determinants of partitioned matrices, a fact that will become apparent in later discussion.

In dealing with determinants it is useful to establish rules for evaluating such quanities for sums or products of matrices. We have

Proposition 25. *Let A, B be two square matrices of order m. Then, in general,*

$$|A + B| \neq |A| + |B|,$$

in the sense that any of the following three relations is possible;

$$|A + B| = |A| + |B|;$$
$$|A + B| > |A| + |B|;$$
$$|A + B| < |A| + |B|.$$

PROOF. We establish the validity of the proposition by a number of examples. Thus, for

$$A = \begin{bmatrix} 2 & 0 \\ 0 & 3 \end{bmatrix}, \qquad B = \begin{bmatrix} -\frac{2}{3} & 0 \\ 0 & 1 \end{bmatrix}$$

we have $|A| = 6, |B| = -\frac{2}{3}, |A + B| = 5\frac{1}{3}$ *and we see that*

$$|A + B| = |A| + |B|.$$

For the matrices

$$A = \begin{bmatrix} 1 & 0 \\ 0 & 1 \end{bmatrix}, \qquad B = \begin{bmatrix} 1 & 0 \\ 0 & 1 \end{bmatrix}$$

we find $|A| = 1, |B| = 1, |A + B| = 4$. Thus $|A + B| > |A| + |B|$.

For the matrices

$$A = \begin{bmatrix} 1 & 0 \\ 0 & 1 \end{bmatrix}, \qquad B = \begin{bmatrix} -1 & 0 \\ 0 & -1 \end{bmatrix}$$

we find $|A| = 1, |B| = 1, |A + B| = 0$. Thus $|A + B| < |A| + |B|$,   q.e.d.

Proposition 26. *Let A, B be two square matrices of order m. Then*

$$|AB| = |A||B|.$$

PROOF. Define the $2m \times 2m$ matrix

$$C = \begin{bmatrix} A & 0 \\ -I & B \end{bmatrix}.$$

Multiply the last *m* rows of *C* on the left by *A* (i.e., take *m* linear combinations of such rows) and add them to the first rows. The resulting matrix is

$$C^* = \begin{bmatrix} 0 & AB \\ -I & B \end{bmatrix}.$$

By Proposition 22 and Remark 17 we have

$$|C| = |C^*|. \tag{20}$$

Expand $|C^*|$ by the method of Proposition 24 and note that, using m-order minors involving the last m rows, their associated complementary cofactors will vanish (since they involve the determinant of a matrix containing a zero column), except for the one corresponding to $-I$. The complementary cofactor for that minor is

$$(-1)^{\sum_{i=1}^{m}(i+m+i)}|AB| = (-1)^{m^2+m^2+m}|AB|.$$

Moreover,

$$|-I| = (-1)^m.$$

Hence

$$|C^*| = (-1)^{2m^2+2m}|AB| = |AB|. \tag{21}$$

Similarly, expand $|C|$ by the same method using m-order minors involving the *first* m rows. Notice now that all m-order minors involving the first m rows of C have a zero determinant save for the one corresponding to A, whose determinant is, evidently,

$$|A|.$$

Its associated complementary cofactor is

$$(-1)^{\sum_{i=1}^{m}(i+i)}|B| = (-1)^{2[m(m+1)/2]}|B| = |B|.$$

Hence we have

$$|C| = |A||B|.$$

But this result together with (20) and (21) imply

$$|AB| = |A||B|. \quad \text{q.e.d.}$$

Corollary 2. *Let A be an invertible matrix of order m. Then*

$$|A^{-1}| = \frac{1}{|A|}.$$

Proof. By definition

$$|AA^{-1}| = |I| = 1,$$

the last equality following immediately from the fundamental definition of the determinant. Since

$$|AA^{-1}| = |A||A^{-1}|,$$

we have

$$|A^{-1}| = \frac{1}{|A|} \quad \text{q.e.d.}$$

We conclude this section by introducing

Definition 24. Let A be a square matrix of order m. Let A_{ij} be the cofactor of the (i, j) element of A, a_{ij}, and define

$$B = (A_{ij}), \qquad i, j, = 1, 2, \ldots, m.$$

The *adjoint of* A, denoted by adj A, is defined by

$$\text{adj } A = B'.$$

2.6 Computation of the Inverse

In earlier discussion we defined the inverse of a matrix, say A, to be another matrix, say B, having the properties

$$AB = BA = I.$$

While this describes the essential property of the inverse it does not provide a useful way in which the elements of B can be determined. In the preceding section we have laid the foundation for providing a practicable method for determining the elements of the inverse of a given matrix. We have

Proposition 27. *Let A be an invertible square matrix of order m. Then its inverse, denoted by A^{-1}, is given by*

$$A^{-1} = \frac{\text{adj } A}{|A|}$$

PROOF. In the standard notation for inverses denote the (i, j) element of A^{-1} by a^{ij}; then the proposition asserts that

$$a^{ij} = \frac{A_{ji}}{|A|},$$

where A_{ji} is the cofactor of the element in the jth row and ith column of A. Let us now verify the validity of the assertion by determining the typical element of AA^{-1}. It is given by

$$\sum_{k=1}^{m} a_{ik} a^{kj} = \frac{1}{|A|} \sum_{k=1}^{m} a_{ik} A_{jk}. \tag{22}$$

Now for $i = j$ we have the expansion by cofactors along the ith row of A. Hence all diagonal elements of AA^{-1} are unity. For $i \neq j$ we may evaluate the quantity in (22) as follows. Strike out the jth row of A and replace it by the ith row. The resulting matrix has two identical rows and as such its determinant is zero. Now expand by cofactors along the jth row. The cofactors of such elements are plainly A_{jk} since the other rows of A have not been disturbed. Thus, expanding by cofactors along the jth row we conclude

$$\sum_{k=1}^{m} a_{ik} A_{jk} = 0.$$

This is so since we have above a representation of the determinant of a matrix with two identical rows. It follows then that

$$AA^{-1} = I.$$

Similarly, consider the typical element of $A^{-1}A$, i.e.,

$$\sum_{k=1}^{m} a^{ik} a_{kj} = \frac{1}{|A|} \sum_{k=1}^{m} a_{kj} A_{ki}. \tag{23}$$

Again for $j = i$ we have the determinant of A evaluated by an expansion along the ith column. Hence, all diagonal elements of $A^{-1}A$ are unity. For $i \neq j$ consider the matrix obtained when we strike out the ith column of A and replace it by its jth column. The resulting matrix has two identical columns and hence its determinant is zero. Evaluating its determinant by expansion along the ith column we note that the cofactors are given by A_{ki} since the other columns of A have not been disturbed. But then we have

$$\sum_{k=1}^{m} a_{kj} A_{ki} = 0, \qquad i \neq j,$$

and thus we conclude that

$$A^{-1}A = I. \quad \text{q.e.d.}$$

Proposition 28. *Let A, B be two invertible matrices of order m. Then*

$$(AB)^{-1} = B^{-1}A^{-1}.$$

PROOF. We verify

$$(AB)(B^{-1}A^{-1}) = A(BB^{-1})A^{-1} = AA^{-1} = I$$
$$(B^{-1}A^{-1})(AB) = B^{-1}A^{-1}AB = B^{-1}B = I. \quad \text{q.e.d.}$$

Remark 21. For any two conformable and invertible matrices A, B, we have

$$(A + B)^{-1} \neq A^{-1} + B^{-1},$$

and, indeed, $A + B$ need not be invertible. For example, suppose $B = -A$. Then even though A^{-1}, B^{-1} exist, $A + B = 0$, which is, evidently, not invertible since its determinant is zero.

2.7 Partitional Matrices

Frequently, we find it convenient to deal with partitioned matrices. In this section, we shall derive certain useful results that will facilitate operations with such matrices. Let A be $m \times n$ and write

$$A = \begin{bmatrix} A_{11} & A_{12} \\ A_{21} & A_{22} \end{bmatrix},$$

where A_{11} is $m_1 \times n_1$, A_{22} is $m_2 \times n_2$ $(m_1 + m_2 = m, \; n_1 + n_2 = n)$, A_{12} is $m_1 \times n_2$, A_{21} is $m_2 \times n_1$. The above is said to be a *partition* of the matrix A.

Now, let B be also $m \times n$ and partition it conformably with A, i.e., put

$$B = \begin{bmatrix} B_{11} & B_{12} \\ B_{21} & B_{22} \end{bmatrix},$$

where B_{11} is $m_1 \times n_1$, B_{22} is $m_2 \times n_2$, and so on.

Addition of (conformably) partitioned matrices is defined by

$$A + B = \begin{bmatrix} A_{11} + B_{11} & A_{12} + B_{12} \\ A_{21} + B_{21} & A_{22} + B_{22} \end{bmatrix}.$$

If A is $m \times n$, C is $n \times q$, and A is partitioned as above, let

$$C = \begin{bmatrix} C_{11} & C_{12} \\ C_{21} & C_{22} \end{bmatrix},$$

where C_{11} is $n_1 \times q_1$, C_{22} is $n_2 \times q_2$, and so on.

Multiplication of two (conformably) partitioned matrices is defined by

$$\begin{aligned} AC &= \begin{bmatrix} A_{11} & A_{12} \\ A_{21} & A_{22} \end{bmatrix} \begin{bmatrix} C_{11} & C_{12} \\ C_{21} & C_{22} \end{bmatrix} \\ &= \begin{bmatrix} A_{11}C_{11} + A_{12}C_{21} & A_{11}C_{12} + A_{12}C_{22} \\ A_{21}C_{11} + A_{22}C_{21} & A_{21}C_{12} + A_{22}C_{22} \end{bmatrix}. \end{aligned}$$

In general, and for either matrix addition or matrix multiplication, readers will not commit an error if, upon (conformably) partitioning two matrices, they proceed to regard the partition blocks as ordinary elements and apply the usual rules. Thus, for example, consider

$$A = \begin{bmatrix} A_{11} & A_{12} & \cdots & A_{1s} \\ A_{21} & A_{22} & \cdots & A_{2s} \\ \vdots & \vdots & & \vdots \\ A_{s1} & A_{s2} & \cdots & A_{ss} \end{bmatrix},$$

where A_{ij} is $m_i \times n_j$, the matrix A is $m \times n$, and

$$\sum_{i=1}^{s} m_i = m, \qquad \sum_{j=1}^{s} n_j = n.$$

Similarly, consider

$$B = \begin{bmatrix} B_{11} & B_{12} & \cdots & B_{1s} \\ B_{21} & B_{22} & \cdots & B_{2s} \\ \vdots & \vdots & & \vdots \\ B_{s1} & B_{s2} & \cdots & B_{ss} \end{bmatrix},$$

where B_{ij} is $m_i \times n_j$ as above. Then A and B are conformably partitioned with respect to matrix addition and their sum is simply

$$A + B = \begin{bmatrix} A_{11} + B_{11} & \cdots & A_{1s} + B_{1s} \\ A_{21} + B_{21} & \cdots & A_{2s} + B_{2s} \\ \vdots & & \vdots \\ A_{s1} + B_{s1} & \cdots & A_{ss} + B_{ss} \end{bmatrix},$$

If, instead of being $m \times n$, B is $n \times q$ and its partition blocks B_{ij} are $n_i \times q_j$ matrices such that

$$\sum_{i=1}^{s} n_i = n, \qquad \sum_{j=1}^{s} q_j = q,$$

then A and B are conformably partitioned with respect to multiplication and their product is given by

$$AB = \begin{bmatrix} \sum_{r=1}^{s} A_{1r}B_{r1} & \cdots & \sum_{r=1}^{s} A_{1r}B_{rs} \\ \sum_{r=1}^{s} A_{2r}B_{r1} & \cdots & \sum_{r=1}^{s} A_{2r}B_{rs} \\ \vdots & & \vdots \\ \sum_{r=1}^{s} A_{sr}B_{r1} & \cdots & \sum_{r=1}^{s} A_{sr}B_{rs} \end{bmatrix}$$

where the (i, j) block of AB, viz.,

$$\sum_{r=1}^{s} A_{ir}B_{rj},$$

is a matrix of dimension $m_i \times q_j$.

For inverses and determinants of partitioned matrices we may prove certain useful results.

Proposition 29. *Let A be a square matrix of order m. Partition*

$$A = \begin{bmatrix} A_{11} & A_{12} \\ A_{21} & A_{22} \end{bmatrix}$$

and let A_{ij} be $m_i \times m_j$, $i, j = 1, 2$, $m_1 + m_2 = m$. Also, let

$$A_{21} = 0.$$

Then

$$|A| = |A_{11}||A_{22}|.$$

PROOF. This follows immediately from the Laplace expansion by noting that if we expand along the last m_2 rows the only $m_2 \times m_2$ minor with non-vanishing determinant is A_{22}. Its complementary factor is

$$(-1)^{\sum_{i=m_1+1}^{m_1+m_2}(i+i)}|A_{11}| = |A_{11}|.$$

Consequently

$$|A| = |A_{11}||A_{22}|. \quad \text{q.e.d.}$$

Corollary 3. *If, instead, we had assumed*

$$A_{12} = 0,$$

then

$$|A| = |A_{11}||A_{22}|.$$

PROOF. Obvious from the preceding.   q.e.d.

Definition 25. A matrix of the form

$$A = \begin{bmatrix} A_{11} & A_{12} \\ 0 & A_{22} \end{bmatrix}$$

of Proposition 29 is said to be an *upper block triangular matrix*. A matrix of the form

$$A = \begin{bmatrix} A_{11} & 0 \\ A_{21} & A_{22} \end{bmatrix}$$

is said to be a *lower block triangular matrix*.

Definition 26. Let A be as in Proposition 29, but suppose

$$A_{12} = 0, \qquad A_{21} = 0,$$

i.e., A is of the form

$$A = \begin{bmatrix} A_{11} & 0 \\ 0 & A_{22} \end{bmatrix},$$

Then A is said to be a *block diagonal matrix* and is denoted by

$$A = \text{diag}(A_{11}, A_{22}).$$

Corollary 4. *Let A be a block diagonal matrix as above. Then*

$$|A| = |A_{11}||A_{22}|.$$

PROOF. Obvious.   q.e.d.

Remark 22. Note that in the definitions of block triangular (block diagonal) matrices the blocks A_{11}, A_{22} need not be triangular (diagonal) matrices.

Proposition 30. *Let A be a partitioned square matrix of order m,*

$$A = \begin{bmatrix} A_{11} & A_{12} \\ A_{21} & A_{22} \end{bmatrix},$$

where the A_{ii} are nonsingular square matrices of order m_i, $i = 1, 2, m_1 + m_2 = m$. Then

$$|A| = |A_{22}||A_{11} - A_{12}A_{22}^{-1}A_{21}|,$$

and

$$|A| = |A_{11}||A_{22} - A_{21}A_{11}^{-1}A_{12}|.$$

PROOF. Consider the matrix

$$A_* = \begin{bmatrix} I & -A_{12}A_{22}^{-1} \\ 0 & I \end{bmatrix} A = \begin{bmatrix} A_{11} - A_{12}A_{22}^{-1}A_{21} & 0 \\ A_{21} & A_{22} \end{bmatrix}.$$

By Proposition 29, and Corollary 3

$$\det \begin{bmatrix} I & -A_{12}A_{22}^{-1} \\ 0 & I \end{bmatrix} = 1,$$

whence we conclude

$$|A_*| = |A|.$$

Again by Proposition 29 the determinant of A_* may be evaluated as

$$|A_*| = |A_{22}||A_{11} - A_{12}A_{22}^{-1}A_{21}|.$$

Hence, we conclude

$$|A| = |A_{22}||A_{11} - A_{12}A_{22}^{-1}A_{21}|.$$

Similarly, consider

$$A^* = \begin{bmatrix} I & 0 \\ -A_{21}A_{11}^{-1} & I \end{bmatrix} A = \begin{bmatrix} A_{11} & A_{12} \\ 0 & A_{22} - A_{21}A_{11}^{-1}A_{12} \end{bmatrix}$$

and thus conclude

$$|A| = |A_{11}||A_{22} - A_{21}A_{11}^{-1}A_{12}|. \quad \text{q.e.d.}$$

Occasionally, we shall have to evaluate the determinants of matrices of the form

$$A + \alpha b d',$$

where α is a *scalar*, A is a square matrix of order m, and b and d are m-element column vectors. Thus, we have

Proposition 31. *Let A be a square invertible matrix of order m; let α be a scalar and b, d be two m-element vectors. Then*

$$|A + \alpha bd'| = |A||1 + \alpha d'A^{-1}b|.$$

Proof. Since

$$A + \alpha bd' = (I + \alpha bd'A^{-1})A$$

we need only evaluate a determinant of the form

$$|I + ac'|,$$

where a, c are m-element (column) vectors. To this effect, observe that

$$\begin{bmatrix} 1 & c' \\ -a & I \end{bmatrix}\begin{bmatrix} 1 & 0 \\ a & I \end{bmatrix} = \begin{bmatrix} 1 + c'a & c' \\ 0 & I \end{bmatrix}, \tag{24}$$

$$\begin{bmatrix} 1 & 0 \\ a & I \end{bmatrix}\begin{bmatrix} 1 & c' \\ -a & I \end{bmatrix} = \begin{bmatrix} 1 & c' \\ 0 & I + ac' \end{bmatrix}. \tag{25}$$

Since

$$\det\begin{bmatrix} 1 & 0 \\ a & I \end{bmatrix} = 1,$$

we conclude from (24) and (25) that

$$1 + c'a = |I + ac'| \tag{26}$$

If in (26) we take

$$c' = \alpha d'A^{-1}, \qquad b = a,$$

we see that

$$|A + \alpha bd'| = |(I + \alpha bd'A^{-1})A| = |A|(1 + \alpha d'A^{-1}b). \quad \text{q.e.d.}$$

For inverses of partitioned matrices we have the following useful results.

Proposition 32. *Let A be a square nonsingular matrix and partition*

$$A = \begin{bmatrix} A_{11} & A_{12} \\ A_{21} & A_{22} \end{bmatrix}$$

so that A_{ii}, $i = 1, 2$, is a nonsingular matrix of order m_i ($m_1 + m_2 = m$). Then

$$A^{-1} = B = \begin{bmatrix} B_{11} & B_{12} \\ B_{21} & B_{22} \end{bmatrix},$$

where

$$B_{11} = (A_{11} - A_{12}A_{22}^{-1}A_{21})^{-1}, \qquad B_{12} = -A_{11}^{-1}A_{12}(A_{22} - A_{21}A_{11}^{-1}A_{12})^{-1},$$

$$B_{21} = -A_{22}^{-1}A_{21}(A_{11} - A_{12}A_{22}^{-1}A_{21})^{-1}, \qquad B_{22} = (A_{22} - A_{21}A_{11}^{-1}A_{12})^{-1}.$$

PROOF. By definition of the inverse B,

$$AB = \begin{bmatrix} A_{11}B_{11} + A_{12}B_{21} & A_{11}B_{12} + A_{12}B_{22} \\ A_{21}B_{11} + A_{22}B_{21} & A_{21}B_{12} + A_{22}B_{22} \end{bmatrix} = \begin{bmatrix} I & 0 \\ 0 & I \end{bmatrix},$$

which implies the equations

$$A_{11}B_{11} + A_{12}B_{21} = I, \qquad A_{11}B_{12} + A_{12}B_{22} = 0,$$

$$A_{21}B_{11} + A_{22}B_{21} = 0, \qquad A_{21}B_{12} + A_{22}B_{22} = I.$$

Solving these by substitution we have the proposition.   q.e.d.

Finally, we have

Proposition 33. *Let A be a nonsingular matrix of order m, and consider*

$$B = A + \alpha ab',$$

where α is a scalar and a, b are two m-element vectors. Then

$$B^{-1} = A^{-1} - cA^{-1}ab'A^{-1},$$

where

$$c = \frac{\alpha}{1 + \alpha b'A^{-1}a}$$

provided

$$1 + \alpha b'A^{-1}a \neq 0.$$

PROOF. We shall verify that

$$BB^{-1} = I$$

But we have

$$[A + \alpha ab'][A^{-1} - cA^{-1}ab'A^{-1}]$$
$$= I - cab'A^{-1} + \alpha ab'A^{-1} - c\alpha ab'A^{-1}ab'A^{-1}$$
$$= I + [\alpha - c(1 + \alpha b'A^{-1}a)]ab'A^{-1}. \tag{27}$$

Noting that

$$c(1 + \alpha b'A^{-1}a) = \alpha$$

we conclude that the last member of equation of (27) is simply the identity matrix.   q.e.d.

2.8 Kronecker Products of Matrices

Definition 27. Let A be $m \times n$, B be $p \times q$. The *Kronecker product* of the two matrices, denoted by

$$A \otimes B$$

is defined by

$$A \otimes B = \begin{bmatrix} a_{11}B & a_{12}B & \cdots & a_{1n}B \\ a_{21}B & a_{22}B & \cdots & a_{2n}B \\ \vdots & \vdots & & \vdots \\ a_{m1}B & a_{m2}B & \cdots & a_{mn}B \end{bmatrix}.$$

Often, it is written more compactly as

$$A \otimes B = [a_{ij}B]$$

and is a matrix of dimension $(mp) \times (nq)$.

One operates with Kronecker products as follows.

Matrix addition. Let A_1, A_2 be matrices of dimension $m \times n$ and B_1, B_2, be matrices of dimension $p \times q$, and put

$$D_i = (A_i \otimes B_1), \qquad i = 1, 2.$$

Then

$$D_1 + D_2 = (A_1 + A_2) \otimes B_1.$$

Similarly, if

$$E_i = (A_1 \otimes B_i), \qquad i = 1, 2$$

then

$$E_1 + E_2 = A_1 \otimes (B_1 + B_2).$$

Scalar Multiplication. Let

$$C_i = (A_i \otimes B_i), \qquad i = 1, 2,$$

with the A_i, B_i as above, and let α be a scalar. Then

$$\alpha C_i = (\alpha A_i \otimes B_i) = (A_i \otimes \alpha B_i).$$

Matrix multiplication. Let C_i, $i = 1, 2$, be two Kronecker product matrices,

$$C_i = A_i \otimes B_i, \qquad i = 1, 2,$$

and suppose that A_1 is $m \times n$, A_2 is $n \times r$, B is $p \times q$, B_2 is $q \times s$. Then

$$C_1 C_2 = A_1 A_2 \otimes B_1 B_2.$$

Matrix inversion. Let C be a Kronecker product

$$C = A \otimes B$$

and suppose that A, B are invertible matrices of order m and n respectively. Then

$$C^{-1} = A^{-1} \otimes B^{-1}.$$

All of the above can be verified directly either from the rules for operating with partitioned matrices or from other appropriate definitions.

We have

Proposition 34. *Let A, B be square matrices of orders m, n respectively. Then*

$$\operatorname{tr}(A \otimes B) = \operatorname{tr}(A)\operatorname{tr}(B).$$

PROOF. Obvious from the definitions of the trace and the Kronecker product. q.e.d.

Proposition 35. *Let A, B be nonsingular matrices of orders m, n respectively. Then*

$$|A \otimes B| = |A|^n |B|^m.$$

PROOF. Denote by A_i the matrix obtained when we suppress the first i rows and columns of A. Similarly, denote by $a_{i.}$ the ith row of A after its first i elements are suppressed and by $a_{.i}$ the ith column of A after its first i elements have been suppressed. *Note that $a_{.i}$ is the first column of A_i and $a_{i.}$ its first row.*

Now partition

$$A = \begin{bmatrix} a_{11} & a_{1.} \\ a_{.1} & A_1 \end{bmatrix}$$

and write the Kronecker product in the partitioned form

$$A \otimes B = \begin{bmatrix} a_{11}B & a_{1.} \otimes B \\ a_{.1} \otimes B & A_1 \otimes B \end{bmatrix}$$

Apply Proposition 30 to the partitioned matrix above to obtain

$$|A \otimes B| = |a_{11}B - (a_{1.} \otimes B)(A_1 \otimes B)^{-1}(a_{.1} \otimes B)||A_1 \otimes B|.$$

Since A is nonsingular, we shall assume that A_1 is also nonsingular,[2] as well as A_2, A_3, ..., etc. Thus, we may evaluate

$$(a_{1.} \otimes B)(A_1 \otimes B)^{-1}(a_{.1} \otimes B) = a_{1.}A_1^{-1}a_{.1} \otimes B,$$

---

[2] This involves some loss of generality but makes a proof by elementary methods possible. The results stated in the proposition are valid without these restrictive assumptions.

where, evidently, $a_1 . A_1^{-1} a_{.1}$ is a scalar. Consequently,

$$|a_{11} B - (a_1 . \otimes B)(A_1 \otimes B)^{-1}(a_{.1} \otimes B)| = |(a_{11} - a_1 . A_1^{-1} a_{.1}) \otimes B|$$
$$= (a_{11} - a_1 . A_1^{-1} a_{.1})^n |B|.$$

Applying Proposition 30 to the partition of A above we note that

$$|A| = (a_{11} - a_1 . A_1^{-1} a_{.1})|A_1|,$$

and so we find

$$|A \otimes B| = |A|^n |A_1|^{-n} |B| |A_1 \otimes B|.$$

Applying the same procedure we also find

$$|A_1 \otimes B| = |A_1|^n |A_2|^{-n} |B| |A_2 \otimes B|,$$

and thus

$$|A \otimes B| = |A|^n |A_2|^{-n} |B|^2 |A_2 \otimes B|$$

Continuing in this fashion $m - 1$ times we have

$$|A \otimes B| = |A|^n |A_{m-1}|^{-n} |B|^{m-1} |A_{n-1} \otimes B|.$$

But

$$A_{m-1} = a_{mm}$$

and

$$|A_{m-1} \otimes B| = |a_{mm} B| = a_{mm}^n |B|.$$

Since

$$|A_{m-1}| = a_{mm}$$

we conclude

$$|A \otimes B| = |A|^n |B|^m. \quad \text{q.e.d.}$$

2.9 Characteristic Roots and Vectors

Definition 28. Let A be a square matrix of order m; let λ, x be, respectively, a scalar and an m-element nonnull vector. If

$$Ax = \lambda x,$$

λ is said to be *a characteristic root* of A and x its associated *characteristic vector*.

Remark 23. Characteristic vectors are evidently not unique. If c is a nonnull scalar, cx is also a characteristic vector, if x is.

Proposition 36. *Let A be a square matrix of order m and let Q be an invertible matrix of order m. Then*

$$B = Q^{-1}AQ$$

has the same characteristic roots as A, and if x is a characteristic vector of A then $Q^{-1}x$ is a characteristic vector of B.

PROOF. Let (λ, x) be any pair of characteristic root and associated characteristic vector of A. Then they satisfy

$$Ax = \lambda x.$$

Premultiply by Q^{-1} to obtain

$$Q^{-1}Ax = \lambda Q^{-1}x.$$

But we may also write

$$Q^{-1}A = Q^{-1}AQQ^{-1}.$$

Thus we obtain

$$Q^{-1}AQ(Q^{-1}x) = \lambda(Q^{-1}x),$$

which shows that the pair $(\lambda, Q^{-1}x)$ is a characteristic root and associated characteristic vector of B.   q.e.d.

Remark 24. If A and Q are as above then

$$B = Q^{-1}AQ$$

and A are said to be *similar matrices*. It is clear that if there exists a matrix P such that

$$P^{-1}AP = D,$$

where D is a diagonal matrix, then P must be the matrix of the characteristic vectors of A, and D the matrix of its characteristic roots. This is so since the equation above implies

$$AP = DP$$

and the columns of this relation read

$$Ap_{\cdot i} = d_i p_{\cdot i}, \qquad i = 1, 2, \ldots, m,$$

thus defining $(d_i, p_{\cdot i})$ as a characteristic root and its associated characteristic vector. The question then arises as to when a matrix A is similar to a diagonal matrix. We have

Proposition 37. *Let A be a square matrix of order m and suppose*

$$r_i, \qquad i = 1, 2, \ldots, n, n \leq m,$$

are the distinct *characteristic roots of A. If*

$$\{x_{\cdot i}: i = 1, 2, \ldots, n\}$$

is the set of associated characteristic vectors, then it is a linearly independent set.

PROOF. Put

$$X = (x_{\cdot 1}, x_{\cdot 2}, \ldots, x_{\cdot n})$$

and note that X is $m \times n$, $n \leq m$. Suppose the columns of X are not linearly independent. Then there exists a nonnull vector

$$b = (b_1, b_2, \ldots, b_n)'$$

such that

$$Xb = 0. \tag{28}$$

Let

$$R = \mathrm{diag}(r_1, r_2, \ldots, r_n)$$

and note that, since

$$AX = XR,$$

multiplying (28) by A we have

$$0 = AXb = XRb.$$

Repeating this j times we find

$$XR^j b = 0, \qquad j = 1, 2, \ldots, n-1, \tag{29}$$

where

$$R^j = \mathrm{diag}(r_1^j, r_2^j, \ldots, r_n^j).$$

Consider now the matrix whose jth column is (29), on the understanding that for $j = 0$ we have (28). In view of (29) this is the null matrix. But note also that

$$0 = (Xb, XRb, \ldots, XR^{n-1}b) = XBV, \tag{30}$$

where

$$B = \mathrm{diag}(b_1, b_2, \ldots, b_n)$$

and V is the so-called *Vandermonde matrix*

$$V = \begin{bmatrix} 1 & r_1 & r_1^2 & \cdots & r_1^{n-1} \\ 1 & r_2 & r_2^2 & \cdots & r_2^{n-1} \\ \vdots & \vdots & \vdots & & \vdots \\ 1 & r_n & r_n^2 & \cdots & r_n^{n-1} \end{bmatrix}.$$

It may be shown (see Proposition 38) that if the r_i are distinct, V is non-singular. Hence from (30) we conclude

$$XB = 0.$$

But this means

$$b_i x_{\cdot i} = 0, \qquad i = 1, 2, \ldots, n.$$

Thus, unless

$$b_i = 0, \qquad i = 1, 2, \ldots, n,$$

we must have, for some i, say i_0,

$$x_{\cdot i_0} = 0.$$

This is a contradiction, and shows that (28) cannot hold for nonnull b, and hence that the characteristic vectors corresponding to distinct characteristic roots are linearly independent. q.e.d.

Proposition 38. *Let*

$$V = \begin{bmatrix} 1 & r_1 & r_1^2 & \cdots & r_1^{n-1} \\ 1 & r_2 & r_2^2 & \cdots & r_2^{n-1} \\ \vdots & \vdots & \vdots & & \vdots \\ 1 & r_n & r_n^2 & \cdots & r_n^{n-1} \end{bmatrix}$$

and suppose the r_i, $i = 1, 2, \ldots, n$, are distinct. Then

$$|V| \neq 0.$$

PROOF. Expand $|V|$ by cofactors along the first row to obtain

$$|V| = a_0 + a_1 r_1 + a_2 r_1^2 + \cdots + a_{n-1} r_1^{n-1},$$

where a_i is the cofactor of r_1^i. This shows $|V|$ to be a polynomial of degree $n - 1$ in r_1; it is immediately evident that $r_2, r_3, \ldots, r_n$ are its roots since if for r_1 we substitute r_i, $i \geq 2$, we have the determinant of a matrix with two identical rows. From the fundamental theorem of algebra we can thus write

$$|V| = a_{n-1} \prod_{j=2}^{n} (r_1 - r_j).$$

But

$$a_{n-1} = (-1)^{n+1} |V_1|,$$

where V_1 is the matrix obtained by striking out the first row and nth column of V. Hence, we can also write

$$|V| = |V_1| \prod_{j=2}^{n} (r_j - r_1).$$

But V_1 is of exactly the same form as V except that it is of dimension $n - 1$ and does not contain r_1.

Applying a similar procedure to V_1 we find

$$|V_1| = |V_2| \prod_{j=3}^{n} (r_j - r_2),$$

where V_2 is evidently the matrix obtained when we strike out the first and second rows as well as columns n and $n - 1$ of V. Continuing in this fashion we find

$$|V| = \prod_{i=1}^{n-1} \prod_{j_i=i+1}^{n} (r_{j_i} - r_i).$$

Since $r_{j_i} \neq r_i$ it is evident that

$$|V| \neq 0. \quad \text{q.e.d.}$$

An immediate consequence of Proposition 37 is

Proposition 39. *Let A be a square matrix of order m and suppose all its roots are distinct. Then A is similar to a diagonal matrix.*

PROOF. Let $(\lambda_i, x_{\cdot i})$, $i = 1, 2, \ldots, m$, be the characteristic roots and associated characteristic vectors of A. Let

$$\Lambda = \text{diag}(\lambda_1, \lambda_2, \ldots, \lambda_m), \qquad X = (x_{\cdot 1}, x_{\cdot 2}, \ldots, x_{\cdot m}),$$

and note that the relationship

$$A x_{\cdot i} = \lambda_i x_{\cdot i}, \qquad i = 1, 2, \ldots, m,$$

between A and its characteristic roots and vectors may be written compactly as

$$AX = X\Lambda.$$

By Proposition 37, X is nonsingular; hence

$$X^{-1}AX = \Lambda. \quad \text{q.e.d.}$$

The usefulness of this proposition is enhanced by the following approximation result.

Proposition 40. *Let A be a square matrix of order m. Then there exists a square matrix of order m, say B, such that B has distinct roots and*

$$\sum_{i,j=1}^{m} |a_{ij} - b_{ij}| < \varepsilon$$

where ε is any preassigned positive quantity howsoever small.

PROOF. The proof of this result lies entirely outside the scope of this volume. The interested reader is referred to Bellman [4, pp. 199 ff]. q.e.d.

In the preceding we have established a number of properties regarding the characteristic roots and their associated characteristic vectors without explaining how such quantities may be obtained. It is thus useful to deal with these aspects of the problem before we proceed. By the definition of characteristic roots and vectors of a square matrix A, we have

$$Ax = \lambda x,$$

or more revealingly

$$(\lambda I - A)x = 0, \tag{31}$$

where λ is a characteristic root and x the associated characteristic vector. We recall that we require

$$x \neq 0. \tag{32}$$

Clearly (31) together with (32) imply that the columns of $\lambda I - A$ are linearly dependent. Hence we can find all λ's for which (31) and (32) may be satisfied for appropriate x's by finding the λ's for which

$$|\lambda I - A| = 0. \tag{33}$$

Definition 29. Let A be square matrix of order m. The relation in (33) regarded as an equation in λ is said to be the *characteristic equation* of the matrix A.

From the basic definition of a determinant we easily see that (33) represents a polynomial of degree m in λ. This is so since in evaluating a determinant we take the sum of all possible products involving the choice of one element from each row and column. In this case the largest power of λ occurs in the term involving the choice of the diagonal elements of $\lambda I - A$. This term is

$$(\lambda - a_{11})(\lambda - a_{22})\cdots(\lambda - a_{mm})$$

and we easily see that the highest power of λ occurring is λ^m and its coefficient is unity. Moreover, collecting terms involving λ^j, $j = 0, 1, 2, \ldots, m$, we can write

$$|\lambda I - A| = \lambda^m + b_{m-1}\lambda^{m-1} + b_{m-2}\lambda^{m-2} + \cdots + b_0, \tag{34}$$

and it is also clear from this discussion that

$$b_0 = |-A|. \tag{35}$$

The fundamental theorem of algebra assures us that over the field of complex numbers the polynomial of degree m in (34) has m roots, which we may number in order, say, of decreasing magnitude $\lambda_1, \lambda_2, \lambda_3, \ldots, \lambda_m$; moreover, we can write

$$|\lambda I - A| = \prod_{i=1}^{m} (\lambda - \lambda_i). \tag{36}$$

Now the roots of the characteristic equation of A are the characteristic roots of A. Since the columns of $\lambda_i I - A$ are linearly dependent, if λ_i is a characteristic root of A it follows that there exists at least one nonnull vector, say, $x_{.i}$, such that

$$(\lambda_i I - A)x_{.i} = 0.$$

But this means that the pair

$$(\lambda_i, x_{.i})$$

represents a characteristic root and its associated characteristic vector.

Thus, obtaining characteristic roots involves solving a polynomial equation of degree m and obtaining characteristic vectors involves solving a system of m linear equations. An immediate consequence of the preceding discussion is

Proposition 41. *Let A be a square matrix of order m. Let $\lambda_1, \lambda_2, \ldots, \lambda_m$ be its characteristic roots. Then*

$$|A| = \prod_{i=1}^{m} \lambda_i$$

PROOF. If in (36) we compute the constant term of the polynomial on the right side we find

$$\prod_{i=1}^{m} (-\lambda_i) = (-1)^m \prod_{i=1}^{m} \lambda_i.$$

From (35) we see that

$$b_0 = |-A| = (-1)^m |A|.$$

Since (36) and (34) are two representations of the same polynomial we conclude

$$|A| = \prod_{i=1}^{m} \lambda_i. \quad \text{q.e.d.}$$

Remark 25. The preceding proposition implies that if A is a singular matrix, then at least one of its roots is zero. It also makes clear the terminology *distinct* and *repeated* characteristic roots. In particular, let $s < m$ and suppose (36) turns out to be of the form

$$|\lambda I - A| = \prod_{j=1}^{s} (\lambda - \lambda_{(j)})^{m_j},$$

where

$$\sum_{j=1}^{s} m_j = m, \qquad \lambda_{(j)} \neq \lambda_{(i)} \quad \text{for } i \neq j.$$

Then we may say that A has s *distinct* roots, viz., the roots $\lambda_{(1)}, \lambda_{(2)}, \ldots, \lambda_{(s)}$, and the root $\lambda_{(i)}$ is *repeated* m_i times since the factor corresponding to it in the characteristic equation is raised to the m_i power.

Remark 26. It may further be shown, but will not be shown here, that if A is a square matrix of order m and rank $r \leq m$ then it has r nonzero roots and $m - r$ zero roots, i.e., the zero root is repeated $m - r$ times or, alternatively, its characteristic equation is of the form

$$|\lambda I - A| = \lambda^{m-r} f(\lambda), \tag{37}$$

where

$$f(\lambda_i) = 0, \qquad i = 1, 2, \ldots, r,$$

and

$$\lambda_i \neq 0, \qquad i = 1, 2, \ldots, r.$$

From the method for obtaining characteristic roots we may easily deduce

Proposition 42. *Let A be a square matrix of order m and let*

$$\lambda_i, \qquad i = 1, 2, \ldots, m,$$

be its characteristic roots. Then

(a) *the characteristic roots of A' are exactly those of A, and*
(b) *if A is also nonsingular, the characteristic roots of A^{-1} are given by*

$$\mu_i = \frac{1}{\lambda_i}, \qquad i = 1, 2, \ldots, m.$$

PROOF. The characteristic roots of A are simply the solution of

$$|\lambda I - A| = 0.$$

The characteristic roots of A' are obtained by solving

$$|vI - A'| = 0.$$

We note that

$$vI - A' = (vI - A)'.$$

By Proposition 17, the determinant of $(vI - A)'$ is exactly the same as the determinant of $vI - A$. Hence, if by v_i, $i = 1, 2, \ldots, m$, we denote the characteristic roots of A', we conclude

$$v_i = \lambda_i, \qquad i = 1, 2, \ldots, m,$$

which proves part (a).

For part (b) we have that

$$|\mu I - A^{-1}| = 0$$

is the characteristic equation for A^{-1}, and moreover

$$\mu I - A^{-1} = A^{-1}(\mu A - I) = -\mu A^{-1}\left(\frac{1}{\mu} I - A\right).$$

Thus

$$|\mu I - A^{-1}| = (-1)^m \mu^m |A^{-1}||\lambda I - A|, \qquad \lambda = \frac{1}{\mu},$$

and we see that since $\mu = 0$ is *not* a root,

$$|\mu I - A^{-1}| = 0$$

if and only if

$$|\lambda I - A| = 0,$$

where

$$\lambda = \frac{1}{\mu}.$$

Hence, if μ_i are the roots of A^{-1} we must have

$$\mu_i = \frac{1}{\lambda_i}, \qquad i = 1, 2, \ldots, m. \quad \text{q.e.d.}$$

Another important result that may be derived through the characteristic equation is

Proposition 43. *Let A, B be two square matrices of order m. Then, the characteristic roots of AB are exactly the characteristic roots of BA.*

PROOF. The characteristic roots of AB and BA are, respectively, the solutions of

$$|\lambda I - AB| = 0, \qquad |\lambda I - BA| = 0.$$

We shall show that

$$|\lambda I - AB| = |\lambda I - BA|,$$

thus providing the desired result. For some square matrix C of order m consider

$$\psi(t) = |\lambda I + tC|,$$

where t is an indeterminate. Quite clearly, $\psi(t)$ is a polynomial of degree m. As such it may be represented by a Taylor series expansion about $t = 0$.

If the expansion contains $m + 1$ terms the resulting representation will be exact. Doing so we find

$$\psi(t) = \psi(0) + \psi'(0)t + \tfrac{1}{2}\psi''(0)t^2 + \cdots + \frac{1}{m!}\psi^{(m)}(0)t^m.$$

By the usual rules for differentiating determinants (see Section 4.3) we easily find that

$$\psi(0) = \lambda^m, \qquad \psi'(0) = \lambda^{m-1} \operatorname{tr} C$$

and, in general,

$$\frac{1}{j!}\psi^{(j)}(0) = \lambda^{m-j} h_j(C),$$

where $h_j(C)$ depends only on $\operatorname{tr} C, \operatorname{tr} C^2, \ldots, \operatorname{tr} C^j$. Thus, the characteristic equation for C is given by

$$\psi(-1) = \lambda^m - \lambda^{m-1} \operatorname{tr} C + \lambda^{m-2} h_2(C) - \lambda^{m-3} h_3(C) + \cdots + (-1)^m h_m(C)$$

Let

$$C_1 = AB, \qquad C_2 = BA$$

and note that

$$\operatorname{tr} C_1 = \operatorname{tr} C_2, \qquad \operatorname{tr} C_1^2 = \operatorname{tr} C_2^2,$$

and, in general,

$$\operatorname{tr} C_1^j = \operatorname{tr} C_2^j.$$

This is so since

$$C_1^j = (AB)(AB) \cdots (AB),$$

$$C_2^j = (BA)(BA) \cdots (BA) = B\overbrace{(AB)(AB) \cdots (AB)}^{j-1 \text{ terms}}A = B C_1^{j-1} A.$$

Thus,

$$\operatorname{tr} C_2^j = \operatorname{tr} B C_1^{j-1} A = \operatorname{tr} C_1^{j-1} AB = \operatorname{tr} C_1^j.$$

Consequently we see that

$$h_j(C_1) = h_j(C_2)$$

and, moreover,

$$|\lambda I - AB| = |\lambda I - BA|. \quad \text{q.e.d.}$$

Corollary 5. Let A, B be, respectively, $m \times n$ and $n \times m$ matrices, where $m \leq n$. Then the characteristic roots of BA (an $n \times n$ matrix) consist of $n - m$ zeros and the m characteristic roots of AB (an $m \times m$ matrix).

PROOF. Define the matrices

$$A_* = \begin{pmatrix} A \\ 0 \end{pmatrix}, \qquad B_* = (B, 0)$$

such that A_* and B_* are $n \times n$ matrices. By Proposition 43 the characteristic roots of $A_* B_*$ are exactly those of $B_* A_*$. But

$$A_* B_* = \begin{bmatrix} AB & 0 \\ 0 & 0 \end{bmatrix}, \qquad B_* A_* = BA.$$

Thus

$$\lambda I - A_* B_* = \begin{bmatrix} \lambda I - AB & 0 \\ 0 & \lambda I \end{bmatrix}$$

and, consequently,

$$|\lambda I - BA| = |\lambda I - B_* A_*| = |\lambda I - A_* B_*| = \lambda^{n-m}|\lambda I - AB|. \quad \text{q.e.d.}$$

Corollary 6. *Let A be a square matrix of order m, and let $\lambda_i, i = 1, 2, \ldots, m$, be its characteristic roots. Then*

$$\text{tr } A = \sum_{i=1}^{m} \lambda_i.$$

PROOF. From the proof of Proposition 43 we have that

$$|\lambda I - A| = \lambda^m - \lambda^{m-1} \text{ tr } A + \lambda^{m-2} h_2(A) + \cdots + (-1)^m h_m(A).$$

From the factorization of polynomials we have

$$|\lambda I - A| = \prod_{i=1}^{m} (\lambda - \lambda_i) = \lambda^m - \lambda^{m-1} \left(\sum_{i=1}^{m} \lambda_i \right) + \cdots + (-1)^m \prod_{i=1}^{m} \lambda_i$$

Equating the coefficients for λ^{m-1} we find

$$\text{tr } A = \sum_{i=1}^{m} \lambda_i. \quad \text{q.e.d.}$$

Proposition 44. *Let A be a square matrix of order m. Then A is diagonalizable, i.e., it is similar to a diagonal matrix, if and only if for each characteristic root λ of A the multiplicity of λ is equal to the nullity of $\lambda I - A$.*

PROOF. Suppose A is diagonalizable. Then we can write

$$Q^{-1} A Q = \Lambda, \qquad \Lambda = \text{diag}(\lambda_1, \lambda_2, \ldots, \lambda_m).$$

Now, suppose the *distinct* roots are $\lambda_{(i)}, i = 1, 2, \ldots, s, s \leq m$. Let the multiplicity of $\lambda_{(i)}$ be m_i, where

$$\sum_{i=1}^{s} m_i = m.$$

It is clear that $\lambda_{(i)}I - \Lambda$ has m_i zeros on its diagonal and hence it is of rank

$$r[\lambda_{(i)}I - \Lambda] = m - m_i = \sum_{j \neq i} m_j.$$

But

$$\lambda_{(i)}I - A = \lambda_{(i)}I - Q\Lambda Q^{-1} = Q(\lambda_{(i)}I - \Lambda)Q^{-1}$$

and it follows that

$$r(\lambda_{(i)}I - A) = r(\lambda_{(i)}I - \Lambda) = \sum_{j \neq i} m_j.$$

But $\lambda_{(i)}I - A$ is an $m \times m$ matrix and by Proposition 5 its nullity obeys

$$n[\lambda_{(i)}I - A] = m - r[\lambda_{(i)}I - A] = m_i,$$

which is the multiplicity of $\lambda_{(i)}$.

Conversely, suppose that the nullity of $\lambda_{(i)}I - A$ is m_i and $\sum_{i=1}^{s} m_i = m$. Choose the basis

$$\xi_{\cdot 1}, \xi_{\cdot 2}, \ldots, \xi_{\cdot m_1}$$

for the null space of $\lambda_{(1)}I - A$,

$$\xi_{\cdot m_1 + 1}, \ldots, \xi_{\cdot m_1 + m_2}$$

for the null space of $\lambda_{(2)}I - A$, and so on until the null space of $\lambda_{(s)}I - A$.

Thus, we have m, m-element, vectors

$$\xi_{\cdot 1}, \xi_{\cdot 2}, \ldots, \xi_{\cdot m},$$

and each *appropriate* subset of m_i vectors, $i = 1, 2, \ldots, s$, is linearly independent. We claim that this set of m vectors is linearly independent. For, suppose not. Then we can find a set of scalars a_i, not all of which are zero, such that

$$\sum_{k=1}^{m} \xi_{\cdot k} a_k = 0.$$

We can also write the equation above as

$$\sum_{i=1}^{s} \zeta_{\cdot i} = 0, \qquad \zeta_{\cdot i} = \sum_{j=m_1 + \cdots + m_{i-1}+1}^{m_1 + \cdots + m_i} \xi_{\cdot j} a_j, \qquad i = 1, 2, \ldots, s, \qquad (38)$$

it being understood that $m_0 = 0$. Because of the way in which we have chosen the $\xi_{\cdot k}$, $k = 1, 2, \ldots, m$, the second equation of (38) implies that the $\zeta_{\cdot i}$ obey

$$(\lambda_{(i)}I - A)\zeta_{\cdot i} = 0,$$

i.e., that they are characteristic vectors of A corresponding to the distinct roots $\lambda_{(i)}$, $i = 1, 2, \ldots, s$. The first equation of (38) then implies that the $\zeta_{\cdot i}$

are linearly dependent. This is a contradiction, however, by Proposition 37. Hence

$$a_k = 0, \qquad k = 1, 2, \ldots, m,$$

and the $\xi_{\cdot i}$, $i = 1, 2, \ldots, m$ are a linearly independent set. Let

$$X = (\xi_{\cdot 1}, \xi_{\cdot 2}, \ldots, \xi_{\cdot m})$$

and arrange the (distinct) roots

$$|\lambda_{(1)}| > |\lambda_{(2)}| > \cdots > |\lambda_{(s)}|.$$

Putting

$$\Lambda = \operatorname{diag}(\lambda_{(1)} I_{m_1}, \lambda_{(2)} I_{m_2}, \ldots, \lambda_{(s)} I_{m_s})$$

we must have

$$AX = X\Lambda.$$

Since X is nonsingular we conclude

$$X^{-1}AX = \Lambda. \quad \text{q.e.d.}$$

2.10 Orthogonal Matrices

Definition 30. Let a, b be two m-element vectors. They are said to be (mutually) *orthogonal* if

$$a'b = 0. \tag{39}$$

They are said to be *orthonormal* if (39) holds and in addition

$$a'a = 1, \qquad b'b = 1.$$

Definition 31. Let Q be a square matrix of order m. It is said to be *orthogonal* if its columns are orthonormal.

An immediate consequence of the definition is

Proposition 45. *Let Q be an orthogonal matrix. Then it is nonsingular.*

PROOF. We shall show that its columns are linearly independent. Suppose there exist scalars c_i, $i = 1, 2, \ldots, m$, such that

$$\sum_{i=1}^{n} c_i q_{\cdot i} = 0, \tag{40}$$

the $q_{\cdot i}$ being the (orthonormal) columns of Q.

Premultiply (40) by $q'_{.j}$ and note that we obtain

$$c_j q'_{.j} q_{.j} = 0.$$

But since $q'_{.j} q_{.j} = 1$ we conclude that (40) implies

$$c_j = 0, \qquad j = 1, 2, \ldots, m. \quad \text{q.e.d.}$$

A further consequence is

Proposition 46. *Let Q be an orthogonal matrix of order m. Then*

$$Q' = Q^{-1}.$$

PROOF. By the definition of an orthogonal matrix,

$$Q'Q = I.$$

By Proposition 45 its inverse exists. Multiplying on the right by Q^{-1} we find

$$Q' = Q^{-1}. \quad \text{q.e.d.}$$

Proposition 47. *Let Q be an orthogonal matrix of order m. Then:*

(a) $$|Q| = 1 \quad or \quad |Q| = -1;$$

(b) *if λ_i, $i = 1, 2, \ldots, m$, are the characteristic roots of Q, then*

$$\lambda_i = \pm 1, i = 1, 2, \ldots, m.$$

PROOF. The validity of (a) follows immediately from $Q'Q = I$, which implies

$$|Q|^2 = 1, \qquad |Q| = \pm 1.$$

For (b) we note that by Proposition 42, the characteristic roots of Q' are exactly those of Q and the characteristic roots of Q^{-1} are $1/\lambda_i$, $i = 1, 2, \ldots, m$, where the λ_i are the characteristic roots of Q. Since for an orthogonal matrix $Q' = Q^{-1}$ we conclude $\lambda_i = 1/\lambda_i$, which implies

$$\lambda_i = \pm 1. \quad \text{q.e.d.}$$

It is interesting that given a set of linearly independent vectors we can transform them into an orthonormal set. This procedure, known as Gram–Schmidt orthogonalization, is explained below.

Proposition 48 (Gram–Schmidt orthogonalization). *Let $\xi_{.i}$, $i = 1, 2, \ldots, m$, be a set of m linearly independent m-element (column) vectors. Then they can be transformed into a set of orthonormal vectors.*

Proof. We shall first transform the $\xi_{\cdot i}$ into an orthogonal set and then simply divide each resulting vector by the square root of its length to produce the desired orthonormal set. Thus, define

$$y_{\cdot 1} = \xi_{\cdot 1}$$
$$y_{\cdot 2} = a_{12}\xi_{\cdot 1} + \xi_{\cdot 2}$$
$$y_{\cdot 3} = a_{13}\xi_{\cdot 1} + a_{23}\xi_{\cdot 2} + \xi_{\cdot 3}$$
$$\vdots$$
$$y_{\cdot m} = a_{1m}\xi_{\cdot 1} + a_{2m}\xi_{\cdot 2}\cdots + a_{m-1,m}\xi_{\cdot m-1} + \xi_{\cdot m}$$

The condition for defining the a_{ij} is that

$$y'_{\cdot i}y_{\cdot j} = 0, \qquad i = 1, 2, \ldots, j-1. \tag{41}$$

But since $y_{\cdot i}$ depends *only* on $\xi_{\cdot 1}, \xi_{\cdot 2}, \ldots, \xi_{\cdot i}$, a condition equivalent to (41) is

$$\xi'_{\cdot i}y_{\cdot j} = 0, \qquad i = 1, 2, \ldots, j-1.$$

To make the notation compact, put

$$X_j = (\xi_{\cdot 1}, \xi_{\cdot 2}, \ldots, \xi_{\cdot j-1}), \qquad a_{\cdot j} = (a_{1j}, a_{2j}, \ldots, a_{j-1,j})',$$

and note that the y's may be written compactly as

$$y_{\cdot 1} = \xi_{\cdot 1}$$
$$y_{\cdot j} = X_j a_{\cdot j} + \xi_{\cdot j}, \qquad j = 2, \ldots, m.$$

We wish the $y_{\cdot j}$ to satisfy

$$X'_j y_{\cdot j} = X'_j X_j a_{\cdot j} + X'_j \xi_{\cdot j} = 0. \tag{42}$$

The matrix $X'_j X_j$ is nonsingular, however, since the columns of X_j are linearly independent;[3] hence

$$a_{\cdot j} = -(X'_j X_j)^{-1} X'_j \xi_{\cdot j}, \qquad j = 2, 3, \ldots, m, \tag{43}$$

and we see we can define the desired orthogonal set by

$$y_{\cdot 1} = \xi_{\cdot 1}$$
$$y_{\cdot i} = \xi_{\cdot i} - X_i (X'_i X_i)^{-1} X'_i \xi_{\cdot i}.$$

Then put

$$\zeta_{\cdot i} = \frac{y_{\cdot i}}{(y'_{\cdot i}y_{\cdot i})^{1/2}}, \qquad i = 1, 2, \ldots, m,$$

and note that

$$\zeta'_{\cdot i}\zeta_{\cdot i} = 1, \qquad i = 1, 2, \ldots, m.$$

---

[3] A simple proof of this is as follows. Suppose there exists a vector c such that $X'_j X_j c = 0$. But then $c'X'_j X_j c = 0$, which implies $X_j c = 0$, which in turn implies $c = 0$.

The set

$$\{\zeta_{\cdot i} : i = 1, 2, \ldots, m\}$$

is the desired orthonormal set.   q.e.d.

A simple consequence is

Proposition 49. *Let a be an m-element nonnull (column) vector with unit length. Then there exists an orthogonal matrix with a as the first column.*

PROOF. Given a there certainly exist m-element vectors $\xi_{\cdot 2}, \xi_{\cdot 3}, \ldots, \xi_{\cdot m}$ such that the set

$$a, \xi_{\cdot 2}, \ldots, \xi_{\cdot m}$$

is linearly independent. The desired matrix is then obtained by applying Gram–Schmidt orthogonalization to this set.   q.e.d.

Remark 27. Evidently, Propositions 48 and 49 are applicable to row vectors.

2.11 Symmetric Matrices

In this section we shall establish certain useful properties of symmetric matrices.

Proposition 50. *Let S be a symmetric matrix of order m whose elements are real. Then its characteristic roots are also real.*

PROOF. Let λ be any characteristic root of S and let z be its associated characteristic vector. Put

$$\lambda = \lambda_1 + i\lambda_2, \qquad z = x + iy$$

so that we allow that λ, z *may* be complex. Since they form a pair of a characteristic root and its associated characteristic vector they satisfy

$$Sz = \lambda z. \tag{44}$$

Premultiply by $\bar{z}'$, $\bar{z}$ being the complex conjugate of z. We find

$$\bar{z}'Sz = \lambda \bar{z}'z. \tag{45}$$

We note that since z is a characteristic vector

$$\bar{z}'z = x'x + yy > 0.$$

In (44) take the complex conjugate to obtain

$$S\bar{z} = \bar{\lambda}\bar{z} \tag{46}$$

since the elements of S are real. Premultiplying (46) by z' we find

$$z'S\bar{z} = \bar{\lambda}z'\bar{z}. \tag{47}$$

Since $\bar{z}'Sz$ is a scalar (a 1×1 "matrix"),

$$(z'S\bar{z}) = (z'S\bar{z})' = \bar{z}'Sz$$

and, moreover,

$$\bar{z}'z = z'\bar{z}.$$

Subtracting (47) from (45) we find

$$0 = (\lambda - \bar{\lambda})\bar{z}'z.$$

Since $\bar{z}'z > 0$ we conclude

$$\lambda = \bar{\lambda}.$$

But

$$\lambda = \lambda_1 + i\lambda_2, \qquad \bar{\lambda} = \lambda_1 - i\lambda_2,$$

which implies

$$\lambda_2 = -\lambda_2 \quad \text{or} \quad \lambda_2 = 0.$$

Hence

$$\lambda = \lambda_1,$$

and the characteristic root is real.   q.e.d.

A further important property is

Proposition 51. *Let S be a symmetric matrix of order m. Let its distinct roots be $\lambda_{(i)}, i = 1, 2, \ldots, s, s \leq m$, and let the multiplicity of $\lambda_{(i)}$ be m_i, $\sum_{i=1}^{s} m_i = m$. Then corresponding to the root $\lambda_{(i)}$ there exist m_i (linearly independent) orthonormal characteristic vectors.*

PROOF. Since $\lambda_{(i)}$ is a characteristic root of S, let $q_{\cdot 1}$ be its associated characteristic vector (of unit length). By Proposition 49 there exist vectors

$$p_{\cdot j}^{(1)}, \qquad j = 2, 3, \ldots, m,$$

such that

$$Q_1 = (q_{\cdot 1}, p_{\cdot 2}^{(1)}, p_{\cdot 3}^{(1)}, \ldots, p_{\cdot m}^{(1)})$$

is an orthogonal matrix. Consider

$$S_1 = Q_1'SQ_1 = \begin{bmatrix} \lambda_{(i)} & 0 \\ 0 & A_1 \end{bmatrix},$$

where A_1 is a matrix whose i, j element is

$$p_{\cdot i}^{(1)\prime} S p_{\cdot j}^{(1)}, \qquad i, j = 2, 3, \ldots, m.$$

But S and S_1 have exactly the same roots. Hence, if $m_i \geq 2$,

$$|\lambda I - S| = |\lambda I - S_1| = \begin{bmatrix} \lambda - \lambda_{(i)} & 0 \\ 0 & \lambda I_{m-1} - A_1 \end{bmatrix}$$

$$= (\lambda - \lambda_{(i)})|\lambda I_{m-1} - A_1| = 0$$

implies that $\lambda_{(i)}$ is also a root of

$$|\lambda I_{m-1} - A_1| = 0.$$

Hence the nullity of $\lambda_{(i)} I - S$ is at least two, i.e.,

$$n(\lambda_{(i)} I - S) \geq 2,$$

and we can thus find another vector, say, $q_{\cdot 2}$, satisfying

$$(\lambda_{(i)} I - S)q_{\cdot 2} = 0$$

and such that $q_{\cdot 1}, q_{\cdot 2}$ are linearly independent and of unit length, and such that the matrix

$$Q_2 = [q_{\cdot 1}, q_{\cdot 2}, p_{\cdot 3}^{(2)}, p_{\cdot 4}^{(2)}, \ldots, p_{\cdot m}^{(2)}]$$

is orthogonal.

Define

$$S_2 = Q_2' S Q_2$$

and note that S_2 has exactly the same roots as S. Note further that

$$|\lambda I - S| = |\lambda I - S_2| = \begin{bmatrix} \lambda - \lambda_{(i)} & 0 & \\ 0 & \lambda - \lambda_{(i)} & 0 \\ & 0 & \lambda I_{m-2} - A_2 \end{bmatrix}$$

$$= (\lambda - \lambda_{(i)})^2 |\lambda I_{m-2} - A_2| = 0$$

and $m_i > 2$ implies

$$|\lambda_{(i)} I_{m-2} - A_2| = 0.$$

Hence

$$n(\lambda_{(i)} I - S) \geq 3$$

and consequently we can choose another characteristic vector, $q_{\cdot 3}$, of unit length orthogonal to $q_{\cdot 1}, q_{\cdot 2}$ and such that

$$Q_3 = [q_{\cdot 1}, q_{\cdot 2}, q_{\cdot 3}, p_{\cdot 4}^{(3)}, p_{\cdot 5}^{(3)}, \ldots, p_{\cdot m}^{(3)}]$$

is an orthogonal matrix.

Continuing in this fashion we can choose m_i orthonormal vectors

$$q_{.1}, q_{.2}, \ldots, q_{.m_i}$$

corresponding to $\lambda_{(i)}$ whose multiplicity is m_i. It is clear that we cannot choose more than m_i such vectors since, after the choice of $q_{.m_i}$, we shall be dealing with

$$|\lambda I - S| = |I - S_{m_i}| = \begin{bmatrix} (\lambda - \lambda_{(i)})I_{m_i} & 0 \\ 0 & \lambda I_{m_i^*} - A_{m_i} \end{bmatrix}$$

$$= |(\lambda - \lambda_{(i)})^{m_i}| \lambda I_{m_i^*} - A_{m_i}| = 0,$$

where $m_i^* = m - m_i^*$. It is evident that

$$|\lambda I - S| = (\lambda - \lambda_{(i)})^{m_i}| \lambda I_{m_i^*} - A_{m_i}| = 0$$

implies

$$|\lambda_{(i)} I_{m_i^*} - A_{m_i}| \neq 0. \tag{48}$$

For, if not, the multiplicity of $\lambda_{(i)}$ would exceed m_i. In turn (48) means that

$$r(\lambda_{(i)} I - S) = m - m_i$$

and thus

$$n(\lambda_{(i)} I - S) = m_i.$$

Since we have chosen m_i linearly independent characteristic vectors corresponding to $\lambda_{(i)}$ they form a basis for the null space of $\lambda_{(i)} I - S$ and, thus, a larger number of such vectors would form a linearly dependent set.   q.e.d.

Corollary 7. *If S is as in Proposition 51, then the multiplicity of the root $\lambda_{(i)}$ is equal to the nullity of*

$$\lambda_{(i)} I - S.$$

PROOF. Obvious from the proof of the proposition above.

An important consequence of the preceding is

Proposition 52. *Let S be a symmetric matrix of order m. Then, the characteristic vectors of S can be chosen to be an orthonormal set, i.e., there exists a matrix Q such that*

$$Q'SQ = \Lambda,$$

or equivalently S is orthogonally similar to a diagonal matrix.

PROOF. Let the distinct characteristic roots of S be $\lambda_{(i)}, i = 1, 2, \ldots, s, s \leq m$, where $\lambda_{(i)}$ is of multiplicity m_i, and $\sum_{i=1}^{s} m_i = m$. By Corollary 7 the nullity of

$\lambda_{(i)}I - S$ is equal to the multiplicity m_i of the root $\lambda_{(i)}$. By Proposition 51 there exist m_i orthonormal characteristic vectors corresponding to $\lambda_{(i)}$. By Proposition 37 characteristic vectors corresponding to district characteristic roots are linearly independent. Hence the matrix

$$Q = (q_{.1}, q_{.2}, \ldots, q_{.m}),$$

where the first m_1 columns are the characteristic vectors corresponding to $\lambda_{(1)}$, the next m_2 columns are those corresponding to $\lambda_{(2)}$, and so on, is an orthogonal matrix. Define

$$\Lambda = \operatorname{diag}(\lambda_{(1)}I_{m_1}, \lambda_{(2)}I_{m_2}, \ldots, \lambda_{(s)}I_{m_s})$$

and note that we have

$$SQ = Q\Lambda.$$

Consequently

$$Q'SQ = \Lambda. \quad \text{q.e.d.}$$

Proposition 53 (Simultaneous diagonalization). *Let A, B be two symmetric matrices of order m; then there exists an orthogonal matrix Q such that*

$$Q'AQ = D_1, \qquad Q'BQ = D_2,$$

where the D_i, $i = 1, 2$, are diagonal matrices if and only if

$$AB = BA.$$

PROOF. The first part of the proposition is trivial since if such an orthogonal matrix exists then

$$Q'AQQ'BQ = D_1D_2,$$
$$Q'BQQ'AQ = D_2D_1 = D_1D_2.$$

But the two equations above imply

$$AB = QD_1D_2Q',$$
$$BA = QD_1D_2Q',$$

which shows that

$$AB = BA. \tag{49}$$

For the second part, suppose (49) holds. Since A is symmetric, let Λ be the diagonal matrix containing its (real) characteristic roots and let Q_1 be the matrix of associated characteristic vectors. Thus

$$Q_1'AQ_1 = \Lambda.$$

Define

$$C = Q_1'BQ_1$$

and note that

$$\begin{aligned}
AC &= Q_1' A Q_1 Q_1' B Q_1 \\
&= Q_1' A B Q_1 \\
&= Q_1' B A Q_1 = Q_1' B Q_1 Q_1' A Q_1 = C\Lambda
\end{aligned} \tag{50}$$

If the roots of A are all distinct we immediately conclude from (50) that

$$C = Q_1' B Q_1$$

is a diagonal matrix. Thus, taking

$$D_1 = \Lambda, \qquad D_2 = C,$$

the proof is completed. If not, let

$$\lambda_{(i)}, \qquad i = 1, 2, \ldots, s,$$

be the distinct roots of A and let $\lambda_{(i)}$ be of multiplicity m_i, where $\sum_{i=1}^{s} m_i = m$. We may write

$$\Lambda = \begin{bmatrix} \lambda_{(1)} I_{m_1} & 0 & \cdots & 0 \\ 0 & \lambda_{(2)} I_{m_2} & & \vdots \\ \vdots & & \ddots & 0 \\ 0 & \cdots & 0 & \lambda_{(s)} I_{m_s} \end{bmatrix}$$

Partition C conformably with Λ, i.e.,

$$C = \begin{bmatrix} C_{11} & C_{12} & \cdots & C_{1s} \\ C_{21} & C_{22} & \cdots & C_{2s} \\ \vdots & \vdots & & \vdots \\ C_{s1} & C_{s2} & \cdots & C_{ss} \end{bmatrix},$$

so that C_{ij} is a matrix with dimensions $m_i \times m_j$. From (50) we thus conclude that

$$\lambda_{(i)} C_{ij} = \lambda_{(j)} C_{ij}. \tag{51}$$

But for $i \neq j$ we have

$$\lambda_{(i)} \neq \lambda_{(j)}$$

and (51) implies

$$C_{ij} = 0, \qquad i \neq j. \tag{52}$$

Thus, C is the block diagonal matrix

$$C = \operatorname{diag}(C_{11}, C_{22}, \ldots, C_{ss}).$$

Clearly, the C_{ii}, $i = 1, 2, \ldots, s$, are symmetric matrices. Thus, there exist orthogonal matrices, say,

$$Q_i^*, \qquad i = 1, 2, 3, \ldots, s,$$

that diagonalize them, i.e.,

$$Q_i^{*\prime} C_{ii} Q_i^* = D_i^*, \qquad i = 1, 2, \ldots, m,$$

the D_i^* being diagonal matrices. Define

$$Q_2 = \operatorname{diag}(Q_1^*, Q_2^*, \ldots, Q_s^*)$$

and note that Q_2 is an othogonal matrix such that

$$D_2 = Q_2' C Q_2 = Q_2' Q_1' B Q_1 Q_2$$

with

$$D_2 = \operatorname{diag}(D_1^*, D_2^*, \ldots, D_s^*).$$

Evidently, D_2 is a diagonal matrix. Define

$$Q = Q_1 Q_2$$

and note:

(i) $Q'Q = Q_2' Q_1' Q_1 Q_2 = Q_2' Q_2 = I$, so that Q is an orthogonal matrix;
(ii) $Q'AQ = Q_2' \Lambda Q_2 = \Lambda$, which follows from the construction of Q_2;
(iii) $Q'BQ = D_2$.

Taking $D_1 = \Lambda$ we see that

$$Q'AQ = D_1, \qquad Q'BQ = D_2. \quad \text{q.e.d.}$$

Corollary 8. *Let A, B be two symmetric matrices of order m such that*

$$AB = 0.$$

Then there exists an orthogonal matrix Q such that

$$Q'AQ = D_1, \qquad Q'BQ = D_2,$$

and, moreover,

$$D_1 D_2 = 0.$$

PROOF. Since A, B are symmetric and

$$AB = 0$$

we see that

$$0 = (AB)' = B'A' = BA = AB.$$

By Proposition 53 there exists an orthogonal matrix Q such that

$$Q'AQ = D_1, \qquad Q'BQ = D_2.$$

Moreover,

$$D_1 D_2 = Q'AQQ'BQ = Q'ABQ = 0. \quad \text{q.e.d.}$$

2.12 Idempotent Matrices

We recall, from Definition 8, that A is said to be idempotent if A is square and

$$AA = A.$$

An easy consequence of the definition is

Proposition 54. Let A be a square matrix of order m; suppose further that A is idempotent. Then its characteristic roots are either zero or one.

PROOF. Let λ, x be a pair consisting of a characteristic root and its associated (normalized) characteristic vector. Thus

$$Ax = \lambda x. \tag{53}$$

Premultiplying by A we find

$$Ax = AAx = \lambda Ax = \lambda^2 x. \tag{54}$$

But (53) and (54) imply, after premultiplication by x',

$$\lambda = \lambda^2.$$

However, this is satisfied only by

$$\lambda = 0 \quad \text{or} \quad \lambda = 1. \quad \text{q.e.d.}$$

Remark 28. In idempotent matrices we have a nonobvious example of a matrix with repeated roots.

Proposition 55. *Let A be an idempotent matrix of order m and rank r. Then*

$$\operatorname{tr} A = r(A).$$

PROOF. Consider the characteristic equation of A

$$|\lambda I - A| = \prod_{i=1}^{m} (\lambda - \lambda_i), \tag{55}$$

where $\lambda_i, i = 1, 2, \ldots, m$, are the characteristic roots of A. By Proposition 43 we can write also

$$|\lambda I - A| = \lambda^m - \lambda^{m-1} \operatorname{tr} A + \lambda^{m-2} h_2(A) + \cdots + (-1)^m h_m(A) \tag{56}$$

But in (55) the coefficient of λ^{m-1} is simply $-\sum_{i=1}^{m} \lambda_i$. From (56) we thus conclude

$$\operatorname{tr} A = \sum_{i=1}^{m} \lambda_i.$$

By Proposition 54,

$$\lambda_i = 0 \quad \text{or} \quad \lambda_i = 1.$$

Hence

$$\text{tr } A = \text{number of nonzero roots}$$

or

$$\text{tr } A = r(A). \quad \text{q.e.d.}$$

2.13 Semidefinite and Definite Matrices

Definition 32. Let A be a square matrix of order m and let x be an m-element vector. Then A is said to be *positive semidefinite* if and only if *for all vectors x*

$$x'Ax \geq 0.$$

The matrix A is said to be *positive definite* if *for nonnull x*

$$x'Ax > 0.$$

Definition 33. Let A be a square matrix of order m. Then A is said to be *negative (semi)definite* if and only if $-A$ is positive (semi)definite.

Remark 29. It is clear that we need only study the properties of positive (semi)definite matrices, since the properties of negative (semi)definite matrices can easily be derived therefrom.

Remark 30. A definite or semidefinite matrix B need not be symmetric. However, since the defining property of such matrices involves the *quadratic form*

$$x'Bx,$$

we see that if we put

$$A = \tfrac{1}{2}(B + B')$$

we have

$$x'Ax = x'Bx$$

with A symmetric. Thus, whatever properties may be ascribed to B by virtue of the fact that for any x, say

$$x'Bx \geq 0,$$

can also be ascribed to A. Thus, we sacrifice no generality *if we always take definite or semidefinite matrices to be also symmetric. In subsequent discussion it should be understood that if we say that A is positive (semi)definite we also mean that A is symmetric.*

Certain properties follow immediately from the definition of definite and semidefinite matrices.

Proposition 56. *Let A be a square matrix of order m. If A is positive definite it is also positive semidefinite. The converse, however, is not true.*

PROOF. The first part is obvious from the definition since if x is *any* m-element vector and A is positive definite then

$$x'Ax \geq 0,$$

so that A is also positive semidefinite.

That the converse is not true is established by an example. Take

$$A = \begin{bmatrix} 1 & 1 \\ 1 & 1 \end{bmatrix}.$$

For any vector $x = (x_1, x_2)'$,

$$x'Ax = (x_1 + x_2)^2 \geq 0,$$

so that A is positive semidefinite. For the choice $x_1 = -x_2, x_2 \neq 0$, we have

$$x'Ax = 0,$$

which shows that A is *not* positive definite.   q.e.d.

Proposition 57. *Let A be a positive definite matrix of order m. Then*

$$a_{ii} > 0, \qquad i = 1, 2, \ldots, m.$$

If A is only positive semidefinite then

$$a_{ii} \geq 0, \qquad i = 1, 2, \ldots, m.$$

PROOF. Let $e_{\cdot i}$ be the m-element vector all of whose elements are zero save the ith, which is unity. If A is positive definite, since $e_{\cdot i}$ is not the null vector, we must have

$$e'_{\cdot i} A e_{\cdot i} > 0, \qquad i = 1, 2, \ldots, m.$$

But

$$e'_{\cdot i} A e_{\cdot i} = a_{ii}, \qquad i = 1, 2, \ldots, m.$$

If A is positive semidefinite but *not* positive definite then repeating the argument above we find

$$a_{ii} = e'_{\cdot i} A e_{\cdot i} \geq 0, \qquad i = 1, 2, \ldots, m.   \text{q.e.d.}$$

Another interesting property is given by

Proposition 58. *Let A be a positive definite matrix of order m. Then there exists a lower triangular matrix T such that*

$$A = TT'.$$

PROOF. Let

$$
T = \begin{bmatrix}
t_{11} & 0 & \cdots & & 0 \\
t_{21} & t_{22} & & & \vdots \\
t_{31} & t_{32} & t_{33} & & \\
& & & \ddots & 0 \\
t_{m1} & t_{m2} & t_{m3} & \cdots & t_{mm}
\end{bmatrix}
$$

Setting

$$ A = TT' $$

we obtain the equations (by equating the (i, j) elements of A and TT')

$$ t_{11}^2 = a_{11}, \quad t_{11}t_{21} = a_{12}, \quad t_{11}t_{31} = a_{13}, \quad \ldots, \quad t_{11}t_{m1} = a_{1m} $$
$$ t_{21}t_{11} = a_{21}, \quad t_{21}^2 + t_{22}^2 = a_{22}, \quad t_{21}t_{31} + t_{22}t_{32} = a_{23}, \quad \ldots, $$
$$ t_{21}t_{m1} + t_{22}t_{m2} = a_{2m} $$
$$ \vdots $$
$$ t_{m1}t_{11} = a_{m1}, \quad t_{m1}t_{21} + t_{m2}t_{22} = a_{m2}, \quad \ldots, \quad \sum_{i=1}^{m} t_{mi}^2 = a_{mm}. $$

In solving the equations as they are arranged, line by line, we see that we are dealing with a recursive system. From the first line we have

$$ t_{11} = \pm\sqrt{a_{11}}, \quad t_{21} = \frac{a_{21}}{t_{11}}, \quad t_{31} = \frac{a_{21}}{t_{11}}, \quad \ldots, \quad t_{m1} = \frac{a_{1m}}{t_{11}}. $$

From the second line we have

$$ t_{21} = \frac{a_{21}}{t_{11}}, \quad t_{22} = \pm\left(\frac{a_{22}a_{11} - a_{21}^2}{a_{11}}\right)^{1/2}, $$

and in general

$$ t_{i2} = \frac{a_{2i} - t_{21}t_{i1}}{t_{22}}, \quad i = 3, 4, \ldots, m. $$

Similarly, in the third line we shall find

$$ t_{33} = \pm\left(a_{33} - \frac{a_{31}^2}{t_{31}^2} - \frac{(a_{23} - t_{21}t_{31})^2}{t_{22}^2}\right)^{1/2}, $$

$$ t_{i3} = \frac{a_{3i} - t_{31}t_{i1} - t_{32}t_{i2}}{t_{33}}, \quad i = 4, 5, \ldots, m, $$

and so on.   q.e.d.

Remark 31. Evidently, the lower triangular matrix above is not unique. In particular, we see that for t_{11} we have the choice

$$ t_{11} = \sqrt{a_{11}} \quad \text{or} \quad t_{11} = -\sqrt{a_{11}}. $$

Similarly, for t_{22} we have the choice

$$t_{22} = \left(\frac{a_{22} a_{11} - a_{21}^2}{a_{11}} \right)^{1/2} \quad \text{or} \quad t_{22} = -\left(\frac{a_{22} a_{11} - a_{21}^2}{a_{11}} \right)^{1/2},$$

and so on. The matrix T can be rendered unique if we specify, say, that all diagonal elements must be positive.

Notice further that the same argument as in Proposition 58 can establish the existence of an *upper triangular* matrix T^* such that

$$A = T^* T^{*'}.$$

The properties of characteristic roots of (semi)definite matrices are established in

Proposition 59. *Let A be a symmetric matrix of order m and let λ_i, $i = 1, 2, \ldots, m$, be its (real) characteristic roots. If A is* positive definite, *then*

$$\lambda_i > 0, \qquad i = 1, 2, \ldots, m.$$

If it is only positive semidefinite *then*

$$\lambda_i \geq 0, \qquad i = 1, 2, \ldots, m.$$

PROOF. Let $x_{\cdot i}$ be the normalized characteristic vector corresponding to the root, λ_i, of A. If A is positive definite, then

$$x'_{\cdot i} A x_{\cdot i} = \lambda_i > 0, \qquad i = 1, 2, \ldots, m.$$

If A is merely positive semidefinite then we can only assert

$$x'_{\cdot i} A x_{\cdot i} = \lambda_i \geq 0, \qquad i = 1, 2, \ldots, m. \quad \text{q.e.d.}$$

By now the reader should have surmised that positive definite matrices are nonsingular and positive semidefinite matrices (which are *not* also positive definite) are singular matrices. This is formalized in

Proposition 60. *Let A be a symmetric matrix of order m. If A is positive definite then*

$$r(A) = m.$$

*If A is merely positive semidefinite (i.e., it is **not** positive definite) then*

$$r(A) < m.$$

PROOF. Since A is symmetric, let Λ denote the diagonal matrix of its (real) characteristic roots and Q the associated (orthogonal) matrix of characteristic vectors. We have

$$AQ = Q\Lambda.$$

By Proposition 59, if A is positive definite,

$$\Lambda^{-1} = \text{diag}\left(\frac{1}{\lambda_1}, \frac{1}{\lambda_2}, \ldots, \frac{1}{\lambda_m}\right).$$

Consequently, the inverse of A exists and is given by

$$A^{-1} = Q\Lambda^{-1}Q'.$$

This establishes the first part of the proposition; for the second part suppose A is *only* positive semidefinite. From Proposition 59 we merely know that $\lambda_i \geq 0, i = 1, 2, \ldots, m$. We shall now establish that at least one root must be zero, thus completing the proof of the proposition.

We have the representation

$$Q'AQ = \Lambda.$$

Consequently, for any vector y,

$$y'Q'AQy = \sum_{i=1}^{m} \lambda_i y_i^2.$$

Now, if x is *any nonnull* vector by the semidefiniteness of A we have

$$0 \leq x'Ax = x'QQ'AQQ'x = x'Q\Lambda Q'x = \sum_{i=1}^{m} \lambda_i y_i^2, \tag{57}$$

where now we have put

$$y = Q'x.$$

Since x is nonnull, y is also nonnull (why?).

If none of the λ_i is zero, (57) implies that for *any* nonnull x

$$x'Ax > 0,$$

thus showing A to be positive definite. Consequently, at least one of the λ_i, $i = 1, 2, \ldots, m$, must be zero and there must exist at least one nonnull x such that

$$x'Ax = \sum_{i=1}^{m} \lambda_i y_i^2 = 0.$$

But this shows that

$$r(A) < m. \quad \text{q.e.d.}$$

Remark 32. Positive definite and semidefinite matrices correspond, roughly, to positive and nonnegative numbers in the usual number system. The reader's intuitive comprehension would be aided if he thinks of them as a sort of matrix generalization of positive and nonnegative real numbers. Just as a nonnegative number can always be written as the square of some other number we have an analogous characterization of definite and semi-definite matrices.

Proposition 61. *Let A be a symmetric matrix, of order m. Then A is positive definite if and only if there exists a matrix S of dimension $n \times m$ and rank m $(n \geq m)$ such that*

$$A = S'S.$$

It is positive semidefinite if and only if

$$r(S) < m.$$

PROOF. If A is positive (semi)definite then, as in Proposition 59, we have the representation

$$A = Q\Lambda Q'.$$

Taking

$$S = \Lambda^{1/2}Q'$$

we have

$$A = S'S.$$

If A is positive definite Λ is nonsingular and thus

$$r(S) = m.$$

If A is merely positive semidefinite then $r(\Lambda) < m$ and hence

$$r(S) < m.$$

On the other hand suppose

$$A = S'S$$

and S is $n \times m$ $(n \geq m)$ of rank m. Let x be any nonnull vector and note

$$x'Ax = x'S'Sx.$$

The right side of the equation above is a sum of squares and thus is zero if and only if

$$Sx = 0. \tag{58}$$

If the rank of S is m, (58) can be satisfied only with *null* x. Hence A is positive definite.

Evidently, for any x

$$x'Ax = x'S'Sx \geq 0,$$

and if S is of rank less then m, there exists at least one nonnull x such that

$$Sx = 0.$$

Consequently, there exists at least one nonnull x such that

$$x'Ax = 0,$$

which shows that A is positive semidefinite but *not* positive definite.   q.e.d.

An obvious consequence of the previous discussion is

Corollary 9. *Let A be a positive definite matrix; then*

$$|A| > 0, \qquad \operatorname{tr}(A) > 0.$$

PROOF. Let $\lambda_i, i = 1, 2, \ldots, m$, be the characteristic roots of A. Since

$$|A| = \prod_{i=1}^{m} \lambda_i, \qquad \operatorname{tr}(A) = \sum_{i=1}^{m} \lambda_i,$$

the result follows immediately from Proposition 59.   q.e.d.

Corollary 10. *Let A be a positive semidefinite but **not** a positive definite matrix. Then*

$$|A| = 0, \qquad \operatorname{tr}(A) \geq 0,$$

and

$$\operatorname{tr}(A) = 0$$

if and only if A is the null matrix.

PROOF. From the representation

$$|A| = \prod_{i=1}^{m} \lambda_i$$

we conclude that $|A| = 0$ by Proposition 60.
   For the second part we note that

$$A = Q\Lambda Q' \tag{59}$$

and

$$\operatorname{tr}(A) = 0$$

if and only if

$$\operatorname{tr}(\Lambda) = 0.$$

But

$$\operatorname{tr}(\Lambda) = \sum_{i=1}^{m} \lambda_i = 0$$

implies

$$\lambda_i = 0, \qquad i = 1, 2, \ldots, m.$$

If this holds, then (59) implies

$$A = 0.$$

Consequently *if A is not a null matrix*

$$\operatorname{tr}(A) > 0. \quad \text{q.e.d.}$$

Corollary 11. *Let A be a positive definite matrix of order m. Then there exists a nonsingular matrix W such that*

$$A = W'W.$$

PROOF. Obvious from Propositions 58, 60, and 61. q.e.d.

Hithertofore when considering characteristic roots and characteristic vectors we have done so in the context of the characteristic equation

$$|\lambda I - A| = 0.$$

Often it is more convenient to broaden the definition of characteristic roots and vectors as follows.

Definition 34. Let A, B be two matrices of order m, where B is nonsingular. The *characteristic roots of A in the metric of B*, and their associated characteristic vectors, are connected by the relation

$$Ax = \lambda Bx,$$

where λ is a characteristic root and x is the associated (nonnull) characteristic vector.

Remark 33. It is evident that the *characteristic roots of A in the metric of B* are found by solving the polynomial equation

$$|\lambda B - A| = 0.$$

It is also clear that this is a simple generalization of the ordinary definition of characteristic roots where the role of B is played by the identity matrix.

Definition 34 is quite useful in dealing with differences of positive (semi)-definite matrices and particularly in determining whether such differences are positive (semi)definite or not. This is intimately connected with the question of relative efficiency of two estimators. We have

Proposition 62. *Let B be a positive definite matrix and let A be positive (semi)-definite. Then the characteristic roots of A in the metric of B, say, λ_i, obey*

$$\lambda_i > 0, \qquad i = 1, 2, \ldots, m,$$

if A is positive definite and

$$\lambda_i \geq 0, \qquad i = 1, 2, \ldots, m,$$

if A is positive semidefinite.

PROOF. Consider

$$|\lambda B - A| = 0.$$

Since B is positive definite, by Corollary 11 there exists a nonsingular matrix P such that

$$B = P'^{-1}P^{-1}.$$

Consequently,

$$0 = |\lambda B - A| = |\lambda P'^{-1}P^{-1} - A| = |\lambda I - P'AP||P|^{-2}.$$

Thus, the *characteristic roots of A in the metric of B are* simply the usual characteristic roots of $P'AP$. i.e., the solution of

$$|\lambda I - P'AP| = 0.$$

If A is positive definite then $P'AP$ is also positive definite; if A is *only* positive semidefinite then $P'AP$ is *only* positive semidefinite. Hence in the former case

$$\lambda_i > 0, \qquad i = 1, 2, \ldots, m,$$

while in the latter case

$$\lambda_i \geq 0, \qquad i = 1, 2, \ldots, m. \quad \text{q.e.d.}$$

A very useful result in this context is

Proposition 63 (Simultaneous decomposition). *Let B be a positive definite matrix and A be positive (semi)definite. Let*

$$\Lambda = \text{diag}(\lambda_1, \lambda_2, \ldots, \lambda_m)$$

be the diagonal matrix of the characteristic roots of A in the metric of B. Then there exists a nonsingular matrix W such that

$$B = W'W, \qquad A = W'\Lambda W.$$

PROOF. From Proposition 62 we have that the λ_i are also the (ordinary) characteristic roots of $P'AP$, where P is such that

$$B = P'^{-1}P^{-1}.$$

Let Q be the (orthogonal) matrix of (ordinary) characteristic vectors of $P'AP$. Thus, we have

$$P'APQ = Q\Lambda. \tag{60}$$

From (60) we easily establish

$$A = P'^{-1}Q\Lambda Q'P^{-1}.$$

Putting

$$W = Q'P^{-1}$$

we have

$$A = W'\Lambda W,$$

$$B = W'W. \quad \text{q.e.d.}$$

From the preceding two propositions flow a number of useful results regarding differences of positive (semi)definite matrices. Thus,

Proposition 64. *Let B be a positive definite matrix and A be positive (semi)-definite. Then B − A is positive (semi)definite if and only if*

$$\lambda_i < 1 \qquad (\lambda_i \le 1),$$

where the λ_i are the characteristic roots of A in the metric of B.

PROOF. From Proposition 63, there exists a nonsingular matrix W such that

$$B = W'W, \qquad A = W'\Lambda W.$$

Hence

$$B - A = W'(I - \Lambda)W.$$

Let x be any m-element vector and note

$$x'(B - A)x = y'(I - \Lambda)y = \sum_{i=1}^{m} (1 - \lambda_i)y_i^2, \tag{61}$$

where $y = Wx$. If for nonnull x

$$x'(B - A)x > 0$$

then we must have

$$1 - \lambda_i > 0,$$

or

$$\lambda_i < 1, \qquad i = 1, 2, \ldots, m. \tag{62}$$

Conversely, if (62) holds then, for any nonnull x, y is also nonnull, and from (61) we see that $B - A$ is positive definite. If on the other hand $B - A$ is only positive semidefinite then for at least one index i we must have

$$\lambda_i = 1,$$

and conversely. q.e.d.

Another useful result easily obtained from the simultaneous decomposition of matrices is given in

Proposition 65. *Let A, B be two positive definite matrices. If B − A is positive definite then so is*

$$A^{-1} - B^{-1}.$$

PROOF. We may write

$$B = W'W, \qquad A = W'\Lambda W,$$

and by Proposition 64 the diagonal elements of Λ (i.e., the roots of A in the metric of B) are less than unity. Hence

$$B^{-1} = W^{-1}W'^{-1}, \qquad A^{-1} = W^{-1}\Lambda^{-1}W'^{-1}.$$

Thus

$$A^{-1} - B^{-1} = W^{-1}(\Lambda^{-1} - I)W'^{-1}.$$

The diagonal elements of $\Lambda^{-1} - I$ are given by

$$\frac{1}{\lambda_i} - 1 > 0, \qquad i = 1, 2, \ldots, m,$$

and thus

$$A^{-1} - B^{-1}$$

is positive definite by Proposition 61.   q.e.d.

Finally, we have

Proposition 66. *Let B be positive definite and A be positive (semi)definite. If*

$$B - A$$

is positive (semi)definite then

$$|B| > |A| \qquad (|B| \geq |A|),$$

$$\mathrm{tr}(B) > \mathrm{tr}(A), \qquad (\mathrm{tr}(B) \geq \mathrm{tr}(A)).$$

PROOF. As in Proposition 65 we can write

$$B = W'W, \qquad A = W'\Lambda W,$$

and by Proposition 64 we know that the diagonal elements of Λ, viz., the λ_i, obey

$$\lambda_i < 1, \qquad (\lambda_i \leq 1).$$

Consequently

$$|B| = |W|^2, \qquad |A| = |W|^2|\Lambda|.$$

Thus, if $B - A$ is positive definite,

$$|\Lambda| < 1$$

and, hence,

$$|B| > |A|.$$

Evidently, the inequality above is automatically satisfied if A itself is merely positive semidefinite. Moreover

$$\text{tr}(B) - \text{tr}(A) = \text{tr}(B - A) > 0.$$

On the other hand, if $B - A$ is merely positive semidefinite then we can only assert

$$|\Lambda| \leq 1,$$

and hence we conclude that

$$|B| \geq |A|, \qquad \text{tr}(B) \geq \text{tr}(A). \quad \text{q.e.d.}$$

Corollary 12. *In Proposition* 66 *the strict inequalities will hold unless*

$$B = A.$$

PROOF. Since

$$\frac{|A|}{|B|} = |\Lambda| = \prod_{i=1}^{m} \lambda_i,$$

we see that

$$|A| = |B|$$

implies

$$\lambda_i = 1, \qquad i = 1, 2, \ldots, m,$$

hence

$$B = A.$$

Similarly,

$$\text{tr}(B) = \text{tr}(A)$$

implies

$$0 = \text{tr}(B - A) = \text{tr } W'(I - \Lambda)W.$$

But this means

$$W'(I - \Lambda)W = 0,$$

which in turn implies

$$\Lambda = I,$$

and consequently

$$B = A. \quad \text{q.e.d.}$$

Linear Systems of Equations and Generalized Inverses of Matrices 3

3.1 Introduction

Consider the linear system of equations

$$Ax = b, \tag{63}$$

where A is $m \times n$ and b is an m-element vector. The meaning of (63), as a system of equations, is that we seek an n-element vector x satisfying (63). If $m = n$ and if A is nonsingular there exists the unique solution

$$x^* = A^{-1}b. \tag{64}$$

If A is singular, i.e., if $r(A) < m$, or if $n > m$, it is clear that more than one solution may exist.

Moreover, if A is $m \times n$ but $m > n$, the system may be *inconsistent*, i.e., there may not exist a vector x satisfying all the conditions (equations) specified in (63). In such a case we may wish to derive "approximate" solutions. In this connection, and by way of motivation, note that we may characterize estimation in the context of the general linear model (GLM) in the following terms. Find a vector b such that

$$y = Xb \tag{65}$$

is satisfied approximately, where X is a $T \times (n + 1)$ matrix of observations on the explanatory variables and y is a T-element vector of observations on the dependent variable. Typically

$$T > n + 1,$$

and the system in (65) is evidently inconsistent, since no vector b exists strictly satisfying (65). It is such considerations that prompt us to study various types of pseudoinverses.

3.2 Conditional, Least Squares, and Generalized Inverses of Matrices

Definition 35. Let A be $m \times n\ (m \leq n)$; the $n \times m$ matrix A_c is said to be the conditional inverse (c-inverse) of A if and only if

$$AA_c A = A.$$

Remark 34. Note that if A is a nonsingular matrix then clearly

$$AA^{-1}A = A,$$

so that the c-inverse satisfies only this property. Evidently, the proper inverse also satisfies other conditions such that, e.g.,

$$A^{-1}A, \qquad A^{-1}A$$

are symmetric matrices. But this is not necessarily satisfied by the conditional inverse.

We now show that the c-inverse is not a vacuous concept.

Proposition 67. *Let A be $m \times m$ and let B be a nonsingular matrix such that*

$$BA = H$$

and H is in (upper) Hermite form. Then B is a c-inverse of A.

PROOF. It is clear by Proposition 9 that such a nonsingular matrix B exists. From Proposition 13 H is an idempotent matrix. Hence

$$H = HH = BABA.$$

Premultiplying by B^{-1} we find

$$ABA = B^{-1}H = A. \quad \text{q.e.d.}$$

For rectangular (nonsquare) matrices we have

Proposition 68. *Let A be $m \times n\ (m \leq n)$ and*

$$A_0 = \begin{pmatrix} A \\ 0 \end{pmatrix},$$

where A_0 is $n \times n$. Let B_0 be a nonsingular $n \times n$ matrix such that

$$B_0 A_0 = H_0$$

and H_0 is in (upper) Hermite form. Partition

$$B_0 = (B, B_1),$$

where B is $n \times m$. Then B is a c-inverse of A.

PROOF. By Proposition 67, B_0 is a c-inverse of A_0. Hence we have

$$A_0 B_0 A_0 = A_0.$$

But

$$\begin{pmatrix} A \\ 0 \end{pmatrix} = A_0 = A_0 B_0 A_0 = \begin{bmatrix} AB & AB_1 \\ 0 & 0 \end{bmatrix} \begin{pmatrix} A \\ 0 \end{pmatrix} = \begin{pmatrix} ABA \\ 0 \end{pmatrix},$$

which shows that

$$A = ABA. \quad \text{q.e.d.}$$

Remark 35. A similar result is obtained if $m \geq n$. One has only to deal with the transpose of A in Proposition 68 and note that if A_c is the c-inverse of A, then A_c' is the c-inverse of A'.

Evidently c-inverses are not unique, since the matrix B reducing a given matrix A to Hermite form is not unique. This is perhaps best made clear by an example.

EXAMPLE 4. Let

$$A = \begin{bmatrix} 2 & 3 & 1 \\ 4 & 5 & 1 \\ 1 & 1 & 0 \end{bmatrix}$$

and observe that both

$$B_1 = \begin{bmatrix} -2 & 1 & 1 \\ 0 & 1 & -4 \\ -1 & 1 & -2 \end{bmatrix}, \quad B_2 = \begin{bmatrix} -3 & 2 & -1 \\ -1 & 2 & -6 \\ -2 & 2 & -4 \end{bmatrix}$$

have the property of reducing A to Hermite form, i.e.,

$$B_1 A = B_2 A = \begin{bmatrix} 1 & 0 & -1 \\ 0 & 1 & 1 \\ 0 & 0 & 0 \end{bmatrix}.$$

A somewhat more stringent set of requirements defines the so-called least squares inverse.

Definition 36. Let A be $m \times n$ ($m \leq n$). The $n \times m$ matrix A_s is said to be a *least squares inverse (s-inverse)* of A if and only if

(i) $AA_s A = A$,
(ii) AA_s is symmetric.

Remark 36. Evidently, if A_s is an s-inverse, it is also a c-inverse. The converse, however, is not true.

That the class of s-inverses is not a vacuous one is shown by

Proposition 69. *Let A be $m \times n$ ($m \leq n$). Then*

$$A_s = (A'A)_c A'$$

is an s-inverse of A.

PROOF. We show that

$$AA_s A = A(A'A)_c A'A = A$$

and

$$AA_s = A(A'A)_c A'$$

is symmetric. The second claim is evident. For the first we note that since $(A'A)_c$ is a c-inverse it satisfies

$$(A'A)(A'A)_c A'A = A'A. \tag{66}$$

Let A have rank $r \leq m$. From Proposition 15 there exist matrices C_1, C_2 of rank r and dimensions $m \times r, r \times n$ respectively such that (rank factorization)

$$A = C_1 C_2.$$

Thus we may write (66) as

$$C_2' C_1' C_1 C_2 (A'A)_c C_2' C_1' C_1 C_2 = C_2' C_1' C_1 C_2.$$

Premultiply by

$$C_1(C_1'C_1)^{-1}(C_2 C_2')^{-1}C_2$$

to obtain

$$C_1 C_2 (A'A)_c C_2' C_1' C_1 C_2 = C_1 C_2. \tag{67}$$

Bearing in mind the definition of A_s and the rank factorization of A we see that (67) can also be written as

$$AA_s A = A. \quad \text{q.e.d.}$$

Remark 37. Evidently, since the c-inverse is not unique the s-inverse is also not unique.

A unique pseudoinverse is defined in the following manner.

Definition 37. Let A be $m \times n$; the $n \times m$ matrix A_g is said to be a generalized inverse (g-inverse) of A if and only if it satisfies

(i) $AA_g A = A$,
(ii) AA_g is symmetric,
(iii) $A_g A$ is symmetric,
(iv) $A_g AA_g = A_g$.

Remark 38. Note that the g-inverse mimics the analogous conditions satisfied by a proper inverse. For if A is nonsingular and A^{-1} is its inverse,

$AA^{-1}A = A,$
$AA^{-1} = I$ is symmetric,
$A^{-1}A = I$ is symmetric,
$A^{-1}AA^{-1} = A^{-1}.$

Moreover, c-inverses, s-inverses and g-inverses are more generally referred to as *pseudoinverses*.

3.3 Properties of the Generalized Inverse

In this section we shall examine a number of useful properties of the g-inverse. We begin with the existence and uniqueness properties.

Proposition 70. *Let A be any $m \times n$ matrix. Then the following statements are true*:

(1) *there exists a unique matrix A_g satisfying the conditions of Definition 37;*
(2) *$A_g A$, $A_g A$ are idempotent matrices;*
(3) *the g-inverse of A' is A'_g.*

PROOF. It is clear that if A is the null matrix, then the $n \times m$ null matrix *is the* g-*inverse of A.* Thus suppose rank$(A) = r > 0$. By Proposition 15 there exist two matrices, namely C_1, which is $m \times r$ of rank r, and C_2, which is $r \times n$ of rank r, such that

$$A = C_1 C_2.$$

Define

$$A_g = C_2'(C_2 C_2')^{-1}(C_1' C_1)^{-1} C_1'$$

and observe that

$$AA_g = C_1 C_2 C_2'(C_2 C_2')^{-1}(C_1' C_1)^{-1} C_1' = C_1(C_1' C_1)^{-1} C_1',$$

$$A_g A = C_2'(C_2 C_2')^{-1}(C_1' C_1)^{-1} C_1' C_1 C_2 = C_2'(C_2 C_2')^{-1} C_2.$$

This shows AA_g *and* $A_g A$ to be symmetric idempotent matrices, thus satisfying (ii) and (iii) of Definition 37 and proving part (2) of the proposition. Moreover,

$$AA_g A = C_1 C_2 C_2'(C_2 C_2')^{-1} C_2 = C_1 C_2 = A,$$

$$A_g AA_g = C_2'(C_2 C_2')^{-1}(C_1' C_1)^{-1} C_1' C_1(C_1' C_1)^{-1} C_1' = A_g,$$

which shows the existence of the g-inverse. To show uniqueness, suppose B_g is another g-inverse of A. We shall show $A_g = B_g$, thus completing the proof of part (1) of the proposition.

Now, $AA_g A = A$. Postmultiplying by B_g we have

$$AA_g AB_g = AB_g.$$

Since AB_g and AA_g are both symmetric, we have

$$AB_g = AB_g AA_g = AA_g.$$

Similarly,

$$B_g A = B_g AA_g A = A_g AB_g A = A_g A.$$

Premultiplying the relation $AB_g = AA_g$ by B_g we have

$$B_g = B_g AB_g = B_g AA_g = A_g AA_g = A_g.$$

To show the validity of part (3) we simply note that if A_g is a g-inverse of A, then transposing the four conditions of Definition 37 yields the conclusion that A'_g is the g-inverse of A', which completes the proof of the proposition. q.e.d.

Let us now establish some other useful properties of the g-inverse.

Proposition 71. *Let A be an $m \times m$ symmetric matrix of rank r ($r \leq m$). Let D_r be the diagonal matrix containing its nonzero characteristic roots (in decreasing order of magnitude), and let P_r be the $m \times r$ matrix whose columns are the (orthonormal) characteristic vectors corresponding to the nonzero roots of A. Then*

$$A_g = P_r D_r^{-1} P_r$$

PROOF. By the definition of characteristic roots and vectors,

$$AP = PD,$$

where P is the orthogonal matrix of characteristic vectors of A and D is the diagonal matrix of the latter's characteristic roots arranged in decreasing order of magnitude. Because A is of rank r, D can be written as

$$D = \begin{bmatrix} D_r & 0 \\ 0 & 0 \end{bmatrix}.$$

Partition P by

$$P = (P_r, P_*),$$

where P_r is $m \times r$, and note

$$A = PDP' = (P_r, P_*) \begin{bmatrix} D_r & 0 \\ 0 & 0 \end{bmatrix} \begin{pmatrix} P'_r \\ P'_* \end{pmatrix} = P_r D_r P'_r.$$

We verify

(i) $AA_g A = (P_r D_r P_r')(P_r D_r^{-1} P_r')(P_r D_r P_r') = P_r D_r P_r' = A,$
(ii) $AA_g = (P_r D_r P_r')(P_r D_r^{-1} P_r') = P_r P_r',$ which is symmetric,
(iii) $A_g A = (P_r D_r^{-1} P_r')(P_r D_r P_r') = P_r P_r',$
(iv) $A_g A A_g = P_r P_r' P_r D_r^{-1} P_r' = P_r D_r^{-1} P_r = A_g.$

This shows that A_g, above, is a g-inverse of A.   q.e.d.

Corollary 13. *If* A *is symmetric* **and** *idempotent, then*

$$A_g = A.$$

PROOF. If A is symmetric *and* idempotent, then its characteristic roots are either zero or one. Hence in the representation above

$$D_r = I_r.$$

Thus

$$A = P_r P_r', \qquad A_g = P_r P_r'. \quad \text{q.e.d.}$$

We have already seen that if A_g is the g-inverse of A, then $(A_g)'$ is the g-inverse of A'. We now examine a number of other properties of the g-inverse that reveal it to be analogous to the ordinary inverse of a nonsingular matrix.

Proposition 72. *Let* A *be* $m \times n$. *Then the following statements are true*:

(i) $(A_g)_g = A$;
(ii) $\text{rank}(A_g) = \text{rank}(A)$;
(iii) $(A'A)_g = A_g A_g'$;
(iv) $(AA_g)_g = AA_g$;
(v) if $m = n$ and A is nonsingular then $A^{-1} = A_g$.

PROOF. Since A_g is the g-inverse of A it satisfies:

$A_g A A_g = A_g$;
$A_g A$ is symmetric;
AA_g is symmetric;
$AA_g A = A$.

But Definition 37 indicates that the above define A as the g-inverse of A_g.

To prove (ii) we note that $AA_g A = A$ implies $\text{rank}(A) \leq \text{rank}(A_g)$, while $A_g A A_g = A_g$ implies $\text{rank}(A_g) \leq \text{rank}(A)$. Together, these show that

$$\text{rank}(A) = \text{rank}(A_g).$$

To prove (iii) we verify that $A_g A_g'$ is, indeed, the g-inverse of $A'A$. But

$(A'A)A_g A_g'(A'A) = (A'A_g' A')(AA_g A) = A'A,$
$(A'A)A_g A_g' = A'A_g' A'A_g' = (A_g A)'$ is symmetric,
$A_g A_g'(A'A) = (A_g A)$ is symmetric,
$A_g A_g'(A'A)A_g A_g' = (A_g A A_g)(A_g' A' A_g') = A_g A_g'.$

Statement (iv) may be proved by noting that AA_g is a symmetric idempotent matrix and that Corollary 13 states that $(AA_g)_g = AA_g$.

To prove (v) we note that

$$AA^{-1}A = A,$$
$$AA^{-1} = A^{-1}A = I \text{ is symmetric,}$$
$$A^{-1}AA^{-1} = A^{-1},$$

which completes the proof.   q.e.d.

Corollary 14. *Let A be an $m \times n$ matrix; let P be $m \times m$ and Q be $n \times n$, and let both be orthogonal. Then*

$$(PAQ)_g = Q'A_g P'.$$

PROOF. We have

$$(PAQ)(PAQ)_g = PAQQ'A_g P' = PAA_g P',$$
$$(PAQ)_g(PAQ) = Q'A_g P'PAQ = Q'A_g AQ.$$

The symmetry of the matrices above follows from the symmetry of AA_g and $A_g A$ respectively. Moreover,

$$(PAQ)_g(PAQ)(PAQ)_g = Q'A_g P'PAQQ'A_g P' = Q'A_g AA_g P' = (PAQ)_g$$
$$(PAQ)(PAQ)_g(PAQ) = PAQQ'A_g P'PAQ = PAA_g AQ = PAQ,$$

which completes the proof.   q.e.d.

Remark 39. It is worth entering a note of caution here. It is well known that if A, B are conformable nonsingular matrices, then

$$(AB)^{-1} = B^{-1}A^{-1}.$$

The results in (iii) and (iv) of the preceding proposition may suggest that the same is true of g-inverses. Unfortunately, it is not *generally* true that if A, B are $m \times n$ and $n \times q$ respectively then

$$(AB)_g = B_g A_g.$$

This is true for the matrices in (iii) and (iv) of the preceding proposition as well as for those in Corollary 14.

In the following discussion we shall consider a number of other instances in which the relation above is valid.

Proposition 73. *Let D be a diagonal matrix,*

$$D = \begin{bmatrix} C & 0 \\ 0 & 0 \end{bmatrix},$$

where the diagonal elements of C are nonzero. Then

$$D_g = \begin{bmatrix} C^{-1} & 0 \\ 0 & 0 \end{bmatrix}.$$

PROOF. Obvious.   q.e.d.

Corollary 15. *Let D, E be two diagonal matrices and put F = DE. Then*

$$F_g = E_g D_g.$$

PROOF. If either E or D is the zero matrix the result holds; thus, let us assume that neither D nor E are null. Without loss of generality, put

$$D = \begin{bmatrix} C_1 & 0 & 0 \\ 0 & 0 & 0 \\ 0 & 0 & 0 \end{bmatrix}, \qquad E = \begin{bmatrix} E_1 & 0 & 0 \\ 0 & E_2 & 0 \\ 0 & 0 & 0 \end{bmatrix}$$

where it is implicitly assumed that E contains more nonnull elements than D.

$$F = \begin{bmatrix} C_1 E_1 & 0 & 0 \\ 0 & 0 & 0 \\ 0 & 0 & 0 \end{bmatrix}$$

By Proposition 73,

$$F_g = \begin{bmatrix} E_1^{-1} C_1^{-1} & 0 & 0 \\ 0 & 0 & 0 \\ 0 & 0 & 0 \end{bmatrix} = \begin{bmatrix} E_1^{-1} & 0 & 0 \\ 0 & E_2^{-1} & 0 \\ 0 & 0 & 0 \end{bmatrix} \begin{bmatrix} C_1^{-1} & 0 & 0 \\ 0 & 0 & 0 \\ 0 & 0 & 0 \end{bmatrix} = E_g D_g. \quad \text{q.e.d.}$$

Proposition 74. *Let A be m × n (m ≤ n) of rank m. Then*

$$A_g = A'(AA')^{-1}, \qquad AA_g = I.$$

PROOF. We verify that A_g is the g-inverse of A. First we note that AA' is $m \times m$ of rank m; hence the inverse exists.

$AA_g A = AA'(AA')^{-1}A = A,$
$AA_g = AA'(AA')^{-1} = I$ is symmetric,
$A_g A = A'(AA')^{-1}A$ is symmetric (and idempotent),
$A_g AA_g = A_g.$ q.e.d.

Corollary 16. *Let A be m × n (m ≥ n) of rank n. Then*

$$A_g = (A'A)^{-1}A', \qquad A_g A = I.$$

PROOF. Obvious.   q.e.d.

Proposition 75. *Let B be m × r, C be r × n, and let both be of rank r. Then*

$$(BC)_g = C_g B_g.$$

PROOF. By Proposition 74,

$$B_g = (B'B)^{-1}B', \qquad C_g = C'(CC')^{-1}.$$

Putting

$$A = BC$$

we may verify that

$$A_g = C_g B_g. \quad \text{q.e.d.}$$

A further useful result is

Proposition 76. *Let A be m × n. Then the following statements are true:*

(i) $I - AA_g, I - A_g A$ are symmetric, idempotent;
(ii) $(I - AA_g)A = 0, A_g(I - AA_g) = 0$;
(iii) $(I - AA_g)AA_g = AA_g(I - AA_g) = 0$;
(iv) $(I - A_g A)A_g A = A_g A(I - A_g A) = 0$.

PROOF. Proposition 70 states that $AA_g, A_g A$ are both symmetric, idempotent. Hence

$$(I - AA_g)(I - AA_g) = I - AA_g - AA_g + AA_g AA_g = I - AA_g.$$

Similarly,

$$(I - A_g A)(I - A_g A) = I - A_g A - A_g A + A_g AA_g A = I - A_g A,$$

which proves (i).
   Since

$$AA_g A = A, \qquad A_g AA_g = A_g,$$

the proof of (ii) is obvious.
   The proof of (iii) and (iv) follows easily from that of (ii).   q.e.d.

   To conclude this section we give some additional results for certain special types of matrices.

Proposition 77. Let B, C be, respectively, $m \times s, n \times s$, such that $BC' = 0$ and

$$A = \begin{pmatrix} B \\ C \end{pmatrix}.$$

Then

$$A_g = (B_g, C_g).$$

PROOF. We verify that A_g is the g-inverse of A. We have to show that

$$A_g A = B_g B + C_g C$$

is symmetric. Now, B_g, C_g are, respectively, the g-inverses of B, C. Thus $B_g B$, $C_g C$ are both symmetric matrices, and consequently so is $A_g A$. Also

$$AA_g = \begin{bmatrix} BB_g & BC_g \\ CB_g & CC_g \end{bmatrix}.$$

We note that

$$BC_g = BC_g CC_g = BC'C'_g C_g = 0,$$

the last equality being valid by the condition $BC' = 0$. Similarly,

$$CB_g = CB_g BB_g = CB'B'_g B = 0,$$

which shows that

$$AA_g = \begin{bmatrix} BB_g & 0 \\ 0 & CC_g \end{bmatrix},$$

which is, clearly, a symmetric matrix. Moreover,

$$AA_g A = \begin{bmatrix} BB_g & 0 \\ 0 & CC_g \end{bmatrix} \binom{B}{C} = \binom{BB_g B}{CC_g C} = \binom{B}{C} = A,$$

$$A_g AA_g = (B_g, C_g) \begin{bmatrix} BB_g & 0 \\ 0 & CC_g \end{bmatrix} = (B_g BB_g, C_g CC_g) = (B_g, C_g) = A_g,$$

thus completing the proof.   q.e.d.

Proposition 78. *If B, C are any matrices, and*

$$A = \begin{bmatrix} B & 0 \\ 0 & C \end{bmatrix},$$

then

$$A_g = \begin{bmatrix} B_g & 0 \\ 0 & C_g \end{bmatrix}.$$

PROOF. Obvious by direct verification.   q.e.d.

Finally, we observe that if B, C are any matrices and

$$A = B \otimes C$$

then

$$A_g = B_g \otimes C_g$$

The proof of this is easily obtained by direct verification.

3.4 Solutions of Linear Systems of Equations and Pseudoinverses

What motivated our exploration of the theory of pseudoinverses was the desire to characterize the class of solutions to the linear system

$$Ax = b,$$

where A is $m \times n$ and, generally, $m \neq n$. When A is not a square matrix the question that naturally arises is *whether the system is consistent, i.e., whether there exists at least one vector, x_*, that satisfies the system,* and, if consistent, how many solutions there are and how they may be characterized.

The first question is answered by

Proposition 79. *Let A be $m \times n$; then the system of equations*

$$Ax = b$$

is consistent, if and only if for some c-inverse of A,

$$AA_c b = b.$$

PROOF. Suppose the system is consistent and let x_0 be a solution, i.e.,

$$b = Ax_0.$$

Premultiply by AA_c to obtain

$$AA_c b = AA_c Ax_0 = Ax_0 = b,$$

which establishes necessity.

Now assume that for some c-inverse,

$$AA_c b = b.$$

Take

$$x = A_c b$$

and observe that this *is a solution*, thus completing the proof.   q.e.d.

The question now arises as to how many solutions there are, given that there is at least one solution (i.e., the system is consistent). This is answered by

Proposition 80. *Let A be $m \times n$ $(m \geq n)$ and suppose*

$$Ax = b$$

is a consistent system. Then, for any arbitrary vector d,

$$x = A_c b + (I - A_c A)d$$

is a solution. Conversely, if x is a solution there exists a d such that x can be written in the form above.

PROOF. Since the system *is consistent* we have

$$AA_cb = b.$$

Let

$$x = A_cb + (I - A_cA)d$$

and observe

$$Ax = AA_cb = b,$$

which shows x to be a solution.

Conversely, suppose x is *any* solution, i.e., it satisfies

$$b - Ax = 0.$$

Premultiply by A_c to obtain

$$A_cb - A_cAx = 0.$$

Adding x to both sides of the equation we have

$$x = A_cb + (I - A_cA)x,$$

which is of the desired form with $d = x$, thus completing the proof.   q.e.d.

Corollary 17. *The statements of the proposition are true if A_c is replaced by A_g.*

PROOF. Clearly, for a consistent system

$$x = A_gb + (I - A_gA)d$$

is a solution, where d is arbitrary.

Conversely, if x is any solution, so is

$$x = A_gb + (I - A_gA)x,$$

which completes the proof.   q.e.d.

Corollary 18. *The solution to the system above is unique if and only if*

$$A_gA = I.$$

PROOF. If

$$A_gA = I$$

then the general solution of the corollary above,

$$x = A_gb + (I - A_gA)d,$$

becomes

$$x = A_gb,$$

which is unique.

Conversely, if the general solution above is unique for every vector d then

$$A_g A = I. \quad \text{q.e.d.}$$

Corollary 19. *The solution of the (consistent) system above is unique if and only if*

$$\text{rank}(A) = n.$$

PROOF. From Corollary 16 if $\text{rank}(A) = n$ then $A_g A = I$. Corollary 18 then shows that the solution is unique.

Conversely, suppose the solution *is unique*. Then

$$A_g A = I$$

shows that

$$n \leq \text{rank}(A).$$

But the rank of A cannot possibly exceed n. Thus

$$\text{rank}(A) = n. \quad \text{q.e.d.}$$

It is clear from the preceding that there are, in general, infinitely many solutions to the system considered above. Thus, for example, if $x._i$ is a solution, $i = 1, 2, \ldots, k$, then

$$x = \sum_{i=1}^{k} \gamma_i x._i$$

is also a solution provided $\sum_{i=1}^{k} \gamma_i = 1$.

This prompts us to inquire as to how many linearly independent solutions there are; if we determine this, then all solutions can be expressed in terms of this linearly independent set. We have

Proposition 81. *Let the system*

$$Ax = b$$

be such that A is $m \times n$ $(m \geq n)$ of rank $0 < r \leq n$, $b \neq 0$. Then, there are $n - r + 1$ linearly independent solutions.

PROOF. Recall that, since

$$A_g A A_g = A_g, \qquad A A_g A = A,$$

we have

$$\text{rank}(A_g A) = \text{rank}(A) = r.$$

Now, the general solution of the system can be written as

$$x = A_g b + (I - A_g A)d$$

for arbitrary d.

Consider, in particular, the vectors

$$x_{\cdot i} = A_g b + (I - A_g A) d_{\cdot i}, \qquad i = 0, 1, 2, \ldots, n,$$

where the $d_{\cdot i}$ are n-element vectors such that for $i = 0$, $d_{\cdot i} = 0$, while for $i \neq 0$ all the elements of $d_{\cdot i}$ are zero save the ith, which is unity. Write

$$X = (x_{\cdot 0}, x_{\cdot 1}, \ldots, x_{\cdot n}) = (A_g b, I - A_g A) \begin{bmatrix} 1 & e' \\ 0 & I \end{bmatrix},$$

where e is an n-element column vector all of whose elements are unity. Since the upper triangular matrix in the right member above is nonsingular, we conclude

$$\operatorname{rank}(X) = \operatorname{rank}(A_g b, I - A_g A) = 1 + n - r.$$

The last equality follows since $A_g b$ is orthogonal to $I - A_g A$, and thus the two are linearly independent. In addition,

$$\operatorname{rank}(I - A_g A) = n - \operatorname{rank}(A_g A) = n - r.$$

Thus, we see that the number of linearly independent solutions cannot exceed n (since we deal with the case $r > 0$)—and at any rate it is exactly $n - r + 1$. q.e.d.

Remark 40. It should be pointed out that the $n - r + 1$ linearly independent solutions above *do not constitute a vector space* since the vector 0 is not a solution—because of the condition $b \neq 0$.

Since there are many solutions to the typical system considered here, the question arises as to whether there are (linear) functions of the solutions that are invariant to the particular choice of solution. This is answered by

Proposition 82. *Let A be m $\times$ n; the linear transformation*

$$Gx,$$

where x is a solution to the (consistent) system

$$Ax = b,$$

is unique if and only if G lies in the space spanned by the rows of A.

PROOF. The general form of the solution is

$$x = A_g b + (I - A_g A) d$$

for arbitrary d. Thus,

$$Gx = G A_g b + G(I - A_g A) d$$

is unique if and only if

$$G = G A_g A.$$

But, if the above is satisfied G lies in the row space of A. Conversely, suppose G lies in the row space of G. Then, there exists a matrix C such that

$$G = CA.$$

Consequently,

$$GA_g A = CAA_g A = CA = G. \quad \text{q.e.d.}$$

3.5 Approximate Solutions of Systems of Linear Equations and Pseudoinverses

In the previous section we examined systems of linear equations and gave necessary and sufficient conditions for their consistency, i.e., for the existence of solutions. Moreover, we gave a characterization of such solutions. Here, we shall examine *inconsistent systems*. Thus, a system

$$Ax = b$$

may have *no solution*, and thus it may better be expressed as

$$r(x) = Ax - b.$$

Nonetheless, we may wish to determine a vector x_* that is an approximate solution in the sense that

$$r(x_*)$$

is "small." The precise meaning of this terminology will be made clear below.

Definition 38. Let A be $m \times n$ and consider the system

$$r(x) = Ax - b.$$

A solution x_* is said to be a least squares (LS) approximate solution if and only if for all n-element vectors x

$$r(x)'r(x) \geq r(x_*)'r(x_*).$$

Remark 41. If the system $r(x) = Ax - b$ of Definition 38 is *consistent*, then any LS approximate solution corresponds to a solution in the usual sense of the previous section.

The question of when does an LS approximate solution exist and how it may be arrived at is answered by

Proposition 83. *Consider the system*

$$r(x) = Ax - b.$$

The vector

$$x_* = Bb$$

is an LS solution to the system above if B is an s-inverse of A, i.e., it obeys

(i) $ABA = A$,
(ii) AB *is symmetric.*

PROOF. We observe that for any n-element vector x,

$$
\begin{aligned}
&(b - Ax)'(b - Ax) \\
&= [(b - ABb) + (ABb - Ax)]'[(b - ABb) + (ABb - Ax)] \\
&= b'(I - AB)'(I - AB)b + (Bb - x)'A'A(Bb - x) \\
&= b'(I - AB)b + (Bb - x)'A'A(Bb - x).
\end{aligned}
$$

This is so since B obeys (i) and (ii) and the cross terms vanish, i.e.,

$$
\begin{aligned}
(b - ABb)'(ABb - Ax) &= b'(A - B'A'A)(Bb - x) \\
&= b'(A - ABA)(Bb - x) = 0.
\end{aligned}
$$

Because $A'A$ is (at least) positive semidefinite, the quantity

$$(b - Ax)'(b - Ax)$$

is minimized only if we take

$$x = Bb. \quad \text{q.e.d.}$$

Corollary 20. *The quantity*

$$b'(I - AB)b$$

is a lower bound for

$$(b - Ax)'(b - Ax).$$

PROOF. Obvious.   q.e.d.

Corollary 21. *If B is a matrix that defines an LS solution then*

$$AB = AA_g.$$

PROOF. We have

$$AB = AA_g AB = A_g' A'B'A' = A_g' A' = AA_g. \quad \text{q.e.d.}$$

We may now ask: what is the connection between s-inverses, LS solutions to inconsistent systems, and our discussion in the previous section. In part, this is answered by

Proposition 84. *An n-element (column) vector x_* is an LS solution to an (inconsistent) system*

$$r(x) = Ax - b,$$

where A is $m \times n$ if and only if x_* is a solution to the consistent system

$$Ax = AA_g b.$$

PROOF. We note, first, that since A_g is also a c-inverse of A, Proposition 79 shows that

$$Ax = AA_g b$$

is, indeed, a consistent system. Because it is a consistent system, Corollary 17 shows that the general form of the solution is

$$x_* = A_g(AA_g b) + (I - A_g A)d = A_g b + (I - A_g A)d$$

for any arbitrary vector d.

Now, with x_* as just defined we have

$$b - Ax_* = b - AA_g b$$

and, consequently,

$$(b - Ax_*)'(b - Ax_*) = b'(I - AA_g)b.$$

Corollary 20, then, shows that x_*, as above, is an LS solution to the (inconsistent) system

$$r(x) = Ax - b.$$

Conversely, suppose that x_* is *any* LS solution to the system above. It must, then, satisfy the condition

$$(Ax_* - b)'(Ax_* - b) = b'(I - AA_g)b.$$

Put

$$q = x_* - A_g b \quad \text{or} \quad x_* = q + A_g b.$$

Substitute in the equation above to obtain

$$\begin{aligned} b'(I - AA_g)b &= (Ax_* - b)'(Ax_* - b) \\ &= (Aq + AA_g b - b)'(Aq + AA_g b - b) \\ &= b'(I - AA_g)b + q'A'Aq, \end{aligned}$$

which immediately implies

$$Aq = 0.$$

Thus,

$$Ax_* = Aq + AA_g b = AA_g b,$$

which completes the proof.   q.e.d.

Remark 42. The import of Proposition 84 is that an LS solution to a (possibly) inconsistent system

$$r(x) = Ax - b$$

can be found by solving the associated (consistent) system

$$Ax = AA_g b.$$

The general class of solutions to this system was determined in Proposition 81. Thus, we see that there may be multiple (or infinitely many) LS solutions. If uniqueness is desired it is clear that the solution must be made to satisfy additional conditions. This leads to

Definition 39. Consider the (possibly) inconsistent system

$$r(x) = Ax - b,$$

where A is $m \times n$. An n-element vector x_* is said to be a *minimum norm least squares* (MNLS) *approximate solution* if and only if

(i) for all n-element vectors x

$$(b - Ax)'(b - Ax) \geq (b - Ax_*)'(b - Ax_*),$$

(ii) for those x for which

$$(b - Ax)'(b - Ax) = (b - Ax_*)'(b - Ax_*)$$

we have

$$x'x > x'_* x_*.$$

This leads to the important

Proposition 85. *Let*

$$r(x) = Ax - b$$

be a (possibly) inconsistent system, where A is $m \times n$. The MNLS *(approximate) solution is given by*

$$x_* = A_g b$$

and is unique.

PROOF. First, we note that

$$(b - AA_g b)'(b - AA_g b) = b'(I - AA_g)b,$$

which shows x_* to be an LS solution—because it attains the lower bound of

$$(b - Ax)'(b - Ax).$$

We must now show that this solution has minimum norm and that it is unique. Now, if x is *any* LS solution, it must satisfy

$$Ax = AA_g b.$$

Premultiply by A_g to obtain

$$A_g Ax = A_g AA_g b = A_g b.$$

Thus, any LS solution x also satisfies

$$x = A_g b - A_g Ax + x = A_g b + (I - A_g A)x.$$

Consequently, for *any* LS *solution* we have

$$x'x = b'A_g' A_g b + (x - A_g Ax)'(x - A_g Ax).$$

But

$$A_g Ax = A_g b$$

if x is any LS solution. Consequently,

$$x'x = b'A_g' A_g b + (x - A_g b)'(x - A_g b),$$

which shows that if

$$x \neq x_* = A_g b$$

then

$$x'x > x_*' x_*.$$

Uniqueness is an immediate consequence of the argument above. Thus, let x_0 be another MNLS solution and suppose $x_0 \neq x_*$. But x_0 must satisfy

$$x_0' x_0 = b'A_g' A_g b + (x_0 - A_g b)'(x_0 - A_g b).$$

Since we assume $x_0 \neq x_*$ we have

$$x_0' x_0 > x_*' x_*,$$

which is a contradiction. Moreover, A_g is unique, which thus completes the proof of the proposition.   q.e.d.

Remark 43. It is now possible to give a summary description of the role of the various pseudoinverses. Thus, the c-inverse is useful in broadly describing the class of solutions to the (consistent) system

$$Ax = b,$$

where A is $m \times n$.

The s-inverse is useful in describing the class of LS solutions to the possibly inconsistent system

$$r(x) = Ax - b,$$

i.e., in the case where no vector x may exist such that $r(x) = 0$. Neither the c-inverse nor the s-inverse of a matrix A is necessarily unique.

The g-inverse serves to characterize the solutions to both types of problems. Particularly, however, it serves to define the MNLS (approximate) solution to the inconsistent system

$$r(x) = Ax - b.$$

This means that of all possible least squares solutions to the inconsistent system above *the g-inverse chooses a unique vector by imposing the additional requirement that the solution vector exhibit minimal norm.* This aspect should always be borne in mind in dealing with econometric applications of the g-inverse, since there is no particular economic reason to believe that the estimator of a vector exhibiting minimal norm is of any extraordinary significance.

4 Vectorization of Matrices and Matrix Functions: Matrix Differentiation

4.1 Introduction

It is frequently more convenient to write a matrix in *vector* form. For lack of a suitable term we have coined for this operation the phrase "vectorization of matrices." For example, if A is a matrix of parameters and $\tilde{A}$ the corresponding matrix of estimators it is often necessary to consider the distribution of

$$\tilde{A} - A.$$

We have a convention to handle what we wish to mean by the expectation of a random matrix, but there is no convention regarding the "covariance matrix" of a matrix. Similarly, we have developed some (limited aspects of) distribution theory for vectors but not for matrices.

In another area, viz., differentiation with a view to obtaining the conditions that may define a wide variety of estimators, vectorization of matrices offers great convenience as well. We now turn to the establishment of the proper notation and the derivation of useful results regarding matrix differentiation.

4.2 Vectorization of Matrices

We begin with

Convention 2. Let A be an $n \times m$ matrix; the notation vec(A) will mean the nm-element column vector whose first n elements are the first column of A, $a_{.1}$; the second n elements, the second column of A, $a_{.2}$, and so on. Thus

$$\text{vec}(A) = (a'_{.1}, a'_{.2}, \ldots, a'_{.m})'.$$

An immediate consequence of Convention 2 is

Proposition 86. *Let A, B be $n \times m$, $m \times q$ respectively. Then*

$$\text{vec}(AB) = (B' \otimes I)\ \text{vec}(A)$$
$$= (I \otimes A)\ \text{vec}(B)$$

PROOF. For the second representation we note that the jth column AB is simply

$$Ab_{.j}.$$

Thus, when AB is vectorized we find

$$\text{vec}(AB) = (I \otimes A)\ \text{vec}(B).$$

To show the validity of the first representation we note that the jth column of AB can also be written as

$$\sum_{i=1}^{m} a_{.i}b_{ij}.$$

We then easily verify that the jth subvector (of n elements) of $\text{vec}(AB)$ is

$$\sum_{i=1}^{m} a_{.i}b_{ij}.$$

Consequently,

$$\text{vec}(AB) = (B' \otimes I)\ \text{vec}(A). \quad \text{q.e.d.}$$

Vectorization of products involving more than two matrices is easily obtained by repeated application of Proposition 86. We shall give a few such results explicitly. Thus,

Corollary 22. *Let A_1, A_2, A_3 be suitably dimensioned matrices. Then*

$$\text{vec}(A_1 A_2 A_3) = (I \otimes A_1 A_2)\ \text{vec}(A_3)$$
$$= (A_3' \otimes A_1)\ \text{vec}(A_2)$$
$$= (A_3' A_2' \otimes I)\ \text{vec}(A_1).$$

PROOF. By Proposition 86, taking

$$A = A_1 A_2, \qquad B = A_3,$$

we have

$$\text{vec}(A_1 A_2 A_3) = (I \otimes A_1 A_2)\ \text{vec}(A_3).$$

Taking

$$A_1 = A, \qquad A_2 A_3 = B,$$

we have

$$\text{vec}(A_1 A_2 A_3) = (A_3' A_2' \otimes I)\, \text{vec}(A_1),$$

as well as

$$\text{vec}(A_1 A_2 A_3) = (I \otimes A_1)\, \text{vec}(A_2 A_3)$$

Applying Proposition 86 again we find

$$\text{vec}(A_2 A_3) = (A_3' \otimes I)\, \text{vec}(A_2),$$

and hence

$$\text{vec}(A_1 A_2 A_3) = (A_3' \otimes A_1)\, \text{vec}(A_2), \quad \text{q.e.d.}$$

Corollary 23. *Let A_1, A_2, A_3, A_4 be suitably dimensioned matrices. Then*

$$\begin{aligned}
\text{vec}(A_1 A_2 A_3 A_4) &= (I \otimes A_1 A_2 A_3)\, \text{vec}(A_4) \\
&= (A_4' \otimes A_1 A_2)\, \text{vec}(A_3) \\
&= (A_4' A_3' \otimes A_1)\, \text{vec}(A_2) \\
&= (A_4' A_3' A_2' \otimes I)\, \text{vec}(A_1).
\end{aligned}$$

PROOF. This follows if we apply Proposition 86, taking

$$A = A_1 A_2 A_3, \qquad B = A_4,$$

and then apply Corollary 22.   q.e.d.

Remark 44. The reader should note the pattern involved in these relations; thus, if we wish to vectorize the product of the n conformable matrices

$$A_1 A_2 A_3 \cdots A_n$$

by vectorizing A_i we shall obtain

$$(A_n' A_{n-1}' \cdots A_{i+1}' \otimes A_1 A_2 \cdots A_{i-1})\, \text{vec}(A_i),$$

so that the matrices appearing to the right of A_i appear on the left of the Kronecker product sign ($\otimes$) in transposed form and order and those appearing on the left of A_i appear on the right of the Kronecker product sign in the original form and order.

We further have

Proposition 87. *Let A, B be $m \times n$. Then*

$$\text{vec}(A + B) = \text{vec}(A) + \text{vec}(B).$$

PROOF. Obvious from Convention 2.   q.e.d.

Corollary 24. *Let A, B, C, D be suitably dimensioned matrices. Then*

$$\begin{aligned}
\text{vec}[(A + B)(C + D)] &= [(I \otimes A) + (I \otimes B)][\text{vec}(C) + \text{vec}(D)] \\
&= [(C' \otimes I) + (D' \otimes I)][\text{vec}(A) + \text{vec}(B)]
\end{aligned}$$

PROOF. By Proposition 86,

$$\text{vec}[(A + B)(C + D)] = [I \otimes (A + B)] \, \text{vec}(C + D)$$
$$= [(C + D)' \otimes I] \, \text{vec}(A + B).$$

Apply then Proposition 87 and the properties of Kronecker product matrices to obtain the result of the Corollary.   q.e.d.

We now turn our attention to the representation of the trace of products of matrices in terms of various functions of vectorized matrices. Thus,

Proposition 88. *Let A, B be suitably dimensioned matrices. Then*

$$\text{tr}(AB) = \text{vec}(A')' \, \text{vec}(B)$$
$$= \text{vec}(B')' \, \text{vec}(A).$$

PROOF. By definition,

$$\text{tr}(AB) = \sum_{i=1}^{m} a_{i.} b_{.i}, \tag{68}$$

where $a_{i.}$ is the ith row of A and $b_{.i}$ is the ith column of B. But $a_{i.}$ is simply the ith column of A' written in row form, and (68) then shows that

$$\text{tr}(AB) = \text{vec}(A')' \, \text{vec}(B).$$

Moreover, since

$$\text{tr}(AB) = \text{tr}(BA) = \sum_{j=1}^{q} b_{j.} a_{.j} \tag{69}$$

we see that

$$\text{tr}(AB) = \text{vec}(B')' \, \text{vec}(A).   \text{q.e.d.}$$

It is an easy consequence of Propositions 86 and 88 to establish a "vectorized representation" of the trace of the product of more than two matrices.

Proposition 89. *Let A_1, A_2, A_3 be suitably dimensioned matrices. Then*

$$\text{tr}(A_1 A_2 A_3) = \text{vec}(A_1')'(A_3' \otimes I) \, \text{vec}(A_2)$$
$$= \text{vec}(A_1')'(I \otimes A_2) \, \text{vec}(A_3)$$
$$= \text{vec}(A_2')'(I \otimes A_3) \, \text{vec}(A_1)$$
$$= \text{vec}(A_2')'(A_1' \otimes I) \, \text{vec}(A_3)$$
$$= \text{vec}(A_3')'(A_2' \otimes I) \, \text{vec}(A_1)$$
$$= \text{vec}(A_3')'(I \otimes A_1) \, \text{vec}(A_2).$$

PROOF. From Proposition 88, taking

$$A = A_1, \qquad B = A_2 A_3,$$

we have

$$\text{tr}(A_1 A_2 A_3) = \text{vec}(A_1')' \, \text{vec}(A_2 A_3). \tag{70}$$

Using Proposition 86 we have

$$\text{vec}(A_2 A_3) = (I \otimes A_2) \, \text{vec}(A_3)$$
$$= (A_3' \otimes I) \, \text{vec}(A_2)$$

This together with (70) establishes

$$\text{tr}(A_1 A_2 A_3) = \text{vec}(A_1')'(A_3' \otimes I) \, \text{vec}(A_2)$$
$$= \text{vec}(A_1')'(I \otimes A_2) \, \text{vec}(A_3).$$

Noting that

$$\text{tr}(A_1 A_2 A_3) = \text{tr}(A_2 A_3 A_1)$$

and using exactly the same procedure as above shows

$$\text{tr}(A_1 A_2 A_3) = \text{vec}(A_2')'(I \otimes A_3) \, \text{vec}(A_1)$$
$$= \text{vec}(A_2')'(A_1' \otimes I) \, \text{vec}(A_3).$$

Finally, since

$$\text{tr}(A_1 A_2 A_3) = \text{tr}(A_3 A_1 A_2),$$

we find by the same argument

$$\text{tr}(A_1 A_2 A_3) = \text{vec}(A_3')'(A_2' \otimes I) \, \text{vec}(A_1)$$
$$= \text{vec}(A_3')'(I \otimes A_1) \, \text{vec}(A_2). \quad \text{q.e.d.}$$

Remark 45. The representation of the trace of the product of more than three matrices is easily established by using the methods employed in the proof of Proposition 89. For example,

$$\text{tr}(A_1 A_2 A_3 A_4) = \text{vec}(A_1')'(A_4' A_3' \otimes I) \, \text{vec}(A_2)$$
$$= \text{vec}(A_1')'(A_4' \otimes A_2) \, \text{vec}(A_3)$$
$$= \text{vec}(A_1')'(I \otimes A_2 A_3) \, \text{vec}(A_4)$$
$$= \text{vec}(A_2')'(I \otimes A_3 A_4) \, \text{vec}(A_1)$$
$$= \text{vec}(A_2')'(A_1' A_4' \otimes I) \, \text{vec}(A_3)$$
$$= \text{vec}(A_2')'(A_1' \otimes A_3) \, \text{vec}(A_4)$$
$$= \text{vec}(A_3')'(A_2' \otimes A_4) \, \text{vec}(A_1)$$
$$= \text{vec}(A_3')'(I \otimes A_4 A_1) \, \text{vec}(A_2)$$
$$= \text{vec}(A_3')'(A_2' A_1' \otimes I) \, \text{vec}(A_4)$$
$$= \text{vec}(A_4')'(A_3' A_2' \otimes I) \, \text{vec}(A_1)$$
$$= \text{vec}(A_4')'(A_3' \otimes A_1) \, \text{vec}(A_2)$$
$$= \text{vec}(A_4')'(I \otimes A_1 A_2) \, \text{vec}(A_3).$$

This example also shows why it is not possible to give all conceivable representations of the trace of the product of an arbitrary number of matrices.

4.3 Vector and Matrix Differentiation

Frequently we need to differentiate quantities like $\mathrm{tr}(AX)$ with respect to the elements of X, or quantities like Ax, $z'Ax$ with respect to the elements of (the vectors) x and/or z.

Although no new concept is involved in carrying out such operations, they involve cumbersome manipulations and, thus, it is desirable to derive such results and have them easily available for reference.

We begin with

Convention 3. Let

$$y = \psi(x),$$

where y, x are, respectively, m- and n-element column vectors. The symbol

$$\frac{\partial y}{\partial x} = \left[\frac{\partial y_i}{\partial x_j}\right], \qquad \begin{array}{l} i = 1, 2, \ldots, m, \\ j = 1, 2, \ldots, n, \end{array}$$

will denote the matrix of first-order partial derivatives (Jacobian matrix) of the transformation from x to y such *that the ith row contains* the derivatives of the ith element of y with respect to the elements of x, viz.,

$$\frac{\partial y_i}{\partial x_1}, \frac{\partial y_i}{\partial x_2}, \ldots, \frac{\partial y_i}{\partial x_n}.$$

Remark 46. Notice that if y, above, is a *scalar* then Convention 3 implies that $\partial y/\partial x$ is a *row* vector. If we wish to represent it as a column vector we may do so by writing $\partial y/\partial x'$, or $(\partial y/\partial x)'$.

We now derive several useful results.

Proposition 90. *If*

$$y = Ax$$

then

$$\frac{\partial y}{\partial x} = A,$$

where A is $m \times n$, and does not depend on x.

PROOF. Since the ith element of y is given by

$$y_i = \sum_{k=1}^{n} a_{ik} x_k,$$

it follows that

$$\frac{\partial y_i}{\partial x_j} = a_{ij},$$

hence that

$$\frac{\partial y}{\partial x} = A. \quad \text{q.e.d.}$$

If the vector x above is a function of another set of variables, say those contained in the r-element column vector α, then we have

Proposition 91. *Let*

$$y = Ax$$

be as in Proposition 90 but suppose that x is a function of the r-element vector α, while A is independent of α. Then

$$\frac{\partial y}{\partial \alpha} = \frac{\partial y}{\partial x} \frac{\partial x}{\partial \alpha} = A \frac{\partial x}{\partial \alpha}.$$

PROOF. Since $y_i = \sum_{k=1}^{n} a_{ik} x_k$,

$$\frac{\partial y_i}{\partial \alpha_j} = \sum_{k=1}^{n} a_{ik} \frac{\partial x_k}{\partial \alpha_j}.$$

But the right side of the above is simply the (i, j) element of $A(\partial x/\partial \alpha)$. Hence

$$\frac{\partial y}{\partial \alpha} = A \frac{\partial x}{\partial \alpha}. \quad \text{q.e.d.}$$

Remark 47. Convention 3 enables us to define routinely the first-order derivative of one vector with respect to another but it is not sufficient to enable us to obtain second-order derivatives. This is so since it is not clear what we would mean by the derivative of a matrix with respect to a vector. From Convention 3 we see that the *rows* of $\partial y/\partial x$ are of the form $\partial y_i/\partial x$. Hence the *columns* of $(\partial y/\partial x)'$ can be written as $\partial y_i/\partial x'$. This suggests

Convention 4. Let

$$y = \psi(x)$$

be as in Convention 3. By the symbol

$$\frac{\partial^2 y}{\partial x \, \partial x'}$$

we shall mean

$$\frac{\partial^2 y}{\partial x \, \partial x'} = \frac{\partial}{\partial x} \text{vec}\left[\left(\frac{\partial y}{\partial x}\right)' \right],$$

so that it is a matrix of dimension $(mn) \times n$. In general, if Y is a matrix and x is a vector, by the symbol

$$\frac{\partial Y}{\partial x}$$

we shall mean the matrix

$$\frac{\partial Y}{\partial x} = \frac{\partial}{\partial x} \, \text{vec}(Y).$$

An easy consequence of Convention 4 is

Proposition 92. *Let*

$$y = Ax$$

be as in Proposition 91. Then

$$\frac{\partial^2 y}{\partial \alpha \, \partial \alpha'} = (A \otimes I) \frac{\partial^2 x}{\partial \alpha \, \partial \alpha'}.$$

PROOF. By Proposition 91

$$\frac{\partial y}{\partial \alpha} = A \frac{\partial x}{\partial \alpha}.$$

By Convention 4 and Proposition 86,

$$\begin{aligned}
\frac{\partial^2 y}{\partial \alpha \, \partial \alpha'} &= \frac{\partial}{\partial \alpha} \, \text{vec} \left[\left(\frac{\partial y}{\partial \alpha} \right)' \right] \\
&= \frac{\partial}{\partial \alpha} \, \text{vec} \left[\left(\frac{\partial x}{\partial \alpha} \right)' A' \right] \\
&= \frac{\partial}{\partial \alpha} (A \otimes I) \, \text{vec} \left[\left(\frac{\partial x}{\partial \alpha} \right)' \right] \\
&= (A \otimes I) \frac{\partial}{\partial \alpha} \, \text{vec} \left[\left(\frac{\partial x}{\partial \alpha} \right)' \right] = (A \otimes I) \frac{\partial^2 x}{\partial \alpha \, \partial \alpha'}. \quad \text{q.e.d.}
\end{aligned}$$

Convention 4 is also useful in handling the case where A depends on the vector α. In particular, we have

Proposition 93. *Let*

$$y = Ax,$$

where y is $m \times 1$, A is $m \times n$, x is $n \times 1$, and both A and x depend on the r-element vector α. Then

$$\frac{\partial y}{\partial \alpha} = (x' \otimes I_m) \frac{\partial A}{\partial \alpha} + A \frac{\partial x}{\partial \alpha}.$$

PROOF. We may write

$$y = \sum_{i=1}^{n} a_{\cdot i} x_i,$$

where $a_{.i}$ is the ith column of A. Hence

$$\frac{\partial y}{\partial \alpha} = \sum_{i=1}^{n} \frac{\partial a_{.i}}{\partial \alpha} x_i + \sum_{i=1}^{n} a_{.i} \frac{\partial x_i}{\partial \alpha}$$

$$= (x' \otimes I_m) \frac{\partial A}{\partial \alpha} + A \frac{\partial x}{\partial \alpha}. \quad \text{q.e.d.}$$

Next we consider the differentiation of bilinear and quadratic forms.

Proposition 94. *Let*

$$y = z'Ax,$$

where z is $m \times 1$, A is $m \times n$, x is $n \times 1$, and A is independent of z and x. Then

$$\frac{\partial y}{\partial z} = x'A', \qquad \frac{\partial y}{\partial x} = z'A.$$

PROOF. Define

$$z'A = c'$$

and note that

$$y = c'x.$$

Hence, by Proposition 90 we have that

$$\frac{\partial y}{\partial x} = c' = z'A.$$

Similarly, we can write

$$y = x'A'z,$$

and employing the same device we obtain

$$\frac{\partial y}{\partial z} = x'A'. \quad \text{q.e.d.}$$

For the special case where y is given by the quadratic form

$$y = x'Ax$$

we have

Proposition 95. *Let*

$$y = x'Ax,$$

where x is $n \times 1$, and A is $n \times n$ and independent of x. Then

$$\frac{\partial y}{\partial x} = x'(A + A').$$

PROOF. By definition

$$y = \sum_{j=1}^{m} \sum_{i=1}^{n} a_{ij} x_i x_j.$$

Differentiating with respect to the kth element of x we have

$$\frac{\partial y}{\partial x_k} = \sum_{j=1}^{n} a_{kj} x_j + \sum_{i=1}^{n} a_{ik} x_i, \qquad k = 1, 2, \ldots, n,$$

and consequently

$$\frac{\partial y}{\partial x} = x'A' + x'A = x'(A' + A). \quad \text{q.e.d.}$$

Corollary 25. *For the special case where A is a symmetric matrix and*

$$y = x'Ax$$

we have

$$\frac{\partial y}{\partial x} = 2x'A.$$

PROOF. Obvious from Proposition 95.   q.e.d.

Corollary 26. *Let A, y, and x be as in Proposition 95; then*

$$\frac{\partial^2 y}{\partial x \, \partial x'} = A' + A,$$

and, for the special case where A is symmetric,

$$\frac{\partial^2 y}{\partial x \, \partial x'} = 2A.$$

PROOF. Obvious if we note that

$$\frac{\partial y}{\partial x'} = (A' + A)x. \quad \text{q.e.d.}$$

For the case where z and/or x are functions of another set of variables we have

Proposition 96. *Let*

$$y = z'Ax,$$

where z is m × 1, A is m × n, x is n × 1, and both z and x are a function of the r-element vector α, while A is independent of α. Then

$$\frac{\partial y}{\partial \alpha} = x'A' \frac{\partial z}{\partial \alpha} + z'A \frac{\partial x}{\partial \alpha},$$

$$\frac{\partial^2 y}{\partial \alpha \, \partial \alpha'} = \left(\frac{\partial z}{\partial \alpha}\right)' A \left(\frac{\partial x}{\partial \alpha}\right) + \left(\frac{\partial x}{\partial \alpha}\right)' A' \left(\frac{\partial z}{\partial \alpha}\right) + (x'A' \otimes I) \frac{\partial^2 z}{\partial \alpha \, \partial \alpha'}$$

$$+ (z'A \otimes I) \frac{\partial^2 x}{\partial \alpha \, \partial \alpha'}.$$

PROOF. We have

$$\frac{\partial y}{\partial \alpha} = \frac{\partial y}{\partial z} \frac{\partial z}{\partial \alpha} + \frac{\partial y}{\partial x} \frac{\partial x}{\partial \alpha} = x'A' \frac{\partial z}{\partial \alpha} + z'A \frac{\partial x}{\partial \alpha},$$

which proves the first part. For the second, note that

$$\frac{\partial^2 y}{\partial \alpha \, \partial \alpha'} = \frac{\partial}{\partial \alpha} \left(\frac{\partial y}{\partial \alpha'}\right) = \frac{\partial}{\partial \alpha} \left(\frac{\partial y}{\partial \alpha}\right)'.$$

But

$$\frac{\partial y}{\partial \alpha'} = \left(\frac{\partial z}{\partial \alpha}\right)' Ax + \left(\frac{\partial x}{\partial \alpha}\right)' A'z,$$

and, by the results of Proposition 93,

$$\frac{\partial}{\partial \alpha} \left(\frac{\partial z}{\partial \alpha}\right)' Ax = (x'A' \otimes I) \frac{\partial^2 z}{\partial \alpha \, \partial \alpha'} + \left(\frac{\partial z}{\partial \alpha}\right)' A \left(\frac{\partial x}{\partial \alpha}\right)$$

$$\frac{\partial}{\partial \alpha} \left(\frac{\partial x}{\partial \alpha}\right)' A'z = (z'A \otimes I) \frac{\partial^2 x}{\partial \alpha \, \partial \alpha'} + \left(\frac{\partial x}{\partial \alpha}\right)' A' \left(\frac{\partial z}{\partial \alpha}\right),$$

which proves the validity of the proposition.   q.e.d.

Remark 48. Note that despite appearances the matrix $\partial^2 y/\partial \alpha \, \partial \alpha'$ is symmetric as required. This is so since, for example, $(x'A' \otimes I) \, \partial^2 z/\partial \alpha \, \partial \alpha'$ is of the form

$$\sum_{i=1}^{m} \frac{\partial^2 z_i}{\partial \alpha \, \partial \alpha'} c_i,$$

where c_i is the ith element of $x'A'$ and evidently the matrices

$$\frac{\partial^2 z_i}{\partial \alpha \, \partial \alpha'}, \qquad i = 1, 2, \ldots, m,$$

are all symmetric.

Corollary 27. *Consider the quadratic form*

$$y = x'Ax$$

where x is $n \times 1$, A is $n \times n$, and x is a function of the r-element vector α, while A is independent of α. Then

$$\frac{\partial y}{\partial \alpha} = x'(A' + A)\frac{\partial x}{\partial \alpha},$$

$$\frac{\partial^2 y}{\partial \alpha \, \partial \alpha'} = \left(\frac{\partial x}{\partial \alpha}\right)'(A' + A)\left(\frac{\partial x}{\partial \alpha}\right) + (x'(A' + A) \otimes I)\frac{\partial^2 x}{\partial \alpha \, \partial \alpha'}.$$

PROOF. Since

$$\frac{\partial y}{\partial \alpha} = \frac{\partial y}{\partial x}\frac{\partial x}{\partial \alpha},$$

Proposition 95 guarantees the validity of the first part. For the second part, applying the arguments of Proposition 96 we see that

$$\frac{\partial^2 y}{\partial \alpha \, \partial \alpha'} = \frac{\partial}{\partial \alpha}\left(\frac{\partial y}{\partial \alpha'}\right) = \frac{\partial}{\partial \alpha}\left(\frac{\partial y}{\partial \alpha}\right)'.$$

But

$$\left(\frac{\partial y}{\partial \alpha}\right)' = \left(\frac{\partial x}{\partial \alpha}\right)'(A' + A)x.$$

Thus

$$\frac{\partial}{\partial \alpha}\left(\frac{\partial y}{\partial \alpha}\right)' = \left(\frac{\partial x}{\partial \alpha}\right)'(A' + A)\left(\frac{\partial x}{\partial \alpha}\right) + (x'(A' + A) \otimes I)\frac{\partial^2 x}{\partial \alpha \, \partial \alpha'} \quad \text{q.e.d.}$$

Corollary 28. *Consider the same situation as in Corollary 27 but suppose in addition that A is symmetric. Then*

$$\frac{\partial y}{\partial \alpha} = 2x'A\frac{\partial x}{\partial \alpha},$$

$$\frac{\partial^2 y}{\partial \alpha \, \partial \alpha'} = 2\left(\frac{\partial x}{\partial \alpha}\right)'A\left(\frac{\partial x}{\partial \alpha}\right) + (2x'A \otimes I)\frac{\partial^2 x}{\partial \alpha \, \partial \alpha'}.$$

PROOF. Obvious from Corollary 26.   q.e.d.

Let us now turn our attention to the differentiation of the trace of matrices. In fact, the preceding discussion has anticipated most of the results to be derived below. We begin with

Convention 5. If it is desired to differentiate, say, $\mathrm{tr}(AB)$ with respect to the elements of A, the operation involved will be interpreted as the "rematricization" of the vector

$$\frac{\partial\,\mathrm{tr}(AB)}{\partial\,\mathrm{vec}(A)},$$

i.e., we shall first obtain the vector

$$\frac{\partial\,\mathrm{tr}(AB)}{\partial\,\mathrm{vec}(A)}$$

and then put the resulting vector in matrix form, thus obtaining

$$\frac{\partial\,\mathrm{tr}(AB)}{\partial A}.$$

With this in mind we establish

Proposition 97. *Let A be a square matrix of order m. Then*

$$\frac{\partial\,\mathrm{tr}(A)}{\partial A} = I.$$

If the elements of A are functions of the r-element vector α, then

$$\frac{\partial\,\mathrm{tr}(A)}{\partial\alpha} = \frac{\partial\,\mathrm{tr}(A)}{\partial\,\mathrm{vec}(A)}\frac{\partial\,\mathrm{vec}(A)}{\partial\alpha} = \mathrm{vec}(I)'\frac{\partial\,\mathrm{vec}(A)}{\partial\alpha}.$$

PROOF. We note that

$$\mathrm{tr}(A) = \mathrm{tr}(A \cdot I),$$

where I is the identity matrix of order m. From Proposition 88 we have

$$\mathrm{tr}(A) = \mathrm{vec}(I)'\,\mathrm{vec}(A).$$

Thus

$$\frac{\partial\,\mathrm{tr}(A)}{\partial\,\mathrm{vec}(A)} = \mathrm{vec}(I)'.$$

Rematricizing this vector we obtain

$$\frac{\partial\,\mathrm{tr}(A)}{\partial A} = I,$$

which proves the first part. For the second part we note that Proposition 91 implies

$$\frac{\partial\,\mathrm{tr}(A)}{\partial\alpha} = \frac{\partial\,\mathrm{tr}(A)}{\partial\,\mathrm{vec}(A)}\frac{\partial\,\mathrm{vec}(A)}{\partial\,\mathrm{vec}(\alpha)} = \mathrm{vec}(I)'\frac{\partial\,\mathrm{vec}(A)}{\partial\alpha}. \qquad \text{q.e.d.}$$

We shall now establish results regarding differentiation of the trace of products of a number of matrices. We have

Proposition 98. *Let A be $m \times n$, and X be $n \times m$; then*

$$\frac{\partial \operatorname{tr}(AX)}{\partial X} = A'.$$

If X is a function of the elements of the vector α, then

$$\frac{\partial \operatorname{tr}(AX)}{\partial \alpha} = \frac{\partial \operatorname{tr}(AX)}{\partial \operatorname{vec}(X)} \frac{\partial \operatorname{vec}(X)}{\partial \alpha} = \operatorname{vec}(A')' \frac{\partial \operatorname{vec}(X)}{\partial \alpha}.$$

PROOF. By Proposition 88,

$$\operatorname{tr}(AX) = \operatorname{vec}(A')' \operatorname{vec}(X).$$

Thus, by Proposition 90,

$$\frac{\partial \operatorname{tr}(AX)}{\partial \operatorname{vec}(X)} = \operatorname{vec}(A')'.$$

Rematricizing this result we have

$$\frac{\partial \operatorname{tr}(AX)}{\partial X} = A',$$

which proves the first part. For the second part, we have, by Proposition 91,

$$\frac{\partial \operatorname{tr}(AX)}{\partial \alpha} = \operatorname{vec}(A')' \frac{\partial \operatorname{vec}(X)}{\partial \alpha}. \quad \text{q.e.d.}$$

Proposition 99. *Let A be $m \times n$, X be $n \times m$, and B be $m \times m$; then*

$$\frac{\partial}{\partial X} \operatorname{tr}(AXB) = A'B'.$$

If X is a function of the r-element vector α then

$$\frac{\partial}{\partial \alpha} \operatorname{tr}(AXB) = \operatorname{vec}(A'B')' \frac{\partial X}{\partial \alpha}.$$

PROOF. We note that

$$\operatorname{tr}(AXB) = \operatorname{tr}(BAX).$$

But then Proposition 98 implies

$$\frac{\partial \operatorname{tr}(AXB)}{\partial \operatorname{vec}(X)} = \operatorname{vec}(A'B')',$$

and thus

$$\frac{\partial \operatorname{tr}(AXB)}{\partial X} = A'B'.$$

For the second part, it easily follows that

$$\frac{\partial \operatorname{tr}(AXB)}{\partial X} = \operatorname{vec}(A'B')' \frac{\partial \operatorname{vec}(X)}{\partial \alpha}. \quad \text{q.e.d.}$$

Proposition 100. *Let A be $m \times n$, X be $n \times q$, B be $q \times r$, and Z be $r \times m$; then*

$$\frac{\partial \operatorname{tr}(AXBZ)}{\partial X} = A'Z'B',$$

$$\frac{\partial \operatorname{tr}(AXBZ)}{\partial Z} = B'X'A'.$$

If X and Z are functions of the r-element vector α, then

$$\frac{\partial \operatorname{tr}(AXBZ)}{\partial \alpha} = \operatorname{vec}(A'Z'B')' \frac{\partial \operatorname{vec}(X)}{\partial \alpha} + \operatorname{vec}(B'X'A')' \frac{\partial \operatorname{vec}(Z)}{\partial \alpha}$$

PROOF. Since

$$\operatorname{tr}(AXBZ) = \operatorname{tr}(BZAX),$$

Proposition 98 implies

$$\frac{\partial \operatorname{tr}(AXBZ)}{\partial X} = A'Z'B',$$

$$\frac{\partial \operatorname{tr}(AXBZ)}{\partial Z} = B'X'A'.$$

For the second part we note

$$\frac{\partial \operatorname{tr}(AXBZ)}{\partial \alpha} = \frac{\partial \operatorname{tr}(AXBZ)}{\partial \operatorname{vec}(X)} \frac{\partial \operatorname{vec}(X)}{\partial \alpha} + \frac{\partial \operatorname{tr}(AXBZ)}{\partial \operatorname{vec}(Z)} \frac{\partial \operatorname{vec}(Z)}{\partial \alpha},$$

and from Proposition 98 we also see that

$$\frac{\partial \operatorname{tr}(AXBZ)}{\partial \alpha} = \operatorname{vec}(A'Z'B')' \frac{\partial \operatorname{vec}(X)}{\partial \alpha} + \operatorname{vec}(B'X'A')' \frac{\partial \operatorname{vec}(Z)}{\partial \alpha}. \quad \text{q.e.d.}$$

Finally, we have

Proposition 101. *Let A be $m \times m$, X be $q \times m$, B be $q \times q$; then*

$$\frac{\partial \operatorname{tr}(AX'BX)}{\partial X} = B'XA' + BXA.$$

If X is a function of the r-element vector α, then

$$\frac{\partial \operatorname{tr}(AX'BX)}{\partial \alpha} = \operatorname{vec}(X)'[(A' \otimes B) + (A \otimes B')] \frac{\partial \operatorname{vec}(X)}{\partial \alpha}.$$

PROOF. From Remark 45, we see that

$$\operatorname{tr}(AX'BX) = \operatorname{vec}(X)'(A' \otimes B) \operatorname{vec}(X),$$

and from Proposition 95 we conclude

$$\frac{\partial \operatorname{tr}(AX'BX)}{\partial \operatorname{vec}(X)} = \operatorname{vec}(X)'[(A' \otimes B) + (A \otimes B')].$$

Matricizing this vector we have, from Corollary 22 and Proposition 87,

$$\frac{\partial \operatorname{tr}(AX'BX)}{\partial X} = B'XA' + BXA.$$

The second part of the proposition follows immediately from Corollary 27 and the preceding result.   q.e.d.

Remark 49. What the results above indicate is that differentiating the trace of products of matrices with respect to the elements of one of the matrix factors is a special case of differentiation of linear, bilinear, and quadratic forms. For this reason it is not necessary to derive second-order derivatives since the latter are easily derivable from the corresponding results regarding linear, bilinear, and quadratic forms, i.e., quantities of the form

$$a'x, \qquad z'Ax, \qquad x'Ax,$$

where a is a vector, A a matrix, and z and x appropriately dimensioned vectors.

Certain other aspects of differentiation of functions of matrices are also important and to these we now turn.

Proposition 102. *Let A be a square matrix of order m; then*

$$\frac{\partial |A|}{\partial A} = A^*,$$

where A^* is the matrix of cofactors (of the elements of A). If the elements of A are functions of the r elements of the vector α, then

$$\frac{\partial |A|}{\partial \alpha} = \operatorname{vec}(A^*)' \frac{\partial \operatorname{vec}(A)}{\partial \alpha}.$$

PROOF. To prove the first part of the proposition, it is sufficient to obtain the typical (i, j) element of the matrix $\partial |A|/\partial A$. The latter is given by

$$\frac{\partial |A|}{\partial a_{ij}}.$$

Expand the determinant by the elements of the ith row and find, by Proposition 23,

$$|A| = \sum_{k=1}^{m} a_{ik} A_{ik},$$

where A_{ik} is the cofactor of a_{ik}. Evidently, A_{ik} does *not* contain a_{ik}. Consequently

$$\frac{\partial |A|}{\partial a_{ij}} = A_{ij},$$

and thus

$$\frac{\partial |A|}{\partial A} = A^*,$$

as was to be proved.

For the second part we note that

$$\frac{\partial |A|}{\partial \alpha} = \frac{\partial |A|}{\partial \text{vec}(A)} \frac{\partial \text{vec}(A)}{\partial \alpha}.$$

But it is easy to see, from Convention 5, that

$$\frac{\partial |A|}{\partial \text{vec}(A)} = \text{vec}\left(\frac{\partial |A|}{\partial A}\right)' = \text{vec}(A^*)'.$$

Hence

$$\frac{\partial |A|}{\partial \alpha} = \text{vec}(A^*)' \frac{\partial \text{vec}(A)}{\partial \alpha}. \quad \text{q.e.d.}$$

Corollary 29. *Assume, in addition to the conditions of Proposition* 102, *that A is nonsingular, and let* $B = A^{-1}$. *Then*

$$\frac{\partial |A|}{\partial A} = |A| B',$$

$$\frac{\partial |A|}{\partial \alpha} = |A| \text{vec}(B')' \frac{\partial \text{vec}(A)}{\partial \alpha}.$$

PROOF. In the proof of Proposition 102 note that

$$A_{ik} = |A| b_{ki},$$

where b_{ki} is the (k, i) element of B.   q.e.d.

Corollary 30. *If in Proposition* 102 α *is assumed to be a scalar then*

$$\frac{\partial |A|}{\partial \alpha} = \text{tr}\left(A^{*'} \frac{\partial A}{\partial \alpha}\right),$$

and if A is nonsingular then

$$\frac{\partial |A|}{\partial \alpha} = |A| \text{tr}\left(B \frac{\partial A}{\partial \alpha}\right).$$

PROOF. If α is a scalar then

$$\frac{\partial \, \text{vec}(A)}{\partial \alpha} = \text{vec}\left(\frac{\partial A}{\partial \alpha}\right),$$

where, obviously,

$$\frac{\partial A}{\partial \alpha} = \left[\frac{\partial a_{ij}}{\partial \alpha}\right].$$

Using Propositions 88 and 102 we see that

$$\frac{\partial |A|}{\partial \alpha} = \text{vec}(A^*)' \, \text{vec}\left(\frac{\partial A}{\partial \alpha}\right) = \text{tr}\left(A^{*\prime} \frac{\partial A}{\partial \alpha}\right).$$

If A is nonsingular then

$$A^{*\prime} = |A|B$$

so that, in this case,

$$\frac{\partial |A|}{\partial \alpha} = |A| \, \text{tr}\left(B \frac{\partial A}{\partial \alpha}\right). \quad \text{q.e.d.}$$

Corollary 31. *If in Proposition 102 A is a symmetric matrix, then*

$$\frac{\partial |A|}{\partial A} = 2A^* - \text{diag}(A_{11}, \ldots, A_{mm}).$$

PROOF. In Corollary 30 take α to be successively $a_{11}, a_{21}, \ldots, a_{m1}$; $a_{22}, a_{32}, \ldots, a_{m2}; a_{33}, a_{43}, \ldots, a_{m3}, \ldots, a_{mm}$, and thus determine that

$$\frac{\partial |A|}{\partial a_{ij}} = 2A_{ij}, \quad i \neq j,$$

$$\frac{\partial |A|}{\partial a_{ii}} = A_{ii}.$$

Hence

$$\frac{\partial |A|}{\partial A} = 2A^* - \text{diag}(A_{11}, A_{22}, \ldots, A_{mm}). \quad \text{q.e.d.}$$

Another useful result is

Proposition 103. *Let X be $n \times m$ and B be $n \times n$, and put*

$$A = X'BX.$$

Then

$$\frac{\partial |A|}{\partial X} = \left[\text{tr}\left(A^{*\prime} \frac{\partial A}{\partial x_{ik}}\right)\right] = BXA^{*\prime} + B'XA^*,$$

where A^* is the matrix of cofactors of A.

PROOF. We shall prove this result by simply deriving the (i, k) element of the matrix $\partial |A| / \partial X$. By the usual chain rule for differentiation we have

$$\frac{\partial |A|}{\partial x_{ik}} = \sum_{r,s=1}^{m} \frac{\partial |A|}{\partial a_{rs}} \frac{\partial a_{rs}}{\partial x_{ik}}.$$

But

$$\frac{\partial |A|}{\partial a_{rs}} = A_{rs}$$

so that

$$\frac{\partial |A|}{\partial x_{ik}} = \sum_{r,s=1}^{m} A_{rs} \frac{\partial a_{rs}}{\partial x_{ik}} = \operatorname{tr}\left(A^{*\prime} \frac{\partial A}{\partial x_{ik}}\right),$$

which proves the first part of the representation.

Next we note that, formally, we can put

$$\frac{\partial A}{\partial x_{ik}} = \frac{\partial X'}{\partial x_{ik}} BX + X'B \frac{\partial X}{\partial x_{ik}}.$$

But

$$\frac{\partial X'}{\partial x_{ik}} = e_{\cdot k} e'_{\cdot i}, \qquad \frac{\partial X}{\partial x_{ik}} = e_{\cdot i} e'_{\cdot k},$$

where $e_{\cdot k}$ is m-element (column) vector all of whose elements are zero save the kth, which is unity, and $e_{\cdot i}$ is an n-element (column) vector all of whose elements are zero save the ith, which is unity.

For simplicity of notation only, put

$$A^{*\prime} = |A| A^{-1}$$

and note that

$$\operatorname{tr}\left(A^{*\prime} \frac{\partial A}{\partial x_{ik}}\right) = |A| \operatorname{tr} A^{-1}(e_{\cdot k} e'_{\cdot i} BX + X'B e_{\cdot i} e'_{\cdot k})$$

$$= |A| [\operatorname{tr}(a^{\cdot k} b_{i\cdot} X) + \operatorname{tr}(X' b_{\cdot i} a^{k\cdot})] = |A| (b_{i\cdot} X a^{\cdot k} + b'_{\cdot i} X a^{k\cdot\prime})$$

where $b_{\cdot i}$ is the ith column of B, $b_{i\cdot}$ is the ith row of B, $a^{\cdot k}$ is the kth column of A^{-1} and $a^{k\cdot}$ its kth row. Thus, the (i, k) element of $\partial |A| / \partial X$ is given by

$$|A| (b_{i\cdot} X a^{\cdot k} + b'_{\cdot i} X a^{k\cdot\prime}).$$

But this is, of course, the (i, k) element of

$$|A| (BXA^{-1} + B'XA'^{-1}) = BXA^{*\prime} + B'XA^{*}.$$

Consequently,

$$\frac{\partial |A|}{\partial X} = BXA^{*\prime} + B'XA^{*}. \quad \text{q.e.d.}$$

Corollary 32. *If in Proposition 103 A is nonsingular then*

$$\frac{\partial |A|}{\partial X} = |A|(BXA^{-1} + B'XA'^{-1}).$$

PROOF. Evident from Proposition 103.   q.e.d.

Corollary 33. *If in Proposition 103 A is nonsingular, then*

$$\frac{\partial \ln |A|}{\partial X} = BXA^{-1} + B'XA'^{-1}.$$

PROOF. We have

$$\frac{\partial \ln |A|}{\partial X} = \frac{\partial \ln |A|}{\partial |A|}\frac{\partial |A|}{\partial X} = \frac{1}{|A|}\frac{\partial |A|}{\partial X}$$

and the conclusion follows from Corollary 32.   q.e.d.

Corollary 34. *If in Proposition 103 B is symmetric, then*

$$\frac{\partial |A|}{\partial X} = 2BXA^*.$$

PROOF. Obvious since if B is symmetric so is A and thus

$$A^{*'} = A^*.\quad\text{q.e.d.}$$

Proposition 104. *Let X be m × n and B be m × m, and suppose that the elements of X are functions of the elements of the vector α. Put*

$$A = X'BX.$$

Then

$$\frac{\partial |A|}{\partial \alpha} = \text{vec}(X)'[(A^{*'} \otimes B') + (A^* \otimes B)]\frac{\partial \text{vec}(X)}{\partial \alpha}.$$

PROOF. By the usual chain rule of differentiation we have

$$\frac{\partial |A|}{\partial \alpha} = \frac{\partial |A|}{\partial \text{vec}(X)}\frac{\partial \text{vec}(X)}{\partial \alpha}.$$

But

$$\frac{\partial |A|}{\partial \text{vec}(X)} = \left[\text{vec}\left(\frac{\partial |A|}{\partial X}\right)\right]'.$$

From Corollary 22 we then obtain, using Proposition 103,

$$\text{vec}\left(\frac{\partial |A|}{\partial X}\right) = [(A^* \otimes B) + (A^{*'} \otimes B')]\text{vec}(X).$$

Thus, we conclude

$$\frac{\partial |A|}{\partial \alpha} = \text{vec}(X)'[(A^* \otimes B) + (A^{*'} \otimes B')] \frac{\partial \text{ vec}(X)}{\partial \alpha}. \quad \text{q.e.d.}$$

Corollary 35. *If in Proposition 104 A is nonsingular, then*

$$\frac{\partial |A|}{\partial \alpha} = |A| \text{ vec}(X)'[(A'^{-1} \otimes B) + (A^{-1} \otimes B')] \frac{\partial \text{ vec}(X)}{\partial \alpha}.$$

PROOF. Obvious if we note that

$$|A|A^{-1} = A^{*'}. \quad \text{q.e.d.}$$

Corollary 36. *If in Proposition 104 A is nonsingular, then*

$$\frac{\partial \ln |A|}{\partial \alpha} = \text{vec}(X)'[(A'^{-1} \otimes B) + (A^{-1} \otimes B')] \frac{\partial \text{ vec}(X)}{\partial \alpha}.$$

PROOF. Obvious.   q.e.d.

Corollary 37. *If in Proposition 104 B is symmetric, then*

$$\frac{\partial |A|}{\partial \alpha} = 2 \text{ vec}(X)'[A^* \otimes B] \frac{\partial \text{ vec}(X)}{\partial \alpha}.$$

PROOF. Obvious since if B is symmetric so is A, and thus

$$A^{*'} = A^*. \quad \text{q.e.d.}$$

The results above exhaust those aspects of differentiation of determinants that are commonly found useful in econometrics. The final topic of interest is the differentiation of inverses. We have

Proposition 105. *Let A be m × m and nonsingular. Then the "derivative of the inverse" $\partial A^{-1}/\partial A$ is given by*

$$\frac{\partial \text{ vec}(A^{-1})}{\partial \text{ vec}(A)} = -(A'^{-1} \otimes A^{-1}).$$

If the elements of A are functions of the elements of the vector α, then

$$\frac{\partial \text{ vec}(A^{-1})}{\partial \alpha} = -(A'^{-1} \otimes A^{-1}) \frac{\partial \text{ vec}(A)}{\partial \alpha}.$$

PROOF. We begin by taking the derivative of A^{-1} with respect to an element of A. From the relation

$$A^{-1}A = I$$

we easily see that

$$0 = \frac{\partial A^{-1}}{\partial a_{rs}} A + A^{-1} \frac{\partial A}{\partial a_{rs}}, \tag{71}$$

whence we obtain

$$\frac{\partial A^{-1}}{\partial a_{rs}} = -A^{-1} \frac{\partial A}{\partial a_{rs}} A^{-1}. \tag{72}$$

But $\partial A / \partial a_{rs}$ is a matrix all of whose elements are zero save the (r, s) element, which is unity.

Consequently,

$$\frac{\partial A}{\partial a_{rs}} = e_{\cdot r} e'_{\cdot s}, \tag{73}$$

where $e_{\cdot j}$ is an m-element vector all of whose elements are zero save the jth, which is unity. Using (73) in (72) we find

$$\frac{\partial A^{-1}}{\partial a_{rs}} = -a^{\cdot r} a^{s \cdot}, \qquad r, s = 1, 2, \ldots, m, \tag{74}$$

where $a^{\cdot r}$ is the rth column and $a^{s \cdot}$ is the sth row of A^{-1}. Vectorizing (74) yields, by Proposition 86,

$$\frac{\partial \, \mathrm{vec}(A^{-1})}{\partial a_{rs}} = -(a^{s \cdot \prime} \otimes a^{\cdot r}), \qquad r, s = 1, 2, \ldots, m. \tag{75}$$

From (75) we see, for example, that

$$\frac{\partial \, \mathrm{vec}(A^{-1})}{\partial a_{11}} = -(a^{1 \cdot \prime} \otimes a^{\cdot 1}), \qquad \frac{\partial \, \mathrm{vec}(A^{-1})}{\partial a_{21}} = -(a^{1 \cdot \prime} \otimes a^{\cdot 2})$$

and so on; thus

$$\frac{\partial \, \mathrm{vec}(A^{-1})}{\partial a_{\cdot 1}} = -(a^{1 \cdot \prime} \otimes A^{-1})$$

or, in general,

$$\frac{\partial \, \mathrm{vec}(A^{-1})}{\partial a_{\cdot s}} = -(a^{s \cdot \prime} \otimes A^{-1}), \qquad s = 1, 2, \ldots, m. \tag{76}$$

But (76) implies

$$\frac{\partial \, \mathrm{vec}(A^{-1})}{\partial \, \mathrm{vec}(A)} = -(A'^{-1} \otimes A^{-1}), \tag{77}$$

which proves the first part of the proposition. For the second part we note that by the chain rule of differentiation

$$\frac{\partial \, \mathrm{vec}(A^{-1})}{\partial \alpha} = \frac{\partial \, \mathrm{vec}(A^{-1})}{\partial \, \mathrm{vec}(A)} \frac{\partial \, \mathrm{vec}(A)}{\partial \alpha},$$

and the desired result follows immediately form (77).   q.e.d.

Since the result of the proposition above may not be easily digestible, let us at least verify that it holds for a simple case. We have

Corollary 38. *Suppose in Proposition* 105 α *is a* ***scalar***; *then*

$$\frac{\partial A^{-1}}{\partial \alpha} = -A^{-1}\frac{\partial A}{\partial \alpha}A^{-1}.$$

PROOF. From Proposition 105 we have, formally,

$$\frac{\partial \, \mathrm{vec}(A^{-1})}{\partial \alpha} = -(A'^{-1} \otimes A^{-1})\frac{\partial \, \mathrm{vec}(A)}{\partial \alpha}.$$

Matricizing the vector above, using Corollary 22, yields

$$\frac{\partial A^{-1}}{\partial \alpha} = -A^{-1}\frac{\partial A}{\partial \alpha}A^{-1}. \quad \text{q.e.d.}$$

Systems of Difference Equations with Constant Coefficients 5

5.1 The Scalar Second-order Equation

The reader will recall that the second-order difference equation

$$a_0 y_t + a_1 y_{t+1} + a_2 y_{t+2} = g(t + 2), \tag{78}$$

where y_t is the *scalar* dependent variable, the a_i, $i = 0, 1, 2$, are the (constant) coefficients, and $g(t)$ is the real-valued "forcing function," is solved in two steps. First we consider the homogeneous part

$$a_0 y_t + a_1 y_{t+1} + a_2 y_{t+2} = 0,$$

and find the most general form of its solution, called the *general solution to the homogeneous part*. Then we find just one solution to the equation in (78), called the *particular solution*. The sum of the general solution to the homogeneous part and the particular solution is said to be the *general solution to the equation*. What is meant by the "general solution," denoted, say, by y_t^*, is that y_t^* satisfies (78) and that it can be made to satisfy any prespecified set of "initial conditions." To appreciate this aspect rewrite (78) as

$$y_{t+2} = \bar{g}(t + 2) + \bar{a}_1 y_{t+1} + \bar{a}_0 y_t, \tag{79}$$

where, assuming $a_2 \neq 0$,

$$\bar{g}(t + 2) = \frac{1}{a_2} g(t + 2), \qquad \bar{a}_1 = -\frac{a_1}{a_2}, \qquad \bar{a}_0 = -\frac{a_0}{a_2}.$$

If $g(t)$ and the coefficients are specified and if, further, we are given the values assumed by y_t for $t = 0$, $t = 1$, we can compute y_2 from (79); then given y_2 we can compute y_3 and so on. Thus, given the coefficients and the function $g(\cdot)$ the behavior of y_t depends solely on the "initial conditions" and

541

if these are also specified then the behavior of y_t is completely determined. Thus, for a solution of (78) to be a "general solution" it must be capable of accommodating any prespecified set of initial conditions.

To obtain the solution we begin with the trial solution

$$y_t = c\lambda^t.$$

Substituting in the homogeneous part we find

$$a_0 c\lambda^t + a_1 c\lambda^{t+1} + a_2 c\lambda^{t+2} = 0.$$

It is clear that

$$y_t = 0$$

is always a solution to the homogeneous part, called the *trivial* solution. The trivial solution is of little interest; consequently, we shall ignore it by factoring out $c\lambda^t$ in the equation above and thus considering

$$a_0 + a_1\lambda + a_2\lambda^2 = 0. \tag{80}$$

Solving this equation yields

$$\lambda_1 = \frac{-a_1 + \sqrt{a_1^2 - 4a_2 a_0}}{2a_2}, \qquad \lambda_2 = \frac{-a_1 - \sqrt{a_1^2 - 4a_2 a_0}}{2a_2}.$$

It is clear that if

$$a_1^2 - 4a_2 a_0 \neq 0,$$

the roots are *distinct*; if

$$a_1^2 - 4a_2 a_0 > 0$$

the roots are real; if

$$a_1^2 - 4a_2 a_0 < 0$$

the roots are a pair of *complex conjugates*, i.e., $\lambda_2 = \bar{\lambda}_1$.

If the roots are distinct, then

$$y_t^H = c_1\lambda_1^t + c_2\lambda_2^t$$

is the general solution to the homogeneous part, and is seen to be a linear combination of the solutions induced by the two roots. That this is indeed the general solution is easily seen by prescribing a set of initial conditions and determining the arbitrary constants c_1, c_2. Thus, if the specified initial conditions are $\bar{y}_0, \bar{y}_1$ we have

$$\bar{y}_0 = c_1 + c_2,$$
$$\bar{y}_1 = c_1\lambda_1 + c_2\lambda_2$$

or

$$\bar{y} = \begin{bmatrix} 1 & 1 \\ \lambda_1 & \lambda_2 \end{bmatrix} c,$$

where

$$\bar{y} = (\bar{y}_0, \bar{y}_1)', \qquad c = (c_1, c_2)'.$$

Since the determinant of the matrix above is

$$\lambda_2 - \lambda_1$$

we see that if the roots are *distinct*

$$c = \begin{bmatrix} 1 & 1 \\ \lambda_1 & \lambda_2 \end{bmatrix}^{-1} \bar{y},$$

so that we establish a unique correspondence between the vector of initial conditions $\bar{y}$ and the vector of undetermined coefficients c.

If the roots *are not distinct* the general solution to the homogeneous part is

$$y_t^H = c_1 \lambda^t + c_2 t \lambda^t.$$

We may show that this is the general solution by solving the equation

$$\bar{y} = \begin{bmatrix} 1 & 0 \\ \lambda & \lambda \end{bmatrix} c,$$

thus obtaining

$$c = \frac{1}{\lambda} \begin{bmatrix} \lambda & 0 \\ -\lambda & 1 \end{bmatrix} \bar{y},$$

which shows again that the vector of undetermined constants can be made to satisfy any prespecified set of initial conditions.

In this case it is easily seen that

$$c_1 = \bar{y}_0, \qquad c_2 = \frac{1}{\lambda} y_1 - \bar{y}_0.$$

In the previous (and more typical) case of distinct roots we have

$$c_1 = \frac{1}{\lambda_2 - \lambda_1} (\lambda_2 \bar{y}_0 - \bar{y}_1), \qquad c_2 = \frac{1}{\lambda_2 - \lambda_1} (\bar{y}_1 - \lambda_1 \bar{y}_0).$$

From the results above it is evident that the *general solution to the homogeneous part simply carries forward the influence of initial conditions.* Generally, in economics, we would not want to say that initial conditions are very crucial to the development of a system, but rather that it is the external forces impinging on the system that are ultimately responsible for its development. This introduces the concept of the *stability* of a difference equation. Stability is defined in terms of the asymptotic vanishing of the general solution to the homogeneous part, i.e., by the condition

$$\lim_{t \to \infty} y_t^H = 0. \tag{81}$$

A little reflection will show that (81) is equivalent to

$$|\lambda_1| < 1, \qquad |\lambda_2| < 1.$$

In the literature of econometrics it is customary to write an equation like (78) as

$$\alpha_0 y_t + \alpha_1 y_{t-1} + \alpha_2 y_{t-2} = g(t) \tag{82}$$

since "period t" refers to the present and one wants to say that the past will determine the present rather than say that the present will determine the future—although the two statements are equivalent. In estimation, however, we can only use information from the past and the present. Because of this difference in the representation of difference equations, there is often a slight confusion regarding stability requirements. To see this let us examine the stability requirements for (82). We see that this will yield the equation

$$\alpha_0 \psi^2 + \alpha_1 \psi + \alpha_2 = 0$$

and the two solutions

$$\psi_1 = \frac{-\alpha_1 + \sqrt{\alpha_1^2 - 4\alpha_2\alpha_0}}{2\alpha_0}, \qquad \psi_2 = \frac{-\alpha_1 - \sqrt{\alpha_1^2 - 4\alpha_2\alpha_0}}{2\alpha_0}$$

and as before, of course, stability here requires that

$$|\psi_1| < 1, \qquad |\psi_2| < 1.$$

If in comparing the homogeneous parts of (78) and (82) we set

$$\alpha_0 = a_2, \qquad \alpha_1 = a_1, \qquad \alpha_2 = a_0, \tag{83}$$

then the two equations are perfectly equivalent. If on the other hand we set

$$\alpha_i = a_i, \qquad i = 0, 1, 2, \tag{84}$$

then the two equations are very different indeed! The reader, however, will easily verify that if

$$y_t^H = c_1 \psi_1^t + c_2 \psi_2^t$$

in the solution to the homogeneous part of (82), then

$$(y_t^*)^H = c_1 \psi_1^{-t} + c_2 \psi_2^{-t}$$

is the general solution to the homogeneous part of (78). Consequently, if

$$|\psi_1| < 1, \qquad |\psi_2| < 1$$

so that (82) is stable, then

$$\lambda_1 = \frac{1}{\psi_1}, \qquad \lambda_2 = \frac{1}{\psi_2}$$

and thus (78) is *not* stable when used for the *forward projection* of the dependent variable, i.e., for the representation of y_t for $t > 0$. On the other hand *it is stable for the backward projection of* y_t, i.e., for $t < 0$.

As we have mentioned earlier we need to find a particular solution in order to obtain the general solution to the equation in (78). It will facilitate matters if we introduce the notion of the lag operator L. If $x(t)$ is a function of "time" the *lag operator* L is defined by

$$Lx(t) = x(t - 1). \tag{85}$$

Powers of the operator are defined as successive applications, i.e.,

$$L^2 x(t) = L[Lx(t)] = Lx(t - 1) = x(t - 2)$$

and in general

$$L^k x(t) = x(t - k), \qquad k > 0. \tag{86}$$

For $k = 0$ we have the *identity operator*

$$L^0 \equiv I, \qquad L^0 x(t) = x(t). \tag{87}$$

It is apparent that

$$L^k L^s = L^s L^k = L^{s+k}. \tag{88}$$

Moreover,

$$\begin{aligned}(c_1 L^{s_1} + c_2 L^{s_2})x(t) &= c_1 L^{s_1} x(t) + c_2 L^{s_2} x(t) \\ &= c_1 x(t - s_1) + c_2 x(t - s_2).\end{aligned} \tag{89}$$

One can further show that the set

$$\{I, L, L^2, \ldots\}$$

over the field of real numbers, together with the operations above, induces *a vector space*. But what is of importance to us is that the set of *polynomial operators*, whose typical element is

$$\sum_{i=0}^{n} c_i L^i,$$

induces an algebra that is *isomorphic to the algebra* of *polynomials* in a real or complex indeterminate. What this means is that in order to determine the outcome of a set of operations on polynomials with a lag operator one only has to carry out such operations with respect to an ordinary polynomial in the real or complex indeterminate, ψ, and then substitute for ψ and its powers, L and its powers. Perhaps a few examples will make this clear.

EXAMPLE 5. Let

$$P_1(L) = c_{01} I + c_{11} L + c_{21} L^2,$$
$$P_2(t) = c_{02} I + c_{12} L + c_{22} L^2 + c_{32} L^3,$$

and suppose we desire the product $P_1(L) \cdot P_2(L)$. We consider, by the usual rules of multiplying polynomials,

$$
\begin{aligned}
P_1(\psi)P_2(\psi) = {} & (c_{01}c_{02}) + (c_{11}c_{02} + c_{12}c_{01})\psi \\
& + (c_{11}c_{12} + c_{22}c_{01} + c_{21}c_{02})\psi^2 \\
& + (c_{01}c_{32} + c_{11}c_{22} + c_{12}c_{21})\psi^3 \\
& + (c_{11}c_{32} + c_{22}c_{21})\psi^4 + c_{32}c_{21}\psi^5,
\end{aligned}
$$

and consequently

$$
\begin{aligned}
P_1(L)P_2(L) = {} & c_{01}c_{02}I + (c_{11}c_{02} + c_{12}c_{01})L \\
& + (c_{11}c_{12} + c_{22}c_{01} + c_{21}c_{02})L^2 \\
& + \left(\sum_{i=0}^{2} c_{i1}c_{3-i2} \right)L^3 + (c_{11}c_{32} + c_{22}c_{21})L^4 + c_{32}c_{21}L^5.
\end{aligned}
$$

EXAMPLE 6. Let

$$
P_1(L) = I - \lambda L
$$

and suppose we wish to find its inverse $I/P_1(L)$. To do so we consider the inverse of $1 - \lambda\psi$; if $|\psi| \le 1$ and $|\lambda| < 1$, then we know that

$$
\frac{1}{1 - \lambda\psi} = \sum_{i=0}^{\infty} \lambda^i \psi^i.
$$

Hence, under the condition $|\lambda| < 1$,

$$
\frac{1}{P_1(L)} = \sum_{i=0}^{\infty} \lambda^i L^i.
$$

Although the discussion above is heuristic, and rather sketchy at that, it is sufficient for the purposes we have in mind. The reader interested in more detail is referred to Dhrymes [8].

If we use the apparatus of polynomial lag operators we see that we can write the equation in (78) as

$$
a_0 L^2 y_{t+2} + a_1 L y_{t+2} + a_2 I y_{t+2} = g(t + 2)
$$

or, substituting t for $t + 2$, perhaps more conveniently as

$$
(a_0 L^2 + a_1 L + a_2 I)y_t = g(t). \tag{90}
$$

Consequently we may write the solution, formally, as

$$
y = [A(L)]^{-1}g(t), \tag{91}
$$

where

$$
A(L) = a_0 L^2 + a_1 L + a_2 I.
$$

The question then arises as to the meaning and definition of the inverse of this polynomial operator. In view of the isomorphism referred to earlier we consider the polynomial equation

$$a_0\psi^2 + a_1\psi + a_2 = 0 \qquad (92)$$

and its roots ψ_1, ψ_2. By the fundamental theorem of algebra we can write

$$A(\psi) = a_0(\psi - \psi_1)(\psi - \psi_2) = a_0\psi_1\psi_2\left(1 - \frac{\psi}{\psi_1}\right)\left(1 - \frac{\psi}{\psi_2}\right).$$

Comparing (92) with (80) and bearing in mind the discussion following (82) we see that

$$\lambda_1 = \frac{1}{\psi_1}, \qquad \lambda_2 = \frac{1}{\psi_2},$$

where λ_i, $i = 1, 2$ are the roots of (80). By the stability condition for (78) we must have

$$|\lambda_1| < 1, \qquad |\lambda_2| < 1.$$

Since

$$a_0\psi_1\psi_2 = a_2$$

we see that

$$\frac{1}{A(\psi)} = \frac{1}{a_2}\left(\frac{1}{1 - \lambda_1\psi}\right)\left(\frac{1}{1 - \lambda_2\psi}\right)$$

and, consequently,

$$y_t = \frac{I}{(I - \lambda_1 L)(I - \lambda_2 L)}\left(\frac{g(t)}{a_2}\right)$$

is the desired interpretation of (91). That this interpretation makes sense, i.e., if $g(\cdot)$ is a bounded function then y_t is also a bounded function, is guaranteed by the stability condition one would normally impose on (78).

Since we would normally like to write the solution in terms of specified initial conditions we may choose to write the general solution to the difference equation as

$$y_t = c_1\lambda_1^t + c_2\lambda_2^t + [A(L)]^{-1}(g(t) - \delta_{0t}g(t) - \delta_{1t}g(t)), \qquad (93)$$

where δ_{it} is the Kronecker delta, which is 0 if $i \neq t$ and is 1 if $i = t$, $i = 0, 1$. The constants c_1, c_2 are to be determined from initial conditions, say y_0, y_1. If the initial conditions were to refer, say, to y_i, y_{i+1} then we would write the solution as

$$y_t = c_1\lambda_1^t + c_2\lambda_2^t + [A(L)]^{-1}[g(t) - \delta_{it}g(t) - \delta_{i+1,t}g(t)]. \qquad (94)$$

5.2 Vector Difference Equations

In this section we shall be considerably more formal than earlier since the results of this section are directly relevant to the analysis of simultaneous equations models.

Definition 40. The equation

$$A_0 y_t + A_1 y_{t-1} + A_2 y_{t-2} + \cdots + A_r y_{t-r} = g(t), \tag{95}$$

where A_i, $i = 0, 1, \ldots, r$, are matrices of constants and y_t and $g(t)$ are m-element (column) vectors, is said to be an rth-*order vector difference equation with constant coefficients* (VDECC), provided the matrix A_0 is nonsingular.

Convention 6. Since the matrix A_0 in the VDECC is nonsingular no loss of generality is entailed by taking $A_0 = I$, which we shall do in the discussion to follow. In discussing the characterization of the solution of (95) it is useful to note that *we can always transform it to an equivalent system that is a first-order* VDECC. This is done as follows. Define

$$\zeta_t = (y_t', y_{t-1}', \ldots, y_{t-r+1}')',$$

$$A^* = \begin{bmatrix} -A_1 & -A_2 & \cdots & -A_r \\ I & 0 & \cdots & 0 \\ 0 & I & \cdots & 0 \\ 0 & 0 & \cdots & I & 0 \end{bmatrix},$$

and notice that the homogeneous part can be written as

$$\zeta_t = A^* \zeta_{t-1}.$$

Indeed, (95) can be written as

$$\zeta_t = A^* \zeta_{t-1} + e_{\cdot 1} \otimes g(t) \tag{96}$$

where $e_{\cdot 1}$ is an r-element (column) vector all of whose elements are zero save the first, which is unity. Thus, in the sequel we shall only discuss the first-order VDECC

$$y_t = A y_{t-1} + g(t), \tag{97}$$

where y_t, $g(t)$ are m-element column vectors and A is an $m \times m$ matrix of constants. We shall further impose the

Condition 1. *The matrix A in (97) is diagonalizable.*

Remark 50. The need for Condition 1 will be fully appreciated below. For the moment let us note that in view of Propositions 39, 40 and 44, Condition 1 is only mildly restrictive.

Definition 41. A *solution to the* VDECC in (97) is a vector ζ_t such that ζ_t satisfies (97) together with a properly specified set of initial conditions, say

$$y_0 = \bar{y}_0.$$

As in the simple scalar case we split the problem in two, finding the general solution to the homogeneous part and a particular solution. We begin with the solution to the homogeneous part

$$y_t = Ay_{t-1}, \tag{98}$$

and we attempt the trial vector

$$y_t = c\lambda^t,$$

where c is a nonnull m-element column vector and λ is a scalar. Substituting in (98) and factoring out $c\lambda^{t-1}$ we have

$$(\lambda I - A)c\lambda^{t-1} = 0.$$

Cancelling the factor λ^{t-1} (since, where relevant, this corresponds to the trivial solution $y_t = 0$) we have

$$(\lambda I - A)c = 0.$$

This condition defines the vector c and the scalar λ. Upon reflection, however, we note that this is nothing more than the condition defining the characteristic roots and vectors of the matrix A.

We have

Definition 42. Consider the VDECC as exhibited in (97). Then

$$|\lambda I - A| = 0$$

is said to be the *characteristic equation* of the VDECC.

The problem of finding a solution for the homogeneous part of VDECC has an obvious formulation. Thus, if

$$\{(\lambda_i, c_{\cdot i}): i = 1, 2, \ldots, m\}$$

are associated pairs of characteristic roots and vectors of A, then

$$y_t^H = \sum_{i=1}^{m} d_i c_{\cdot i} \lambda_i^t \tag{99}$$

is the general solution to the homogeneous part. This is so since, evidently, (99) satisfies (98). Thus, we need only check whether the constants d_i, $i = 1, 2, \ldots, m$, can be so chosen as to satisfy any prespecified set of initial conditions. Putting

$$C = (c_{\cdot 1}, c_{\cdot 2}, \ldots, c_{\cdot m}), \qquad \Lambda = \text{diag}(\lambda_1, \lambda_2, \ldots, \lambda_m),$$
$$d = (d_1, d_2, \ldots, d_m)',$$

we can write (99) more conveniently as

$$y_t^H = C\Lambda^t d. \tag{100}$$

If we specify the initial condition

$$y_0 = \bar{y}_0$$

the elements of D must be so chosen as to satisfy

$$\bar{y}_0 = C\Lambda^0 d = Cd.$$

Since A is diagonalizable C is nonsingular. Consequently,

$$d = C^{-1}\bar{y}_0$$

and the solution to the homogeneous part, with initial condition $\bar{y}_0$, can be written as

$$y_t^H = C\Lambda^t C^{-1}\bar{y}_0. \tag{101}$$

The particular solution is found by noting that formally we can put

$$y_t = (I - AL)^{-1}g(t),$$

where we have written (97) as

$$(I - AL)y_t = g(t)$$

and have formally inverted the *matrix* of polynomial operators $I - AL$. By the isomorphism we have noted earlier we can obtain the desired inverse by operating on $I - A\psi$, where ψ is a real or complex indeterminate. By the definition of inverse we note that we first need the *adjoint* of $I - A\psi$, i.e., the transpose of the matrix of cofactors. Since each cofactor is the determinant of an $(m - 1) \times (m - 1)$ matrix, the adjoint is a matrix, say

$$B(\psi) = [b_{ij}(\psi)],$$

the individual elements $b_{ij}(\psi)$ being polynomials of degree at most $m - 1$ in ψ.

We also need to evaluate the determinant of $I - A\psi$. But we note that for $\lambda = \psi^{-1}$

$$
\begin{aligned}
|I - A\psi| &= |\psi(\lambda I - A)| \\
&= \psi^m |\lambda I - A| \\
&= \psi^m \prod_{i=1}^{m} (\lambda - \lambda_i) \\
&= \psi^m \lambda^m \prod_{i=1}^{m} (1 - \psi\lambda_i) = \prod_{i=1}^{m} (1 - \psi\lambda_i).
\end{aligned}
$$

Consequently, we obtain

$$a(L) = |I - AL| = \prod_{i=1}^{m} (I - \lambda_i L).\qquad(102)$$

If the system in (97) is a *stable one*, i.e., if

$$|\lambda_i| < 1, \qquad i = 1, 2, \ldots, m,$$

then the inverse of $I - AL$ is well defined and is given by

$$(I - AL)^{-1} = \left[\frac{b_{ij}(L)}{a(L)}\right],\qquad(103)$$

so that each element is the *ratio of two polynomials*; the numerator polynomial is of degree, at most, $m - 1$, while the denominator polynomial is of degree m. Thus, the (general) solution to the VDECC in (97), with initial conditions $\bar{y}_0$, is given by

$$y_t = C\Lambda^t C^{-1} \bar{y}_0 + (I - AL)^{-1}[(1 - \delta_{0t})g(t)].$$

Bibliography

[1] Anderson, T. W., "On the theory of testing serial correlation," *Skandinavisk Aktuarietidskrift* **31** (1948), 88–116.

[2] Anderson, T. W., *The Statistical Analysis of Time series* (1971; Wiley, New York).

[3] Bassmann, R., "A generalized classical method of linear coefficients in a structural equation," *Econometrica* **25** (1957), 77–83.

[4] Bellman, Richard, *Introduction to Matrix Analysis* (1960; McGraw-Hill, New York).

[5] Beyer, W. H. (ed.), *Handbook of Tables for Probability and Statistics* (1960; The Chemical Rubber Co., Cleveland).

[6] Chung, K. L., *A Course in Probability Theory* (1968; Harcourt, Brace and World, New York).

[7] Cochrane, D., and G. H. Orcutt, "Applications of least squares to relations containing autocorrelated error terms," *Journal of the American Statistical Association* **44** (1949), 32–61.

[8] Dhrymes, Phoebus J., *Distributed Lags: Problems of Estimation and Formulation* (1971; Holden–Day, San Francisco).

[9] Dhrymes, Phoebus J., *Econometrics: Statistical Foundations and Applications* (1974; Springer–Verlag, New York).

[10] Dhrymes, Phoebus J., "Econometric models," in *Encyclopedia of Computer Science and Technology* (J. Beizer, A. G. Holzman, and A. Kent, eds.), Vol. 8 (1977; Marcel Dekker, Inc., New York).

[11] Durbin, J., and G. S. Watson, "Testing for serial correlation in least squares regression, I," *Biometrika* **37** (1950), 408–428.

[12] Durbin, J., and G. S. Watson, "Testing for serial correlation in least squares regression, II," *Biometrika* **38** (1951), 159–178.

[13] Durbin, J., and G. S. Watson, "Testing for serial correlation in least squares regression, III," *Biometrika* **58** (1970) 1–19.

[14] Durbin, J., "Testing for serial correlation in least squares regressions when some of the regressors are lagged dependent variables," *Econometrica* **38** (1970), 410–421.

[15] Durbin, J., "An alternative to the bounds test for testing for serial correlation in least squares regression," *Econometrica* **38** (1970), 422–429.

[16] Fisher, F. M., *The Identification Problem in Econometrics* (1966; McGraw–Hill, New York).
[17] Geary, R. C., and C. E. V. Leser, "Significance tests in multiple regression," *The American Statistician* **22** (1968), 20–21.
[18] Hadley, George, *Linear Algebra* (1961; Addison–Wesley, Reading, Mass.).
[19] Hoerl, A. E., "Optimum solution of many variables equations," *Chemical Engineering Progress* **55** (1959), 69–78.
[20] Hoerl, A. F., "Application of ridge analysis to regression problems," *Chemical Engineering Progress* **58** (1962), 54–59.
[21] Hoerl, A. E., and R. W. Kennard, "Ridge regression: biased estimation of non-orthogonal problems," *Technometrics* **12** (1970), 55–69.
[22] Hoerl, A. E., and R. W. Kennard, "Ridge regression: applications to non-orthogonal problems," *Technometrics* **12** (1970), 69–82.
[23] Koopmans, T. C., and W. C. Hood, "The estimation of simultaneous linear economic relationships," in *Studies in Econometric Method* (W. C. Hood and T. C. Koopmans, eds.), Chapter 6 (1953; Wiley, New York).
[24] Marquardt, D. W., "Generalized inverses, ridge regression, biased linear and nonlinear estimation," *Technometrics* **12** (1970), 591–612.
[25] McDonald, G. C., and D. I. Galarneau, "A Monte Carlo evaluation of some ridge-type estimators" (1973; General Motors Research Laboratory Report, GMR-1322-B, Marven, Michigan).
[26] Newhouse, J. P., and S. D. Oman, "An evaluation of ridge estimators" (1971; Rand Corporation Report No. R-716-PR, Santa Monica, California).
[27] Raifa, H., and R. Scheifer, *Applied Statistical Decision Theory* (1961; Harvard Business School, Boston).
[28] Sargan, J. D., "Wages and prices in the United Kingdom: a study in econometric methodology," in *Econometric Analysis for National Economic Planning* (P.E. Hart *et al.*, eds.) (1964; Butterworths, London).
[29] Scheffè, H., "A method for judging all contrasts in the analysis of variance," *Biometrika* **40** (1973), 87–104.
[30] Scheffè, H., *The Analysis of Variance* (1959; Wiley, New York).
[31] Scheffè, H., "A note on a reformulation of the S-method of multiple comparison," *Journal of the American Statistical Association* **72** (1977), 145–146. [See also in this issue R. A. Olshen's comments, together with Scheffè's rejoinders.]
[32] Theil, H., *Estimation and Simultaneous Correlation in Complete Equation Systems* (1953; Central Plan Bureau, The Hague).
[33] Vinod, H. D., "Ridge regression of signs and magnitudes of individual regression coefficients" (undated mimeograph).
[34] Vinod, H. D., "Generalization of the Durbin–Watson statistic for higher order autoregressive processes," *Communications in Statistics* **2** (1973), 115–144.
[35] Wald, A., "The fitting of straight lines if both variables are subject to error," *Annals of Mathematical Statistics* **11** (1940), 284–300.
[36] Zellner, Arnold, *Introduction to Bayesian Inference in Econometrics* (1971; Wiley, New York).

Index

Econometrics

Statistical Foundations and Applications

By **P.J. Dhrymes**

1974. xv, 592p. paper
(Springer Study Edition; Revised Reprint of the lst Edition)

Written primarily for the graduate student in econometrics, this book provides a comprehensive review of the techniques employed in econometric research. Assuming statisical competence of only an elementary level on the part of the reader, the book covers multivariate normal distribution before going on to the more sophisticated spectral analysis methods. Both practicing econometricians and students will find the integrated presentation of simultaneous equations estimation theory and spectral analysis a convenient and valued reference.

Contents: Elementary Aspects of Multivariate Analysis. Applications of Multivariate Analysis. Probability Limits, Asymptotic Distributions, and Properties of Maximum Likelihood Estimators. Estimation of Simultaneous Equations Systems. Applications of Classical and Simultaneous Equations Techniques and Related Problems. Alternative Estimation Methods; Recursive Systems. Maximum Likelihood Methods. Relations Among Estimators; Monte Carlo Methods. Spectral Analysis. Cross-Spectral Analysis. Approximate Sampling Distributions and Other Statistical Aspects of Spectral Analysis. Applications of Spectral Analysis to Simultaneous Equations Systems.

Applications of Mathematics

Vol. 1
**Deterministic and Stochastic Optimal
  Control**
By **W.H. Fleming** and **R.W. Rishel**
1975. ix, 222p. 4 illus. cloth

Vol. 2
Methods of Numerical Mathematics
By **G.I. Marchuk**
1975. xii, 316p. 10 illus. cloth

Vol. 3
Applied Functional Analysis
By **A.V. Balakrishnan**
1976. x, 309p. cloth

Vol. 4
Stochastic Processes in Queueing Theory
By **A.A. Borovkov**
1976. xi, 280p. 14 illus. cloth

Vol. 5
Statistics of Random Processes I
General Theory
By **R.S. Liptser** and **A.N. Shiryayev**
1977. x, 394p. cloth

Vol. 6
Statistics of Random Processes II
Applications
By **R.S. Liptser** and **A.N. Shiryayev**
1978. x, 339p. cloth

Vol. 7
Game Theory
Lectures for Economists and Systems
  Scientists
By **N.N. Vorob'ev**
1977. xi, 178p. 60 illus. cloth

Vol. 8
Optimal Stopping Rules
By **A.N. Shiryayev**
1978. x, 217p. 7 illus. cloth

Vol. 9
Gaussian Random Processes
By **I.A. Ibragimov** and **Y.A. Rosanov**
1978. approx. 290p. cloth